Advanced Machine Learning and Deep Learning Approaches for Remote Sensing II

Advanced Machine Learning and Deep Learning Approaches for Remote Sensing II

Editor

Gwanggil Jeon

MDPI

Basel • Beijing • Wuhan • Barcelona • Belgrade • Novi Sad • Cluj • Manchester

Editor
Gwanggil Jeon
Incheon National University
Incheon
Republic of Korea

Editorial Office
MDPI
St. Alban-Anlage 66
4052 Basel, Switzerland

This is a reprint of articles from the Special Issue published online in the open access journal *Remote Sensing* (ISSN 2072-4292) (available at: https://www.mdpi.com/journal/remotesensing/special_issues/MVH19UZ2J0).

For citation purposes, cite each article independently as indicated on the article page online and as indicated below:

Lastname, A.A.; Lastname, B.B. Article Title. *Journal Name* **Year**, *Volume Number*, Page Range.

ISBN 978-3-7258-0771-0 (Hbk)
ISBN 978-3-7258-0772-7 (PDF)
doi.org/10.3390/books978-3-7258-0772-7

Contents

About the Editor

Gwanggil Jeon

Gwanggil Jeon received the B.S., M.S., and Ph.D. (summa cum laude) degrees from the Department of Electronics and Computer Engineering, Hanyang University, Seoul, Korea, in 2003, 2005, and 2008, respectively. From 2009.09 to 2011.08, he was with the School of Information Technology and Engineering, University of Ottawa, Ottawa, ON, Canada, as a Post-Doctoral Fellow. From 2011.09 to 2012.02, he was with the Graduate School of Science and Technology, Niigata University, Niigata, Japan, as an Assistant Professor. From 2014.12 to 2015.02 and 2015.06 to 2015.07, he was a Visiting Scholar at Centre de Mathématiques et Leurs Applications (CMLA), École Normale Supérieure Paris-Saclay (ENS-Cachan), France. From 2019 to 2020, he was a Prestigious Visiting Professor at Dipartimento di Informatica, Università degli Studi di Milano Statale, Italy. From 2019 to 2020 and 2023 to 2024, he was a Visiting Professor at Faculdade de Ciência da Computação, Universidade Federal de Uberlândia, Brasil. He is currently a professor at Incheon National University, Incheon, Korea. He was a general chair of IEEE SITIS 2023, and served as a workshop chairs in numerous conferences.

Dr. Jeon is an Associate Editor of IEEE Transactions on Circuits and Systems for Video Technology (TCSVT), Elsevier Sustainable Cities and Society, IEEE Access, Springer Real-Time Image Processing, Journal of System Architecture, and Wiley Expert Systems.

Dr. Jeon was a recipient of the IEEE Chester Sall Award in 2007, ACM's Distinguished Speaker in 2022, the ETRI Journal Paper Award in 2008, and Industry-Academic Merit Award by Ministry of SMEs and Startups of Korea Minister in 2020.

remote sensing

MDPI

Article

Simulation of the Ecological Service Value and Ecological Compensation in Arid Area: A Case Study of Ecologically Vulnerable Oasis

Jiamin Liu [1,2], Xiutong Pei [1,2], Wanyang Zhu [1,2] and Jizong Jiao [1,2,3,*]

[1] College of Earth and Environmental Sciences, Lanzhou University, Lanzhou 730000, China; 120220909740@lzu.edu.cn (J.L.); 220220947421@lzu.edu.cn (X.P.); 220210946551@lzu.edu.cn (W.Z.)
[2] The Key Laboratory of Western China's Environmental Systems, Ministry of Education (MOE), Lanzhou 730000, China
[3] Institute of Tibet Plateau Human Environment Research, Lanzhou University, Lanzhou 730000, China
* Correspondence: jiaojz@lzu.edu.cn; Tel.: +86-136-6938-5107

Abstract: In recent years, the delicate balance between economic development and ecological environment protection in ecologically fragile arid areas has gradually become apparent. Although previous research has mainly focused on changes in ecological service value caused by land use, a comprehensive understanding of ecology–economy harmony and ecological compensation remains elusive. To address this, we employed a coupled deep learning model (convolutional neural network-gated recurrent unit) to simulate the ecological service value of the Wuwei arid oasis over the next 10 years. The ecology–economy harmony index was used to determine the priority range of ecological compensation, while the GeoDetector analyzed the potential impact of driving factors on ecological service value from 2000 to 2030. The results show the following: (1) The coupled model, which extracts spatial features in the neighborhood of historical data using a convolutional neural network and adaptively learns time features using the gated recurrent unit, achieved an overall accuracy of 0.9377, outperforming three other models (gated recurrent unit, convolutional neural network, and convolutional neural network—long short-term memory); (2) Ecological service value in the arid oasis area illustrated an overall increasing trend from 2000 to 2030, but urban expansion still caused a decrease in ecological service value; (3) Historical ecology–economy harmony was mainly characterized by low conflict and potential crisis, while future ecology–economy harmony will be characterized by potential crisis and high coordination. Minqin and Tianzhu in the north and south have relatively high coordination between ecological environment and economic development, while Liangzhou and Guluang in the west and east exhibited relatively low coordination, indicating a greater urgency for ecological compensation; (4) Geomorphic, soil, and digital elevation model emerged as the most influential natural factor affecting the spatial differentiation of ecological service value in the arid oasis area. This study is of great significance for balancing economic development and ecological protection and promoting sustainable development in arid areas.

Keywords: convolutional neural network; gated recurrent unit; ecological service value; ecological–economic harmony; driving mechanism

Citation: Liu, J.; Pei, X.; Zhu, W.; Jiao, J. Simulation of the Ecological Service Value and Ecological Compensation in Arid Area: A Case Study of Ecologically Vulnerable Oasis. *Remote Sens.* **2023**, *15*, 3927. https://doi.org/10.3390/rs15163927

Academic Editor: Prasad S. Thenkabail

Received: 23 June 2023
Revised: 29 July 2023
Accepted: 5 August 2023
Published: 8 August 2023

1. Introduction

The benefits that people derive from multiple processes and ecosystem functions can be described as ecosystem services [1]. Driven by the growth of urban demand, land use change has led to serious degradation of global ecosystems [2,3]. On the one hand, the invasion of large areas of ecological land has resulted in irreversible biodiversity loss [4]. On the other hand, local climate change, the urban heat island effect, and changes in precipitation have contributed to the decline in the Ecological service value (ESV) [5]. With population growth and economic development, global ecosystems have been seriously damaged, and the imbalance between economic development and ecological environment protection has gradually become prominent, especially in ecologically fragile arid areas [6,7]. In arid regions, characterized by harsh climatic conditions, soil salinization and alkalization, and the sustainability of ecosystem services has always been a focus of attention [8,9]. Oasis ecosystems play an essential role in social and economic stability and development in arid areas, but their ecological fragility is particularly pronounced due to the low precipitation and high evaporation rates [10,11]. Wuwei Oasis is situated in the Shiyang River Basin, an important inland river in Northwest China's ecologically fragile area, and its ecological environment quality has a serious impact on the entire basin [12,13]. Therefore, focusing on the ESV and EEH in the arid oasis area of Wuwei is of great significance for promoting sustainable development and achieving a balance between economic growth and ecological protection [14].

Ecological services are characterized by complex interconnections and strong scale effects, with changes in ESV often being determined by multiple ecosystem services [15] The benefit transfer method can not only rapidly assess the individual ecological benefits of multiple ecosystem services but also evaluate their overall ecological benefits, and therefore, it has been widely used in ESV evaluation [16]. However, the benefit transfer method relies on equivalent factor coefficients to characterize the relationship between different land use types and ESV, which is subjective. In addition, there is spatiotemporal heterogeneity in land use distribution. Thus, it is necessary to adjust the coefficient value of multiple ecosystem services according to the natural and socioeconomic characteristics of the area to improve the accuracy of ESV estimates.

Assessing the impact of future land use changes on ESV and Ecology–Economy Harmony (EEH) can provide scientific policy recommendations for ecosystem management [17]. Li, et al. [18] employed the InVEST and SLEUTH models to evaluate the impact of land use changes on habitat quality. However, existing models often have difficulty in reliably predicting future land use changes, leading to significant errors in evaluation results [19]. Deep learning has recently emerged as a powerful tool for time-series object modeling, demonstrating excellent performance in various domains [20]. It can not only extract implicit spatial features from datasets with multiple variables to improve feature representation ability [21] but also exploit long-term time dependencies among large amounts of time-series data to establish accurate feature maps [22]. Among the various deep learning models, convolutional neural networks (CNN) have been extensively utilized in the dynamic simulation of time-series data. Zhai, et al. [23] fused CNN and vector-based cellular automata to extract high-level features of irregularly shaped cells in the neighborhood and simulate land use changes, achieving higher simulation accuracy than other models such as Random Forest and Artificial Neural Networks. Qian, et al. [24] also validated the effectiveness of deep learning models such as CNN applying land use data from Shanghai from 2000 to 2015. However, existing studies exploring neighborhood effects in transformation rules have only considered the extraction of spatial features in historical data dimensions, ignoring the significant long-term time dependencies in neighborhood interactions, resulting in low simulation accuracy [25]. A gated recurrent unit (GRU) network is a deep learning model used to extract time-dimension features. Compared with traditional recurrent neural networks, it can improve memory capacity and training performance and better solve overfitting, gradient vanishing, and explosion problems. Cao, et al. [26] predicted grain loss and waste rates based on a multi-task multi-gate recurrent unit autoencoder method, and the results indicated that the accuracy of this method was higher than that of existing models. Chen, et al. [27] applied the GRU network to predict long-term degradation trends based on available data on degradation features. In light of the excellent performance of the GRU in time feature extraction, we coupled the CNN-GRU model to complement the deficiencies in existing time-series data simulation research.

Ecological compensation is a widely recognized economic approach to improving water yield, soil and water conservation, intensive and efficient use of water resources, ecological, environmental protection, and pollution control by coordinating the relationships between different stakeholders [28]. There exist various methods for evaluating ecological compensation, including the willingness-to-pay, opportunity cost, ecological footprint, and value theory methods [29,30]. While the willingness-to-pay method relies on subjective survey data [31], the opportunity cost method tends to undervalue ecosystem services by focusing on cost-benefit analysis [32]. Similarly, the ecological footprint method determines the sustainability of ecological compensation by evaluating the supply and demand relationship between humans and ecological resources, but its sustainability is weak [33]. The ecological service value method, which is based on the theory of externalities, bridges the gap between natural ecosystems and economic systems by quantifying the direct or indirect available ecological value used to produce ecosystem services [34]. It quantifies ecological compensation by comparing the non-market ESV per unit area with the GDP per unit area of the area. Although ESV is complex and unstable at cross-regional scales, it can be corrected by incorporating various regional data, such as food and GDP, and is applicable to a wide range of research scales [35]. Consequently, ESV evaluation appears to be a more suitable method for ecological compensation. In addition, EEH is a critical foundation for setting reasonable ecological compensation standards and accurately quantifying ecological compensation, which has often been frequently overlooked in previous research. ESV comprises various ecosystem services, including supply, regulation, support, and cultural services, and exploring the ESV represented by different ecosystem services is necessary to fully express the EEH of the area, serving ecological compensation and sustainable development.

A thorough analysis of the influencing factors and mechanisms of ESV is a crucial basis for guiding ecological protection decision-making [36]. Wu et al. [17] quantitatively analyzed the impact of rapid urbanization on ecosystem services in Kunshan from 2006 to 2030. Chen, et al. [37] utilized cellular automata and geographically weighted regression to simulate the ESV loss caused by land use changes in Chongqing. Previous studies have primarily focused on the rise or fall of ESV caused by land use changes, but little is known about the driving mechanisms of ESV, particularly in arid areas [38,39]. Research methods for ESV and its driving factors have primarily included principal component and correlation analysis [40], regression models, and grey relational analysis [41]. Although these methods can explain the contribution of influencing factors to a certain extent, they fail to capture the interaction and joint effects between influencing factors and cannot fully express the complex spatial correlation and spatiotemporal differentiation characteristics within ESV [42]. GeoDetector can further reveal the spatial distribution relationship and interaction mechanism between independent and dependent variables from a statistical perspective by converting qualitative data into quantitative data [43,44]. Therefore, this study utilizes GeoDetector to quantitatively analyze the explanatory power of each driving factor for spatial variable distribution characteristics and explore the interaction between two factors [45].

The main contributions of this study are as follows:

(1) Proposed a new CNN-GRU model, which integrates both temporal and spatial neighborhood features, for simulating the dynamic process of land use change. This approach outperforms three other models, including GRU, CNN, and CNN long short-term memory (LSTM), and provides higher accuracy in predicting land use change;

(2) Revealed the impact of land use change on ESV in the arid oasis area of Northwest China;

(3) Determined EEH in the historical period and the next 10 years in the arid oasis area, as well as the priority for ecological compensation;

(4) Employed the GeoDetector to explore the driving mechanism of ESV;

2. Study Area and Data Sources

2.1. Study Area

Wuwei (Figure 1) (36°29′~39°27′N, 101°49′~104°16′E) is located in Northwest China, at the intersection of the Loess Plateau, the Qinghai–Tibet Plateau, and the Mongolian

Plateau [46]. The terrain is complex, with the southern area belonging to the Qilian Mou tains, and the climate is suitable for the development of forestry and animal husbandry. Th central area is a flat oasis area with fertile land and is an important agricultural productic base in China. The northern area is a desert area with low precipitation [12]. Wuwei spar 326 km in length and 204 km in width and has natural landscapes, such as snow-covere highlands, oases, and deserts. The permanent population was 1.825×10^4 at the end 2019 [13].

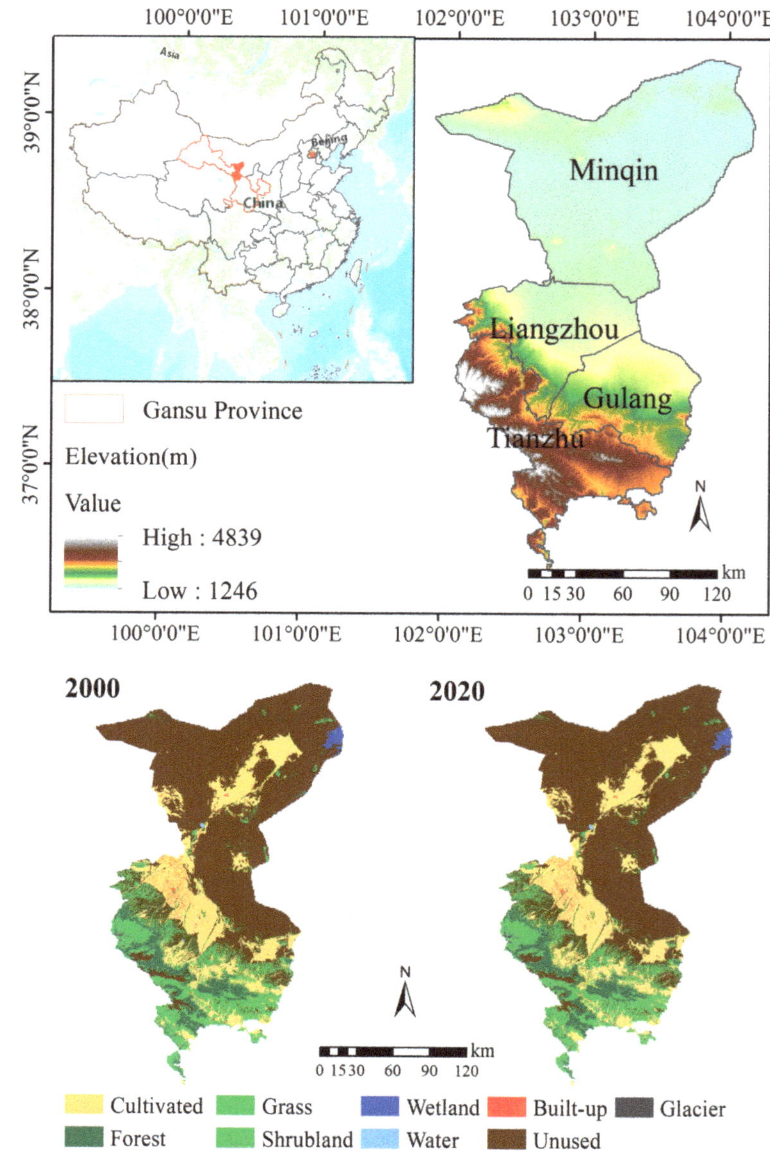

Figure 1. Study area and land use spatial distribution. Wuwei belongs to a warm-temperate continer tal arid climate with an average annual temperature of 7.8 °C and a precipitation range of 60–610 mm In terms of administrative divisions, it includes one district, two counties, and one autonomou county, with a total area of 3.32×10^4 km^2.

2.2. Data Sources

The land use datasets for 2000, 2010, and 2020 were obtained from the Global Geographic Information Products Platform for this study. The driving factors were categorized into four types: transportation accessibility; socioeconomic conditions; terrain conditions; and climate conditions, consisting of 14 categories. Data sources for each category are presented in Table A1 of the attached Appendix A. Transportation accessibility variables included major roads, railways, rivers, residential areas, and ecological function protection areas, while socioeconomic variables included nighttime lights, GDP, population, and NPP. Terrain conditions included elevation, slope, and faults, while conditions included precipitation and temperature. All data were resampled to a spatial resolution of 30 m and normalized to ensure consistency across variables. In the initial phase, remote sensing images were acquired, and an extensive data preprocessing pipeline was implemented. This preprocessing encompassed radiometric calibration, atmospheric correction, geometric correction, image mosaicking, and cropping. These rigorous steps were undertaken to rectify image distortions, geometric irregularities, and atmospheric interferences arising from sensor characteristics, spatial variations, atmospheric absorption, scattering, and other influential factors. Subsequently, we leveraged a land use remote sensing dataset to obtain comprehensive land use classification data. Additionally, key remote sensing variables, such as nighttime lights, were strategically integrated as driving factors into the CNN-GRU algorithm. This integration facilitated the acquisition of spatiotemporal features, thereby enabling the model to effectively learn and process complex temporal dynamics and land use patterns.

3. Methods

The research framework is illustrated in Figure 2.

3.1. Land Use Modeling

3.1.1. CNN

The CNN architecture typically comprises convolutional layers, pooling layers, activation functions, and fully connected layers [47]. Convolutional layers extract the spatial features of the input image by using filters learned from the training data set. Usually, an activation function is used after the convolutional layers to introduce nonlinearity into the network and capture the complex relationship between the input and output [48]. After the activation function, a pooling layer is added to retain the main features of the convolutional layer while reducing parameters. Finally, the objective of the fully connected layer is to predict the output value based on a nonlinear combination of a series of feature maps from convolutional and pooling layers. The core of this study is to use CNN to extract the complex spatial features of the data and pooling layers are omitted to prevent the loss of relevant features [49].

3.1.2. GRU

GRU calculates the probability distribution of the time series data by employing the encoder and decoder [50]. Initially, the conditional distribution on a variable-length output sequence given another variable-length sequence is learned (e.g., $p(y_1, \ldots, y_{T'} | x_{1, \ldots}, x_T)$, where T and T' are the input and output sequences, respectively. Secondly, the encoder reads the temporal features of the input sequence x in order. The hidden state $h_{(t)}$ changes with the time step (Equation (1)). Upon reading the sequence end, $h_{(t)}$ is the summary of the entire input sequence c. The decoder is trained to generate the output sequence by predicting the time dimension feature y_t of the next neighboring unit. The hidden state at time t is determined by Equation (2). Using the softmax activation function to predict the probability distribution of the next neighboring unit learning sequence (Equation (3)), the output of each time step t is the conditional distribution $p(x_t | x_{t-1, \ldots}, x_1)$. By combining the probability of each neighboring unit, the probability of sequence x is calculated by

Equation (4). Therefore, the conditional distribution of the time dimension feature of the next neighboring unit is Equation (5) [51].

$$h_{(t)} = f\left(h_{(t-1)}, x_t\right) \tag{1}$$

$$h_{(t)} = f\left(h_{(t-1)}, y_{(t-1)}, c\right) \tag{2}$$

$$p(x_{(t,j)} = 1 | x_{t-1}, \ldots, x_1) = \frac{exp\left(w_j h_{(t)}\right)}{\sum_{j'=1}^{K} exp\left(w_{j'} h_{(t)}\right)} \tag{3}$$

$$p(x) = \prod_{t=1}^{T} p_t(x | x_{t-1}, \ldots, x_1) \tag{4}$$

$$p\left(y_t \middle| y_{(t-1)}, y_{(t-2)}, \ldots, y_1, c\right) = g\left(h_{(t)}, y_{(t-1)}, c\right) \tag{5}$$

Here, f is a non-linear activation function; w_j is the row of weight matrix w. For a given activation function g, it must generate effective probabilities.

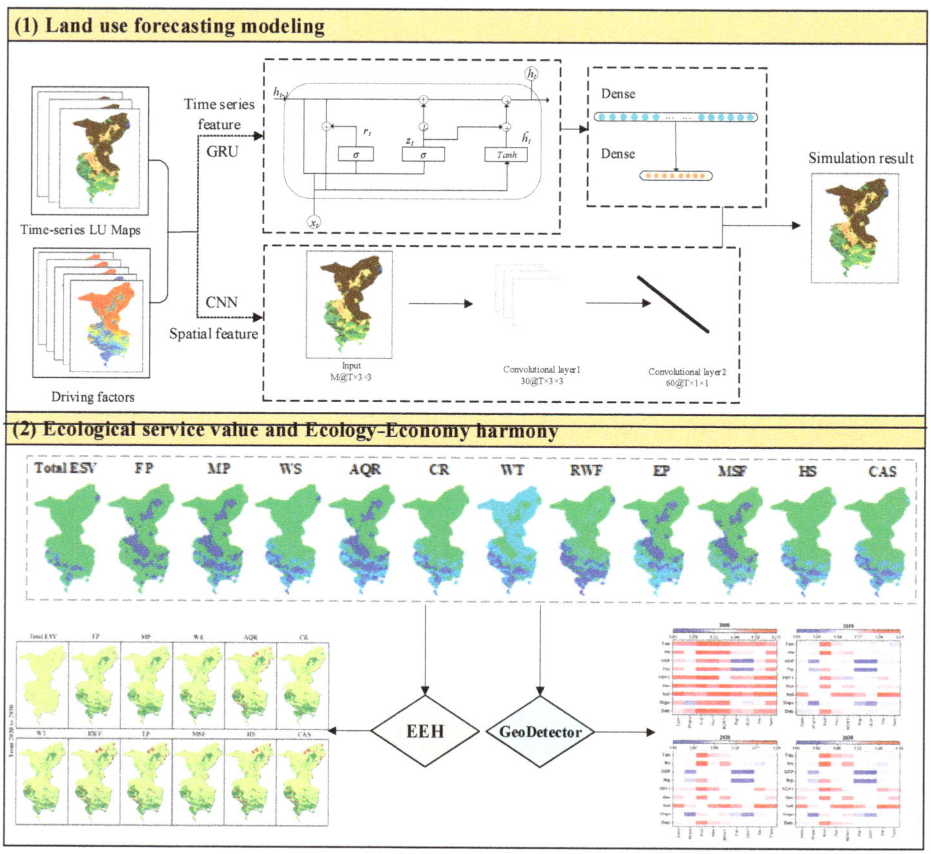

Figure 2. Research framework. (ESV: ecological service value; FP: food production; MP: material production; WS: water supply; AQR: air quality regulation; CR: climate regulation; WT: waste treatment; RWF: regulation of water flows; EP: erosion prevention; MSF: maintenance of soil fertility; HS: habitat services; CAS: cultural and amenity services).

The activation calculation for the jth hidden unit is given by:

$$r_j = \sigma\left(\left[W_r x\right]_j + \left[U_r h_{(t-1)}\right]_j\right) \tag{6}$$

$$z_j = \sigma\left(\left[W_z x\right]_j + \left[U_z h_{(t-1)}\right]_j\right) \tag{7}$$

$$h_j^{(t)} = z_j h_j^{(t-1)} + (1 - z_j)\overset{\sim(t)}{h_j} \tag{8}$$

$$\overset{\sim(t)}{h_j} = f([Wx]_j + r_j[Uh_{(t-1)}]) \tag{9}$$

Here, r_j represents the reset gate; z_j represents the update gate; h_j represents the actual activation of the unit; σ is the sigmoid function; $[.]_j$ represents the jth element of the vector; x and $h_{(t-1)}$ are the input state and the previous hidden state, and W_r and U_r are weight matrices.

When the reset gate is close to 0, the hidden state is forced to ignore the previous hidden state and only use the current input to reset. The update gate controls the amount of information transferred from the previous hidden state to the current hidden state for the long-term memory [52]. Each hidden unit has separate reset and update gates, so it can learn to capture dependencies at different time scales.

3.1.3. CNN-GRU

To optimize the land-use change simulation research, we constructed a six-layer network structure consisting of two CNN layers, two GRU layers, and two fully connected layers. The two convolutional layers each consist of 14 3×3 convolutional kernels, resulting in a $(N - 2) \times (N - 2) \times 14$ feature map. The data were then formatted with 14 time steps and one input feature per time step. The first GRU layer has 64 cores, with $h_{(t)}$ being passed to the next layer at each time step. The second GRU layer has 94 cores and only outputs $h_{(t)}$ at the final time step. To avoid overfitting, the dropout rate was set to 20% for both GRU layers, and the tanh activation function was chosen to improve model performance. Finally, there are two fully connected layers, with 128 neurons in the first layer and a dropout rate of 20% and 8 neurons in the second layer with a softmax classifier. After continuous iteration, we found that the optimal learning rate for the research area data was 0.002; the batch size was set to 128, and the Adam algorithm was selected as the optimizer. Further, the cross-entropy loss function was introduced to optimize model performance. The number of epochs was set to 50, the loss value decreased rapidly to a certain point, and the iteration process basically converged.

The modeling process consists of four steps: (1) Data preprocessing and model training: preprocessing land use historical data and driving factor variables to prepare for training and conversion rules; (2) Model calibration: utilizing CNN and GRU algorithms to extract spatial and temporal neighborhood features of land use and driving factors, continuously optimizing the model's performance; (3) Model validation: comparing the simulated land use change results in the CNN, GRU, CNN-LSTM, and CNN-GRU models with the actual situation using the same data set; (4) Future prediction: using the calibrated model to simulate future land use, ESV, and EEH changes.

Using Python coding, we calibrated the model parameters with historical data from 2000 to 2010 and generated simulation results for 2020 (Figure 3). To verify the model performance, we compared the results with three sets of indicators, overall accuracy, Kappa coefficient, and figure of merit (FOM). Specifically, we conducted comparisons among (1) The coupled model and single models (CNN-GRU, CNN, and GRU) to examine the importance of spatiotemporal feature extraction, (2) Different recurrent neural networks (CNN-GRU and CNN-LSTM) as feature samplers for comparing the performance of time dimension feature extraction, and (3) Single spatiotemporal models (CNN and GRU) t analyze the impact of temporal and spatial features on time series data simulation.

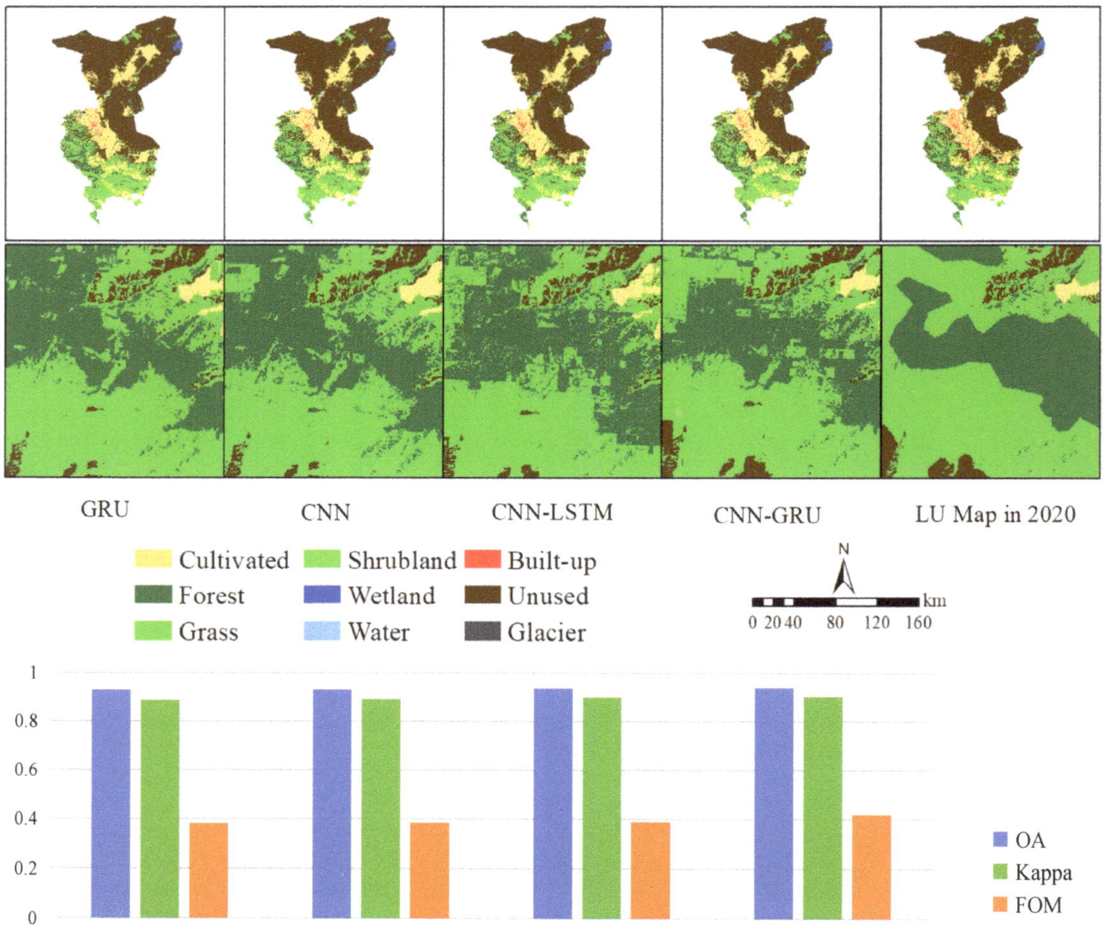

Figure 3. Simulated and actual land use maps for 2020.

3.2. ESV Evaluation

In this study, the ESV of Wuwei Oasis was calculated by exploiting the standard uni results and evaluation method of the ecological service value equivalent factor improved by Xie, et al. [53]. To ensure the applicability of the numerical coefficients in the calculation of ESV at the regional scale, the coefficients were adjusted based on the correction factor fo grain production. The equations applied for calculating ESV are as follows:

$$E_a = \frac{1}{7} \times P \times Y \quad (10)$$

$$E_i = E_a \times q \quad (11)$$

$$ESV = \sum (A_i \times E_i) \quad (12)$$

Here, E_a is the economic value of an ESV equivalent factor; E_i is the ESV of the land ecosystem i per unit area; q is the ESV equivalent factor per unit area; A_i is the area of land ecosystem type I; Y is the crop yield per unit area in Wuwei, and P is the average grain price in 2020.

3.3. Ecology–Economy Harmony

ESV change serves as a pivotal gauge for assessing regional socioeconomic and ecological environment sustainability. Coordinated development between the ecological environment and the economy entails a harmonious interaction and alignment of elements within the environmental and economic subsystems throughout the regional development trajectory, fostering their reciprocal advancement and ultimately elevating the overall developmental status of the region. Through an in-depth analysis of the association between alterations in ESV resulting from land use dynamics and the level of regional socio-economic development, an assessment of the degree of harmony between the regional ecological environment and economic progress can be achieved. The ecological environment and economic development status in arid oasis areas was measured by utilizing the Ecology–Economy Harmony (EEH) index (Table 1), which combines datasets of ESV and GDP of the period from 2000 to 2030. The ecological compensation priority was then determined. Additionally, the 2023 GDP data were obtained through time-series forecasting employing Python 3.9 software.

$$\text{EEH} = \frac{\frac{(ESV_{hj}-ESV_{hi})}{ESV_{hi}}}{\frac{(GDP_{hj}-GDP_{hi})}{GDP_{hi}}} \tag{13}$$

Here, EEH is the ecology–economy harmony index; ESV_{hj} and ESV_{hi} are the ecosystem service values for different periods, and GDP_{hj} and GDP_{hi} are the GDP values for different periods. Coordination and conflict levels are divided based on the regional characteristics of the arid oasis area and the existing literature [54].

3.4. GeoDetector

We employed the Geodetector model to quantify the influence of various factors on the changes in ESV in the Wuwei Oasis area [43]. Geodetector is a spatial statistical method used for identifying driving factors of geographic phenomena, widely applied in the fields of geography, environmental science, and public health, among others. It has the capability to reveal the impact extent and interaction relationships of various factors on specific events or phenomena. The determination of single-factor and two-factor contributions to ESV values ranged from 0 to 1, with higher values denoting a more pronounced influence. Unlike conventional approaches employed in identifying driving factors, Geodetector demonstrates a distinctive advantage in its capacity to investigate the combined impact of two independent variables on the dependent variable. Notably, Geodetector exhibits a high degree of flexibility concerning the incorporation of input data, as it can effectively accommodate both quantitative and qualitative data by means of a reclassification process, enabling their seamless integration into the analytical framework. While previous studies have mainly focused on socioeconomic data as the primary drivers for analysis, it is well recognized that single socioeconomic factors cannot comprehensively predict regional ESV changes. Thus, this study selected a range of factors, including natural factors, such as DEM, slope, soil type, geomorphic type, and NDVI, as well as socioeconomic factors, such as population density and GDP, and climate factors, such as precipitation and temperature. It is important to note that natural environmental factors, climate, and landscape patterns all have a certain impact on ESV in arid oasis areas, making the inclusion of these factors critical for a comprehensive analysis.

Table 1. Classification level of EEH index. At the coordination level, an EEH value greater than or equal to 1 denotes that the growth rate of ESV equals or surpasses the growth rate of GDP. This finding reflects a high degree of synchronization between the ecological environment and economic development within this study's area. Alternatively, it may suggest that the ecological environment experienced significant damage initially but subsequently underwent ecological restoration, resulting in certain limitations on economic development. On the other hand, when the EEH falls within the range of 0 to 1, it indicates that the growth rate of ESV is lower than that of GDP. Despite economic development not directly causing ecological degradation, varying degrees of ecological pressure persist. A higher EEH value indicates enhanced coordination between ecological and economic factors. In the conflict level, negative ESV growth signifies that economic development has detrimental effects on ecological environment conservation, leading to disharmony between the two. A lower EEH value indicates more pronounced conflicts between economic development and ecological protection.

EEH Index	Classification Level	EEH Index	Classification Level
$EEH \geq 1$	high coordination	$-0.5 \leq EEH < 0$	low conflict
$0.5 \leq EEH < 1$	moderate coordination	$-1 \leq EEH < -0.5$	moderate conflict
$0 \leq EEH < 0.5$	potential crisis	$EEH \leq -1$	serious conflict

4. Results

4.1. Model Comparison

4.1.1. Quantitative Analysis

(1) The CNN-GRU model outperformed the single models, highlighting that the extraction of spatial-temporal neighborhood features is crucial in time series data simulation, and ignoring any feature would substantially decrease the model's performance;

(2) The FOM values showed that CNN-GRU was more effective in capturing temporal features than CNN-LSTM. GRU's ability to directly use gate control for linear self-updating in the hidden unit overcomes the impact of short-term memory compared to linear self-updating memory units used by LSTM;

(3) The OA was higher in the single spatiotemporal models (CNN and GRU) than in the coupled CNN-GRU model, suggesting that spatial features have a greater impact on simulation accuracy than temporal features;

(4) The CNN-GRU model, which comprehensively considers both spatial and temporal features, exhibited superior accuracy compared to the other three models, providing strong evidence of the effectiveness and superiority of the coupled model.

4.1.2. Qualitative Analysis

Qualitative evaluation of the simulation results revealed consistency between the predicted land use maps and the actual spatial distribution of Wuwei Oasis in 2020. However, subtle differences were observed between the models (Figure 3). Specifically, the forest and cultivated land ratios of GRU, CNN, and CNN-LSTM were higher than the corresponding proportions in the actual land use map, suggesting insufficient feature extraction. GRU was particularly prone to misjudgment, possibly due to the challenge of accurately capturing feature maps from temporal sequence features alone. Moreover, notable discrepancies were found in the prediction of unused land among the four models. While the predictions generated by GRU and CNN were more dispersed, CNN-LSTM produced a more compact distribution. Nonetheless, CNN-GRU exhibited the highest degree of spatial similarity to the actual land use map, highlighting its exceptional simulation performance in predicting time-series data. As such, we utilized the CNN-GRU model to forecast changes in land use and ESV in 2030.

4.2. ESV Changes from 2000 to 2030

The analysis revealed that the ESV of Wuwei Oasis experienced a decline of 6.96×10^8 from 2000 to 2010, and while the area experienced a partial recovery from 2010 to 2020, the rate of recovery was slower than the decline from 2000 to 2010 (Figures 4 and 5). Furthermore, the ESV of this study's area remained in a state of loss from 2000 to 2020, with a slight increase predicted for 2030.

4.2.1. Contribution of Different Ecosystem Services to ESV

In terms of the contribution of different ecosystem services to ESV, climate regulation and regulation of water flows were the main types of ecosystem services in Wuwei Oasis, accounting for 20.11% and 19.84% of the total ESV, respectively. In contrast, water supply and maintenance of soil fertility had the smallest proportions, only 1.83% and 1.31%, respectively. During the period from 2000 to 2010, all ecosystem services exhibited a decreasing trend, with the highest loss rates for climate regulation services (−0.71%). From 2010 to 2020, except for food production, all ES exhibited an increasing trend, although with a small overall growth rate. Among them, water supply had the highest ESV growth rate of 0.71%, while food production had a loss rate of −0.31%. From 2000 to 2020, except for the regulation of water flows and water supply, all other ecosystem services led to ESV losses. The ESV changes from 2020 to 2030 were consistent with those from 2010 to 2020, with an increasing trend for all ecosystem services except food production. However, the loss rate of food production was low, and the regulation of water flows had the highest growth rate. Qualitatively, the distribution pattern of ESV increased gradually from northeast to southwest, which was attributed to the distribution of land use types from unused land, cultivated land, and grassland to forest from northeast to southwest, with a corresponding increase in vegetation cover. The 11 ESV types exhibited differences and similarities, with similarities in their spatial distribution patterns, while differences mainly reflected the composition of different ecosystem services. High values of the total ESV were relatively scarce, scattered in Tianzhu in the south. The 11 types of ESV could be divided into three categories. The first category included food production, material production, air quality regulation, erosion prevention, and maintenance of soil fertility, with relatively balanced high, medium, and low-value areas. The high-value areas were mainly distributed in the central part of Minqin, the southern parts of Liangzhou and Gulang, and the northern part of Tianzhu. The low-value areas surrounded the high-value areas, and the medium-value areas were only present in the southern part of Tianzhu. The second category included water supply, climate regulation, regulation of water flows, habitat services, and cultural and amenity services, where some high-value areas in the first category were replaced by medium-value areas in the second category, indicating that the functions of the second category of ecosystem services were lower than those of the first category. The third category was water treatment, which differed from the second category in that medium-value areas replaced high-value areas, indicating that water treatment in Minqin had stronger functional capabilities than the above services.

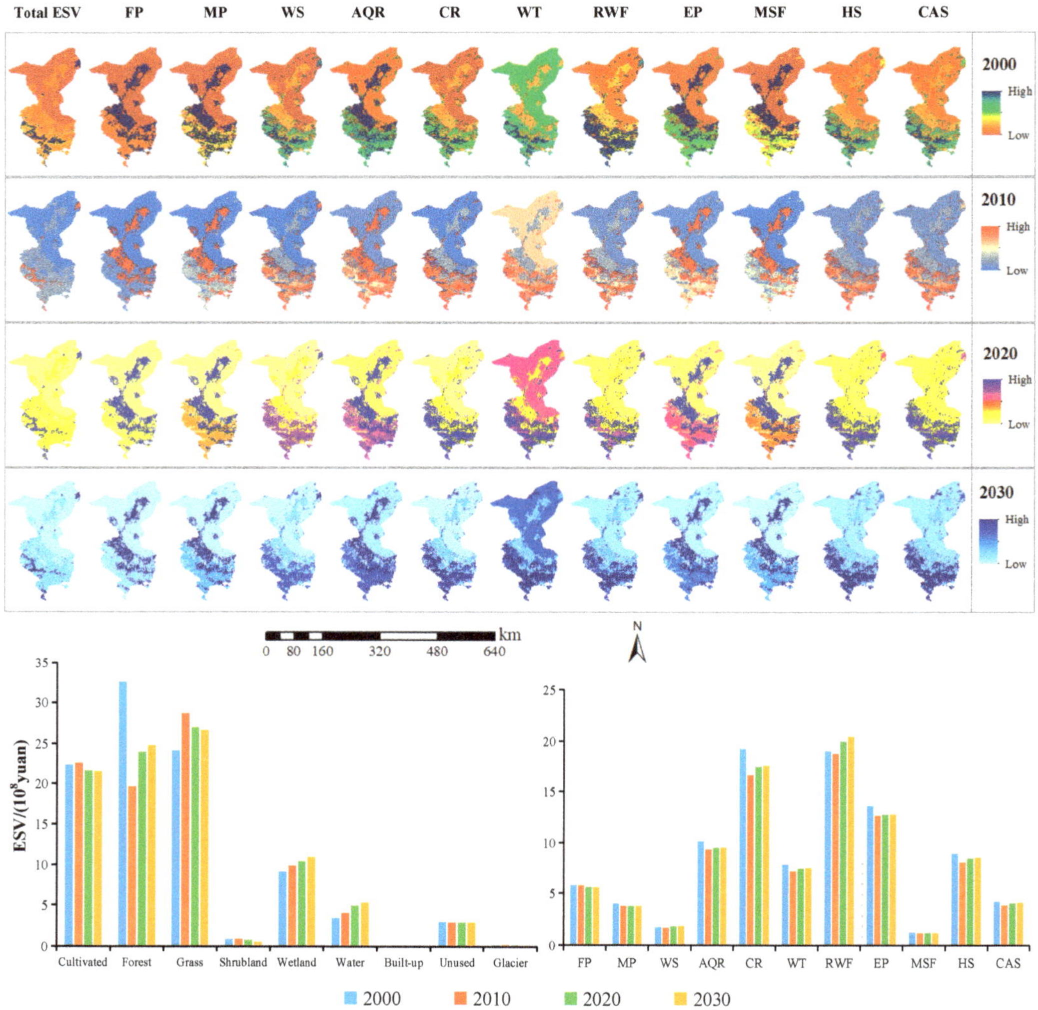

Figure 4. Spatial distribution of total ESV and 11 ESV from 2000 to 2030, and bar graphs are ESV of different land use types and ecosystem service representations from 2000 to 2030. (FP: food production; MP: material production; WS: water supply; AQR: air quality regulation; CR: climate regulation; WT: waste treatment; RWF: regulation of water flows; EP: erosion prevention; MSF: maintenance of soil fertility; HS: habitat services; CAS: cultural and amenity services).

4.2.2. Contribution of Different Land Use Types to ESV

The examination of ESV from the perspective of land use types revealed a notable trend in forestland, wetland, and water, which all showed an increase from 2000 to 2010. However, the ESV represented by the remaining land use types exhibited a decrease, with shrubland and glaciers experiencing the highest loss rate of ESV (−1.37% and −2.38%, respectively). The ESV changes observed from 2010 to 2020 remained consistent with the trend from 2000 to 2010, except for unused land, which changed from a decline to an increase. Over the entire 20-year period from 2000 to 2020, forests experienced the most severe loss of ESV. Looking ahead to 2030, the distribution of ESV in Wuwei Oasis is expected to be the lowest in the edge area of Minqin, which is consistent with the spatial distribution of unused land. Additionally, the forest in Tianzhu contributed the most to ESV

Among the ESVs represented by different land use types, grassland, wetland, and water all showed a decreasing trend, with water having the largest loss rate of ESV (−3.45%) and shrubland having the highest growth rate (6.32%).

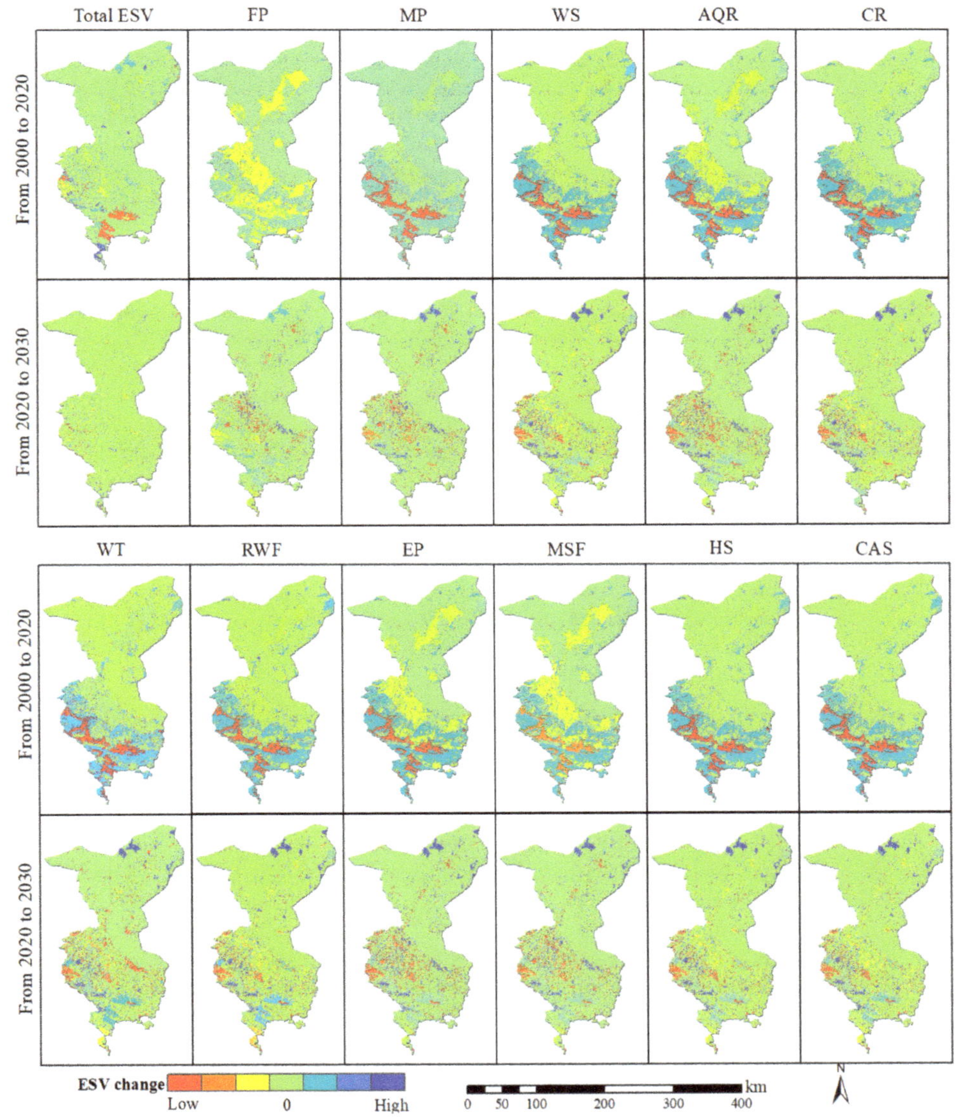

Figure 5. Spatial distribution changes in total ESV and 11 types of ESV in different periods. Lower variation values correspond to more substantial declines in ESV; higher ESV variation values are indicative of a more pronounced increase in ESV. An ESV variation value of 0 denotes a state of stable ESV, indicating no net change in ESV.

4.3. Ecological Compensation Changes from 2000 to 2030

There are significant differences in the temporal and spatial distributions of total EEH and different ecosystem services' EEH (Figure 6). From a temporal perspective, this study identified five types of EEH from 2000 to 2010, including moderate conflict, low conflict, potential crisis, moderate coordination, and high coordination. The overall study area exhibited a potential crisis, with highly coordinated areas in the Northern Minqin and Northwestern Tianzhu, suggesting that the changes in ESV and GDP were positively correlated during this period. In contrast, Southeastern Tianzhu showed low conflict, indicating that economic development had caused some loss of ESV and had an impact on the ecological environment. From 2010 to 2020, the EEH types remained consistent with those from 2000 to 2010, but the proportion of moderate and high conflict and coordination increased. In the 2020–2030 EEH types, high conflict replaced moderate conflict, and coordination shifted toward a negative direction. The ecological and economic status of this study's area underwent a shift from coordination to contradiction and then back to coordination due to the rapid population growth and negative growth of ESV from 2010 to 2020, leading to a lower level of conflict between the ecological environment and economic development.

Although the total EEH changes were relatively peaceful, the EEH changes in different ecosystem services were more intense. Specifically, from 2000 to 2020, the potential crisis turned into low conflict. From 2010 to 2030, the contradiction and coordination status became more apparent. The strong coordination is mainly due to the high altitude of these areas, which partly limited the regional economic development. However, a large amount of water and grass resources provided a higher ESV, indicating a high demand for ecological compensation in the area. Therefore, priority should be given to ecological compensation to promote the common development of ecology and economy. The more apparent the contradiction, the more serious the ecological degradation problem, and resolving the contradiction should be the main way to solve the problem, with ecological compensation as an auxiliary tool. As time passes, the gap in ecological compensation priority is gradually increasing, indicating that the economic development level gap between counties is gradually widening. Thus, focus on addressing the ecological degradation problems related to air quality regulation, regulation of water flows, erosion prevention, habitat services, and cultural and amenity services. Meanwhile, to deal with the widening gap between counties, ecological compensation should be given priority to Tianzhu, followed by Minqin, Gulang, and Liangzhou.

4.4. Driving Mechanism of ESV

We found that the impact trends of driving factors on ESV were consistent across different years, with an overall decreasing trend from 2000 to 2030 (Figure 7). Among the natural, socio-economic, and climatic factors considered, the geomorphic type had the highest q value, followed by soil type and DEM. Geomorphic, soil, and DEM were identified as the primary driving factors affecting regional ESV. This was because this study's area is located in an ecologically sensitive area with significant spatial differences in terrain. Our findings suggest that elevation plays an important role in the spatial distribution of ESV in the arid oasis area. Climate factors mainly affected the material exchange between underground soil and aboveground vegetation through changes in precipitation and temperature, ultimately impacting changes in regional ESV. In contrast, the socio-economic factors had the weakest driving force. The population density represented the degree of disturbance of human activities on ESV. The q value for GDP was the lowest, indicating that its impact on the spatial differentiation of ESV was the smallest. The low contribution rate of population density and GDP in the area was mainly due to the small proportion of urban areas and population distribution in this study's area. This is consistent with previous research, but the driving mechanisms of ESV differ significantly between the arid oasis area and the humid coastal area of Southeast China. In the humid southeast area, socio-economic factors such as GDP and population density are the main driving forces behind the loss of ESV, while in the northwest arid area, the increase in ESV is mainly driven by natural landscape patterns. This corresponds to the significant differences in socio-economic development and natural landscape between the humid southeast area and the arid northwest area.

Figure 6. Changes in the spatial distribution of EEH for different ecosystem services in different periods.

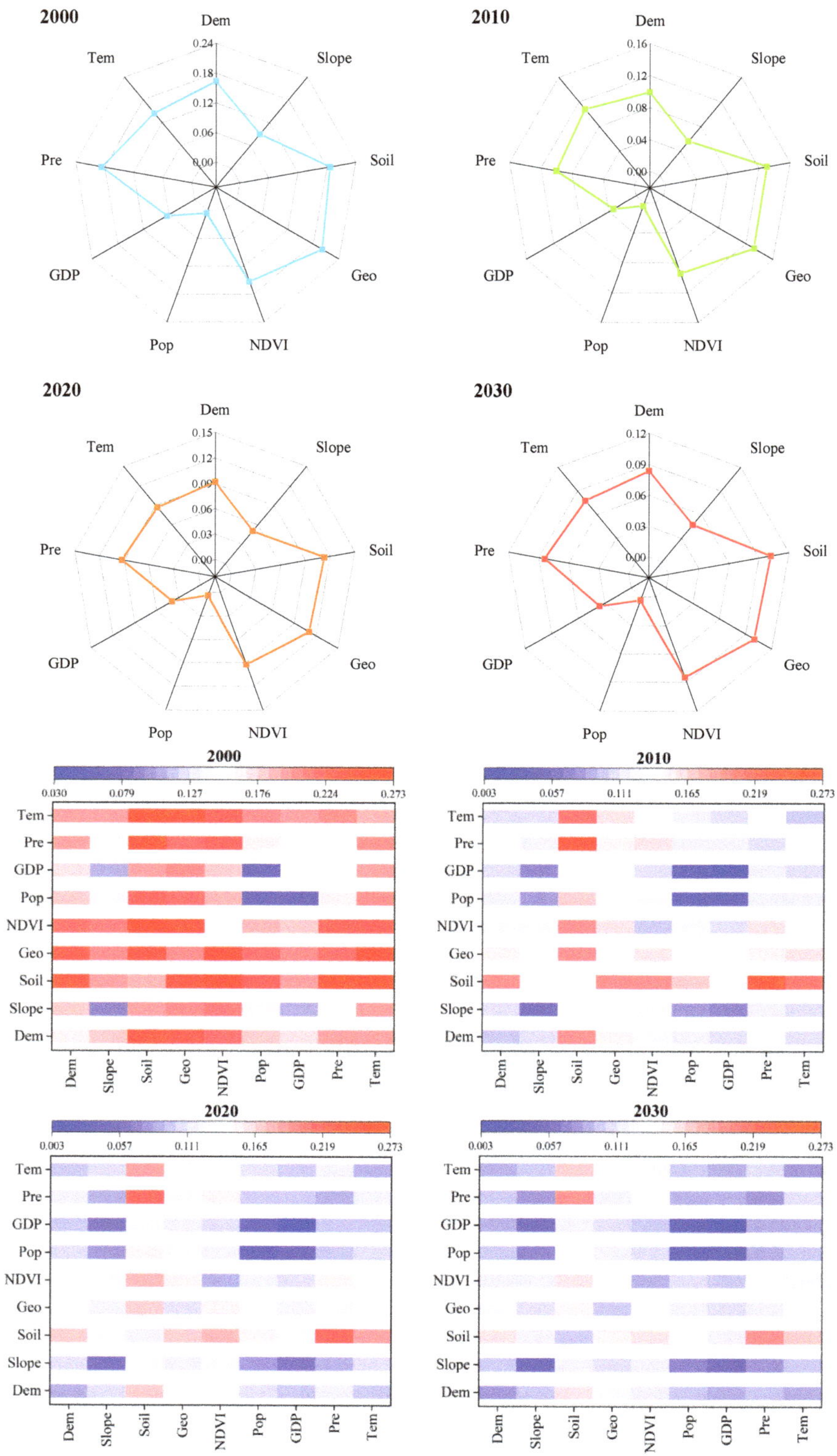

Figure 7. The contribution of single-factor (Radar map) and two-factor (Heat map) to ESV.

The results of exploring the interactions between factors showed that the driving factors had a synergistic and enhancing effect on ESV in the Wuwei Oasis. Specifically, there were two distinct modes of interaction, nonlinear enhancement, and two-factor enhancement. The results of two-factor interactions were consistent with those of single-factor interactions, with q values showing a decreasing trend from 2000 to 2030. However, the contribution of two-factor interactions was significantly higher than that of single-factor interactions, indicating a significant enhancement effect on the spatial differentiation of ESV. Notably, the interaction between natural factors and other factors had the most significant effect on ESV, with the q values exceeding the average value. Specifically, the interaction between geomorphic type and other factors had the greatest driving force.

5. Discussion

5.1. Model Advantages

The neighborhood effect plays a critical role in extracting transfer rules and calculating conversion probabilities in the dynamic simulation of the urban expansion [24,55]. Extracting temporal features is an essential part of this process, which determines the dependency relationships between model variables and parameters by computing gradients and storing them through a time-backward propagation [56]. Recurrent Neural Network (RNN) can pass the output and state of the current time as inputs to the next time, maintaining the data relationship between each time, and has been proven to be an effective deep learning model for processing time-series data [57]. However, RNN faces challenges such as vanishing and exploding gradients, which limit their ability to maintain long-term dependencies. Many excellent evolutionary models have been developed to optimize RNN, such as LSTM and GRU. LSTM adds memory units to address long-term dependency issues [58], while GRU reduces computational tensors by combining forget gates and input gates into a single update gate [52]. GRU also mixes cell states and hidden states, making the model more efficient and faster to train [59]. Applying batch normalization to the model optimizes the distribution width and offset, accelerates the network learning rate, and facilitates the gradient propagation [60]. ReLU, as an activation function, has sparse activation properties, enabling it to learn relatively sparse features from effective data dimensions and automatically decouple features to avoid overfitting [61]. In the context of abundant and comprehensive data, our model effectively harnesses regional land use data, in conjunction with key natural geographic and socio-economic factors, to undergo rigorous learning and training processes, iteratively fine-tuning various hyperparameters to achieve optimal performance. Although this study focused on a specific region, the model's underlying principles and methodologies were designed to be adaptable to different climatic zones. By carefully considering the environmental and socio-economic characteristics of various regions during model calibration and validation, this model's accuracy and reliability can be enhanced for use in diverse geographical contexts. Consequently, the applicability of this model extends beyond arid and semi-arid regions, encompassing a broader spectrum of climate zones, including humid and semi-humid areas.

Previous studies on the spatiotemporal variation of ESV have mainly focused on statistical analysis of quantitative data, with limited investigations on the underlying mechanisms driven by spatial factors. GeoDetector is highly inclusive in their analysis of data features. On the one hand, it can directly analyze quantified numerical values such as temperature and precipitation, which influence ecosystem services by regulating water and heat conditions and affecting biological behavior [62–64]. Additionally, socioeconomic factors such as GDP and population density directly impact ESV through human activities [65]. On the other hand, it quantifies qualitative numerical values before analysis. For example, natural factors such as soil type and geomorphic type, as the background elements of biological habitat in the ecosystem, have important functions in accumulating organic carbon and promoting water cycling. Changes in the background ecological conditions have a substantial impact on ecosystem services such as soil conservation, soil erosion, and biodiversity [66].

5.2. Relationship between Land Use and ESV

Between 2000 to 2020, hydrological regulation remained the dominant function in this study's area. By 2030, climate regulation will surpass hydrological regulation to become the dominant function in the arid oasis area of Wuwei. Despite this change, the proportion of most ecosystem service functions remained stable with no more than a 0.005 change, indicating a relatively stable structure of ecosystem service functions. Food production and soil erosion experienced the most significant decrease in proportion, while hydrological regulation increased by 0.009. This transformation primarily occurred in areas with more intense human activities, which aligns with the resource utilization characteristics in China's arid areas [67,68]. Qualitatively, the ESV of various ecosystem services in this study's area exhibited slight changes between 2000 and 2030, which can be visually observed in Figure 4. By comparison, the spatial distribution of ESV changes in the first 20 years was generally more drastic than those in the future 10 years, with continuous changes in the former and scattered changes in the latter. From 2000 to 2020, the degradation areas of food production, material production, air quality regulation, erosion prevention, and maintenance of soil fertility accounted for the largest proportion. The proportion of ESV losses and gains of other ecosystem services was relatively average. Although the distribution of ESV changes varied slightly among different ecosystem services, the spatial distribution changes in ESV for ecosystem services were generally greater than the overall ESV amplitude. This complexity underscores the importance of studying ESV characterization for different ecosystem services.

Land use change is a complex dynamic process that can have direct or indirect impacts on ecosystem services and ESV [69]. The increase or decrease in ESV in this study area is mainly contributed by farmland, forest, and grassland. Urban expansion, in particular, has occupied a considerable amount of ecological land, leading to a deterioration of the coupling coordination relationship between urban expansion and food production function. This phenomenon has caused varying degrees of damage to the original functions of the ecosystem, resulting in the problem of high-speed and low-quality urban expansion [70]. Due to natural geographic conditions, cultivated land is the most commonly occupied land type during urban expansion. The rapid reduction in cultivated land area disrupts the balanced ecological process and leads to a decline in the ecological system's food production value [71,72]. However, high-ESV land types, such as forests, wetlands, and water bodies, are the main drivers of ESV changes because their ESV per unit area is higher than that of cultivated land.

While previous research by Long, et al. [73] has shown that land use change due to urban expansion in the eastern coastal area of China has severely damaged the ecosystem and resulted in a decrease in ESV; our quantitative analysis of the Wuwei Oasis area's ESV indicates an opposite trend over time. This discrepancy can be attributed to differences in climate, topography, urban expansion speed, and ecological environment between the arid northwest area and the southeastern coastal area. To promote the sustainable development of such eco-fragile cities as Wuwei Oasis, it is essential to plan regional land use reasonably and optimize both economic and ecological benefits. Built-up land can reduce the ESV of this study's area, while the increase in ecological land, such as water bodies, wetlands, and forests, will lead to an increase in ESV. Therefore, optimizing both economic and ecological benefits, reducing ESV losses caused by unregulated development, and protecting land use types with high ESV are the most effective ways to increase ESV [74].

5.3. Insights and Recommendations on Ecological Compensation

In reality, ecological compensation schemes in arid oasis areas are still in their early stages, making the EEH prediction of ESV and its ecological compensation priority practically significant. ESV is a composite measure of diverse ecosystem services, including provisioning, regulating, supporting, and cultural services. Unfortunately, current research has only focused on total ESV policy and has not fully expressed the relationship between ESV and the various ecosystem services [75,76]. Based on the EEH prediction results of

Wuwei in the arid oasis area, we propose suggestions for its sustainable development. Firstly, measures must be taken to alleviate the degradation of air quality regulation, hydrological regulation, soil retention, biodiversity, and cultural services in Tianzhu and Minqin. The main way is to increase vegetation coverage through afforestation to neutralize carbon emissions in the atmosphere. It is also possible to prevent natural disasters such as drought, floods, and debris flow to prevent large-scale soil and water loss and to designate ecological protection zones to prevent the extinction of rare animals and plants. Secondly, to further quantify ecological compensation standards in the arid oasis area, the ESV, characterized by 11 ecosystem services, should be divided into natural contribution, human input, human preference, and natural contribution + human input, based on their importance and differences. Furthermore, our study has explored the driving factors of total ESV; natural and human factors also have certain impacts on ecosystem services. For example, a large terrain relief or excessive rainfall will accelerate surface runoff velocity, enhance soil erosion, and cause soil and water loss. A higher vegetation coverage of forest or grassland with certain canopy closures can reduce soil erosion and increase hydrological regulation and soil conservation ability. Unreasonable land use by humans may destroy surface vegetation and stable terrain, leading to the degradation of ecosystem services. Therefore, targeted exploration of the driving mechanisms of ESV characterized by different ecosystem services should be conducted to promote ecological and economic coordinated sustainable development.

5.4. Limitations and Future Perspectives

In this study, we employed a high-performance deep learning model to simulate the future ESV and ecological compensation in arid regions. The conducted investigation offers valuable practical implications for land use planning and ecological compensation policies. The main findings of this study are attractive for various regions and countries facing similar challenges in land use management and ecological compensation. The deep learning model's transferability can be evaluated by adapting it to different study areas and considering region-specific data and contextual factors [77]. In addition, the advanced land use simulation and geospatial analysis techniques facilitate the identification of ecologically sensitive areas, potential conflicts between economic development and environmental conservation, and opportunities for ecological compensation schemes [78]. The insights gained from our research can be utilized to inform policy development and land use planning in diverse geographic contexts.

However, it is crucial to acknowledge the inherent limitations of our research. Firstly, our land use simulation did not encompass multiple scenarios. While the baseline scenario captures one potential future development trajectory, the implementation of novel ecological and economic policies could exert notable influences on land use dynamics. As a result, future investigations could integrate historical trends of land use changes and pertinent policy considerations to furnish scientific underpinnings for territorial spatial planning and the advancement of sustainable urban development. By accounting for a broader range of scenarios, more comprehensive insights into the complex interplay between human activities and ecological systems can be attained, enhancing the utility and robustness of our findings. Moreover, further efforts in data collection and model refinement could aid in reducing uncertainties and refining the precision of our predictions, ensuring greater accuracy and applicability in decision-making processes and policy formulation.

6. Conclusions

Through the application of deep learning models and spatial analysis methods, this study provides valuable insights into the identification of priority areas for ecological compensation and the driving factors contributing to ESV in arid oasis areas. Results demonstrate that (1) Deep learning models effectively captured the spatiotemporal neighborhood features of land use dynamics, and CNN-GRU exhibited the highest accuracy and most accurately simulated the 2020 land use; (2) The built-up area of Wuwei Oasis is projected to increase by 25.35% from 2000 to 2030, resulting in a significant declin in ESV (−2.38%). Climate regulation was identified as the main contributor to ESV i this study, while the loss rate was also the highest. Wetlands and water bodies were th dominant factors affecting the change in ESV per area unit; (3) In the historical perio EEH was primarily characterized by low conflicts and potential crises, while potentia crises and high coordination will be the main features in the future. The coordinatio of Minqin and Tianzhu in the south and north of this study's area was generally highe than that of Liangzhou and Guluang in the east and west, and the urgency of ecologica compensation was correspondingly higher; (4) Natural factors had the most significan impact on ESV, and the explanatory power of bivariate interaction detection for ESV spatia differentiation increased significantly. Moreover, the contribution of single and multipl factors to ESV showed a decreasing trend from 2000 to 2030. Overall, the findings of thi study provide important insights that can inform strategies for promoting the restoratio of oasis ecosystems and sustainable urban development.

Author Contributions: Conceptualization, J.L. and X.P.; methodology, J.L.; software, J.L.; validatio J.L. and X.P.; formal analysis, J.L.; investigation, J.L.; resources, J.L.; data curation, J.L.; writing– original draft preparation, J.L.; writing—review and editing, X.P., W.Z. and J.J.; visualization, J.L supervision, X.P.; project administration, J.J.; funding acquisition, J.J. All authors have read an agreed to the published version of the manuscript.

Funding: This research was funded by the National Key R&D Program of China, grant numbe 2018YFC1903700.

Data Availability Statement: Not applicable.

Acknowledgments: The authors would like to thank the reviewers for their constructive comment

Conflicts of Interest: The authors declare no conflict of interest.

Appendix A

Table A1. Data Format and Source.

Category	Data	Data Format	Data Sources	Spatial Resolution
Traffic accessibility	distance to the settlement	vector (Point)	National Geographic Information Resource Directory Service System (https://webmap.cn/) accessed on 1 January 2022	30 m
	distance to road	vector (Polyline)	National Geographic Information Resource Directory Service System (https://webmap.cn/) accessed on 1 January 2022	30 m
	distance to railway	vector (Polyline)	National Geographic Information Resource Directory Service System (https://webmap.cn/) accessed on 1 January 2022	30 m
	distance to river	vector (Polyline)	National Geographic Information Resource Directory Service System (https://webmap.cn/) accessed on 1 January 2022	30 m

Table A1. *Cont.*

Category	Data	Data Format	Data Sources	Spatial Resolution
Social and economic conditions	distance to ecological function protection area	vector (Polygont)	Resource and Environmental Science and Data Center, Chinese Academy of Sciences (http://www.resdc.cn/) accessed on 1 January 2022	30 m
	population	raster	Resource and Environmental Science and Data Center, Chinese Academy of Sciences (http://www.resdc.cn/) accessed on 2 January 2022	30 m
	GDP	raster	Resource and Environmental Science and Data Center, Chinese Academy of Sciences (http://www.resdc.cn/) accessed on 2 January 2022	30 m
	nighttime lights	rasterd	Hubei high-resolution earth observation system application platform (http://59.175.109.173:8888) accessed on 2 January 2022	30 m
	NPP	raster	Resource and Environmental Science and Data Center, Chinese Academy of Sciences (http://www.resdc.cn/) accessed on 2 January 2022	30 m
Terrain conditions	elevation	raster	USGS Earth Explorer (https://earthexplorer.usgs.gov/) accessed on 3 January 2022	30 m
	slope	raster	USGS Earth Explorer (https://earthexplorer.usgs.gov/) accessed on 3 January 2022	30 m
	fault	vector (Polyline)	"Hydrogeological Map of Gansu Province" (Gansu Geological and Mineral Bureau Hydrogeological Engineering Geological Survey Institute) (http://www.gssgy.com/) accessed on 3 January 2022	30 m
Climatic conditions	temperature	raster	Resource and Environmental Science and Data Center, Chinese Academy of Sciences (http://www.resdc.cn/) accessed on 4 January 2022	30 m
	precipitation	raster	Resource and Environmental Science and Data Center, Chinese Academy of Sciences (http://www.resdc.cn/) accessed on 4 January 2022	30 m

References

1. Lu, Y.; Fu, B.; Feng, X.; Zeng, Y.; Liu, Y.; Chang, R.; Sun, G.; Wu, B. A Policy-Driven Large Scale Ecological Restoration: Quantifying Ecosystem Services Changes in the Loess Plateau of China. *PLoS ONE* **2012**, *7*, e31782. [CrossRef]
2. Wang, Y.; Li, X.; Zhang, Q.; Li, J.; Zhou, X. Projections of future land use changes: Multiple scenarios -based impacts analysis on ecosystem services for Wuhan city, China. *Ecol. Indic.* **2018**, *94*, 430–445. [CrossRef]
3. He, C.; Liu, Z.; Tian, J.; Ma, Q. Urban expansion dynamics and natural habitat loss in China: A multiscale landscape perspective. *Glob. Chang. Biol.* **2014**, *20*, 2886–2902. [CrossRef] [PubMed]
4. Zhang, D.; Wang, X.; Qu, L.; Li, S.; Lin, Y.; Yao, R.; Zhou, X.; Li, J. Land use/cover predictions incorporating ecological security for the Yangtze River Delta region, China. *Ecol. Indic.* **2020**, *119*, 106841. [CrossRef]
5. Mirbagheri, B.; Alimohammadi, A. Improving urban cellular automata performance by integrating global and geographically weighted logistic regression models. *Trans. Gis* **2017**, *21*, 1280–1297. [CrossRef]
6. Seto, K.C.; Gueneralp, B.; Hutyra, L.R. Global forecasts of urban expansion to 2030 and direct impacts on biodiversity and carbon pools. *Proc. Natl. Acad. Sci. USA* **2012**, *109*, 16083–16088. [CrossRef]
7. McDonald, R.I.; Guneralp, B.; Huan, C.-W.; Seto, K.C.; You, M. Conservation priorities to protect vertebrate endemics from global urban expansion. *Biol. Conserv.* **2018**, *224*, 290–299. [CrossRef]
8. Fensholt, R.; Langanke, T.; Rasmussen, K.; Reenberg, A.; Prince, S.D.; Tucker, C.; Scholes, R.J.; Le, Q.B.; Bondeau, A.; Eastman, R.; et al. Greenness in semi-arid areas across the globe 1981-2007—An Earth Observing Satellite based analysis of trends and drivers. *Remote Sens. Environ.* **2012**, *121*, 144–158. [CrossRef]
9. Tan, Z.; Guan, Q.; Lin, J.; Yang, L.; Luo, H.; Ma, Y.; Tian, J.; Wang, Q.; Wang, N. The response and simulation of ecosystem services value to land use/land cover in an oasis, Northwest China. *Ecol. Indic.* **2020**, *118*, 106711. [CrossRef]
10. Abulizi, A.; Yang, Y.; Mamat, Z.; Luo, J.; Abdulslam, D.; Xu, Z.; Zayiti, A.; Ahat, A.; Halik, W. Land-Use Change and its Effects in Charchan Oasis, Xinjiang, China. *Land Degrad. Dev.* **2017**, *28*, 106–115. [CrossRef]
11. Fu, Q.; Li, B.; Hou, Y.; Bi, X.; Zhang, X. Effects of land use and climate change on ecosystem services in Central Asia's arid regions: A case study in Altay Prefecture, China. *Sci. Total Environ.* **2017**, *607*, 633–646. [CrossRef]
12. Zhang, T.; Du, Z.; Yang, J.; Yao, X.; Ou, C.; Niu, B.; Yan, S. Land Cover Mapping and Ecological Risk Assessment in the Context of Recent Ecological Migration. *Remote Sens.* **2021**, *13*, 1381. [CrossRef]
13. Guan, Q.; Zhao, R.; Pan, N.; Wang, F.; Yang, Y.; Luo, H. Source apportionment of heavy metals in farmland soil of Wuwei, China: Comparison of three receptor models. *J. Clean. Prod.* **2019**, *237*, 117792. [CrossRef]
14. Kindu, M.; Schneider, T.; Doellerer, M.; Teketay, D.; Knoke, T. Scenario modelling of land use/land cover changes in Munessa-Shashemene landscape of the Ethiopian highlands. *Sci. Total Environ.* **2018**, *622*, 534–546. [CrossRef] [PubMed]
15. Zhu, S.; Zhao, Y.; Huang, J.; Wang, S. Analysis of Spatial-Temporal Differentiation and Influencing Factors of Ecosystem Services in Resource-Based Cities in Semiarid Regions. *Remote Sens.* **2023**, *15*, 871. [CrossRef]
16. Zhou, T.; Chen, W.; Wang, Q.; Li, Y. Urbanisation and ecosystem services in the Taiwan Strait west coast urban agglomeration, China, from the perspective of an interactive coercive relationship. *Ecol. Indic.* **2023**, *146*, 109861. [CrossRef]
17. Wu, Y.; Tao, Y.; Yang, G.; Ou, W.; Pueppke, S.; Sun, X.; Chen, G.; Tao, Q. Impact of land use change on multiple ecosystem services in the rapidly urbanizing Kunshan City of China: Past trajectories and future projections. *Land Use Policy* **2019**, *85*, 419–427. [CrossRef]
18. Li, F.; Wang, L.; Chen, Z.; Clarke, K.C.; Li, M.; Jiang, P. Extending the SLEUTH model to integrate habitat quality into urban growth simulation. *J. Environ. Manag.* **2018**, *217*, 486–498. [CrossRef]
19. Zhang, D.; Huang, Q.; He, C.; Wu, J. Impacts of urban expansion on ecosystem services in the Beijing-Tianjin-Hebei urban agglomeration, China: A scenario analysis based on the Shared Socioeconomic Pathways. *Resour. Conserv. Recycl.* **2017**, *125*, 115–130. [CrossRef]
20. Xiao, R.; Yu, X.; Zhang, Z.; Wang, X. Built-up land expansion simulation with combination of naive Bayes and cellular automaton model-A case study of the Shanghai-Hangzhou Bay agglomeration. *Growth Chang.* **2021**, *52*, 1804–1825. [CrossRef]
21. Wu, Q.; Guan, F.; Lv, C.; Huang, Y. Ultra-short-term multi-step wind power forecasting based on CNN-LSTM. *IET Renew. Power Gener.* **2021**, *15*, 1019–1029. [CrossRef]
22. Wang, H.; Zhao, X.; Zhang, X.; Wu, D.; Du, X. Long Time Series Land Cover Classification in China from 1982 to 2015 Based on Bi-LSTM Deep Learning. *Remote Sens.* **2019**, *11*, 1639. [CrossRef]
23. Zhai, Y.; Yao, Y.; Guan, Q.; Liang, X.; Li, X.; Pan, Y.; Yue, H.; Yuan, Z.; Zhou, J. Simulating urban land use change by integrating a convolutional neural network with vector-based cellular automata. *Int. J. Geogr. Inf. Sci.* **2020**, *34*, 1475–1499. [CrossRef]
24. Qian, Y.; Xing, W.; Guan, X.; Yang, T.; Wu, H. Coupling cellular automata with area partitioning and spatiotemporal convolution for dynamic land use change simulation. *Sci. Total Environ.* **2020**, *722*, 137738. [CrossRef]
25. Liu, J.; Xiao, B.; Li, Y.; Wang, X.; Jiao, J. Simulation of Dynamic Urban Expansion under Ecological Constraints Using a Long Short Term Memory Network Model and Cellular Automata. *Remote Sens.* **2021**, *13*, 1499. [CrossRef]
26. Cao, J.; Wang, Y.; He, J.; Liang, W.; Tao, H.; Zhu, G. Predicting Grain Losses and Waste Rate Along the Entire Chain: A Multitask Multigated Recurrent Unit Autoencoder Based Method. *IEEE Trans. Ind. Inform.* **2021**, *17*, 4390–4400. [CrossRef]
27. Chen, Z.; Xia, T.; Li, Y.; Pan, E. A hybrid prognostic method based on gated recurrent unit network and an adaptive Wiener process model considering measurement errors. *Mech. Syst. Signal Process.* **2021**, *158*, 107785. [CrossRef]

8. Gao, X.; Zeng, S.; Shen, J.; Yang, X.; Kang, L.; Chi, C.; Song, R. Predicting payment for ecosystem services regarding land use: A simulation study in China. *Environ. Impact Assess. Rev.* **2023**, *98*, 106972. [CrossRef]
9. Ren, Y.; Lu, L.; Yu, H.; Zhu, D. Game strategies in government-led eco-compensation in the Xin'an River Basin from the perspective of the politics of scale. *J. Geogr. Sci.* **2021**, *31*, 1205–1221. [CrossRef]
0. Jiang, Y.; Zhang, J.; Chen, K.; Xue, X.; Michael, A.U. Moving towards a systematic marine eco-compensation mechanism in China: Policy, practice and strategy. *Ocean. Coast. Manag.* **2019**, *169*, 10–19. [CrossRef]
1. Wu, L.; Jin, L. How eco-compensation contribute to poverty reduction: A perspective from different income group of rural households in Guizhou, China. *J. Clean. Prod.* **2020**, *275*, 122962. [CrossRef]
2. Xie, Y.; Kong, F.; Zhang, J.; Li, Y.; Huang, G.; Zhang, W. Medium- and Long-Term Planning of an Integrated Eco-Compensation System Considering Ecological Water Demand under Uncertainty: A Case Study of Daguhe Watershed in China. *J. Water Resour. Plan. Manag.* **2022**, *148*, 04022049. [CrossRef]
3. Xiao, W.; Qu, L.; Li, K.; Guo, C.; Li, J. An Assessment of the Rational Range of Eco-Compensation Standards: A Case Study in the Nujiang Prefecture, Southwestern China. *Land* **2022**, *11*, 1417. [CrossRef]
4. Liu, R.; Xu, H.; Li, J.; Pu, R.; Sun, C.; Cao, L.; Jiang, Y.; Tian, P.; Wang, L.; Gong, H. Ecosystem service valuation of bays in East China Sea and its response to sea reclamation activities. *J. Geogr. Sci.* **2020**, *30*, 1095–1116. [CrossRef]
5. Wang, Z.; Wang, Y.; Zhou, Z.; Yu, F.; Ma, D.; Li, J. Combining spatial planning and ecosystem services value to assist ecological compensation decision-making-A case study of Yangtze River Delta ecological barrier, China. *Front. Environ. Sci.* **2022**, *10*, 1002014. [CrossRef]
6. Liu, Z.; Wu, R.; Chen, Y.; Fang, C.; Wang, S. Factors of ecosystem service values in a fast-developing region in China: Insights from the joint impacts of human activities and natural conditions. *J. Clean. Prod.* **2021**, *297*, 126588. [CrossRef]
7. Chen, S.; Feng, Y.; Tong, X.; Liu, S.; Xie, H.; Gao, C.; Lei, Z. Modeling ESV losses caused by urban expansion using cellular automata and geographically weighted regression. *Sci. Total Environ.* **2020**, *712*, 136509. [CrossRef] [PubMed]
8. Luo, Q.; Luo, Y.; Zhou, Q.; Song, Y. Does China's Yangtze River Economic Belt policy impact on local ecosystem services? *Sci. Total Environ.* **2019**, *676*, 231–241. [CrossRef]
9. Song, W.; Deng, X. Land-use/land-cover change and ecosystem service provision in China. *Sci. Total Environ.* **2017**, *576*, 705–719. [CrossRef]
0. Zhang, X.; Li, H.; Xia, H.; Tian, G.; Yin, Y.; Lei, Y.; Kim, G. The Ecosystem Services Value Change and Its Driving Forces Responding to Spatio-Temporal Process of Landscape Pattern in the Co-Urbanized Area. *Land* **2021**, *10*, 1043. [CrossRef]
1. Chen, M.; Lu, Y.; Ling, L.; Wan, Y.; Luo, Z.; Huang, H. Drivers of changes in ecosystem service values in Ganjiang upstream watershed. *Land Use Policy* **2015**, *47*, 247–252. [CrossRef]
2. Huang, M.; Fang, B.; Yue, W.; Feng, S. Spatial differentiation of ecosystem service values and its geographical detection in Chaohu Basin during 1995–2017. *Acta Geogr. Sin.* **2019**, *38*, 2790–2803.
3. Wang, J.; Xu, C. Geodetector: Principle and prospective. *Acta Geogr. Sin.* **2017**, *72*, 19.
4. Wang, J.F.; Hu, Y. Environmental health risk detection with GeogDetector. *Environ. Model. Softw.* **2012**, *20*, 114–115. [CrossRef]
5. Liu, X.; Chen, X.; Hua, K.; Wang, Y.; Wang, P.; Han, X.; Ye, J.; Wen, S. Effects of Land Use Change on Ecosystem Services in Arid Area Ecological Migration. *Chin. Geogr. Sci.* **2018**, *28*, 894–906. [CrossRef]
6. Bauer, S. Identification of Water-Reuse Potentials to Strengthen Rural Areas in Water-Scarce Regions-The Case Study of Wuwei. *Land* **2020**, *9*, 492. [CrossRef]
7. Kadulkar, S.; Howard, M.P.; Truskett, T.M.; Ganesan, V. Prediction and Optimization of Ion Transport Characteristics in Nanoparticle-Based Electrolytes Using Convolutional Neural Networks. *J. Phys. Chem. B* **2021**, *125*, 4838–4849. [CrossRef]
8. Fei, X.; Zhang, Y.; Zheng, W. XB-SIM*: A Simulation Framework for Modeling and Exploration of ReRAM-Based CNN Acceleration Design. *Tsinghua Sci. Technol.* **2021**, *26*, 322–334. [CrossRef]
9. Xie, J.; Chen, S.; Zhang, Y.; Gao, D.; Liu, T. Combining generative adversarial networks and multi-output CNN for motor imagery classification. *J. Neural Eng.* **2021**, *18*, 046026. [CrossRef]
0. Rehman, S.U.; Khaliq, M.; Imtiaz, S.I.; Rasool, A.; Shafiq, M.; Javed, A.R.; Jalil, Z.; Bashir, A.K. DIDDOS: An approach for detection and identification of Distributed Denial of Service (DDoS) cyberattacks using Gated Recurrent Units (GRU). *Future Gener. Comput. Syst.* **2021**, *118*, 453–466. [CrossRef]
1. Huang, N.; Nie, F.; Ni, P.; Luo, F.; Gao, X.; Wang, J. NeuralPolish: A novel Nanopore polishing method based on alignment matrix construction and orthogonal Bi-GRU Networks. *Bioinformatics* **2021**, *37*, 3120–3127. [CrossRef] [PubMed]
2. Huang, G.; Li, X.; Zhang, B.; Ren, J. PM2.5 concentration forecasting at surface monitoring sites using GRU neural network based on empirical mode decomposition. *Sci. Total Environ.* **2021**, *768*, 144516. [CrossRef] [PubMed]
3. Xie, G.; Zhang, C.; Zhen, L.; Zhang, L. Dynamic changes in the value of China's ecosystem services. *Ecosyst. Serv.* **2017**, *26*, 146–154. [CrossRef]
4. Wang, X.; Xu, M.; Zhang, Y.; Xu, N.; Zhang, Y. Evaluation of Eco-economy Harmony and Spatial Evolution of the Urban Agglomeration Area in The Great Pearl River Delta. In Proceedings of the 2nd International Workshop on Renewable Energy and Development (IWRED), Guilin, China, 20–22 April 2018. [CrossRef]
5. Liao, J.; Tang, L.; Shao, G.; Su, X.; Chen, D.; Xu, T. Incorporation of extended neighborhood mechanisms and its impact on urban land-use cellular automata simulations. *Environ. Model. Softw.* **2016**, *75*, 163–175. [CrossRef]

56. Aburas, M.M.; Ahamad, M.S.S.; Omar, N.Q. Spatio-temporal simulation and prediction of land-use change using conventiona and machine learning models: A review. *Environ. Monit. Assess.* **2019**, *191*, 205. [CrossRef]
57. Liu, L.; Sun, X.-K. Volcanic Ash Cloud Diffusion From Remote Sensing Image Using LSTM-CA Method. *IEEE Access* **2020**, 54681–54690. [CrossRef]
58. Jia, X.; Khandelwal, A.; Nayak, G.; Gerber, J.; Carlson, K.; West, P.; Kumar, V. *Incremental Dual-Memory LSTM in Land Cove Prediction*; ACM: New York, NY, USA, 2017; pp. 867–876.
59. Yan, L.; Wang, H.; Wang, H.; Liu, Z. An integrated GRU based real-time prognostic method towards uncertainty quantificatio *Meas. Sens.* **2021**, *18*, 100220. [CrossRef]
60. Lian, J.; Dong, P.; Zhang, Y.; Pan, J.; Liu, K. A Novel Data-Driven Tropical Cyclone Track Prediction Model Based on CNN an GRU With Multi-Dimensional Feature Selection. *IEEE Access* **2020**, *8*, 97114–97128. [CrossRef]
61. Zhang, N.; Zhang, N.; Zheng, Q.; Xu, Y.S. Real-time prediction of shield moving trajectory during tunnelling using GRU dee neural network. *Acta Geotech.* **2021**, *17*, 1167–1182. [CrossRef]
62. Mina, M.; Bugmann, H.; Cordonnier, T.; Irauschek, F.; Klopcic, M.; Pardos, M.; Cailleret, M. Future ecosystem services fro European mountain forests under climate change. *J. Appl. Ecol.* **2017**, *54*, 389–401. [CrossRef]
63. Braun, D.; de Jong, R.; Schaepman, M.E.; Furrer, R.; Hein, L.; Kienast, F.; Damm, A. Ecosystem service change caused b climatological and non-climatological drivers: A Swiss case study. *Ecol. Appl. A Publ. Ecol. Soc. Am.* **2019**, *29*, e01901. [CrossRe [PubMed]
64. Prather, C.M.; Pelini, S.L.; Laws, A.; Rivest, E.; Woltz, M.; Bloch, C.P.; Toro, I.D.; Ho, C.K.; Kominoski, J.; Newbold, T.A. Invertebrates, ecosystem services and climate change. *Biol. Rev.* **2013**, *88*, 327–348. [CrossRef]
65. Su, S.; Li, D.; Hu, Y.; Xiao, R.; Zhang, Y. Spatially non-stationary response of ecosystem service value changes to urbanization i Shanghai, China. *Ecol. Indic.* **2014**, *45*, 332–339. [CrossRef]
66. Khl, L.; Oehl, F.; Heijden, M.G.A.V.D. Agricultural practices indirectly influence plant productivity and ecosystem service through effects on soil biota. *Ecol. Appl.* **2014**, *24*, 1842–1853. [CrossRef] [PubMed]
67. Jiang, W. Ecosystem services research in China: A critical review. *Ecosyst. Serv.* **2017**, *26*, 10–16. [CrossRef]
68. He, C.; Li, J.; Zhang, X.; Liu, Z.; Zhang, D. Will rapid urban expansion in the drylands of northern China continue: A scenari analysis based on the Land Use Scenario Dynamics-urban model and the Shared Socioeconomic Pathways. *J. Clean. Prod.* **201** *165*, 57–69. [CrossRef]
69. Tripathi, R.; Moharana, K.C.; Nayak, A.D.; Dhal, B.; Shahid, M.; Mondal, B.; Mohapatra, S.D.; Bhattacharyya, P.; Fitton, N.; Smit P.; et al. Ecosystem services in different agro-climatic zones in eastern India: Impact of land use and land cover change. *Enviro Monit. Assess.* **2019**, *191*, 98. [CrossRef]
70. Salvati, L.; Zambon, I.; Chelli, F.M.; Serra, P. Do spatial patterns of urbanization and land consumption reflect differen socioeconomic contexts in Europe? *Sci. Total Environ.* **2018**, *625*, 722–730. [CrossRef]
71. Msofe, N.K.; Sheng, L.; Li, Z.; Lyimo, J. Impact of Land Use/Cover Change on Ecosystem Service Values in the Kilombero Valle Floodplain, Southeastern Tanzania. *Forests* **2020**, *11*, 109. [CrossRef]
72. Rukundo, E.; Liu, S.; Dong, Y.; Rutebuka, E.; Asamoah, E.F.; Xu, J.; Wu, X. Spatio-temporal dynamics of critical ecosystem service in response to agricultural expansion in Rwanda, East Africa. *Ecol. Indic.* **2018**, *89*, 696–705. [CrossRef]
73. Long, H.; Liu, Y.; Hou, X.; Li, T.; Li, Y. Effects of land use transitions due to rapid urbanization on ecosystem services: Implication for urban planning in the new developing area of China. *Habitat Int.* **2014**, *44*, 536–544. [CrossRef]
74. Chuai, X.; Huang, X.; Wu, C.; Li, J.; Lu, Q.; Qi, X.; Zhang, M.; Zuo, T.; Lu, J. Land use and ecosystems services value changes an ecological land management in coastal Jiangsu, China. *Habitat Int.* **2016**, *57*, 164–174. [CrossRef]
75. Fan, H.; Xu, J.; Gao, S. Modeling the dynamics of urban and ecological binary space for regional coordination: A case of Fuzho coastal areas in Southeast China. *Habitat Int.* **2018**, *72*, 48–56. [CrossRef]
76. Xing, X.; Yang, X.; Guo, J.; Chen, A.; Zhang, M.; Yang, D.; Hou, Z.; Zhang, H.; Wang, X. Response of ecosystem services i Beijing-Tianjin Sandstorm Source Control Project to differing engineering measures scenarios. *J. Clean. Prod.* **2023**, *384*, 13557 [CrossRef]
77. van Duynhoven, A.; Dragicevic, S. Assessing the Impact of Neighborhood Size on Temporal Convolutional Networks fo Modeling Land Cover Change. *Remote Sens.* **2022**, *14*, 4957. [CrossRef]
78. Vaissiere, A.-C.; Meinard, Y. A policy framework to accommodate both the analytical and normative aspects of biodiversity i ecological compensation. *Biol. Conserv.* **2021**, *253*, 108897. [CrossRef]

remote sensing

MDPI

Article

Hybrid-Scale Hierarchical Transformer for Remote Sensing Image Super-Resolution

Jianrun Shang [1], Mingliang Gao [1], Qilei Li [2], Jinfeng Pan [1,*], Guofeng Zou [1] and Gwanggil Jeon [1,3]

1 School of Electrical and Electronic Engineering, Shandong University of Technology, Zibo 255000, China; 21404020515@stumail.sdut.edu.cn (J.S.); mlgao@sdut.edu.cn (M.G.); gfzou@sdut.edu.cn (G.Z.); gjeon@inu.ac.kr (G.J.)
2 School of Electronic Engineering and Computer Science, Queen Mary University of London, London E1 4NS, UK; q.li@qmul.ac.uk
3 Department of Embedded Systems Engineering, Incheon National University, Incheon 22012, Republic of Korea
* Correspondence: pjfbysj@163.com

Abstract: Super-resolution (SR) technology plays a crucial role in improving the spatial resolution of remote sensing images so as to overcome the physical limitations of spaceborne imaging systems. Although deep convolutional neural networks have achieved promising results, most of them overlook the advantage of self-similarity information across different scales and high-dimensional features after the upsampling layers. To address the problem, we propose a hybrid-scale hierarchical transformer network (HSTNet) to achieve faithful remote sensing image SR. Specifically, we propose a hybrid-scale feature exploitation module to leverage the internal recursive information in single and cross scales within the images. To fully leverage the high-dimensional features and enhance discrimination, we designed a cross-scale enhancement transformer to capture long-range dependencies and efficiently calculate the relevance between high-dimension and low-dimension features. The proposed HSTNet achieves the best result in PSNR and SSIM with the UCMecred dataset and AID dataset. Comparative experiments demonstrate the effectiveness of the proposed methods and prove that the HSTNet outperforms the state-of-the-art competitors both in quantitative and qualitative evaluations.

Keywords: super-resolution; remote sensing image; convolutional neural network; transformer; self-similarity

Citation: Shang, J.; Gao, M.; Li, Q.; Pan, J.; Zou, G.; Jeon, G. Hybrid-Scale Hierarchical Transformer for Remote Sensing Image Super-Resolution. *Remote Sens.* **2023**, *15*, 3442. https://doi.org/10.3390/rs15133442

Academic Editors: Prashant K. Srivastava and Salah Bourennane

Received: 19 April 2023
Revised: 21 June 2023
Accepted: 30 June 2023
Published: 7 July 2023

1. Introduction

With the rapid progress of satellite platforms and optical remote sensing technology, remote sensing images (RSIs) have been broadly deployed in civilian and military fields, e.g., disaster prevention, meteorological forecast, military mapping, and missile warning [1,2]. However, due to hardware limitations and environmental restrictions [3,4], RSIs often suffer from low-resolution (LR) and contain some intrinsic noise. Upgrading physical imaging equipment to improve resolution is often plagued by high costs and long development cycles. Therefore, it is of utmost urgency to explore the remote sensing image super-resolution (RSISR).

Single-image super-resolution (SR) is a highly ill-posed visual problem which aims to reconstruct high-resolution (HR) images from corresponding degraded LR images. To this end, many representative algorithms have been proposed, which can be roughly divided into three categories, i.e., interpolation-based methods [5,6], reconstruction-based methods [7,8], and learning-based methods [9,10]. The interpolation-based methods generally utilize different interpolation operations, including bilinear interpolation, bicubic interpolation, and nearest interpolation, to estimate unknown pixel value [11]. These methods are relatively straightforward in practice, while the reconstructed images lack essential details. In contrast, reconstruction-based methods improve image quality by incorporating prior information of the image as constraints into the HR image. These methods can restore high-frequency details with the help of prior knowledge, while they require substantial computational costs, making it difficult for them to be readily applied to RSIs [12]. Learning-based approaches try to produce HR images by learning the mapping relationship established between external LR–HR image training pairs. Compared with the aforementioned two lines of methods, learning-based methods achieve better performance and become the mainstream in this domain due to the powerful feature representation ability provided by convolutional neural networks (CNNs) [13]. However, learning-based methods generally adopt the post-upsampling framework [14], which solely exploits low-dimensional features while ignoring the discriminative high-dimensional feature information after the upsampling process.

In addition to utilizing nonlinear mapping between LR–HR image training pairs, the self-similarity of the image is also employed to improve the performance of SR algorithms. Self-similarity refers to the property of similar patches appear repeatedly in a single image and is broadly adopted in image denoising [15,16], deblurring [17], and SR [18–20]. Self-similarities are also an intrinsic property in RSIs, i.e., internal recursive information. Figure 1 illustrates the self-similarities in RSIs. One can see that the down-scaled image is on the left, and the original one is on the right. Similar highway patches with green box labels appear repeatedly in the same scale image, while the roof of factories with red box labels appear repeatedly across different scales, and these patches with similar edges and textures contain abundant internal recursive information. Previously, Pan et al. [21] employed dictionary learning to capture structural self-similarity features as additional information to improve the performance of the model. However, the sparse representation of SR has a limited ability to leverage the internal recursive information within the entire remote sensing image.

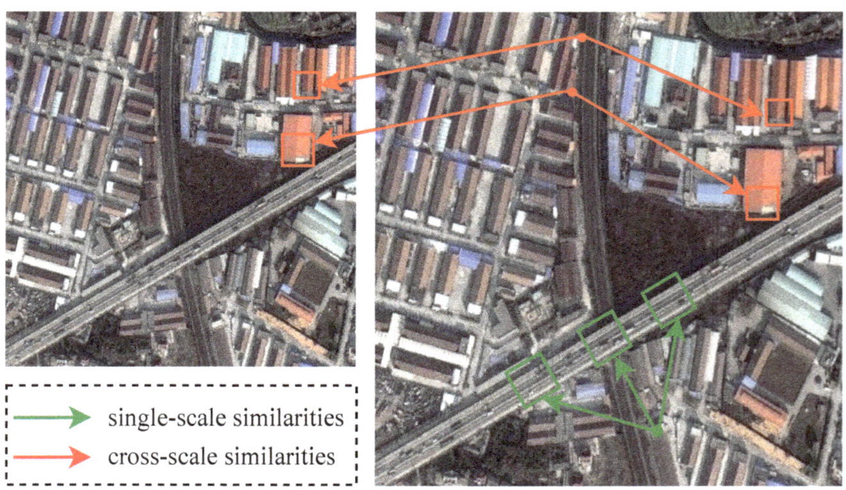

Figure 1. Illustration of self-similarities in RSIs with single-scale (green box) and cross-scale (red box).

In this paper, we propose a Hybrid-Scale Hierarchical Transformer Network (HSTNet) for RSISR. The HSTNet can enhance the representation of the high-dimensional features after upsampling layers and fully utilize the self-similarity information in RSIs. Specifically, we propose a hybrid-scale feature exploitation (HSFE) module to leverage the internal similar information both in single and cross scales within the images. The HSFE module contains two branches, i.e., a single-scale branch and a cross-scale branch. The former is employed to capture the recurrence within the same scale image, and the latter is utilized to learn the feature correlation across different scales. Moreover, we designed a cross-scale enhancement transformer (CSET) module to capture long-range dependencies and efficiently model the relevance between high-dimension and low-dimension features. In the CSET module, the encoders are used to encode low-dimension features from the HSFE module, and the decoder is used to utilize to fuse the multiple hierarchies high-/low-dimensional features so as to enhance the representation ability of high-dimensional features. To sum up, the main contributions of this work are as follows:

1. We propose an HSFE module with two branches to leverage the internal recursive information from both single and cross scales within the images for enriching the feature representations for RSISR.
2. We designed a CSET module to capture long-range dependencies and efficiently calculate the relevance between high-dimension and low-dimension features. It helps the network reconstruct SR images with rich edges and contours.
3. Jointly incorporating the HSFE and CSET modules, we formed the HSTNet for RSISR. Extensive experiments on two challenging remote sensing datasets verify the superiority of the proposed model.

2. Related Literature

2.1. CNN-Based SR Models

Dong et al. [22] pioneered the adoption of an SR convolutional neural network (SR-CNN) that utilizes three convolution layers to establish the nonlinear mapping relationship between LR–HR image training pairs. On the basis of the residual network introduced by He et al. [23], Kim et al. [24] designed a very deep SR convolutional neural network (VDSR) where residual learning is employed to accelerate model training and improve reconstruction quality. Lim et al. [25] built the enhanced deep super-resolution model to simplify the network and improve the computational efficiency via optimizing the initial residual block. Zhang et al. [26] designed a deep residual dense network in which the residual network with dense skip connections is used to transfer intermediate features. Benefiting from the channel attention (CA) module, Zhang et al. [27] presented a deep residual channel attention network to enhance the high-frequency channel feature representation. Dai et al. [28] designed a second-order CA mechanism to guide the model to improve the ability of discriminative learning ability and exploit more conducive features. Li et al. [29] proposed an image super-resolution feedback network (SRFBN) in which a feedback mechanism is adopted to transfer high-level feature information. The SRFBN could leverage high-level features to polish up the representation of low-level features.

Because of the impact of spatial resolution on the final performance of many RSI tasks, including instance segmentation, object detection, and scene classification, RSISR also raises significant research interest. Lei et al. [30] proposed a local–global combined network (LGC-Net) which can enhance the multilevel representations, including local detail features and global information. Haut et al. [31] produced a deep compendium model (DCM), which leverages skip connection and residual unit to exploit more informative features. To fuse different hierarchical contextual features efficiently, Wang et al. [32] designed a contextual transformation network (CTNet) based on a contextual transformation layer and contextual feature aggregation module. Ni et al. [33] designed a hierarchical feature aggregation and self-learning network in which both self-learning and feedback mechanisms are employed to improve the quality of reconstruction images. Wang et al. [34] produced a multiscale fast Fourier transform (FFT)-based attention network (MSFFTAN), which employs a multi-input U-shape structure as the backbone for accurate RSISR. Liang et al. [35] presented a multiscale hybrid attention graph convolution neural network for RSISR in which a hybrid attention mechanism was adopted to obtain more abundant critical high-frequency information. Wang et al. [36] proposed a multiscale enhancement network which utilizes multiscale features of RSIs to recover more high-frequency details. However, the CNN-based methods above generally employ the post-upsampling framework that directly recovers HR images after the upsampling layer, ignoring the discriminative high-dimensional feature information after the upsampling process [14].

2.2. Transformer-Based SR Models

Due to the strong long-range dependence learning ability of transformers, transforme based image SR methods have been studied recently by many scientific researcher Yang et al. [37] produced a texture transformer network for image super-resolution, in whic a learnable texture extractor is utilized to exploit and transmit the relevant textures to L images. Liang et al. [38] proposed SwinIR by transferring the ability of the Swin Tran former, which could achieve competitive performance on three representative tasks, namel image denoising, JPEG compression artifact reduction, and image SR. Fang et al. [39] de signed a lightweight hybrid network of a CNN and transformer that can extract beneficia features for image SR with the help of local and non-local priors. Lu et al. [40] presented hybrid model with a CNN backbone and transformer backbone, namely the efficient supe resolution transformer, which achieved impressive results with low computational cos Yoo et al. [41] introduced an enriched CNN–transformer feature aggregation network i which the CNN branch and transformer branch can mutually enhance each representatio during the feature extraction process. Due to the limited ability of multi-head self-attentio to extract cross-scale information, cross-token attention is adopted in the transforme branch to utilize information from tokens of different scales.

Recently, transformers have also found their way into the domain of RSISR. Lei et al. [14 proposed a transformer-based enhancement network (TransENet) to capture features fron different stages and adopted a multistage-enhanced structure that can integrate features fron different dimensions. Ye et al. [42] proposed a transformer-based super-resolution metho for RSIs, and they employed self-attention to establish dependencies relationships withi local and global features. Tu et al. [43] presented a GAN that draws on the strengths o the CNN and Swin Transformer, termed the SWCGAN. The SWCGAN fully considers th characteristics of large size, a large amount of information, and a strong relevance betwee pixels required for RSISR. He et al. [44] designed a dense spectral transformer to extract th long-range dependence for spectral super-resolution. Although the transformer can improv the long-range dependence learning ability of the model, these methods do not leverage th self-similarity within the entire remote sensing image [45].

3. Methodology

3.1. Overall Framework

The framework of the proposed HSTNet is shown in Figure 2. It is built by the comb nation of three kinds of fundamental modules, i.e., a low-dimension feature extraction (LFE module, a cross-scale enhancement transformer (CSET) module, and an upsample modul Specifically, the LFE module is utilized to extract high-frequency features across differen scales, and the CSET module is employed to capture long-range dependency to enhanc the final feature representation. The upsample module is adopted to transform the featur representation from a low-dimensional space to a high-dimensional space.

Given an LR image I_{LR}, a convolutional layer with a 3×3 kernel is utilized to extrac the initial feature F_0. The process of shallow feature extraction is formulated as

$$F_0 = f_{\mathrm{sf}}(I_{\mathrm{LR}}), \quad (1$$

where $f_{\mathrm{sf}}(\cdot)$ represents the operation of the convolutional operation and F_0 is the sha low feature.

As shown in Figure 3, the LFE module consists of five basic extraction (BE) module and each BE module contains two 3×3 convolution layers and one hybrid-scale featur exploitation (HSFE) module. As the core component of the BE module, the HSFE modul is proposed to model image self-similarity. The whole low-dimensional feature extractio process is formulated as

$$F^i_{\mathrm{LFE}} = f^i_{\mathrm{lfe}}\left(F^{i-1}_{\mathrm{LFE}}\right) = f^i_{\mathrm{lfe}}\left(f^{i-1}_{\mathrm{lfe}}\left(\cdots f^1_{\mathrm{lfe}}(F_0)\cdots\right)\right), \quad i = 1,2,3, \quad (2$$

where $f^i_{\rm lfe}(\cdot)$ and $F^i_{\rm LFE}$ represent the operation of ith LFE module and its output. After the three cascaded LFE modules, a subpixel layer [46] is adopted to transform low-dimensional features into high-dimensional features, which is formulated as

$$F_{\rm up} = {\rm Subpixel}\left(F^3_{\rm LFE}\right), \tag{3}$$

where $F_{\rm up}$ represents the high-dimension feature and Subpixel$(\cdot)$ denotes the function of the subpixel layer. The low-dimension features $F^1_{\rm LFE}$, $F^2_{\rm LFE}$, and $F^3_{\rm LFE}$ and the high-dimension feature $F_{\rm up}$ are fed into three cascaded CSET modules for feature hierarchical enhancement. To reduce the redundancy of the enhanced features, a 1×1 convolution layer is employed to reduce the feature dimension. The complete process including the enhancement and dimension reduction is formulated as

$$F^i_{CSET} = \begin{cases} f^i_{\rm cset}\left(F^i_{\rm LFE}, F^{i+1}_{\rm CSET}\right), & i = 1,2, \\ f^i_{\rm cset}\left(F^i_{\rm LFE}, F_{\rm up}\right), & i = 3, \end{cases} \tag{4}$$

where $f^i_{\rm cset}(\cdot)$ and $F^i_{\rm CSET}$ represent the operation of ith CSET module and its output, respectively. Finally, one convolution layer is employed to obtain SR image $I_{\rm SR}$ from the enhanced features. A conventional L_1 loss function was employed to train the proposed HSTNet model. Given a training set $\{I^i_{\rm LR}, I^i_{\rm HR}\}^N_{i=1}$, the loss function is formulated as:

$$L(\theta) = \frac{1}{N}\sum_{i=1}^{N}\left\|F_{\rm HSTNet}\left(I^i_{\rm LR}\right) - I^i_{\rm HR}\right\|_1 , \tag{5}$$

where $F_{\rm HSTNet}$ denotes the proposed model parameterized by θ and N represents the number of training LR–HR pairs.

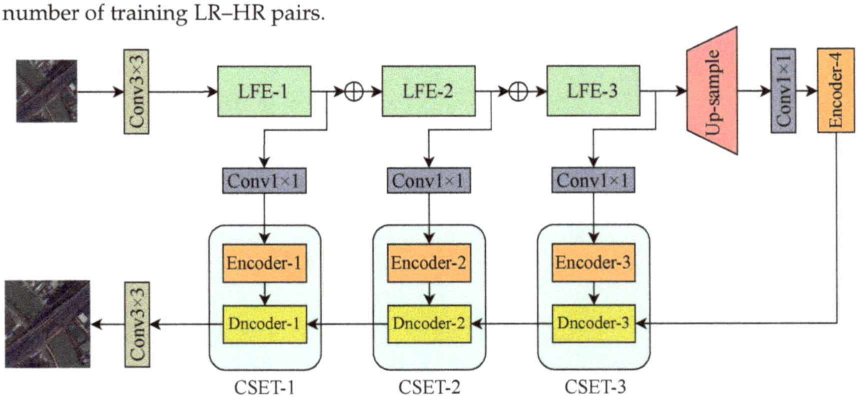

Figure 2. Architecture of the proposed HSTNet for remote sensing image SR.

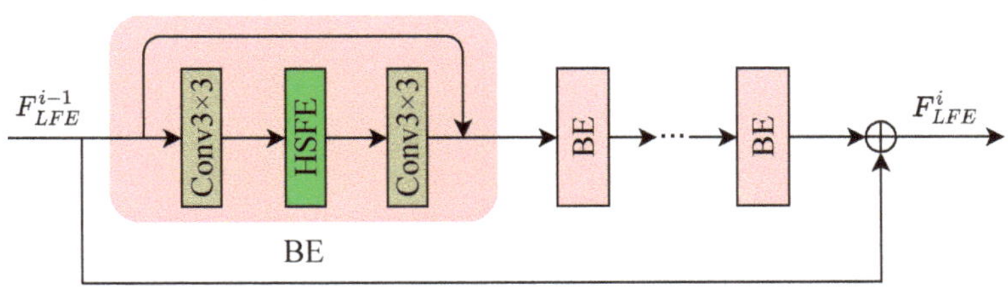

Figure 3. Architecture of the LFE module.

3.2. Hybrid-Scale Feature Exploitation Module

To explore the internal recursive information in single-scale and cross-scale, we propose an HSFE module. Figure 4 exhibits the architecture of the HSFE module, which consists of a single-scale branch and a cross-scale branch. The single-scale branch aims to capture similar features within the same scale, and a non-local (NL) block [47] was utilized to calculate the relevance of these features. The cross-scale branch was applied to capture recursive features of the same image at different scales, and an adjusted non-local (ANL) block [45] was utilized to calculate the relevance of features between two different scales.

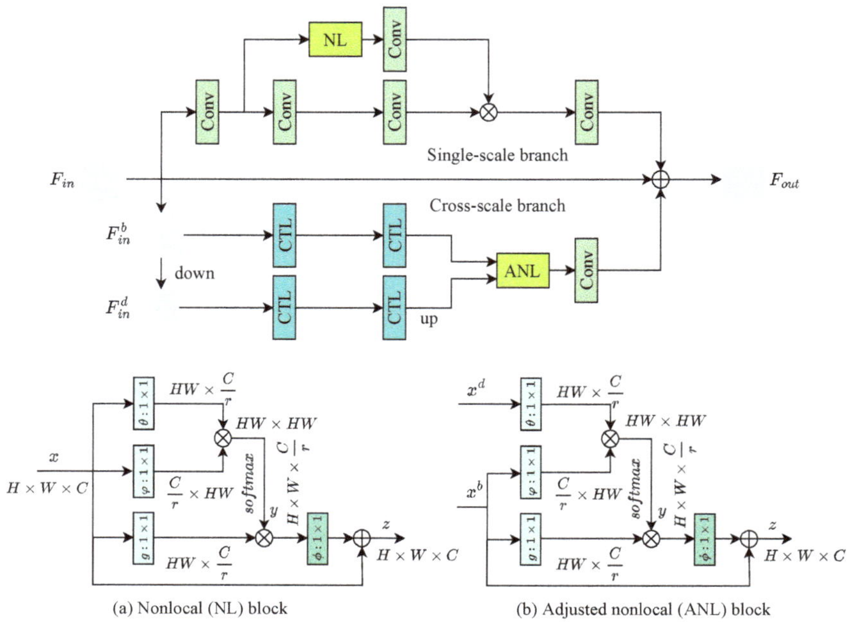

Figure 4. Architecture of the proposed HSFE module.

Single-scale branch: As depicted in Figure 4, we built the single-scale branch to extract single-scale features. Specifically, in the single-scale branch, several convolutional layers are applied to capture recursive features, and an NL block is employed to guide the network to concentrate on informative areas. As shown in Figure 4a, an embedding function is utilized to mine the similarity information as

$$f(x_i, x_j) = e^{(\theta^T(x_i)\varphi(x_j))} = e^{((W_\theta x_i)^T(W_\varphi x_j))}, \tag{6}$$

where i is the index of the output position, j is the index that enumerates all positions, and x denotes the input of the NL block. W_θ and W_φ are the embeddings weight matrix. The non-local function is symbolized as

$$y_i = \left(\sum_{\forall j} f(x_i, x_j) g(x_j)\right) \Big/ \sum_{\forall j} f(x_i, x_j). \tag{7}$$

The relevance between x_i and all x_j can be calculated by pairwise function $f(\cdot)$. The feature representation of x_j can be obtained by the function $g(\cdot)$. Eventually, the output of the NL block is obtained by

$$z_i = W_\phi y_i + x_i, \tag{8}$$

where W_ϕ is a weight matrix.

The convolution layer following the NL block transforms the input into an attention diagram, which is then normalized with a sigmoid function. In addition, the main branch output features are multiplied by the attention diagram, where the activation values for each space and channel location are rescaled.

Cross-scale branch: As depicted in Figure 4, the cross-scale branch is utilized to perform cross-scale feature representation. Specifically, the input of the HSFE module is considered the basic scale feature, which is symbolized as F_{in}^{b}. To exploit the internal recursive information at different scales, the downsampled scale feature F_{in}^{d} is formulated as

$$F_{in}^{d} = f_{\text{down}}^{s}\left(F_{in}^{b}\right), \tag{9}$$

where $f_{\text{down}}^{s}(\cdot)$ denotes the operation of downsampling with scale factor s.

Two contextual transformation layers (CTLs) [48] are employed to extract feature with two different scales F_{in}^{b} and F_{in}^{d}. To align the spatial dimension of the features in different scales, the downsampled feature is firstly upsampled with the scale factor of s. x^{b} and x^{b} represent the output of the basic scale and the downsampled scale through the two CTLs, and this process is formulated as

$$\begin{aligned} x^{b} &= f_{ctl}\left(F_{in}^{b}\right) \\ x^{d} &= f_{up}^{s}\left(f_{ctl}\left(F_{in}^{b}\right)\right), \end{aligned} \tag{10}$$

where $f_{ctl}(\cdot)$ denotes the operation of two CTLs and $f_{up}^{s}(\cdot)$ represents the operation of upsample with scale factor s.

Similar to the single-scale branch, an ANL block [45] was introduced to exploit the feature correlation between two different scales RSIs. As shown in Figure 4b, the ANL block is improved compared to the NL block, and they have different inputs. Thus, z_i in Equation (8) for ANL block can be rewritten as

$$f\left(x_i^{d}, x_j^{b}\right) = e^{\left(\theta^{T}\left(x_i^{d}\right)\varphi\left(x_j^{b}\right)\right)} = e^{\left(\left(W_{\theta}x_i^{d}\right)^{T}\left(W_{\varphi}x_j^{b}\right)\right)}, \tag{11}$$

$$y_i = \left(\sum_{\forall j} f\left(x_i^{d}, x_j^{b}\right) g\left(x_j^{b}\right)\right) \Big/ \sum_{\forall j} f\left(x_i^{b}, x_j^{d}\right) \tag{12}$$

$$z_i = W_{\phi} y_i + x_i. \tag{13}$$

In the cross-scale branch, we employ the ANL block to fuse multiple scale features, therefore fully utilizing the self-similarity information. The HSFE module can be formulated as

$$F_{out} = f_{sin}(F_{in}) + f_{cro}(F_{in}) + F_{in}, \tag{14}$$

where F_{in} is the input of the HSFE module and F_{out} is the output of the HSFE module. $f_{sin}(\cdot)$ and $f_{cro}(\cdot)$ are the operation of the single-scale branch and cross-scale branch, respectively.

3.3. Cross-Scale Enhancement Transformer Module

The cross-scale enhancement transformer module is designed to learn the dependency relationship across long distances between high-dimension and low-dimension features and enhance the final feature representation. The architecture of the CSET module is shown in Figure 5a. Specifically, we introduced the cross-scale token attention (CSTA) module [41] to exploit the internal recursive information within an input image across different scales. Moreover, we use three CSET modules to utilize different hierarchies of feature information. Figure 5a illustrates in detail the procedure of feature enhancement using CSET-3 module as an example.

Transformer encoder: The encoders are used to encode different hierarchies of features from LFE modules. As shown in Figure 5a, the encoder is mainly composed of a multi-headed self-attention (MHSA) block and a feed-forward network (FFN) block, which is similar to the original design in [49]. The FFN block contains two multilayer perceptron (MLP) layers with an expansion ratio r and a GELU activation function [50] in the middle. Moreover, we adopted layer normalization (LN) before the MHSA block and FFN block, and employed a local residual structure to avoid the gradient vanishing or explosion during gradient backpropagation. The entire process of the encoder can be formulated as

$$\begin{aligned} {F_{EN}^{i}}' &= f_{mhsa}\left(f_{ln}\left(F_{LFE}^{i}\right)\right) + F_{LFE}^{i} \\ F_{EN}^{i} &= f_{ffn}\left(f_{ln}\left({F_{EN}^{i}}'\right)\right) + {F_{EN}^{i}}', \end{aligned} \quad (15)$$

where $f_{mhsa}(\cdot)$, $f_{ln}(\cdot)$, and $f_{ffn}(\cdot)$ denote the function of the MHSA block, layer normalization, and FFN block, respectively. ${F_{EN}^{i}}'$ is the intermediate output of the encoder. F_{LFE}^{i} and F_{EN}^{i} are the input and output of the encoder in the ith CSET module.

Transformer decoder: The decoders are utilized to fuse high-/low-dimensional features from multiple hierarchies to enhance the representation ability of high-dimensional features. As shown in Figure 5a, the decoder contains two MHSA blocks and a CSTA block [41]. With the CSTA block, the decoder can exploit the recursive information within an input image across different scales. The operation of the decoder can be formulated as

$$\begin{aligned} {F_{DE}^{i}}'' &= f_{csta}(f_{ln}(F_{up})) + F_{up} \\ {F_{DE}^{i}}' &= f_{mhsa}\left(f_{ln}\left({F_{DE}^{i}}''\right), {F_{EN}^{i}}'\right) + {F_{DE}^{i}}'' \\ F_{CSET}^{i} &= f_{mhsa}\left(f_{ln}\left({F_{DE}^{i}}'\right)\right) + {F_{DE}^{i}}' \end{aligned} \quad (16)$$

where $f_{csta}(\cdot)$ denotes the process of the CSTA block and F_{up} is the output of Encoder-4. Each CSET module has two inputs, and the composition of the inputs is determined by the location of the CSET module. ${F_{DE}^{i}}'$ and ${F_{DE}^{i}}''$ represent the intermediate outputs of the decoder. F_{CSET}^{i} represents the output of ith CSET module.

CSTA block: The CSTA block [41] is introduced to utilize the recurrent patch information of different scales in the input image. The feature information flow of the CSTA module is illustrated in Figure 5b. Specifically, the input token embeddings $T \in \mathbb{R}^{n \times d}$ of the CSTA block are split into $T^{a} \in \mathbb{R}^{n \times \frac{d}{2}}$ and $T^{b} \in \mathbb{R}^{n \times \frac{d}{2}}$ along the channel axis. Then $T^{s} \in \mathbb{R}^{n \times \frac{d}{2}}$ including n tokens from $T^{a}$and $T^{l} \in \mathbb{R}^{n' \times d'}$ including n' tokens by rearranging T^{b} are generated. The number of tokens in T^{l} can be set to $n' = \left[\frac{h-t'}{s'} + 1\right] \times \left[\frac{w-t'}{s'} + 1\right]$ where t' and s' represent the stride and token size. To improve efficiency, T^{s} is replaced by T^{a}, and T^{l} is tokenized with a larger token size and overlapping. Numerous large-size tokens can be obtained by overlapping, enabling the transformer to actively learn patch recurrence across scales.

To effectively exploit self-similarity across different scales, we computed cross-scale attention scores between tokens in both T^{s} and T^{l}. Specifically, the queries $q^{s} \in \mathbb{R}^{n \times \frac{d}{2}}$, keys $k^{s} \in \mathbb{R}^{n \times \frac{d}{2}}$, and values $v^{s} \in \mathbb{R}^{n \times \frac{d}{2}}$ were generated from T^{s}. Similarly, the queries $q^{l} \in \mathbb{R}^{n' \times \frac{d}{2}}$, keys $k^{l} \in \mathbb{R}^{n' \times \frac{d}{2}}$, and values $v^{l} \in \mathbb{R}^{n' \times \frac{d}{2}}$ were generated from T^{l}. The reorganized triples $\left(q^{s}, k^{l}, v^{l}\right)$ and $\left(q^{l}, k^{s}, v^{s}\right)$ were obtained by swapping their key–value pairs to each other. Then, the attention operation was executed using the reorganized triples. It should be noted that the projection of attention operations reduces the last dimension of queries, keys, and values in T^{l} from d' to $d/2$. Subsequently, we re-projected the attention results of T^{l} to the dimension of $n' \times d'$ then transformed to the dimension of $n \times \frac{d}{2}$. Finally, we concatenated the attention results to obtain the output of the CSTA block.

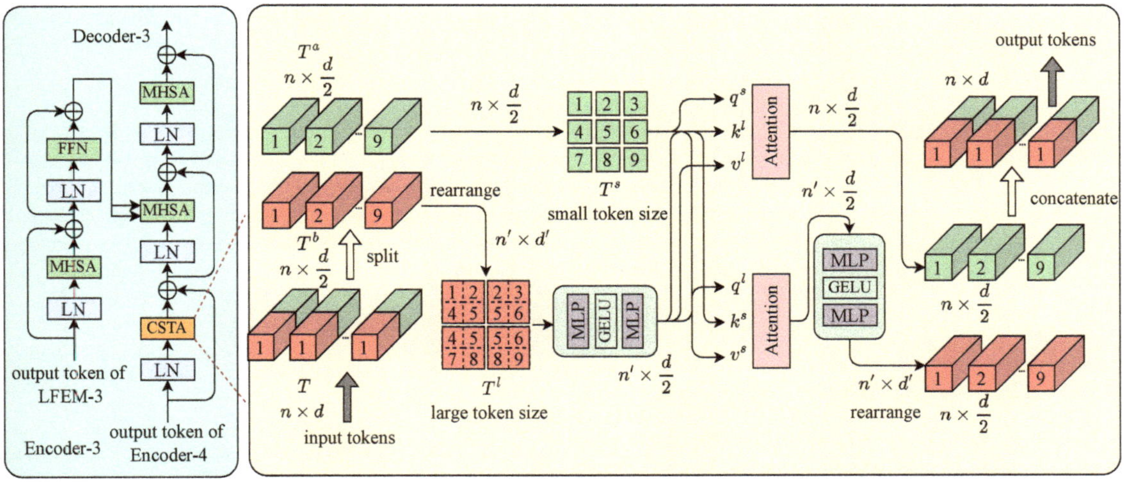

(a) Transformer module (b) Cross-scale token attention (CSTA) block

Figure 5. Architecture of the CSET module.

4. Experiments

4.1. Experimental Dataset and Settings

We evaluate the proposed method on two widely adopted benchmarks [30,31,51], namely the UCMecred dataset [52] and AID dataset [53], to demonstrate the effectiveness of the proposed HSTNet.

UCMerced dataset: This dataset consists of 2100 images belonging to 21 categories of varied remote sensing image scenes. All images exhibit a pixel size of 256 × 256 and a spatial resolution of 0.3 m/pixel. The dataset is divided equally into two distinct sets, one comprising 1050 images for training and the other for testing.

AID dataset: This dataset encompasses 10,000 remote sensing images, spanning 30 unique categories. In contrast to the UCMerced dataset, all images in this dataset have a pixel size of 600 × 600 and spatial resolution of 0.5 m/pixel. A selection of 8000 images from this dataset was randomly chosen for the purpose of training, while the remaining 2000 images were used for testing. In addition, a validation set consisting of five arbitrary images from each category was established.

To verify the generalization of the proposed method, we further adapted the trained model to the real-world images of Gaofen-1 and Gaofen-2 satellites. We downsampled HR images through bicubic operations to obtain LR images. Two mainstream metrics, namely peak signal-to-noise ratio (PSNR) and structural similarity index measure (SSIM), were calculated on the Y channel of the YCbCr space for objective evaluation. They are formulated as

$$PSNR(I_{SR}, I_{HR}) = 10 \cdot \log 10 \times \left(\frac{L^2}{\frac{1}{N}\sum_{i=1}^{N}(I_{SR}(i) - I_{HR}(i))^2} \right), \tag{17}$$

where L represents the maximum pixel, and N denotes the number of all pixels in I_{SR} and I_{HR}.

$$SSIM(x, y) = \frac{2u_x u_y + k_1}{u_x^2 + u_y^2 + k_1} \cdot \frac{\sigma_{xy} + k_2}{\sigma_x^2 + \sigma_y^2 + k_2}, \tag{18}$$

where x and y represent two images. σ_{xy} symbolizes the covariance between x and y. u and σ^2 represent the average value and variance. k_1 and k_2 denote constant relaxation terms. Multi-adds and model parameters were utilized to evaluate the computational complexity [32,54]. In addition, the natural image quality evaluator (NIQE) was adopted to validate the reconstruction of real-world images from Gaofen-1 and Gaofen-2 satellites [55].

4.2. Implementation Details

We conducted experiments on remote sensing image data with scale factors of ×2, ×3, and ×4. During training, we randomly cropped 48 × 48 patches from LR images and extracted ground-truth references from corresponding HR images. We also employed horizontal flipping and random rotation (90°, 180° and 270°) to augment training samples. Table 1 presents the comprehensive hyperparameter setting of the cross-scale enhancement transformer (CSET) module.

Table 1. Parameter setting of the CSET module in the HSTNet.

	Heads	Head Dim	Hidden Size D	MLP Dim	Layers
Transformer Encoder	6	32	512	512	8
Transformer Decoder	6	32	512	512	1

We adopted the Adam optimizer [56] to train the HSTNet with β_1 = 0.9, β_2 = 0.99, and $\epsilon = 10^{-8}$. The initial learning rate was set to 10^{-4}, and the batch size was 16. The proposed model was trained for 800 epochs, and the learning rate decreased by half after 400 epochs. Both the training and testing stages were performed using the PyTorch framework, utilizing CUDAToolkit 11.4, cuDNN 8.2.2, Python 3.7, and two NVIDIA 3090 Ti GPUs.

4.3. Comparison with Other Methods

To verify the effectiveness of the proposed HSTNet, we conducted comparative experiments with some state-of-the-art (SOTA) competitors, namely SC [12], SRCNN [22], FSRCNN [57], VDSR [24], LGCNet [30], DCM [31], CTNet [48], ESRT [40], ACT [41], and TransENet [14]. Among these methods, SC [12], SRCNN [22], FSRCNN [57], VDSR [24], ESRT [40], and ACT [41] are the methods proposed for natural image SR. LGCNet [30], DCM [31], CTNet [48], and TransENet [14] are designed for RSISR. The experimental results for the UCMerced dataset and AID dataset with the scale factors of ×2, ×3 and ×4 are reported in Table 2.

4.3.1. Quantitative Evaluation

Evaluation with UCMerced dataset: Table 2 shows that the proposed HSTNet achieves first place among competitors for the UCMerced dataset for all scale factors. Specifically, the HSTNet improves the PSNR comparatively by 0.71 dB, 0.54 dB, and 0.60 dB for scale factor ×2 for LGCNet [30], DCM [31] and CTNet [48], respectively. The average PSNR values of the proposed HSTNet over the second-best TransENet that employs a transformer module are 0.16 dB, 0.15 dB and 0.12 dB when the scale factors are ×2, ×3 and ×4, respectively. Additionally, the HSTNet outperforms LGCNet [30], DCM [31], and CTNet [48] in terms of SSIM by 0.0183, 0.0027, and 0.0102 for scale factor ×3. Compared to ACT [41], which also uses a transformer structure, the average PSNR obtained by the proposed method increased by 0.31 dB, 0.27 dB, and 0.35 dB at scale factors of ×2, ×3 and ×4, respectively. Moreover, Table 3 lists the mean PSNR of different methods on all 21 classes (All these 21 classes of UCMerced dataset: 1—Agricultural, 2—Airplane, 3—Baseballdiamond, 4—Beach, 5—Buildings, 6—Chaparral, 7—Denseresidential, 8—Forest, 9—Freeway, 10—Golfcourse, 11—Harbor, 12—Intersection, 13—Mediumresidential, 14—Mobilehomepark, 15—Overpass, 16—Parkinglot, 17—River, 18—Runway, 19—Sparseresidential, 20—Storagetanks, and 21—Tenniscourt) of the UCMerced dataset when the scale factor is ×3. One can see that the proposed HSTNet performs best in 14 scene classes, ranks second in 5 scene classes, and third in 2 scene classes. The DCM [31] obtains the best PSNR in the other seven categories. It is worth mentioning that the HSTNet shows more effective performance in some scenes comprising prominent contours and rich edges, such as "Baseballdiamond",

"Buildings", and "Overpass". Overall, the mean PSNR in all 21 class scenes of the proposed HSTNet is 0.55 dB higher than DCM [31].

Table 2. Comparative results for the UCMerced dataset and AID dataset. The best and the second-best results are marked in red and blue, respectively.

Method	Scale	UCMerced Dataset		AID Dataset	
		PSNR	SSIM	PSNR	SSIM
Bicubic	×2	30.76	0.8789	32.39	0.8906
SC [12]	×2	32.77	0.9166	32.77	0.9166
SRCNN [22]	×2	32.84	0.9152	34.49	0.9286
FSRCNN [57]	×2	33.18	0.9196	34.11	0.9228
VDSR [24]	×2	33.47	0.9234	35.05	0.9346
LGCNet [30]	×2	33.48	0.9235	34.80	0.9320
DCM [31]	×2	33.65	0.9274	35.21	0.9366
CTNet [48]	×2	33.59	0.9255	35.13	0.9354
ESRT [40]	×2	33.70	0.9270	35.15	0.9358
ACT [41]	×2	33.88	0.9283	35.17	0.9362
TransENet [14]	×2	34.03	0.9301	35.28	0.9374
Ours	×2	34.19	0.9338	35.35	0.9387
Bicubic	×3	27.46	0.7631	29.08	0.7863
SC [12]	×3	28.26	0.7971	28.26	0.7671
SRCNN [22]	×3	28.66	0.8038	30.55	0.8372
FSRCNN [57]	×3	29.09	0.8167	30.30	0.8302
VDSR [24]	×3	29.34	0.8263	31.15	0.8522
LGCNet [30]	×3	29.28	0.8238	30.73	0.8417
DCM [31]	×3	29.52	0.8394	31.31	0.8561
CTNet [48]	×3	29.44	0.8319	31.16	0.8527
ESRT [40]	×3	29.52	0.8318	31.34	0.8562
ACT [41]	×3	29.80	0.8395	31.39	0.8579
TransENet [14]	×3	29.92	0.8408	31.45	0.8595
Ours	×3	30.07	0.8421	31.61	0.8613
Bicubic	×4	25.65	0.6725	27.30	0.7036
SC [12]	×4	26.51	0.7152	26.51	0.7152
SRCNN [22]	×4	26.78	0.7219	28.40	0.7561
FSRCNN [57]	×4	26.93	0.7267	28.03	0.7387
VDSR [24]	×4	27.11	0.7360	28.99	0.7753
LGCNet [30]	×4	27.02	0.7333	28.61	0.7626
DCM [31]	×4	27.22	0.7528	29.17	0.7824
CTNet [48]	×4	27.41	0.7512	29.00	0.7768
ESRT [40]	×4	27.41	0.7485	29.18	0.7831
ACT [41]	×4	27.54	0.7531	29.19	0.7836
TransENet [14]	×4	27.77	0.7630	29.38	0.7909
Ours	×4	27.89	0.7694	29.57	0.7983

Evaluation with AID dataset: Table 2 reports the averaged evaluation results of the proposed method in comparison to other methods for AID datasets for scale factors of ×2, ×3, and ×4. One can see that the proposed HSTNet outperforms SRCNN [22], FSRCNN [57], and VDSR [24] by 1.17 dB, 1.54 dB, and 0.58 dB for scale factors ×4 in terms of PSNR values. It proves that the HSTNet ranks first with PSNR scores that are higher than LGCNet [30] by 0.55 dB, 0.88 dB, and 0.96 dB for scale factors ×2, ×3, and ×4, respectively. Compared to ESRT [40], which adopts a transformer structure, the average PSNR obtained by the proposed method increased by 0.20 dB, 0.27 dB, and 0.39 dB at scale factors of ×2, ×3, and ×4, respectively. Compared to the second-best method, TransENet [14], the HSTNet achieves a performance improvement of 0.16 dB and 0.0013 in PSNR and SSIM scores, respectively, for scale factor ×3. In contrast to the UCMerced dataset, the AI dataset comprises 30 categories of scenes and a significantly larger number of image Table 4 reports a detailed performance comparison of different methods for scale facto ×4 on all 30 scene classes (All these 30 classes of AID dataset: 1—Airport, 2—Barelan 3—Baseballdiamond, 4—Beach, 5—Bridge, 6—Center, 7—Church, 8—Commercial, 9- Denseresidential, 10—Desert, 11—Farmland, 12—Forest, 13—Industrial, 14—Meadow, 15- Mediumresidential, 16—Mountain, 17—Park, 18—Parking, 19—Playground, 20—Pon 21—Port, 22—Railwaystation, 23—Resort, 24—River, 25—School, 26—Sparseresidentia 27—Square, 28—Stadium, 29—Storagetanks, 30—Viaduct) of the AID dataset. It can b seen that the proposed HSTNet outperforms the other methods in 28 scene classes, whil TransENet [14] obtains the best PSNR scores in the remaining 2 categories. Although th HSTNet ranks second in those two scene classes, its PSNR values are very close to th TransENet [14]. Notably, the HSTNet has an overall average PSNR that is 0.48 dB highe than TransENet [14].

4.3.2. Qualitative Evaluation

To further verify the advantages of the proposed method, the subjective results o SR images reconstructed by the aforementioned methods are shown in Figures 6 and Figure 6 shows the reconstruction results of the above methods for the UCMerced datase by taking "airplane" and "runway" scenes as examples. Figure 7 shows the visual result of the "stadium" and "medium-residential" scenes in the AID dataset. In general, the S results reconstructed by the proposed method possess sharper edges and clearer contour compared with other methods, which verifies the effectiveness of the HSTNet.

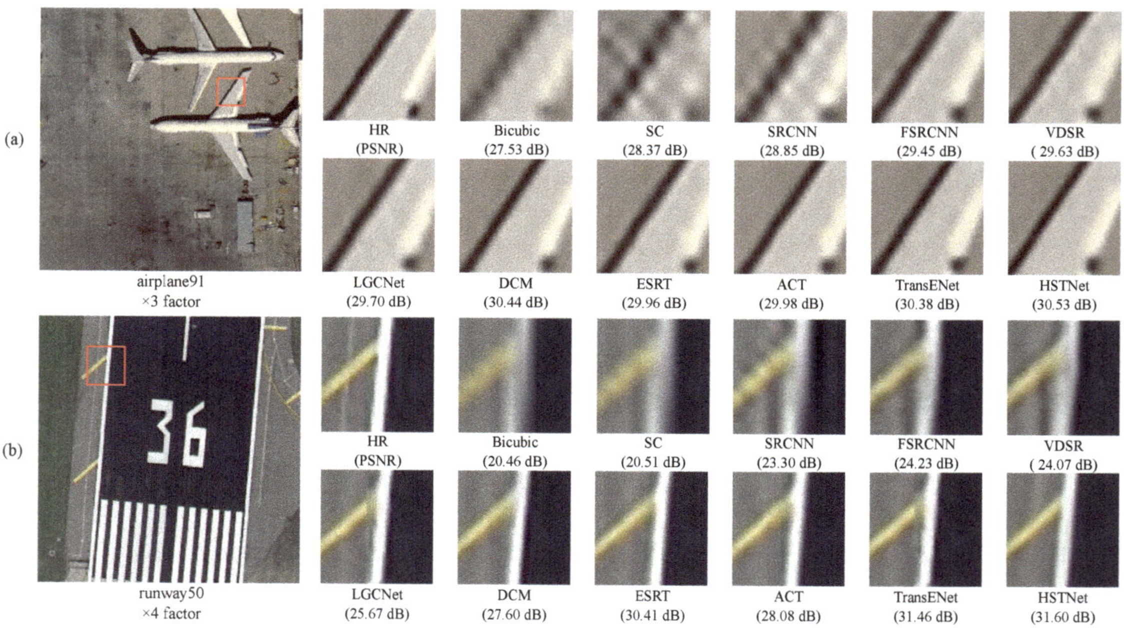

Figure 6. Subjective results for UCMerced dataset: (**a**) "Airplane91" scene with ×3 factor. (**b**) "Runway50" scene with ×4 factor.

Table 3. Average PSNR of per-category for UCMerced dataset with the scale factor of ×3. The best and the second-best results are marked in red and blue, respectively.

Class	Bicubic	SC [12]	SRCNN [22]	FSRCNN [57]	LGCNet [30]	DCM [31]	CTNet [48]	ESRT [40]	ACT [41]	TransENet [14]	Ours
1	26.86	27.23	27.47	27.61	27.66	29.06	28.53	28.13	27.86	28.02	27.93
2	26.71	27.67	28.24	28.98	29.12	30.77	29.22	29.45	29.78	29.94	29.98
3	33.33	34.06	34.33	34.64	34.72	33.76	34.81	34.88	35.05	35.04	35.13
4	36.14	36.87	37.00	37.21	37.37	36.38	37.38	37.45	37.55	37.53	37.76
5	25.09	26.11	26.84	27.50	27.81	28.51	27.99	28.18	28.66	28.81	29.12
6	25.21	25.82	26.11	26.21	26.39	26.81	26.40	26.43	26.62	26.69	26.78
7	25.76	26.75	27.41	28.02.	28.25	28.79	28.42	28.53	28.97	29.11	29.27
8	27.53	28.09	28.24.	28.35	28.44	28.16	28.48	28.47	28.56	28.59	28.65
9	27.36	28.28	28.69	29.27	29.52	30.45	29.60	29.87	30.25	30.38	30.65
10	35.21	35.92	36.15	36.43	36.51	34.43	36.46	36.54	36.63	36.68	36.69
11	21.25	22.11	22.82	23.29	23.63	26.55	23.83	23.87	24.42	24.72	24.91
12	26.48	27.20	27.67	28.06	28.29	29.28	28.38	28.53	28.85	29.03	29.32
13	25.68	26.54	27.06	27.58	27.76	27.21	27.87	27.93	28.30	28.47	28.64
14	22.25	23.25	23.89	24.34	24.59	26.05	24.87	24.92	25.32	25.64	25.74
15	24.59	25.30	25.65	26.53	26.58	27.77	26.89	27.17	27.76	27.83	28.31
16	21.75	22.59	23.11	23.34	23.69	24.95	23.59	23.72	24.11	24.45	24.53
17	28.12	28.71	28.89	29.07	29.12	28.89	29.11	29.14	29.28	29.25	29.32
18	29.30	30.25	30.61	31.01	31.15	32.53	30.60	30.98	31.21	31.25	31.21
19	28.34	29.33	29.40	30.23	30.53	29.81	31.25	31.35	31.55	31.57	31.71
20	29.97	30.86	31.33	31.92	32.17	29.02	32.29	32.42	32.74	32.71	32.98
21	29.75	30.62	30.98	31.34	31.58	30.76	31.74	31.99	32.40	32.51	32.77
AVG	27.46	28.23	28.66	29.09	29.28	29.52	29.41	29.52	29.80	29.92	30.07

Table 4. Average PSNR of per-category for AID dataset with the scale factor of ×4. The best and the second-best results are marked in red and blue, respectively.

Class	Bicubic	SRCNN [22]	FSRCNN [57]	VDSR [24]	LGCNet [30]	DCM [31]	CTNet [48]	ESRT [40]	ACT [41]	TransENet [14]	Ours
1	27.03	28.17	27.70	28.82	28.39	28.99	28.80	28.98	29.01	29.23	29.29
2	34.88	35.63	35.73	35.98	35.78	36.17	36.12	36.15	36.15	36.20	36.45
3	29.06	30.51	29.89	31.18	30.75	31.36	31.15	31.35	31.37	31.59	31.69
4	31.07	31.92	31.79	32.29	32.08	32.45	32.40	32.47	32.45	32.55	32.61
5	28.98	30.41	29.83	31.19	30.67	31.39	31.17	31.42	31.42	31.63	31.75
6	25.26	26.59	25.96	27.48	26.92	27.72	27.48	27.73	27.75	28.03	28.23
7	22.15	23.41	22.74	24.12	23.68	24.29	24.10	24.29	24.32	24.51	24.56
8	25.83	27.05	26.65	27.62	27.24	27.78	27.63	27.78	27.79	27.97	28.06
9	23.05	24.13	23.69	24.70	24.33	24.87	24.70	24.88	24.89	25.13	25.32
10	38.49	38.84	38.84	39.13	39.06	39.27	39.25	39.25	39.24	39.31	39.45
11	32.30	33.48	32.95	34.20	33.77	34.42	34.25	34.41	34.43	34.58	34.59
12	27.39	28.15	28.19	28.36	28.20	28.47	28.47	28.53	28.47	28.56	28.76
13	24.75	26.00	25.49	26.72	26.24	26.92	26.71	26.93	26.94	27.21	27.19
14	32.06	32.57	32.50	32.77	32.65	32.88	32.84	32.89	32.87	32.94	33.26
15	26.09	27.37	26.84	28.06	27.63	28.25	28.06	28.25	28.25	28.45	28.54
16	28.04	28.90	28.70	29.11	28.97	29.18	29.15	29.20	29.18	29.28	29.42
17	26.23	27.25	26.98	27.69	27.37	27.82	27.69	27.84	27.84	28.01	28.34
18	22.33	24.01	23.47	25.21	24.40	25.74	25.27	25.80	25.75	26.40	26.38
19	27.27	28.72	28.09	29.62	29.04	29.92	29.66	29.96	29.96	30.30	30.52
20	28.94	29.85	29.50	30.26	30.00	30.39	30.25	30.39	30.38	30.53	30.79
21	24.69	25.82	25.40	26.43	26.02	26.62	26.41	26.62	26.61	26.91	27.18
22	26.31	27.55	27.12	28.19	27.76	28.38	28.19	28.40	28.40	28.61	28.76
23	25.98	27.12	26.77	27.71	27.32	27.88	27.72	27.90	27.89	28.08	28.22
24	29.61	30.48	30.22	30.82	30.60	30.91	30.83	30.92	30.92	31.00	31.27
25	24.91	26.13	25.66	26.78	26.34	26.94	26.75	26.96	26.99	27.22	27.43
26	25.41	26.16	25.88	26.46	26.27	26.53	26.46	26.55	26.54	26.63	26.87
27	26.75	28.13	27.62	28.91	28.39	29.13	28.94	29.17	29.15	29.39	29.72
28	24.81	26.10	25.50	26.88	26.37	27.10	26.86	27.14	27.10	27.41	27.68
29	24.18	25.27	24.73	25.86	25.48	26.00	25.82	26.01	26.02	26.20	26.43
30	25.86	27.03	26.54	27.74	27.26	27.93	27.67	27.92	27.95	28.21	28.48
AVG	27.3	28.4	28.03	28.99	28.61	29.17	29.03	29.18	29.19	29.38	29.57

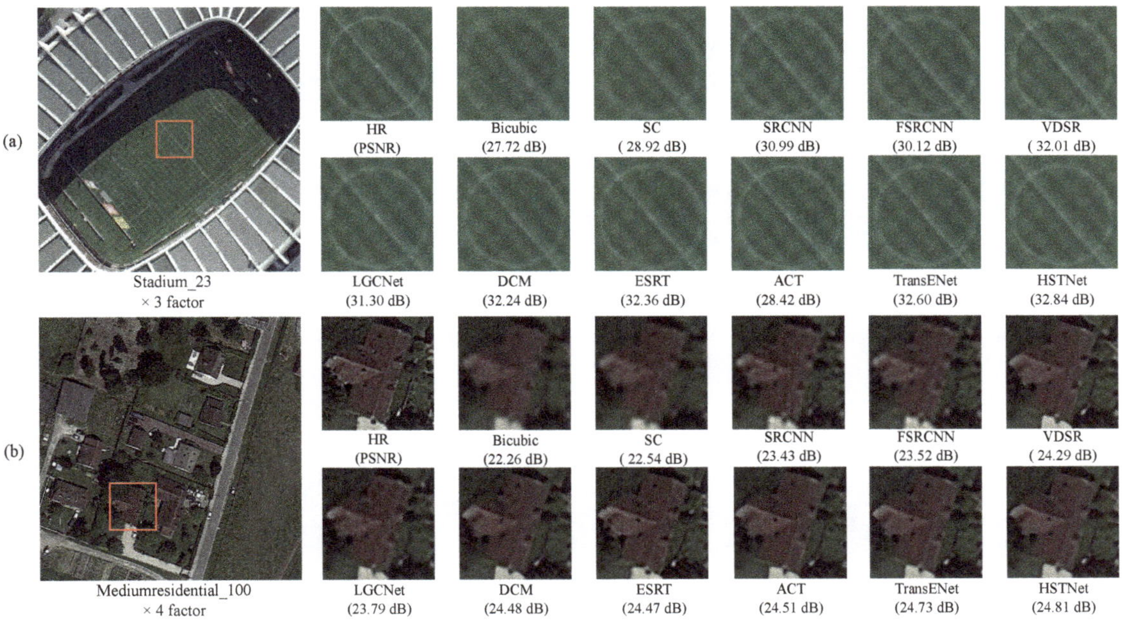

Figure 7. Subjective results for AID dataset: (**a**) "Stadium_23" scene with ×3 factor. (**b**) "Mediumresidential_100" scene with ×4 factor.

4.4. Results on Real Remote Sensing Data

Real images acquired by GaoFen-1 (GF-1) and GaoFen-2 (GF-2) satellites were employed to assess the robustness of the HSTNet. The spatial resolutions of GF-1 and GF-2 are 8 and 3.2 m/pixel, respectively. Three visible bands are selected from GF-1 and GF-2 satellite images to generate the LR inputs. The pre-trained DCM [31], ACT [41], and the proposed HSTNet models for the UCMerced dataset are utilized for SR image reconstruction. Figures 8 and 9 demonstrate the reconstruction results of the aforementioned methods on real data in some common scenes including river, factory, overpass, and paddy fields. One can see that the proposed HSTNet can obtain favorable results. Compared with DCM [31] and ACT [41], the reconstructed images of the proposed HSTNet achieved the lowest NIQE scores in all the four common scenes. Although the pixel size of these input images is different from the LR images in the training set, which are 600 × 600 and 256 × 256 for real-world images and training images, respectively, the HSTNet can still achieve good results in terms of visual perception qualities. It verifies the robustness of the proposed HSTNet.

4.5. Ablation Studies

Ablation studies with the scale factor of ×4 were conducted on the UCMerced dataset to demonstrate the effectiveness of the proposed fundamental modules in the HSTNet model.

4.5.1. Ablation Studies on the LFE Module

Number of LFE and HSFE modules: Table 5 presents a comparative analysis of varying quantities of LFE and HSFE modules. It indicates that when adopting two LFE and 2 HSFE modules, the model has the smallest number of parameters and computation, but the model has the lowest PSNR and SSIM values. The results indicate that the proposed HSTNet achieves the highest PSNR and SSIM when utilizing three LFE and five HSFE modules. When employing three LFE and eight HSFE modules, the model has the largest number of parameters and computation, and its performance is not optimal. Therefore, considering the performance of the model and the computational complexity, we adopte three LFE and five HSFE modules in the proposed method. The results confirm th effectiveness of the LFE and HSFE modules in the proposed model, as well as the rationalit of the number of LFE and HSFE modules.

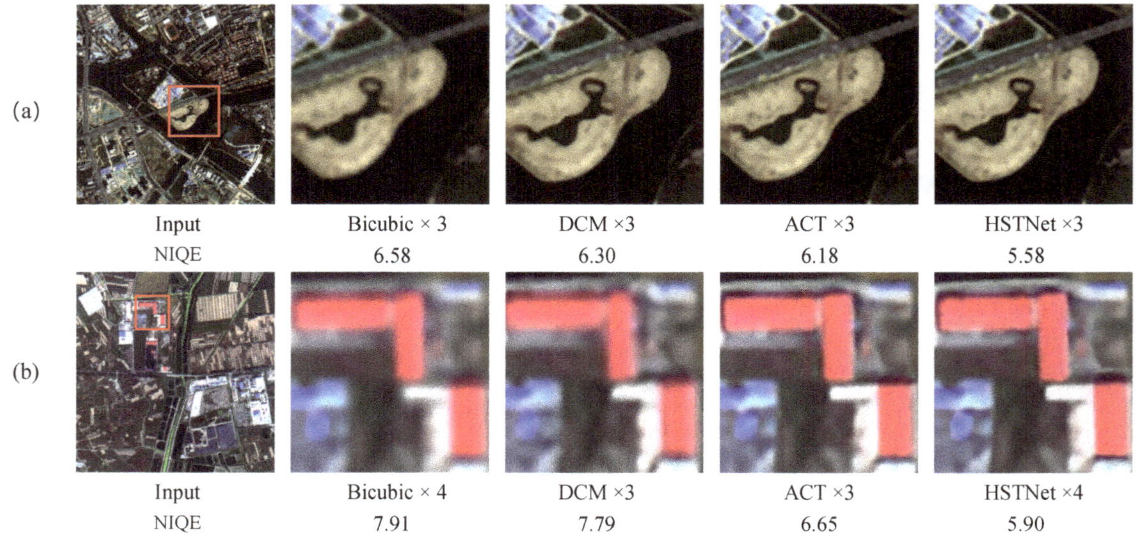

Figure 8. Subjective results on real GaoFen-1 satellite data: (**a**) "River" with ×3 factor. (**b**) "Factory" with ×4 factor.

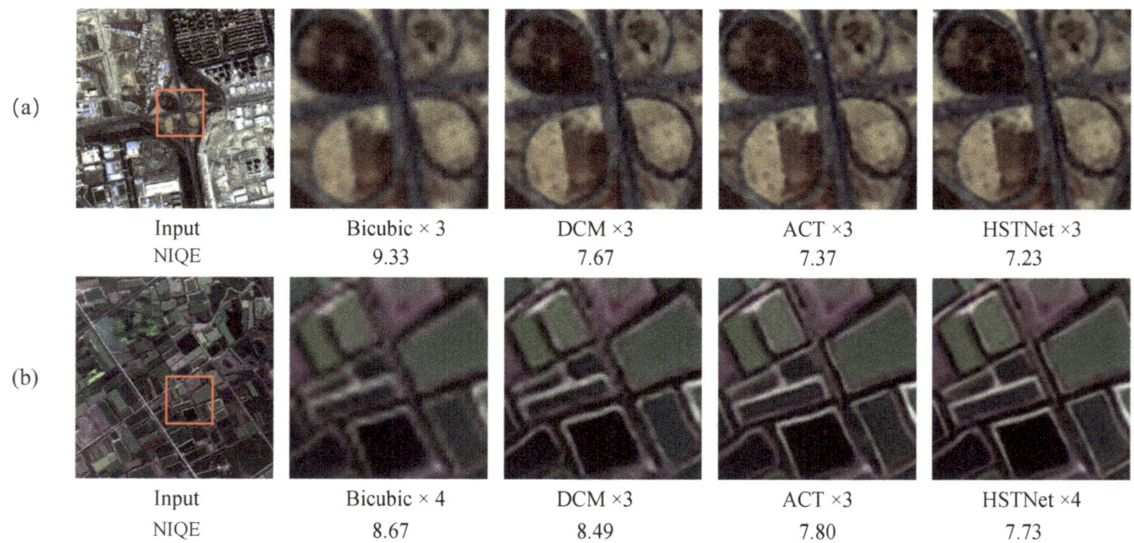

Figure 9. Subjective results on real GaoFen-2 satellite data: (**a**) "Overpass" with ×3 factor. (**b**) "Paddy fields" with ×4 factor.

Effects of the HSFE module: We devised the HSFE module in the proposed LFE module to exploit the recursive information inherent in the image. We conducted further ablation studies by substituting the HSFE module with widely used feature extraction modules in SR algorithms, namely RCAB [27], CTB [48], CB [58], and SSEM [45] to validate the effectiveness of the HSFE module. Among them, SSEM [45] is also used to mine scale information. As presented in Table 6, the HSFE module outperforms the other feature extraction modules in terms of PSNR and SSIM, demonstrating its effectiveness in feature extraction. Meanwhile, it is also competitive in terms of parameter quantity and computational complexity.

Table 5. Ablation analysis of the number of LFE and HSFE modules (the best result is highlighted in bold).

Scale	Numbers of LFE	Numbers of HSFE	PSNR	SSIM	Params	Multi-Adds
×4	2	2	27.57	0.7546	30.2M	73.6G
×4	2	5	27.72	0.7603	31.9M	135.9G
×4	2	8	27.61	0.7566	33.6M	205.1G
×4	3	2	27.58	0.7542	40.8M	95.5G
×4	3	5	**27.89**	**0.7694**	43.4M	194.4G
×4	3	8	27.73	0.7608	46.0M	292.8G

Table 6. Ablation analysis of different feature extraction modules in LFE module (the best result is highlighted in bold).

Scale	RCAB	CTB	CB	SSEM	HSFE	PSNR	SSIM	Params	Multi-Adds
×4	✓	✗	✗	✗	✗	26.33	0.7010	41.2M	112.0G
×4	✗	✓	✗	✗	✗	27.36	0.7451	40.3M	75.1G
×4	✗	✗	✓	✗	✗	27.51	0.7510	45.7M	275.2G
×4	✗	✗	✗	✓	✗	27.61	0.7561	42.5M	160.0G
×4	✗	✗	✗	✗	✓	**27.89**	**0.7694**	43.4M	194.4G

4.5.2. Ablation Studies on the CSET Module

Number of CSET modules: The CSET module is designed to learn the dependency relationship across long distances between features of different dimensions. To validate the effectiveness of the proposed CSET modules, we conducted ablation experiments using varying numbers of CSET modules. Table 7 proves that the configuration of three CSET modules performs the best in terms of PSNR and SSIM. The features of low-dimension space are transmitted more to the high-dimension space, reducing the difficulty of optimization and facilitating the convergence of the deep model. The aforementioned results demonstrate the effectiveness of the CSET module in enhancing the representation of high-dimensional features.

Effects of the CSTA block: The CSTA [41] block is introduced to enable the CSET module to utilize the recurrent patch information of different scales in the input image. To verify the effectiveness of the CSTA module, we analyzed the composition of the transformer. Table 8 presents the comparative results of two different transformers. It proves that the CSTA block is beneficial to improve the performance of the HSTNet.

Table 7. Ablation analysis of different feature extraction modules in the LFE module (the best result is highlighted in bold).

Scale	Transformer-3	Transformer-2	Transformer-1	Transformer-0	PSNR	SSIM
×4	✗	✗	✗	✗	27.54	0.7522
×4	✓	✗	✗	✗	27.61	0.7562
×4	✓	✓	✗	✗	27.73	0.7618
×4	✓	✓	✓	✗	**27.89**	**0.7694**
×4	✓	✓	✓	✓	27.50	0.7509

Table 8. Ablation analysis of the CSTA block. The best performances are highlighted in **bold**.

Transformer	PSNR	SSIM
MHSA + FFN	27.77	0.7630
MHSA + FFN + CSTA	**27.89**	**0.7694**

5. Conclusions and Future Work

In this paper, we present a hybrid-scale hierarchical transformer network (HSTNet) for remote sensing image super-resolution (RSISR). The HSTNet contains two crucial components, i.e., a hybrid-scale feature exploitation (HSFE) module and a cross-scale enhancement transformer (CSET) module. Specifically, the HSFE module with two branches was built to leverage the internal recurrence of information both in single and cross scales within the images. Meanwhile, the CSET module was built to capture long-range dependencies and effectively mine the correlation between high-dimension and low-dimension features. Experimental results on two challenging remote sensing datasets verified the effectiveness and superiority of the proposed HSTNet. In the future, more efforts are expected to simplify the network architecture and design a more effective low-dimensional feature extraction branch to improve RSISR performance.

Author Contributions: Conceptualization, J.S., M.G. and G.J.; methodology, J.S. and M.G.; software, J.S. J.P., and G.Z.; validation, J.S. Q.L. and M.G.; formal analysis, J.S. and M.G.; investigation, J.S. and Q.L.; resources, M.G. and J.S.; writing, J.S. and Q.L.; supervision, M.G. and G.J.; project administration, J.S., M.G. and G.Z. All authors have read and agreed to the published version of the manuscript.

Funding: This work is supported in part by the Natural Science Foundation of Shandong Province of China (ZR2022MF307) and the National Natural Science Foundation of China (Nos. 61601266 and 61801272).

Data Availability Statement: Not applicable.

Acknowledgments: This work is supported in part by the Natural Science Foundation of Shandong Province of China (ZR2022MF307) and the National Natural Science Foundation of China (Nos.61601266 and 61801272).

Conflicts of Interest: The authors declare no conflict of interest.

References

1. Harrie, L.; Oucheikh, R.; Nilsson, Å.; Oxenstierna, A.; Cederholm, P.; Wei, L.; Richter, K.F.; Olsson, P. Label Placement Challenges in City Wayfinding Map Production—Identification and Possible Solutions. *J. Geovisualization Spat. Anal.* **2022**, *6*, 16. [CrossRef]
2. Kokila, S.; Jayachandran, A. Hybrid Behrens-Fisher- and Gray Contrast–Based Feature Point Selection for Building Detection from Satellite Images. *J. Geovisualization Spat. Anal.* **2023**, *7*, 8. [CrossRef]
3. Shen, H.; Zhang, L.; Huang, B.; Li, P. A MAP Approach for Joint Motion Estimation, Segmentation, and Super Resolution. *IEEE Trans. Image Process.* **2007**, *16*, 479–490. [CrossRef] [PubMed]
4. Köhler, T.; Huang, X.; Schebesch, F.; Aichert, A.; Maier, A.K.; Hornegger, J. Robust Multiframe Super-Resolution Employing Iteratively Re-Weighted Minimization. *IEEE Trans. Comput. Imaging* **2016**, *2*, 42–58. [CrossRef]
5. Zhang, L.; Wu, X. An edge-guided image interpolation algorithm via directional filtering and data fusion. *IEEE Trans. Image Process.* **2006**, *15*, 2226–2238. [CrossRef]
6. Hung, K.W.; Siu, W.C. Robust Soft-Decision Interpolation Using Weighted Least Squares. *IEEE Trans. Image Process.* **2012**, *21*, 1061–1069. [CrossRef]
7. Lu, X.; Yuan, H.; Yuan, Y.; Yan, P.; Li, L.; Li, X. Local learning-based image super-resolution. In Proceedings of the 2011 IEEE 13th International Workshop on Multimedia Signal Processing, Hangzhou, China, 17–19 October 2011; pp. 1–5.
8. Zhang, K.; Gao, X.; Tao, D.; Li, X. Single Image Super-Resolution With Non-Local Means and Steering Kernel Regression. *IEEE Trans. Image Process.* **2012**, *21*, 4544–4556. [CrossRef]
9. Schulter, S.; Leistner, C.; Bischof, H. Fast and accurate image upscaling with super-resolution forests. In Proceedings of the 2015 IEEE Conference on Computer Vision and Pattern Recognition (CVPR), Boston, MA, USA, 7–12 June 2015; pp. 3791–3799.
10. Wang, L.; Guo, Y.; Liu, L.; Lin, Z.; Deng, X.; An, W. Deep Video Super-Resolution Using HR Optical Flow Estimation. *IEEE Trans. Image Process.* **2020**, *29*, 4323–4336. [CrossRef]
11. Chang, K.; Ding, P.L.K.; Li, B. Single image super-resolution using collaborative representation and non-local self-similarity. *Signal Process.* **2018**, *149*, 49–61. [CrossRef]

12. Yang, J.; Wright, J.; Huang, T.S.; Ma, Y. Image Super-Resolution Via Sparse Representation. *IEEE Trans. Image Process.* **2010**, *19*, 2861–2873. [CrossRef]
13. Li, Y.; Sixou, B.; Peyrin, F. A review of the deep learning methods for medical images super resolution problems. *Irbm* **2021**, *42*, 120–133. [CrossRef]
14. Lei, S.; Shi, Z.; Mo, W. Transformer-based Multi-Stage Enhancement for Remote Sensing Image Super-Resolution. *IEEE Trans. Geosci. Remote. Sens.* **2021**, *60*, 1–11.
15. Buades, A.; Coll, B.; Morel, J.M. A non-local algorithm for image denoising. In Proceedings of the 2005 IEEE Computer Society Conference on Computer Vision and Pattern Recognition (CVPR'05), San Diego, CA, USA, 20–25 June 2005; Volume 2, pp. 60–65.
16. Xu, J.; Zhang, L.; Zuo, W.; Zhang, D.; Feng, X. Patch Group Based Nonlocal Self-Similarity Prior Learning for Image Denoising. In Proceedings of the 2015 IEEE International Conference on Computer Vision (ICCV), Santiago, Chile, 7–13 December 2015; pp. 244–252.
17. Michaeli, T.; Irani, M. Blind Deblurring Using Internal Patch Recurrence. In Proceedings of the European Conference on Computer Vision, Zurich, Switzerland, 6–12 September 2014.
18. Freedman, G.; Fattal, R. Image and video upscaling from local self-examples. *ACM Trans. Graph.* **2011**, *30*, 12:1–12:11. [CrossRef]
19. Yang, J.; Lin, Z.L.; Cohen, S.D. Fast Image Super-Resolution Based on In-Place Example Regression. In Proceedings of the 2013 IEEE Conference on Computer Vision and Pattern Recognition, Portland, OR, USA, 23–28 June 2013; pp. 1059–1066.
20. Shocher, A.; Cohen, N.; Irani, M. "Zero-Shot" Super-Resolution Using Deep Internal Learning. In Proceedings of the Computer Vision and Pattern Recognition, Honolulu, HI, USA, 21–26 July 2017.
21. Pan, Z.; Yu, J.; Huang, H.; Hu, S.; Zhang, A.; Ma, H.; Sun, W. Super-Resolution Based on Compressive Sensing and Structural Self-Similarity for Remote Sensing Images. *IEEE Trans. Geosci. Remote Sens.* **2013**, *51*, 4864–4876. [CrossRef]
22. Dong, C.; Loy, C.C.; He, K.; Tang, X. Learning a deep convolutional network for image super-resolution. In Proceedings of the European Conference on Computer Vision, Zurich, Switzerland, 6–12 September 2014; pp. 184–199.
23. He, K.; Zhang, X.; Ren, S.; Sun, J. Deep Residual Learning for Image Recognition. In Proceedings of the 2016 IEEE Conference on Computer Vision and Pattern Recognition (CVPR), Las Vegas, NV, USA, 27–30 June 2016; pp. 770–778.
24. Kim, J.; Lee, J.K.; Lee, K.M. Accurate image super-resolution using very deep convolutional networks. In Proceedings of the IEEE Conference on Computer Vision and Pattern Recognition, Las Vegas, NV, USA, 27–30 June 2016; pp. 1646–1654.
25. Lim, B.; Son, S.; Kim, H.; Nah, S.; Lee, K.M. Enhanced Deep Residual Networks for Single Image Super-Resolution. In Proceedings of the 2017 IEEE Conference on Computer Vision and Pattern Recognition Workshops (CVPRW), Honolulu, HI, USA, 21–26 July 2017; pp. 1132–1140.
26. Zhang, Y.; Tian, Y.; Kong, Y.; Zhong, B.; Fu, Y. Residual dense network for image super-resolution. In Proceedings of the IEEE Conference on Computer Vision and Pattern Recognition, Salt Lake City, UT, USA, 18–23 June 2018; pp. 2472–2481.
27. Zhang, Y.; Li, K.; Li, K.; Wang, L.; Zhong, B.; Fu, Y. Image super-resolution using very deep residual channel attention networks. In Proceedings of the European Conference on Computer Vision (ECCV), Munich, Germany, 8–14 September 2018; pp. 286–301.
28. Dai, T.; Cai, J.; Zhang, Y.; Xia, S.T.; Zhang, L. Second-order attention network for single image super-resolution. In Proceedings of the IEEE/CVF Conference on Computer Vision and Pattern Recognition, Long Beach, CA, USA, 15–20 June 2019; pp. 11065–11074.
29. Li, Z.; Yang, J.; Liu, Z.; Yang, X.; Jeon, G.; Wu, W. Feedback network for image super-resolution. In Proceedings of the IEEE/CVF Conference on Computer Vision and Pattern Recognition, Long Beach, CA, USA, 15–20 June 2019; pp. 3867–3876.
30. Lei, S.; Shi, Z.; Zou, Z. Super-Resolution for Remote Sensing Images via Local–Global Combined Network. *IEEE Geosci. Remote Sens. Lett.* **2017**, *14*, 1243–1247. [CrossRef]
31. Haut, J.M.; Paoletti, M.E.; Fernández-Beltran, R.; Plaza, J.; Plaza, A.J.; Li, J. Remote Sensing Single-Image Superresolution Based on a Deep Compendium Model. *IEEE Geosci. Remote Sens. Lett.* **2019**, *16*, 1432–1436. [CrossRef]
32. Wang, X.; Wang, Q.; Zhao, Y.; Yan, J.; Fan, L.; Chen, L. Lightweight Single-Image Super-Resolution Network with Attentive Auxiliary Feature Learning. In Proceedings of the Asian Conference on Computer Vision, Kyoto, Japan, 30 November–4 December 2020.
33. Ni, N.; Wu, H.; Zhang, L. Hierarchical Feature Aggregation and Self-Learning Network for Remote Sensing Image Continuous-Scale Super-Resolution. *IEEE Geosci. Remote Sens. Lett.* **2022**, *19*, 1–5. [CrossRef]
34. Wang, Z.; Zhao, Y.; Chen, J. Multi-Scale Fast Fourier Transform Based Attention Network for Remote-Sensing Image Super-Resolution. *IEEE J. Sel. Top. Appl. Earth Obs. Remote Sens.* **2023**, *16*, 2728–2740. [CrossRef]
35. Liang, G.M.; KinTak, U.; Yin, H.; Liu, J.; Luo, H. Multi-scale hybrid attention graph convolution neural network for remote sensing images super-resolution. *Signal Process.* **2023**, *207*, 108954. [CrossRef]
36. Wang, Y.; Shao, Z.; Lu, T.; Wu, C.; Wang, J. Remote Sensing Image Super-Resolution via Multiscale Enhancement Network. *IEEE Geosci. Remote Sens. Lett.* **2023**, *20*, 1–5. [CrossRef]
37. Yang, F.; Yang, H.; Fu, J.; Lu, H.; Guo, B. Learning Texture Transformer Network for Image Super-Resolution. In Proceedings of the 2020 IEEE/CVF Conference on Computer Vision and Pattern Recognition (CVPR), Seattle, WA, USA, 13–19 June 2020; pp. 5790–5799.
38. Liang, J.; Cao, J.; Sun, G.; Zhang, K.; Gool, L.V.; Timofte, R. SwinIR: Image Restoration Using Swin Transformer. In Proceedings of the 2021 IEEE/CVF International Conference on Computer Vision Workshops (ICCVW), Montreal, BC, Canada, 11–17 October 2021; pp. 1833–1844.

39. Fang, J.; Lin, H.; Chen, X.; Zeng, K. A Hybrid Network of CNN and Transformer for Lightweight Image Super-Resolution. I Proceedings of the 2022 IEEE/CVF Conference on Computer Vision and Pattern Recognition Workshops (CVPRW), New Orlean LA, USA, 19–24 June 2022; pp. 1102–1111.
40. Lu, Z.; Li, J.; Liu, H.; Huang, C.; Zhang, L.; Zeng, T. Transformer for Single Image Super-Resolution. In Proceedings of the 202 IEEE/CVF Conference on Computer Vision and Pattern Recognition Workshops (CVPRW), New Orleans, LA, USA, 19–24 Jun 2022; pp. 456–465.
41. Yoo, J.; Kim, T.; Lee, S.; Kim, S.; Lee, H.S.; Kim, T.H. Enriched CNN-Transformer Feature Aggregation Networks for Supe Resolution. In Proceedings of the 2023 IEEE/CVF Winter Conference on Applications of Computer Vision (WACV), Waikolo HI, USA, 2–7 January 2023; pp. 4945–4954.
42. Ye, C.; Yan, L.; Zhang, Y.; Zhan, J.; Yang, J.; Wang, J. A Super-resolution Method of Remote Sensing Image Using Transformers. I Proceedings of the 2021 11th IEEE International Conference on Intelligent Data Acquisition and Advanced Computing System Technology and Applications (IDAACS), Online, 22–25 September 2021; Volume 2, pp. 905–910.
43. Tu, J.; Mei, G.; Ma, Z.; Piccialli, F. SWCGAN: Generative Adversarial Network Combining Swin Transformer and CNN fc Remote Sensing Image Super-Resolution. *IEEE J. Sel. Top. Appl. Earth Obs. Remote Sens.* **2022**, *15*, 5662–5673. [CrossRef]
44. He, J.; Yuan, Q.; Li, J.; Xiao, Y.; Liu, X.; Zou, Y. DsTer: A dense spectral transformer for remote sensing spectral super-resolutio *Int. J. Appl. Earth Obs. Geoinf.* **2022**, *109*, 102773. [CrossRef]
45. Lei, S.; Shi, Z. Hybrid-Scale Self-Similarity Exploitation for Remote Sensing Image Super-Resolution. *IEEE Trans. Geosci. Remot Sens.* **2022**, *60*, 1–10. [CrossRef]
46. Shi, W.; Caballero, J.; Huszár, F.; Totz, J.; Aitken, A.P.; Bishop, R.; Rueckert, D.; Wang, Z. Real-Time Single Image and Vide Super-Resolution Using an Efficient Sub-Pixel Convolutional Neural Network. In Proceedings of the 2016 IEEE Conference o Computer Vision and Pattern Recognition (CVPR), Las Vegas, NV, USA, 27–30 June 2016; pp. 1874–1883.
47. Wang, X.; Girshick, R.B.; Gupta, A.K.; He, K. Non-local Neural Networks. In Proceedings of the 2018 IEEE/CVF Conference o Computer Vision and Pattern Recognition, Salt Lake City, UT, USA, 18–22 June 2018; pp. 7794–7803.
48. Wang, S.; Zhou, T.; Lu, Y.; Di, H. Contextual Transformation Network for Lightweight Remote Sensing Image Super-Resolutio *IEEE Trans. Geosci. Remote. Sens.* **2021**, *60*, 1–13. [CrossRef]
49. Vaswani, A.; Shazeer, N.M.; Parmar, N.; Uszkoreit, J.; Jones, L.; Gomez, A.N.; Kaiser, L.; Polosukhin, I. Attention is All you Nee *arXiv* **2017**, arXiv: abs/1706.03762.
50. Hendrycks, D.; Gimpel, K. Gaussian Error Linear Units (GELUs). *arXiv* **2016**, arXiv:1606.08415.
51. Qin, M.; Mavromatis, S.; Hu, L.; Zhang, F.; Liu, R.; Sequeira, J.; Du, Z. Remote Sensing Single-Image Resolution Improvemen Using A Deep Gradient-Aware Network with Image-Specific Enhancement. *Remote. Sens.* **2020**, *12*, 758. [CrossRef]
52. Yang, Y.; Newsam, S. Bag-of-visual-words and spatial extensions for land-use classification. In Proceedings of the ACN SIGSPATIAL International Workshop on Advances in Geographic Information Systems, San Jose, CA, USA, 2–5 November 201
53. Xia, G.S.; Hu, J.; Hu, F.; Shi, B.; Bai, X.; Zhong, Y.; Zhang, L.; Lu, X. AID: A Benchmark Data Set for Performance Evaluation o Aerial Scene Classification. *IEEE Trans. Geosci. Remote Sens.* **2016**, *55*, 3965–3981. [CrossRef]
54. Muqeet, A.; Hwang, J.; Yang, S.; Kang, J.H.; Kim, Y.; Bae, S.H. Multi-attention Based Ultra Lightweight Image Super-Resolutio In Proceedings of the ECCV Workshops, Glasgow, UK, 23–28 August 2020.
55. Mittal, A.; Soundararajan, R.; Bovik, A.C. Making a "Completely Blind" Image Quality Analyzer. *IEEE Signal Process. Lett.* **201** *20*, 209–212. [CrossRef]
56. Kingma, D.P.; Ba, J. Adam: A method for stochastic optimization. *arXiv* **2014**, arXiv:1412.6980.
57. Dong, C.; Loy, C.C.; Tang, X. Accelerating the Super-Resolution Convolutional Neural Network. In Proceedings of the Europea Conference on Computer Vision, Amsterdam, The Netherlands, 11–14 October 2016.
58. Zhang, D.; Shao, J.; Li, X.; Shen, H.T. Remote Sensing Image Super-Resolution via Mixed High-Order Attention Network. *IEE Trans. Geosci. Remote Sens.* **2020**, *59*, 5183–5196. [CrossRef]

remote sensing MDPI

Communication

Application of Data Sensor Fusion Using Extended Kalman Filter Algorithm for Identification and Tracking of Moving Targets from LiDAR–Radar Data

Oscar Javier Montañez [1], Marco Javier Suarez [2] and Eduardo Avendano Fernandez [1,*]

[1] School of Electronic Engineering, Pedagogical and Technological University of Colombia, Sogamoso 152210, Colombia; oscarjavier.montanez@uptc.edu.co

[2] School of Systems and Computing Engineering, Pedagogical and Technological University of Colombia, Sogamoso 152210, Colombia

* Correspondence: eduardo.avendano@uptc.edu.co

Abstract: In surveillance and monitoring systems, the use of mobile vehicles or unmanned aerial vehicles (UAVs), like the drone type, provides advantages in terms of access to the environment with enhanced range, maneuverability, and safety due to the ability to move omnidirectionally to explore, identify, and perform some security tasks. These activities must be performed autonomously by capturing data from the environment; usually, the data present errors and uncertainties that impact the recognition and resolution in the detection and identification of objects. The resolution in the acquisition of data can be improved by integrating data sensor fusion systems to measure the same physical phenomenon from two or more sensors by retrieving information simultaneously. This paper uses the constant turn and rate velocity (CTRV) kinematic model of a drone but includes the angular velocity not considered in previous works as a complementary alternative in Lidar and Radar data sensor fusion retrieved using UAVs and applying the extended Kalman filter (EKF) for the detection of moving targets. The performance of the EKF is evaluated by using a dataset that jointly includes position data captured from a LiDAR and a Radar sensor for an object in movement following a trajectory with sudden changes. Additive white Gaussian noise is then introduced into the data to degrade the data. Then, the root mean square error (RMSE) versus the increase in noise power is evaluated, and the results show an improvement of 0.4 for object detection over other conventional kinematic models that do not consider significant trajectory changes.

Keywords: data sensor fusion; extended Kalman filter; lidar; radar

Citation: Montañez, O.J.; Suarez, M.J.; Fernandez, E.A. Application of Data Sensor Fusion Using Extended Kalman Filter Algorithm for Identification and Tracking of Moving Targets from LiDAR–Radar Data. *Remote Sens.* **2023**, *15*, 3396. https://doi.org/10.3390/rs15133396

Academic Editors: Gemine Vivone and Gwanggil Jeon

Received: 30 April 2023
Revised: 4 June 2023
Accepted: 14 June 2023
Published: 4 July 2023

1. Introduction

In surveillance and monitoring systems, the use of unmanned aerial vehicles (UAVs), such as drones or mobile vehicles, provides advantages in terms of access to the environment for exploration like augmented range, maneuverability, and safety due to their omnidirectional displacement capacity. These tasks must be performed autonomously by capturing information from sensors in the environment at scheduled or random points at specific times and areas. The collected data present errors and uncertainties that make object recognition difficult and depend on the resolution of the sensors for detection and identification. Data acquisition resolution can be improved by integrating sensor data fusion systems to measure the same physical phenomenon by capturing information from two or more sensors simultaneously and applying filtering or pattern recognition techniques to obtain better results than those obtained with only one sensor

Sensor data fusion consists of different techniques, inspired by the human cognitive ability to extract information from the environment by integrating different stimuli. In the case of sensor fusion, measurement variables are integrated through a set of sensors, often different from each other, that make inferences that cannot be possible from a single sensor [1].

The fusion of Radar (Radio Detecting and Ranging) and LiDAR (Light Detection and Ranging or Laser Imaging Detection and Ranging) sensor data presents a better response considering two key aspects: (i) the use of two coherent systems that allow an accurate phase capture and (ii) the improvement in the extraction of data from the environment, with the combination of two or more sensors arranged on the mobile vehicle or UAV [1,2]. This integration allows the error to be decreased in the detection of objects in a juxtaposition relationship by determining the distances through the reflection of radio frequency signals in the Radar case and through the reflection of a light beam (photons) for the case of the LiDAR sensor, generating a double observer facing the same event, in this case, the measurement of proximity and/or angular velocity [3,4]. Thus, the choice of Radar and LiDAR sensors requires special care, mainly about technical characteristics and compatibility [5,6], coherence in range, and data acquisition. The above allows a complementary performance to be achieved with its associated element in data fusion, facilitating a better understanding of the three-dimensional environment that feeds the data processing system integrated into the UAV [7] or at a remote site.

A proper sensor fusion of LiDAR and Radar data must rely on the use of estimators to achieve higher consistency in the measurements to mitigate the uncertainties by using three parameters: Radar measurements, LiDAR measurements, and Kalman filtering. This improves the estimation of the measured variable. The Kalman filtering technique allows the description of the real world using linear differential equations to be expressed as a function of state variables. In most real-world problems, the measurements may not be a linear function of the states of the system. However, applying extended Kalman filtering (EKF) techniques counteracts this situation by modeling the phenomenon using a set of nonlinear differential equations, X_k, which describe the dynamics of the system. The EKF allows "projecting" in time the behavior of the system to be filtered, with variables that are non-measurable but are calculable from the measurable variables. Then, by predicting the future data and their deviation concerning the measured data, the Kalman gain, K_k, is calculated, and it continuously adapts to the dynamics of the system. Finally, updating the matrix state $\overline{x}_k$ and the covariance matrix P_k associated with the filtered system. This process is graphically described in Figure 1.

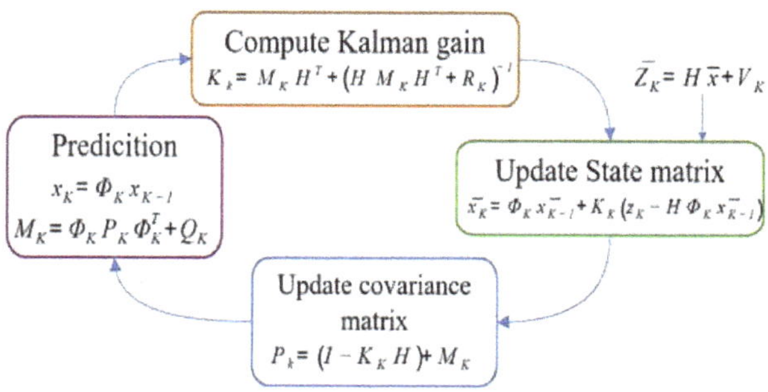

Figure 1. The schematic diagram for an extended Kalman filter.

In this work, sensor data fusion was performed for target tracking from a UAV, using an EKF and taking into consideration the results from data fusions performed in autonomous driving. The kinematic modeling Constant Turn Rate and Velocity (CTRV) [8] was taken as a reference, and this model includes in its description the angular velocity variable, provided by the Radar, a parameter that introduces an improvement in omnidirectional motion detection.

This paper shows the performance of an implementation of data sensor fusion using LiDAR and Radar through an EKF for the tracking of moving targets, taking into account changes in their direction and trajectory, to generate a three-dimensional reconstruction when the information is captured from a UAV.

2. Dynamic Model of UAV

The UAV dynamics were obtained from the 2D CTRV model [8] for vehicle and pedestrian detection on highways. It is assumed that the possible movements of the elements around the UAV are not completely arbitrary and not holonomous, in which case there will be displacements in a bi-dimensional plane. The curvilinear model (CTRV) includes angular velocities and angular movements in its modeling, which allows a better description of the changes in the direction and velocity of an object in a linear model. The CTRV model is shown in Figure 2.

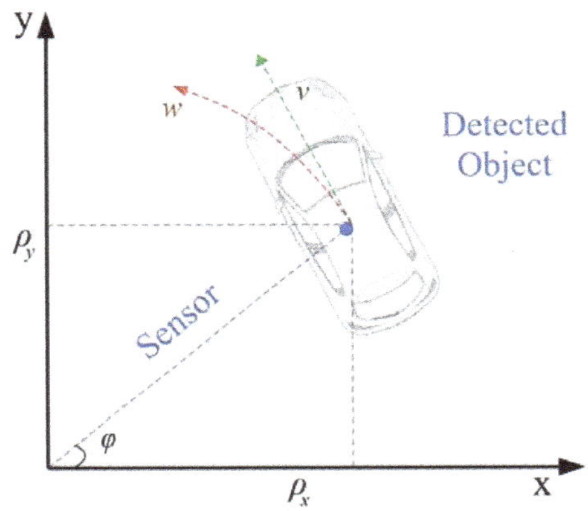

Figure 2. CTRV model for a moving object.

The velocity variable provides the system model the ability to calculate the target's lateral position variations for a correct prediction of the future position of the target, thus starting from initial positions x and y and projecting this location over time, defined as $x + \Delta x$ and $y + \Delta y$ for the target as shown in Figure 3.

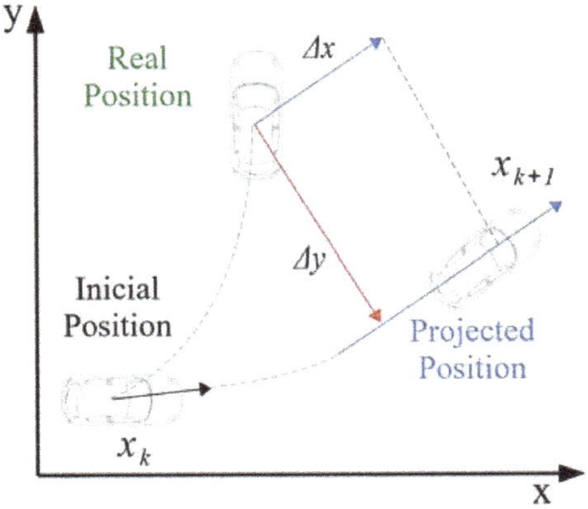

Figure 3. Position prediction through the CTRV model.

The CTRV model for the UAV system's moving target in the three-dimensional case determines the projection of the position of the target x_{i+1} on the axis, starting from the values of the angular frequency w and the angle θ [9–11] for x_i, and equally for y_i and its position projection. Therefore, the variables of interest in the system are the position x

and y; these are calculated by modeling their projection through the frontal velocity v, the angle θ formed between the Radar and the target, the angular frequency w of the target, and finally the angular frequency w_d of the UAV. The set of state variables involved in the system is the following:

$$\overline{x} = [x, y, v, \theta, w, w_d] \quad (1)$$

The kinematic equations describing the change from an initial position of the UAV to a future position are as follows:

$$\overline{x}_{i+1} = x_i + \left[v_{object} - v_{drone}\right] \cdot \Delta T \quad (2)$$

$$\overline{y}_{i+1} = y_i + \left[v_{object} - v_{drone}\right] \cdot \Delta T \quad (3)$$

The state variables are the frontal velocity, the theta angle, the target angular velocity and the angular velocity of the UAV.

$$\begin{gathered} v = w \cdot \Delta T \\ \theta = 0 \\ w = 0 \\ w_d = 0 \end{gathered} \quad (4)$$

Because the data sensor fusion operation is bidimensional, the CTRV model does not include motion in the position around the z-axis in its state variables. To maintain a bi-dimensional analysis, the UAV velocity [10] is projected as

$$v_x = V\cos\phi \quad (5)$$

$$v_y = V\sin\phi \quad (6)$$

In this way, ϕ represents the elevation of the UAV concerning the sensed target, this angle allows the velocities of the drone to be projected in the xz plane, and the $x + \Delta x$ or $y + \Delta y$ to be determined, as shown in Figure 3, concerning the position prediction. To simplify the model and to have a congruence of the LiDAR and Radar models in the sensor data fusion, a data acquisition method is proposed in which the UAV only uses pitch (rotation on the lateral Y axis) and yaw (rotation on the vertical Z axis) movements, and their projection in a three-dimensional coordinate system. These motions are included in the CTRV model through the projection of the UAV velocity v_d, through the angles ϕ and θ as shown below.

$$v_d = \begin{bmatrix} V_{dx} \\ V_{dy} \\ V_{dz} \\ W_{dz} \end{bmatrix} = \begin{bmatrix} V_0\cos\phi\cos\theta \\ V_0\cos\phi\sin\theta \\ V_0\sin\phi \\ 0 \end{bmatrix} \quad (7)$$

The difference between the estimated position and the actual position of the target is determined by the displacement generated by w and θ, i.e., ($\Delta T{\cdot}w + \theta$) [8], so the space and velocity projections are also a function of these variations. The velocity equations are obtained from x_i and y_i, which correspond to the first derivative, such that v_{dx} and v_{dy} are expressed as

$$\begin{gathered} \dot{x}_{i+1} = \tfrac{v}{w}[sin(\Delta T \cdot w + \theta) - sin\theta] - v_{dx}\cos\theta\cos\phi \\ \dot{y}_{i+1} = \tfrac{v}{w}[\cos\theta - \cos(\Delta T \cdot w + \theta)] - v_{dy}\cos\theta\sin\phi \\ a = \dot{w} \\ \dot{\theta} = 0 \\ \dot{w} = 0 \\ \dot{w}_d = 0 \end{gathered} \quad (8)$$

When the target has an initial angular velocity $w = 0$, the expressions change to [8]

$$\begin{aligned}
\overline{x}_{i+1} &= x_i + v\cos\theta\cdot\Delta T - v_{dx}\cos\theta\cos\phi\cdot\Delta T \\
\overline{y}_{i+1} &= y_i + v\ sin\theta\cdot\Delta T - v_{dx}\cos\theta\ sin\phi\cdot\Delta T \\
v &= 0 \\
\theta &= 0 \\
w &= 0 \\
w_d &= 0
\end{aligned} \tag{9}$$

The EKF performs the filtering in a bi-dimensional plane formed by the intersection of the range of the LiDAR and Radar sensor to achieve a three-dimensional reconstruction of the sensed target and a rotation is accomplished on the x-axis of the sensors, using the cylindrical coordinates as orientation. Figure 4 shows the dynamics between the UAV for generating a three-dimensional reconstruction from bi-dimensional data gathered by the sensors and the target in the XYZ plane.

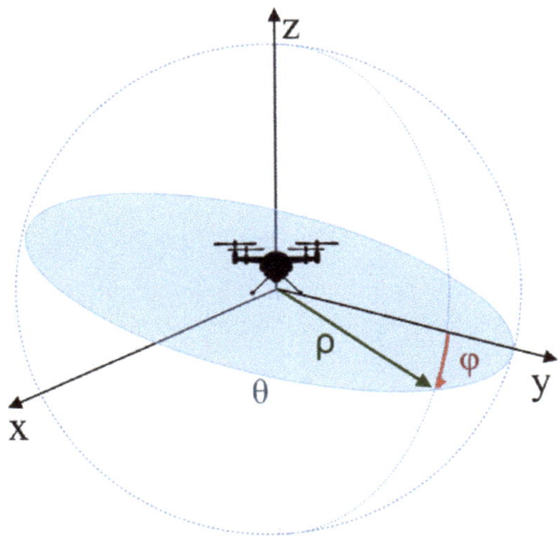

Figure 4. Three-dimensional reconstruction with UAV.

For the data fusion design, the RPLIDAR Slam S1 LiDAR sensor and the Positio2go BGT24MTR12 Radar were used as references. The LiDAR sensor operates in 2D with rotation capability, delivering data for a 360° scan, and the Radar achieves a range of 10 m. The range of the sensors according to the implementation of data fusion in the UAV is shown in Figure 5.

Figure 5. LiDAR and Radar sensor range.

Now, to improve the estimation of the measured variable from the noisy sensors, the Kalman filter is implemented through sequential steps: (i) the estimation or prediction of the system behavior from the nonlinear equations; (ii) the calculation of the Kalman gain to reduce the error of the prediction of the current state versus the previous state, and (iii) the update of the measurement matrix, as well as the covariance associated with the uncertainty of the system. The representation from the selected state variables corresponds to the following equation:

$$\dot{x} = f(x) + w \tag{10}$$

where $\dot{x}$ is the vector of the system states and $f(x)$ is a nonlinear function of the states This state–space model of the system allows us to determine the future states and the output is obtained by filtering the input signal. The Kalman filter performs estimations and corrections iteratively, where the possible errors of the system will be reflected in the covariance values present between the measured values and the values estimated by the filter. The forward projection of the covariance error has the following representation:

$$M_k = \Phi_k P_k \Phi_k^T + Q_k \tag{11}$$

The system update is implemented according to the following equations:

$$K_k = M_k H^T \left(H M_k H^T + R_k\right)^{-1} \tag{12}$$

$$P_k = (I - K_k H) M_k \tag{13}$$

$$\overline{z}_k = H\overline{x} + V_k \tag{14}$$

$$\overline{x}_k = \Phi_k \overline{x}_{k-1} + K_k (z_k - H\Phi_k \overline{x}_{k-1}) \tag{15}$$

Based on these state variables, the fundamental matrix for the extended Kalman filter is calculated and must satisfy the condition

$$F = \left.\frac{\delta f(x)}{\delta x}\right|_{x=\hat{x}} \tag{16}$$

$$\Phi_k = I + F \cdot \Delta T \tag{17}$$

Making $\alpha = \Delta T + \theta$, $\beta = -sin\theta + sin\alpha$, and $\chi = -cos\theta + cos\alpha$ in Equation (18), the fundamental matrix is

$$\Phi_k = \begin{bmatrix} 1 & 0 & \frac{\beta}{w} & \frac{v}{w}\chi & \frac{\Delta T v}{w}\left[cos\,\alpha - \frac{v}{w^2}\beta\right] & -v_d sin\theta cos\Phi \\ 0 & 1 & \frac{-\chi}{w} & \frac{v}{w}\beta & \frac{\Delta T v}{w}\left[sin\,\alpha + \frac{v}{w^2}\chi\right] & v_d cos\theta sin\Phi \\ 0 & 0 & 1 & \Delta T & 0 & 0 \\ 0 & 0 & 0 & 1 & 0 & 0 \\ 0 & 0 & 0 & 0 & 1 & 0 \\ 0 & 0 & 0 & 0 & 0 & 1 \end{bmatrix} \tag{18}$$

When the angular velocity of the target is zero, the fundamental matrix reduces to the following matrix:

$$\Phi_k = \begin{bmatrix} 1 & 0 & \cos\theta \cdot \Delta T & -v \cdot sin\theta \cdot \Delta T & 0 & -v_d sin\theta cos\phi \\ 0 & 1 & sin\theta \cdot \Delta T & v \cdot \cos\theta \cdot \Delta T & 0 & v_d cos\theta sin\phi \\ 0 & 0 & 1 & 0 & 0 & 0 \\ 0 & 0 & 0 & 1 & 0 & 0 \\ 0 & 0 & 0 & 0 & 1 & 0 \\ 0 & 0 & 0 & 0 & 0 & 1 \end{bmatrix} \tag{19}$$

The matrix associated with the system noise Q_k was calculated from the discrete output matrix G_k. The matrix must consider the output variables on which the Kalman filter can act. For the proposed model, the angular acceleration of the target has been taken into account, as well as the angular velocity and acceleration of the UAV. The output matrix G_k for the EKF is presented below.

$$G_k\mu = \int_0^{Ts} \Phi_k(\tau) \cdot G \cdot d(\tau) \quad (20)$$

$$G_k\mu = \begin{bmatrix} \frac{\Delta T^2}{2}\cos\theta & 0 & -\cos\theta\cos\phi \cdot \Delta T \\ \frac{\Delta T^2}{2}sin\theta & 0 & -\cos\theta\sin\phi \cdot \Delta T \\ \Delta T & 0 & 0 \\ 0 & 0 & 0 \\ 0 & \Delta T & 0 \\ 0 & 0 & \Delta T \end{bmatrix} \cdot \begin{bmatrix} \mu_a \\ \mu_w \\ \mu_{wd} \end{bmatrix} \quad (21)$$

The noise matrix from the output matrix is calculated through the following expression:

$$Q_k = G_k \cdot E[\mu \cdot \mu^T] \cdot G_k^T \quad (22)$$

where

$$E\left[\mu \cdot \mu^T\right] = \begin{bmatrix} \sigma_a^2 & 0 & 0 \\ 0 & \sigma_w^2 & 0 \\ 0 & 0 & \sigma_{wd}^2 \end{bmatrix} \quad (23)$$

With $\gamma = cos\theta\ cos\varphi$, $\nu = cos\varphi\ sin\varphi$, $\eta = sin\theta\ sin\varphi$, $\kappa = sin\theta\ cos\theta$, $\sigma = cos\theta\ sin\varphi$, $\lambda = cos\varphi\ sin\varphi$, the noise matrix is defined as

$$Q_k = \begin{bmatrix} \left(\frac{\Delta T^2}{2}\sigma_a\cos\theta\right)^2 + (\Delta T \cdot \sigma_{wd}\gamma)^2 & \frac{\Delta T^4}{4}{\sigma_a}^2\kappa + (\Delta T \cdot \sigma_{wd}\cos\theta)^2\lambda & \frac{\Delta T^3}{4}{\sigma_a}^2\kappa & 0 & 0 & -(\Delta T \cdot \sigma_{wd})^2\gamma \\ \left(\frac{\Delta T^2}{2}\sigma_a sin\theta\right)^2 + (\Delta T \cdot \sigma_{wd}\cos\theta)^2\nu & \frac{\Delta T^4}{4}{\sigma_a}^2\kappa + (\Delta T \cdot \sigma_{wd}\sigma)^2 & \frac{\Delta T^3}{2}{\sigma_a}^2 sin\theta & 0 & 0 & -(\Delta T \cdot \sigma_{wd})^2\eta \\ \frac{\Delta T^3}{2}{\sigma_a}^2\cos\theta & \frac{\Delta T^3}{2}{\sigma_a}^2 sin\theta & \Delta T^2\sigma_a^2 & 0 & 0 & 0 \\ 0 & 0 & 0 & 0 & 0 & 0 \\ 0 & 0 & 0 & 0 & \frac{\Delta T^2}{2}\sigma_{\dot{w}} & 0 \\ -(\Delta T \cdot \sigma_{wd})^2\gamma & -(\Delta T \cdot \sigma_{wd})^2\cos\theta\sin\phi & 0 & 0 & 0 & -(\Delta T \cdot \sigma_{wd})^2 \end{bmatrix} \quad (24)$$

Regarding the variables obtained from the sensors, it should be taken into account that the LiDAR and Radar sensors provide the measurements in different formats. For the LiDAR case, position data are retrieved in rectangular coordinates for x and y that correspond to the first two variables of the state vector. Since x_k has six state variables, the measurement matrix for LiDAR data processing should operate only on the x and y variables, making the product between the state vector x_k and H, conformable, i.e.,

$$H_L = \begin{bmatrix} 1 & 0 & 0 & 0 & 0 & 0 \\ 0 & 1 & 0 & 0 & 0 & 0 \end{bmatrix} \quad (25)$$

For Radar, the measurement matrix changes to

$$H_R = \begin{bmatrix} 1 & 0 & 0 & 0 & 0 & 0 \\ 0 & 1 & 0 & 0 & 0 & 0 \\ 0 & 0 & 1 & 0 & 0 & 0 \\ 0 & 0 & 0 & 1 & 0 & 0 \\ 0 & 0 & 0 & 0 & 1 & 0 \end{bmatrix} \quad (26)$$

The measurement error covariance matrix of the LiDAR sensor, obtained from the statistical analysis of the dataset obtained from this sensor, is as follows:

$$R_{k-L} = \begin{bmatrix} 0.0222 & 0 \\ 0 & 0.0222 \end{bmatrix} \quad (27)$$

R_{k-L} is obtained from the variance in the LiDAR dataset. Likewise, the measurement error covariance matrix of the Radar sensor obtained is

$$R_{k-R} = \begin{bmatrix} 0.088 & 0 & 0 & 0 & 0 \\ 0 & 0.00088 & 0 & 0 & 0 \\ 0 & 0 & 0.088 & 0 & 0 \\ 0 & 0 & 0 & 0.0088 & 0 \\ 0 & 0 & 0 & 0 & 0.08 \end{bmatrix} \quad (28$$

These covariance values are directly related to the resolution and reliability of bot the LiDAR and the Radar sensors. In the LiDAR case, the uncertainty is present in it measurement of the target distance, measured and represented as x and y coordinate while for the Radar, this uncertainty is found in this same measurement, but is represente as a vector distance of magnitude ρ and angle θ. Likewise, the covariance matrix for th Radar includes the estimated velocity at the target.

3. Results and Discussion

The EKF filter was implemented under numerical evaluation using Matlab®. T determine its performance, a dataset combining position measurements from a LiDAR an Radar sensor for a pedestrian and real position measurements for the pedestrian were use and with these results, an estimation of the performance was obtained using the RMSE [12 To evaluate the robustness of the model, the dataset was contaminated with different level of additive white Gaussian noise (AWGN).

The system was initialized by predefining values for the state matrix as shown i Figure 6, as well as the fundamental matrix, the system covariance matrix, and the nois matrix. Each new LiDAR or Radar sensor input triggers the filtering, starting by dete mining the time-lapse DT concerning the previous measurement. Next, the state matri is estimated using the set of Equations (8) or (9) when w = 0, the fundamental matri according to Equations (18) or (19) if w = 0, the noise matrix given by (24), and the systen covariance matrix as given by Equation (11). Next, the configuration of the measuremen and uncertainty matrices, (25) and (27) for the LiDAR case and (26) and (28) for the Rada case, is performed. The Kalman gain given by Equation (13) is determined, and, finally an updating of the measurement matrix given by (15), as well as the system and stat covariance matrix, is achieved.

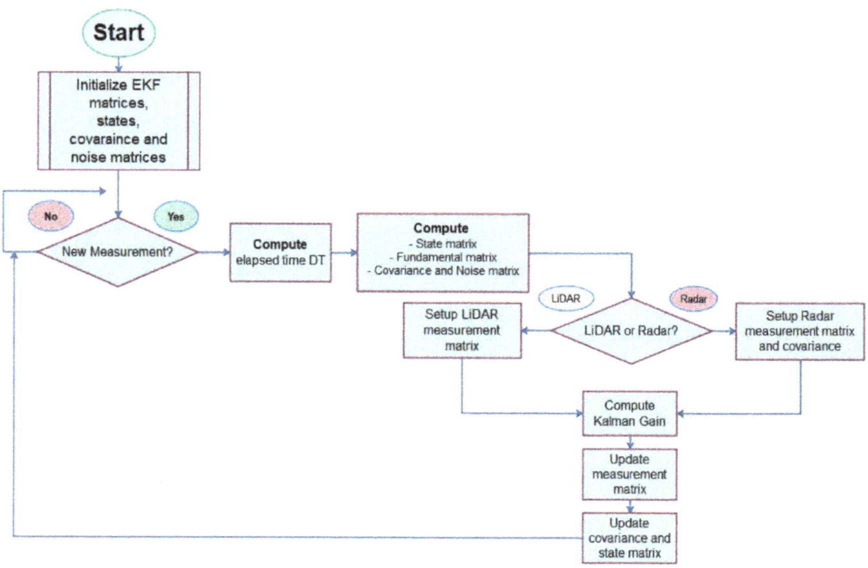

Figure 6. Schematic for EKF implementation.

The CTRV model proposed by [8] in the context of autonomous driving was designed, taking as reference highways and locations commonly used by automotive vehicles. In these scenarios, the tangential velocity changes to the sensors are presented to a lesser extent concerning the same scenario, but with measurements taken from a UAV. This behavior is accentuated when it is necessary to perform a three-dimensional reconstruction of the moving target. The CTRV model developed in this work includes the angular velocity of the drone, modifying the fundamental matrix of the system, as well as the noise matrix associated with the system, and a favorable response of the filter to the newly established changes was observed.

To evaluate the response of the x and y position variables to measurements contaminated with noise, the equations of the CTRV model were implemented in Matlab®, and a sweep of the position variables contaminated with AWGN was performed. The response of the x and y state variables of the EKF to these contaminated measurements is shown in Figure 7.

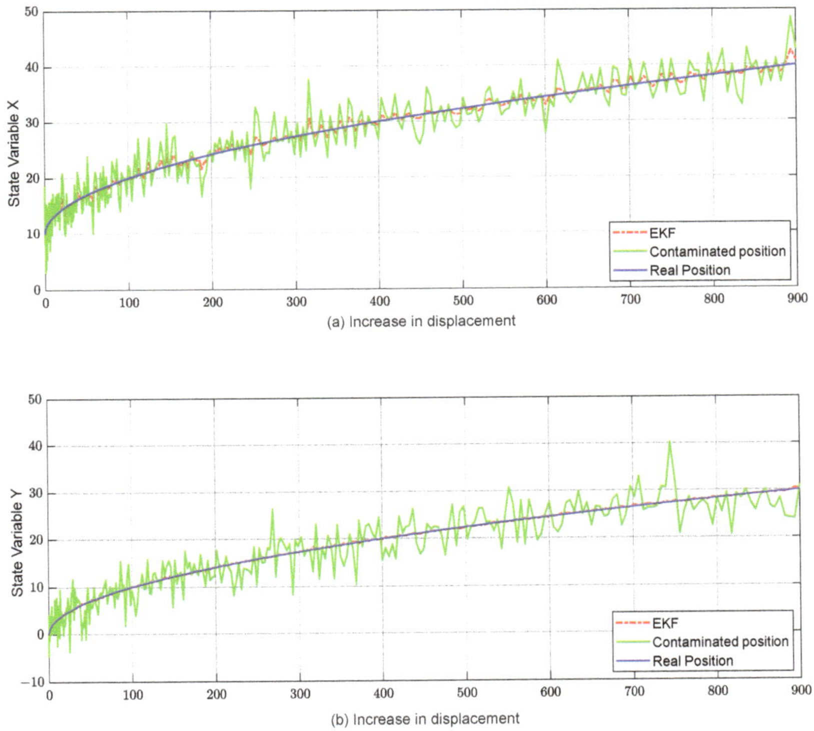

Figure 7. Response of the EKF in the state variables (**a**) x and (**b**) y to measurements contaminated with AWGN. Source: authors.

The response of the EKF to significant changes in the angular velocity of the target, represented as a change in the direction of the trajectory on the x-axis, is presented below. The EKF succeeds in predicting the target (green band), even at the point of greatest deflection. The zoom of the filter's response to the change in trajectory is shown in Figure 8. The EKF was tested with the help of a dataset that provides 1225 positions and angles from simultaneous Radar and LiDAR measurements, where the Radar sensor provides the distance along with the angle of displacement, concerning the horizontal of the Radar, and also the angular velocity detected by the Radar; the LiDAR sensor gives the position in x and y coordinates. This dataset was contaminated with AWGN by increasing the noise power progressively and testing the EKF.

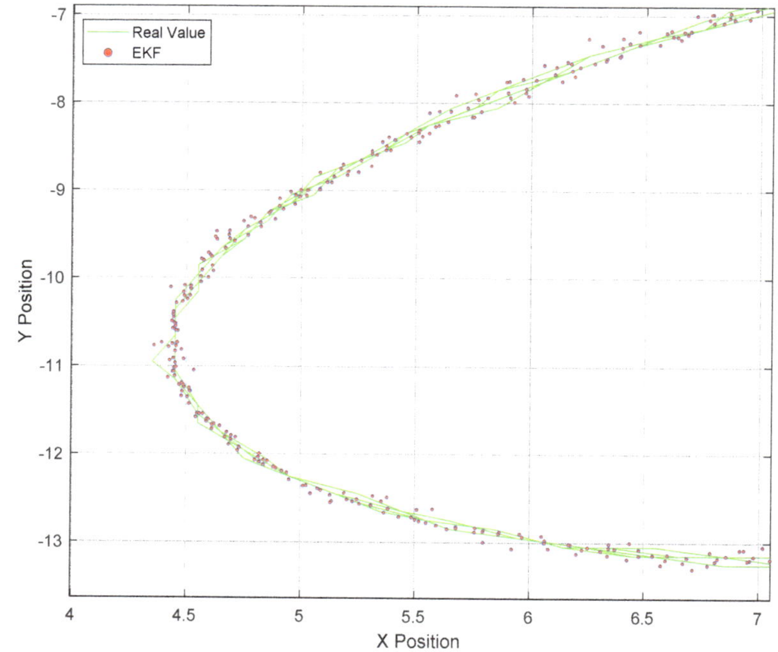

Figure 8. EKF response to trajectory changes in the moving target.

The representation of the real data versus the measured data from the Radar and LiDAR sensors is visualized in Figure 9.

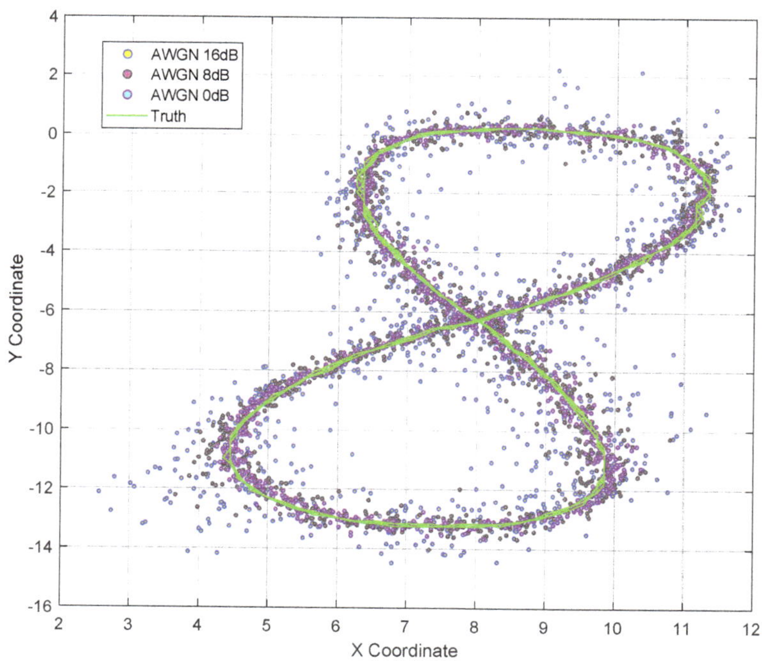

Figure 9. The plot of x and y position values for a 16 dB AWGN-contaminated LiDAR and Radar fusion.

To draw a comparison with other fusion models with Radar data, the root mean square error (RMSE) was computed, and it was possible to validate the effectiveness of the proposed CTRV model against fusions of previously used sensor data. The root mean square error is reduced from 0.21 to 0.163 in terms of the linear model; however, in contrast to the state-of-the-art [13] the unscented Kalman filter (UKF) maintains a better response compared to EKF based on the CTRV model. The response of the EKF to AWGN variations in the input data is presented below.

Figure 10 shows that the EKF acts by reducing the difference between the real values (red signal) and the values contaminated with Gaussian noise (blue signal). In [13], the authors state that the RMSE response can be improved with the unscented Kalman filter; however, it should be noted that this filter implies a higher computational complexity concerning the EKF. On the other hand, the response of the KF with a linear model and increasing AWGN is shown in Figure 11.

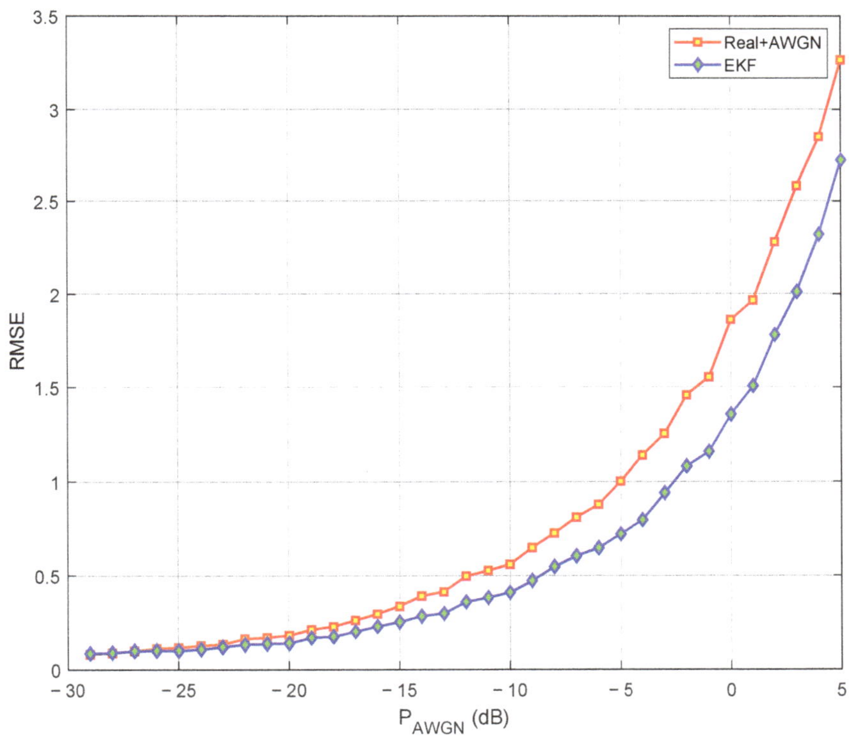

Figure 10. RMSE vs. AWGN in CTRV model.

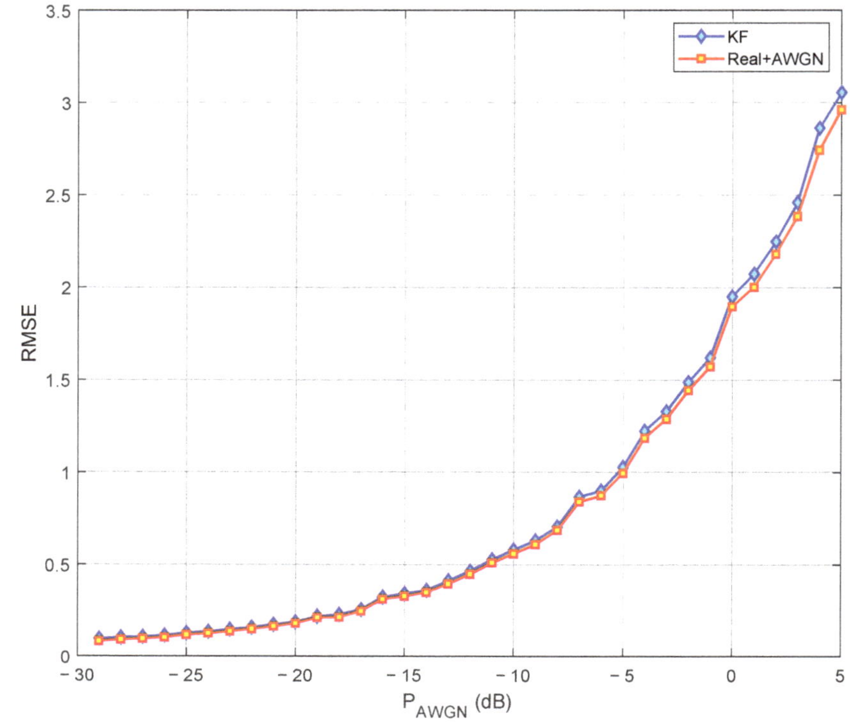

Figure 11. RMSE vs. AWGN in the linear model.

4. Conclusions

An architecture for LiDAR and Radar sensor data fusion through the extended Kalman filter model was implemented based on the CTRV model and the angular velocity projection of a UAV (a parameter not identified in related previous research). The robustness against trajectory changes for a moving target was demonstrated and determined by the angular velocity and angle of the target concerning the UAV provided by the LiDAR and Radar sensor. The evaluation of this model from the dataset allowed an accurate tracking of the target in the face of position changes. In CTRV modeling, the angle of radar and angular velocity of drone, when working together ensure a better response of the EKF. In the review of the state of the art, no references have been found that include the angular velocity of the drone. The projection of the UAV angular velocity on an *xz*-plane allows a bi-dimensional analysis to be performed, as well as the modeling of the drone–moving target system without negatively affecting the EKF response.

The CTRV model proposed in this article for the drone–moving target system was validated by numerical analysis using real data captured from LiDAR and Radar sensors. In future work, when the implemented system in a UAV including the kinematic proposed model requires a performance validation of the data sensor fusion using the EKF, the implementation of the EKF algorithm must be evaluated in a Field-Programmable Gate Array (FPGA) or System-on-Chip module due to their parallel processing capacity.

Author Contributions: Conceptualization, O.J.M., E.A.F. and M.J.S.; methodology, O.J.M.; software O.J.M. and E.A.F.; validation, O.J.M., E.A.F. and M.J.S.; formal analysis, O.J.M.; investigation, O.J.M and E.A.F.; resources O.J.M.; writing—original draft preparation, O.J.M., E.A.F. and M.J.S.; writing—review and editing, O.J.M. and E.A.F.; visualization, E.A.F. and M.J.S.; supervision, E.A.F.; project administration, E.A.F. All authors have read and agreed to the published version of the manuscript

Funding: This study was supported by the UPTC SGI 3139 Clarifier Research Project, as a partner of an International Research Cooperation agreement under the NATO Science for Peace Program—SPS G5888.

Data Availability Statement: The data presented in this study are available on request from the corresponding author.

Acknowledgments: All authors would like to sincerely thank the reviewers and editors for their suggestions and opinions for improving this article.

Conflicts of Interest: The authors declare no conflict of interest.

References

1. Lee, S.; Har, D.; Kum, D. Drone-Assisted Disaster Management: Finding Victims via Infrared Camera and Lidar Sensor Fusion. In Proceedings of the 2016 3rd Asia-Pacific World Congress on Computer Science and Engineering (APWC on CSE), Nadi, Fiji, 5–6 December 2016; pp. 84–89. [CrossRef]
2. Ki, M.; Cha, J.; Lyu, H. Detect and avoid system based on multi-sensor fusion for UAV. In Proceedings of the 2018 International Conference on Information and Communication Technology Convergence (ICTC), Jeju, Korea, 17–19 October 2018; pp. 1107–1109. [CrossRef]
3. Steinbaeck, J.; Steger, C.; Holweg, G.; Druml, N. Design of a low-level radar and time-of-flight sensor fusion framework. In Proceedings of the 2018 21st Euromicro Conference on Digital System Design (DSD), Prague, Czech Republic, 29–31 August 2018; pp. 268–275. [CrossRef]
4. De Silva, V.; Roche, J.; Kondoz, A. Fusion of LiDAR and Camera Sensor Data for Environment Sensing in Driverless Vehicles. *arXiv* **2017**. Available online: http://arxiv.org/abs/1710.06230 (accessed on 21 January 2023).
5. Na, K.; Byun, J.; Roh, M.; Seo, B. Fusion of multiple 2D LiDAR and RADAR for object detection and tracking in all directions. In Proceedings of the 2014 International Conference on Connected Vehicles and Expo (ICCVE), Vienna, Austria, 3–7 November 2014; pp. 1058–1059. [CrossRef]
6. Kwon, S.K.; Son, S.H.; Hyun, E.; Lee, J.H.; Lee, J. Radar-Lidar Sensor Fusion Sheme Using Occluded Depth Generation for Pedestrian Detection. In Proceedings of the 2017 International Conference on Computational Science and Computational Intelligence (CSCI), Las Vegas, NV, USA, 14–16 December 2017; pp. 1811–1812. [CrossRef]
7. Wittmann, D.; Chucholowski, F.; Lienkamp, M. Improving lidar data evaluation for object detection and tracking using a priori knowledge and Sensor fusion. In Proceedings of the 2014 11th International Conference on Informatics in Control, Automation and Robotics (ICINCO), Vienna, Austria, 1–3 September 2014; Volume 1, pp. 794–801. [CrossRef]
8. Wang, Y.; Liu, D.; Matson, E. Accurate Perception for Autonomous Driving: Application of Kalman Filter for Sensor Fusion. In Proceedings of the 2020 IEEE Sensors Applications Symposium (SAS), Kuala Lumpur, Malaysia, 9–11 March 2020. [CrossRef]
9. Lee, H.; Chae, H.; Yi, K. A Geometric Model based 2D LiDAR/Radar Sensor Fusion for Tracking Surrounding Vehicles. *IFAC-PapersOnLine* **2019**, *52*, 277–282. [CrossRef]
10. Kim, B.; Yi, K.; Yoo, H.J.; Chong, H.J.; Ko, B. An IMM/EKF approach for enhanced multitarget state estimation for application to integrated risk management system. *IEEE Trans. Veh. Technol.* **2015**, *64*, 876–889. [CrossRef]
11. Han, J.; Kim, J.; Son, N.S. Persistent automatic tracking of multiple surface vessels by fusing radar and lidar. In Proceedings of the OCEANS 2017-Aberdeen, Aberdeen, UK, 19–22 June 2017; pp. 1–5. [CrossRef]
12. Farag, W. Kalman-filter-based sensor fusion applied to road-objects detection and tracking for autonomous vehicles. *Proc. Inst. Mech. Eng. Part I J. Syst. Control Eng.* **2021**, *235*, 1125–1138. [CrossRef]
13. Tian, K.; Radovnikovich, M.; Cheok, K. Comparing EKF, UKF, and PF Performance for Autonomous Vehicle Multi-Sensor Fusion and Tracking in Highway Scenario. In Proceedings of the 2022 IEEE International Systems Conference (SysCon), Montreal, QC, Canada, 25–28 April 2022; pp. 1–6. [CrossRef]

remote sensing

Article

A Prediction Method of Ionospheric hmF2 Based on Machine Learning

Jian Wang [1,2,3,†], Qiao Yu [1,†], Yafei Shi [1,2] and Cheng Yang [1,2,*]

1 School of Microelectronics, Tianjin University, Tianjin 300072, China; wangjian16@tju.edu.cn (J.W.)
2 Qingdao Institute for Ocean Technology, Tianjin University, Qingdao 266200, China
3 Shandong Engineering Technology Research Center of Ocean Information Awareness and Transmission, Qingdao 266200, China
* Correspondence: ych2041@tju.edu.cn
† These authors contributed equally to this work.

Abstract: The ionospheric F2 layer is the essential layer in the propagation of high-frequency radio waves, and the peak electron density height of the ionospheric F2 layer (hmF2) is one of the important parameters. To improve the predicted accuracy of hmF2 for further improving the ability of HF skywave propagation prediction and communication frequency selection, we present an interpretable long-term prediction model of hmF2 using the statistical machine learning (SML) method. Taking Moscow station as an example, this method has been tested using the ionospheric observation data from August 2011 to October 2016. Only by inputting sunspot number, month, and universal time into the proposed model can the predicted value of hmF2 be obtained for the corresponding time. Finally, we compare the predicted results of the proposed model with those of the International Reference Ionospheric (IRI) model to verify its stability and reliability. The result shows that, compared with the IRI model, the predicted average statistical RMSE decreased by 5.20 km, and RRMSE decreased by 1.78%. This method is expected to provide ionospheric parameter prediction accuracy on a global scale.

Keywords: ionosphere; peak height of F2 layer; hmF2; machine learning; prediction

Citation: Wang, J.; Yu, Q.; Shi, Y.; Yang, C. A Prediction Method of Ionospheric hmF2 Based on Machine Learning. *Remote Sens.* **2023**, *15*, 3154. https://doi.org/10.3390/rs15123154

Academic Editor: Gwanggil Jeon

Received: 17 April 2023
Revised: 26 May 2023
Accepted: 12 June 2023
Published: 16 June 2023

1. Introduction

The ionosphere is the atmosphere between 60 km and 1000 km above the Earth's surface. Due to its electrical and ionized structure, and its complex temporal and spatial variability, it is of significant importance for its high frequency (HF) [1]. It influences sky waves, challenging radio propagation and wireless communication [2]. hmF2 is an important parameter of the F2 layer in the ionosphere, which serves as the basis for predicting the usable frequency [3] by reflecting the ionosphere [4]. Namely, usable frequency and propagation loss [5] are a function of hmF2, which indicates the height characteristics and changes of the ionospheric F2 layer [6]. Therefore, a reliable modeling method of hmF2 will help in propagation prediction, frequency selection, and spectrum management for HF communication systems [7]. Moreover, estimating and predicting the characteristics of hmF2 is vital for identifying adverse space weather [8] and hmF2 is a major aeronautical parameter involved in aeronautical [9] and ionospheric electrodynamic [10] studies. In general, the hmF2 can be observed at the sounding station using ionospheric sounders [11]. Without sounding stations, hmF2 can be predicted based on the ionospheric models that provide helpful empirical values for educators, engineers, and scientists.

As an internationally recognized standard, the International Reference Ionosphere (IRI) provides ionospheric parameters [12] and is often used as a benchmark to evaluate the performance of new ionospheric prediction models. Similar to critical frequency and the propagation factor, the modeling methods and models for predicting hmF2 are continuously developing. In order to improve the prediction performance of the ionospheric model, many new methods have constantly been introduced by experts at home and abroad.

For example, based on the empirical orthogonal function, Zhang et al. [13,14] and Yu et al. [15] constructed a global and Chinese ionospheric hmF2 prediction model, respectively, and the results were superior to the IRI model. Themens [16] proposed an ionospheric empirical model of the Canadian high Arctic. Compared with the IRI model, the prediction error of hmF2 of this model was reduced by 3~9 km. Sai et al. established a two-dimensional ionospheric model based on the artificial neural network (ANN) [17] and improved it [18], which can more accurately predict the detailed changes of hmF2 compared with the IRI model. Li et al. [19] established a global ionospheric model based on the improved ANN technology based on the genetic algorithm, which has better temporal and spatial characteristics of the global or regional ionosphere.

To further improve the hmF2 prediction accuracy, we propose an explicable long-term method of hmF2 based on the statistical machine learning (SML) method and the correlation between hmF2 and the solar activity index. The structure is arranged as follows: firstly, the paper elaborates on the SML method, briefly introduces the data required for modeling, and establishes a long-term prediction model for hmF2 based on this data; next, the model prediction results are analyzed, followed by a discussion of the model and a conclusion of the entire paper.

2. Materials and Methods

2.1. Method

Machine learning is an interdisciplinary field that uses probabilistic models to analyze and predict data based on provided data [20]. The idea behind machine learning's data processing is simple to understand, and the process is straightforward. Unlike the black-box algorithm, the model parameters determined using SML methods have explainable and transparent meanings [21]. For example, SML algorithms can be used to solve specific functional analytic expressions, which deep learning algorithms such as artificial neural networks cannot do [22]. Therefore, statistical machine learning is widely used to model ionospheric parameters. Using SML to reconstruct the ionospheric parameter hmF2 model, it is indispensable to determine the algorithm, strategy, and model with hmF2 data as the core and solve four problems in the process of machine learning:

(1) What data are needed? In machine learning, data are central. The paper is carried out using the data of the median value (the median of each month measured by the hour) calculated from the ionospheric hmF2 observation data of the Moscow station;

(2) How is the model chosen? The selection model finds the mapping relationship between input and output variables. The model should be based on analyzing input and output variables' characteristics. Ionospheric parameters are affected by solar activities, and there are seasonal, semi-annual, annual, and more subtle changes [23]. Therefore, to establish the hmF2 long-term prediction model, it is necessary to find the mapping relationship between hmF2 and solar activity index and time;

(3) How is the model determined? The model needs to be determined based on the relationship between the independent and target-dependent variables. Here, the relationship between hmF2 and solar activity index and time is determined by regression analysis under the least square;

(4) How is the model evaluated? The discrepancy between the sample's real output and the learner's actual predicted output is called an "error". "Training error" or "empirical error" is the learner's error on the training set. "Generalization error" refers to the learner's error on the new sample. Generally, it is desirable to obtain a model with low generalization error. Therefore, the hmF2 data are divided into three parts: training data, verification data, and test data, using training data to train the model and validation data to select and adjust the model. Test data are used to represent the generalization ability of the model [24]. Because of the ionosphere's prominent time-varying characteristics, this paper uses the

relative root mean square error (RRMSE) as the general evaluation standard to evaluat the model.

In brief, this paper uses the RRMSE analysis strategy as the model selection an evaluation criteria to establish the long-term prediction model of hmF2 according to th correlation between the Moscow station's hmF2 median value data and solar activit and time. Finally, the validity and reliability of the prediction model are verified by th observation data and IRI model. According to the learning process of SML, the following i the process of data acquisition, model training, validation, and testing.

Figure 1 shows the hmF2 modeling process according to SML:

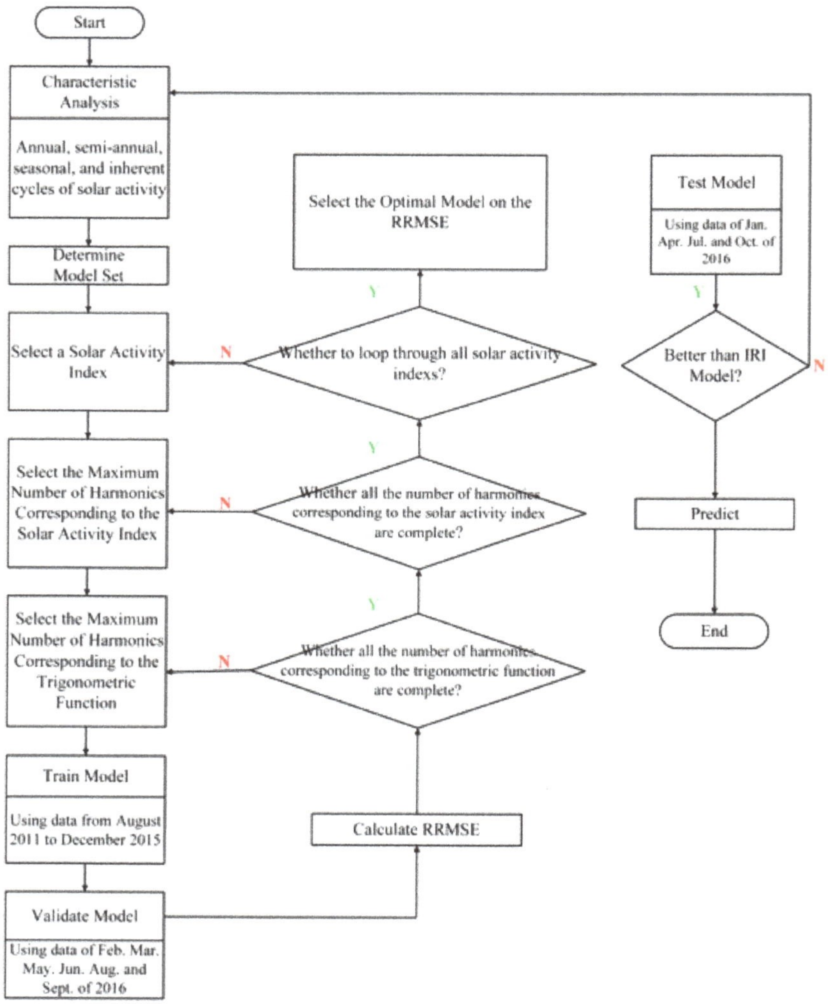

Figure 1. Modeling flowchart based on SML.

(1) The hmF2 are closely related to solar activity, and there are seasonal, semi-annual, and annual variations. According to the above characteristics, this paper determines the training model set.

(2) The solar activity index, including the solar radio wave flux with a wavelength of 10.7 cm, the number of sunspots, and the strongest single line in the ultraviolet band are selected. The highest power index of solar activity parameters and the highest harmonic number in the trigonometric function is also selected.

(3) Data from August 2011 to December 2015 are used to train, and data from February, March, May, June, August, and September 2016 are used to validate.

(4) The relative root mean square error calculated and recording between the verified and actual data are calculated.

(5) Whether all the highest harmonic numbers in the trigonometric function are traversed is checked. If the traversal is completed, step b is entered; otherwise, the remaining highest harmonic number for model training is selected; Whether all the highest power index of solar activity parameters are traversed is checked. If the traversal is completed, step c is entered; otherwise, the remaining highest power index for model training is selected. Whether the index of solar activity has been traversed is judged. If so, the next step is entered; otherwise, the remaining solar activity indices for training are selected.

(6) The prediction model according to RRMSE is determined.

(7) A modeling test is undertaken. According to the division of seasons, the data of January, April, July, and October 2016 are respectively used for testing and compared with the IRI-2016 model. If the model performance is worse than the IRI model, the data characteristics need to be re-analyzed to determine the model set; if the model performance is better than the IRI model, it can be used for engineering prediction.

2.2. Data

Following the modeling technique route and flow identified in the previous section, this section specifies the data required for hmF2 parameter modeling.

2.2.1. Ionospheric hmF2

hmF2 data from the Moscow station (55.9°N, 37.7°E) are selected to train and learn the proposed model and parameters, which can be obtained from http://www.wdcb.ru (accessed on 1 January 2023) and were measured by an instrument known as the ionosonde. The collected data were averaged by month and hour to obtain the corresponding median value, which is the target variable modeled in this paper and referred to as hmF2 monthly median value data. Figure 2 shows the hmF2 monthly value data of the collected station, which was completed from August 2011 to October 2016 and used in this study to train and validate the proposed model.

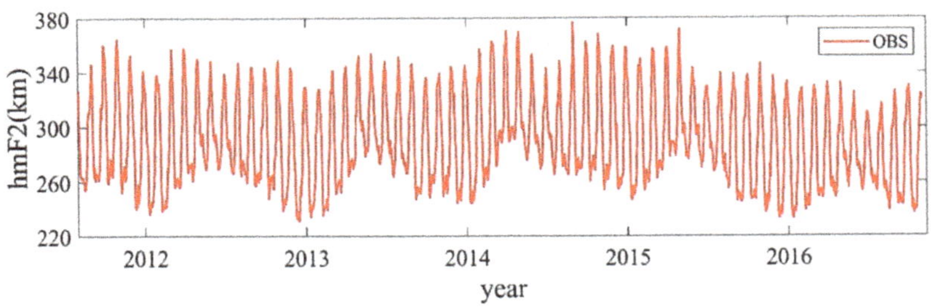

Figure 2. hmF2 monthly median data from the Moscow station.

2.2.2. Solar Activity Index

People usually use the solar activity index to represent the intensity of solar activity. The most commonly used solar activity index includes: (1) *F*10.7 [25], the solar radio wave flux with a wavelength of 10.7 cm affected by the upper atmosphere and chromospheric corona [26], denoted simply as *F*; (2) *R* [27], the number of sunspots affected by the lower chromosphere and the photosphere; (3) Lyman-α [28], the strongest single line in the ultraviolet band, abbreviated as *A*. The three solar activity indices are available from the corresponding forecast websites.

(1) The outer chromosphere and part of the inner corona of the sun's atmosphere emit *F*10.7. Flux (sfu) is the unit of *F*10.7, 1 sfu $=10^{-22}$ $Wm^{-2}Hz^{-1}$ [29]. As a common index of solar activity, *F*10.7 is closely related to the intensity of solar activity [30]. It is mainly

determined by sunspot number groups on the solar surface and can describe the intensity of solar activity [31]. It is widely used in ionospheric parameter prediction models. For example, $F10.7$ is used as the input parameter of the model to fit the ionospheric parameter hmF2 [13,14]. The twelve-monthly smoothed value of $F10.7$ is used here, denoted as F_{12}.

(2) The sunspot number is a swirling airflow caused by the solid solar magnetic field activity located in the solar sphere [32]. The sunspot number is often used to measure the level of solar activity. Changes in the ionosphere are subject to solar activity, and so the sunspot number is used to describe changes in ionospheric parameters and to study prediction models of ionospheric parameters [31]. Li et al. introduced sunspot numbers into the model when predicting ionospheric hmF2 [19]. The twelve-monthly smoothed value of R is used here, denoted as R_{12}.

(3) The Lyman-α line is the hydrogen line in the Lyman series and represents the most vital single line in the outer band. Electron transitions produce it within hydrogen atoms when atomic electrons transition from the first excited state to the ground state Lyman-α is released by hydrogen produced in large quantities in the universe [33] and also participate in the modeling of ionospheric parameters as an input parameter [34]. Here, the twelve-monthly smoothed value of Lyman-α is used, denoted as A_{12}.

The twelve-monthly smoothed value is calculated by the following formula:

$$S_{12} = \frac{1}{12}\left[\sum_{i=n-5}^{n+5} \overline{S}_i + \frac{1}{2}(\overline{S}_{n-6} + \overline{S}_{n+6})\right] \quad (1)$$

where S represents the index of solar activity, $\overline{S}$ represents the solar activity index's monthly mean value, and n represents the month.

Figure 3 shows the changes in F_{12}, R_{12}, and A_{12} over time. The three solar activity indices tend to be the same year by year, but the details are still different.

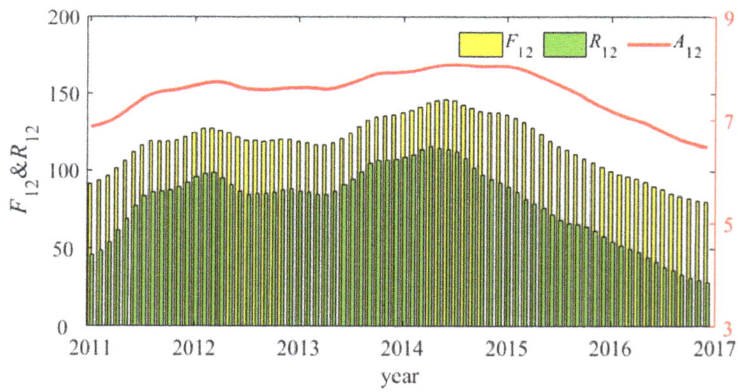

Figure 3. Trend of solar activity index over time.

3. Results

3.1. Model Determination

There is a correlation between ionospheric parameters and the solar activity index, and there are annual, semi-annual, seasonal, and more subtle variations [23] Therefore, for the given local time and geographical coordinates, the Formula (2) shows the general formula for defining the harmonic mapping between the solar cycle variation parameters, year, season, and month, and hmF2:

$$\text{hmF2}(p, m) = \sum_{k=0}^{K} \sum_{j=0}^{J} \left[\gamma_{k,j} p^j \cdot \cos(2\pi k m/12) + \beta_{k,j} p^j \cdot \sin(2\pi k m/12)\right] \quad (2)$$

where p represents F_{12}, R_{12}, A_{12} and m represents the integer of the month. The harmonic number k describes the variation characteristics of annual, semi-annual, seasonal, and monthly cycles. $k = 1, 2, 3$, and 4, respectively, represent one year, half a year, a quarter, and one month. Considering that the increase in the K value does not bring a significant increase in calculation accuracy [35], $K = 1$ and 2 are chosen here. The value of j is directly related to the solar activity index. $J = 1$ and 2 are selected here. Given the solar activity index and the values of K and J, the hyperparameters in the model can be statistically obtained by regression analysis under the least square method. The final determination of the model requires the consideration of RRMSE. The specific solving process is as follows:

$$\left(\mathbf{C}\mathbf{C}^T\right)\begin{bmatrix}\gamma_{0,0}\\ \gamma_{0,1}\\ \vdots\\ \beta_{K,J-1}\\ \beta_{K,J}\end{bmatrix} = \mathbf{C}\begin{bmatrix}\mathrm{hmF2}_1\\ \mathrm{hmF2}_2\\ \vdots\\ \mathrm{hmF2}_{O-1}\\ \mathrm{hmF2}_O\end{bmatrix} \tag{3}$$

where O is the number of hmF2 obtained statistically, and

$$\mathbf{C} = \begin{bmatrix} 1 & 1 & \cdots & 1\\ (p)_1 & (p)_2 & \cdots & (p)_O\\ \vdots & \vdots & \ddots & \vdots\\ (p)_1^{J-1}\sin\left(\frac{2\pi Km}{12}\right) & (p)_2^{J-1}\sin\left(\frac{2\pi Km}{12}\right) & \cdots & (p)_O^{J-1}\sin\left(\frac{2\pi Km}{12}\right)\\ (p)_1^{J}\sin\left(\frac{2\pi Km}{12}\right) & (p)_2^{J}\sin\left(\frac{2\pi Km}{12}\right) & \cdots & (p)_O^{J}\sin\left(\frac{2\pi Km}{12}\right) \end{bmatrix} \tag{4}$$

In the given model set, the data from August 2011 to December 2015 are used for training to obtain the hyperparameters in the model. Then, the data from February, March, May, June, August, and September 2016 are used to verify the trained model. The RRMSE between the verified and observations is calculated, and RRMSE is taken as the evaluation strategy of the model. Formula (5) is the calculation formula of RRMSE:

$$\mathrm{RRMSE} = \sqrt{\frac{1}{N}\sum_{i=1}^{N}\left(\frac{\mathrm{hmF2}_i' - \mathrm{hmF2}_i}{\mathrm{hmF2}_i}\right)^2} \tag{5}$$

where $\mathrm{hmF2}_i'$ is the calculated value of the model, $\mathrm{hmF2}_i$ is the measured statistical value, and N is the total data count.

We calculated the RRMSE value obtained by verification calculation of all training models in the model set. The result shows that: (1) the increase of orders J and K does not improve the model's predictive performance, but improve the calculation of the algorithm; (2) RRMSE is minimum when R is selected as the index of solar activity to participate in the modeling. The minimum RMSE of using F, R, and A is 4.52%, 4.27%, and 4.67% with $J = 1$ and $K = 1$. This is different from the model of ionospheric foF2 [1] and other parameters [35].

3.2. Results Analysis

Based on the training results, the values of p, J, and K were selected as R, 1, and 1, respectively, and then plugged into Equation (2). This led to the derivation of the prediction model for hmF2 at the Moscow station, which is represented by Equation (6):

$$\begin{aligned}\hat{\mathrm{hmF2}}(R,m) &= \sum_{k=0}^{1}\sum_{j=0}^{1}\left[\gamma_{k,j}R^j\cdot\cos\left(\frac{2\pi km}{12}\right) + \beta_{k,j}R^j\cdot\sin\left(\frac{2\pi km}{12}\right)\right]\\ &= \left(\gamma_{0,0}+\gamma_{0,1}R\right) + \left(\gamma_{1,0}+\gamma_{1,1}R\right)\cdot\cos\left(\frac{2\pi m}{12}\right) + \left(\beta_{1,0}+\beta_{1,1}R\right)\cdot\sin\left(\frac{2\pi m}{12}\right)\\ &= \gamma_{0,0}+\gamma_{1,0}\cdot\cos\left(\frac{2\pi m}{12}\right)+\beta_{1,0}\cdot\sin\left(\frac{2\pi m}{12}\right)+\left(\gamma_{0,1}+\gamma_{1,1}\cdot\cos\left(\frac{2\pi m}{12}\right)+\beta_{1,1}\cdot\sin\left(\frac{2\pi m}{12}\right)\right)\cdot R\end{aligned}, \tag{6}$$

The exact value of the hyperparameter in the model can be obtained by substitutin the training data into Equation (6) and using the least square method to fit it. The resul is shown in Figure 4, where the rows represent the world from UT = 0 to UT = 23, th columns represent the names of the hyperparameters, and the colors represent the value of the hyperparameters.

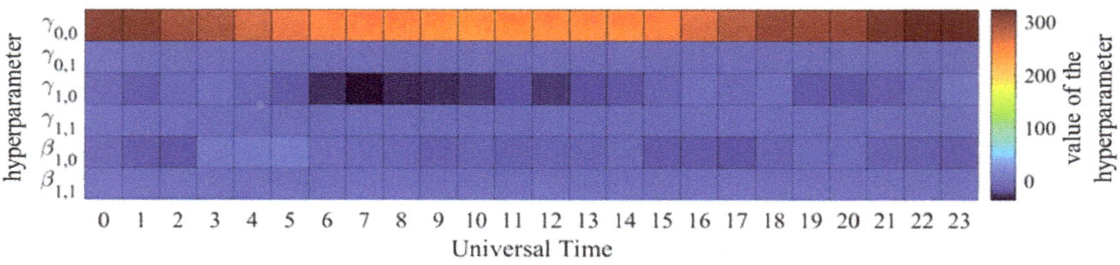

Figure 4. Prediction model hyperparameter distribution diagram.

As can be seen from Figure 4, the change of the hyperparameter changes obviousl with time, and the $\gamma_{0,0}$ is the maximum. With the increase in the order, the value contribu tion of the hyperparameter decreases.

Based on the above conditions, it is only necessary to provide the sunspot numbe month, and universal time corresponding to the time in the model to obtain the predicte value of hmF2. In Figure 5, we present the RRMSE between the observations and th predicted values obtained by using the verification dataset, where OBS is the observe value and VAL is the model validation value.

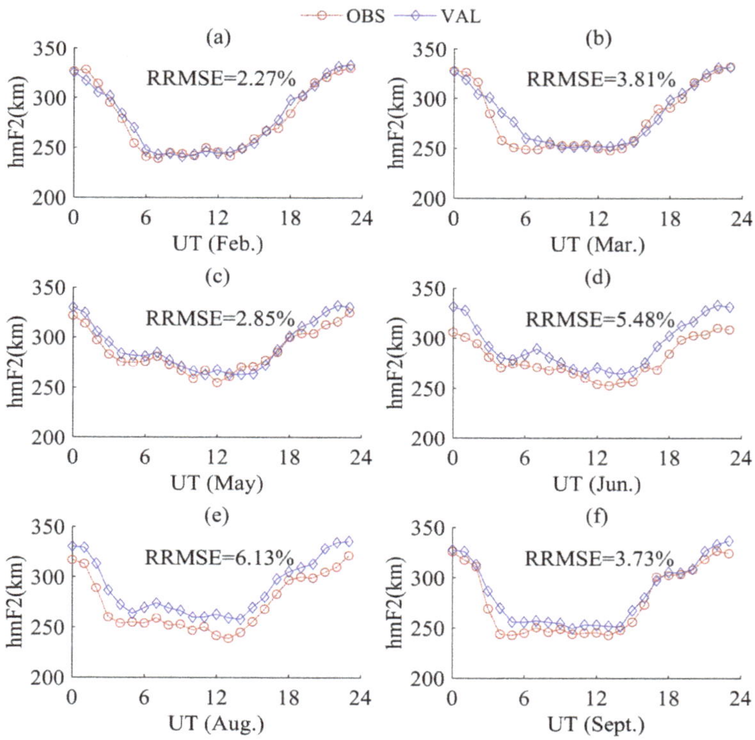

Figure 5. Figure comparing observed and verified values: (**a**) February; (**b**) March; (**c**) May; (**d**) June (**e**) August; (**f**) September.

As shown in Figure 5, the model's predicted ability varies for different months. The RRMSE is less than 3% for February and May, while the predicted ability is relatively poor for July and August, with RRMSE greater than 5%. Furthermore, the predicted values are generally higher than the observations.

4. Discussions

To test the generalization ability of the model, we compared the proposed model (denoted as PRO) with the IRI model [36]. The test data used are hmF2 of January (winter), April (spring), July (summer), and October (autumn) of 2016.

Figure 6 compares the observations, predicted values of the IRI model and PRO model.

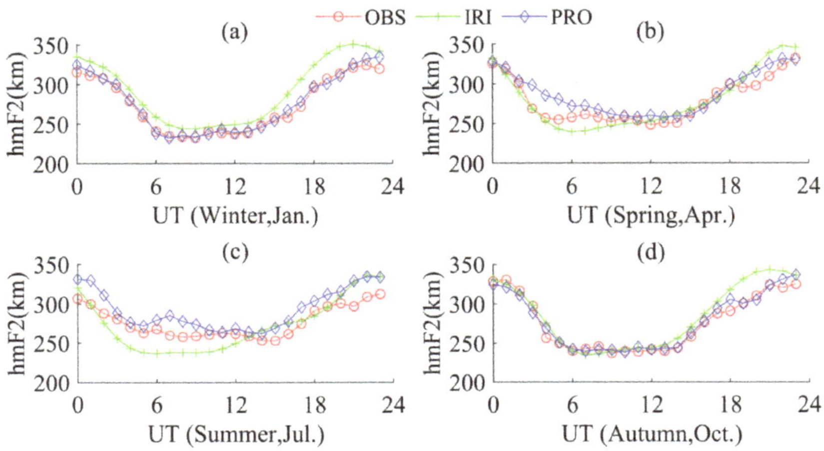

Figure 6. Comparison observations, and predictions of the IRI and PRO models: (**a**) January, Winter; (**b**) April, Spring; (**c**) July, Summer; (**d**) October, Autumn.

Overall, the hmF2 exhibit a trend of being low during the day and high at night. In January and October, the hmF2 values are relatively stable, while, in April and July, they fluctuate to some extent at UT = 7 and UT = 15, respectively. The predicted values of the IRI and PRO models for January and October are relatively close to the observations, but the IRI model tends to overestimate the hmF2 values. However, the IRI and PRO models show significant prediction errors compared with the observations in April and July. Specifically, the predicted values of the IRI model are lower than the observations around UT = 6, while the PRO model's predictions are higher than the observations.

To further evaluate the model's predicted ability and accuracy, RRMSE (Equation (7)) and RRMSE (Equation (4)) is calculated to intuitively analyze the difference between hmF2 predicted by the IRI model and the PRO model and the observed value.

Formula (7) is the calculation formula of RMSE:

$$\text{RMSE} = \sqrt{\frac{1}{N}\sum_{i=1}^{N}(\text{hmF2}'_i - \text{hmF2}_i)^2} \tag{7}$$

where $\text{hmF2}'_i$ is the calculated value of the model, hmF2_i is the measured statistical value, and N is the total data count.

Figure 7 shows the RMSE and RRMSE results of the IRI and PRO models. To conduct a more refined analysis of the results, we also calculated the error of the prediction results in different months. The results are shown as follows:

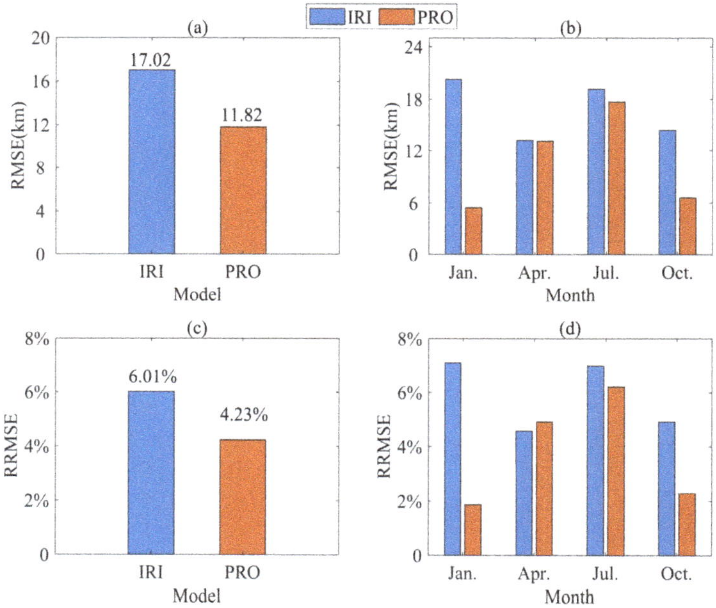

Figure 7. RMSE and RRMSE obtained by the IRI model and PRO model: (**a**) RMSE of the model; (**b**) RMSE calculated by month; (**c**) RRMSE of the model; (**d**) RRMSE calculated by month.

(1) For all the predictions, the RMSE of the IRI model is 17.02 km, and that of the PRO model is 11.82 km. Compared with the IRI model, the RMSE of the PRO model is 5.20 km smaller, proving that the PRO model's stability is better than that of the IRI model.

(2) Specific to each month, the RMSE of the PRO model is smaller than that of the IRI model. In January and October, the RMSE of the PRO model was much smaller than that of the IRI model. In April, the RMSE of the PRO model was only slightly lower than that of the IRI model. Compared with other months, the RMSE of both models was relatively large in July.

(3) The total RRMSE of the IRI model is 6.01%, and that of the PRO model is 4.23%. Compared with the IRI model, the RRMSE of the PRO model is 1.78% smaller, proving that the PRO model's prediction accuracy is better than that of the IRI model.

(4) In each month, except April, the RRMSE of the PRO model is smaller than that of the IRI model in other months. In January and October, the RRMSE of the PRO model decreased most significantly compared with the IRI model. Compared with other months, the RRMSE of both models was relatively large in July.

Generally speaking, the PRO model is superior to the IRI model in both stability and accuracy of hmF2 prediction at the Moscow station. In other words, the PRO model improves the predicted accuracy at the Moscow station. The following are our reflections on the results and prospects for the future:

(1) During January and October, the hmF2 of observation showed relatively stable changes, and the PRO model demonstrated a noticeably better predicted ability for the hmF2 data than the IRI model. However, in April and July, the hmF2 exhibited significant fluctuations, causing a decrease in the predicted ability of both models. This suggests that there is room for improvement room in both models to learn the finer details;

(2) The PRO model utilized $J = 1$ and $K = 1$ as its model parameters, which differs from other research findings. This phenomenon could be attributed to the size of the collected data, which warrants further exploration. This choice could also result in a weaker generalization ability of the model towards hmF2 details, which also requires further investigation;

(3) Due to data collection limitations, the model could only verify hmF2 data collected from the Moscow station. Future efforts will aim to verify hmF2 data from stations in different latitude regions;

(4) In the future, his model can also be compared with other models such as ANN and LSTM.

5. Conclusions

Based on the SML method, this paper proposed an interpretable long-term prediction model for the ionospheric hmF2 median value of the Moscow station. The model only needs to input the sunspot number, month, and universal time to predict the monthly median value data of hmF2 in the corresponding month. In general, compared with the IRI model, the RMSE of the PRO model decreased by 5.20 km and the RRMSE of the PRO model decreased by 1.78%, indicating that the PRO model has certain advantages in predicting hmF2 parameters at this station. Specifically, when predicting hmF2 in January, July, and October, the PRO model has a higher precision prediction. When predicting hmF2 in April, the PRO model has a better predicted degree but lower prediction accuracy than the IRI model. In the future, the applicability of this model needs to be further discussed on other stations in different latitude ranges or other ionospheric parameters. In addition, this model can be compared with other methods such as ANN and LSTM models in the future.

Author Contributions: Conceptualization, J.W. and Q.Y.; methodology, J.W. and Q.Y.; software, J.W., Q.Y., Y.S. and C.Y.; validation, J.W. and Q.Y.; formal analysis, J.W., Q.Y. and Y.S.; investigation, J.W., Q.Y., Y.S. and C.Y.; resources, J.W. and C.Y.; data curation, J.W. and Q.Y.; writing—original draft preparation, J.W. and Q.Y.; writing—review and editing, J.W., Y.S. and C.Y.; visualization, J.W., Q.Y. and C.Y.; supervision, J.W., Q.Y. and Y.S.; project administration, J.W. and C.Y.; funding acquisition, J.W. and C.Y. All authors have read and agreed to the published version of the manuscript.

Funding: This research was funded by the State Key Laboratory of Complex Electromagnetic Environment Effects on Electronics and Information Systems (No. CEMEE2022G0201).

Data Availability Statement: Not applicable.

Conflicts of Interest: The authors declare no conflict of interest.

References

1. Sezen, U.; Sahin, O.; Arikan, F.; Arikan, O. Estimation of hmF2 and foF2 Communication Parameters of Ionosphere F2-Layer Using GPS Data and IRI-Plas Model. *IEEE Trans. Antennas Propag.* **2013**, *61*, 5264–5273. [CrossRef]
2. Tsagouri, I.; Goncharenko, L.; Shim, J.S.; Belehaki, A.; Buresova, D.; Kuznetsova, M.M. Assessment of current capabilities in modeling the ionospheric climatology for space weather applications: foF2 and hmF2. *Space Weather* **2018**, *16*, 1930–1945. [CrossRef]
3. ITU. *ITU-R P.1240, ITU-R Methods of Basic MUF, Operational MUF and Ray-Path Prediction*; ITU: Geneva, Switzerland, 2015.
4. Yan, Z.; Zhang, L.; Rahman, T.; Su, D. Prediction of the HF Ionospheric Channel Stability Based on the Modified ITS Model. *IEEE Trans. Antennas Propag.* **2013**, *61*, 3321–3333. [CrossRef]
5. Yan, Z.; Wang, G.; Tian, G.; Li, W.; Su, D.; Rahman, T. The HF Channel EM Parameters Estimation Under a Complex Environment Using the Modified IRI and IGRF Model. *IEEE Trans. Antennas Propag.* **2011**, *59*, 1778–1783. [CrossRef]
6. Arikan, F.; Sezen, U.; Gulyaeva, T.L.; Cilibas, O. Online, automatic, ionospheric maps: IRI-PLAS-MAP. *Adv. Space Res.* **2015**, *55*, 2106–2113. [CrossRef]
7. Rishbeth, H.; Edwards, R. Modeling the F2 layer peak height in terms of atmospheric pressure. *Radio Sci.* **1990**, *25*, 757–769. [CrossRef]
8. Rao, T.V.; Sridhar, M.; Ratnam, D.V.; Harsha, P.B.S.; Srivani, I. A Bidirectional Long Short-Term Memory-Based Ionospheric foF2 and hmF2 Models for a Single Station in the Low Latitude Region. *IEEE Geosci. Remote Sens. Lett.* **2022**, *19*, 1–5. [CrossRef]
9. Perrone, L.; Mikhailov, A.V.; Scotto, C.; Sabbagh, D. Testing of the Method Retrieving a Consistent Set of Aeronomic Parameters with Millstone Hill ISR Noontime hmF2 Observations. *IEEE Geosci. Remote Sens. Lett.* **2021**, *18*, 1698–1700. [CrossRef]
10. Zhang, B.; Wang, Z.; Shen, Y.; Li, W.; Xu, F.; Li, X. Evaluation of foF2 and hmF2 Parameters of IRI-2016 Model in Different Latitudes over China under High and Low Solar Activity Years. *Remote Sens.* **2022**, *14*, 860.
11. Wang, J.; Shi, Y.; Yang, C. Investigation of Two Prediction Models of Maximum Usable Frequency for HF Communication Based on Oblique- and Vertical-Incidence Sounding Data. *Atmosphere* **2022**, *13*, 1122. [CrossRef]
12. Bilitza, D.; Mckinnell, L.A.; Reinisch, B.; Rowell, T.F. The international reference ionosphere today and in the future. *J. Geod.* **2011**, *85*, 909–920. [CrossRef]

13. Zhang, M.L.; Liu, C.; Wan, W.; Liu, L.; Ning, B. A global model of the ionospheric F2 peak height based on EOF analysis. *An Geophys.* **2009**, *27*, 3203–3212. [CrossRef]
14. Zhang, M.L.; Liu, C.; Wan, W.; Liu, L.; Ning, B. Evaluation of global modeling of M(3000)F2 and hmF2 based on alternativ empirical orthogonal function expansions. *Adv. Space Res.* **2010**, *46*, 1024–1031. [CrossRef]
15. Yu, Y.; Wan, W.; Xiong, B.; Ren, Z.; Zhao, B.; Zhang, Y.; Ning, B.; Liu, L. Modeling Chinese ionospheric layer parameters based o EOF analysis. *Space Weather* **2015**, *13*, 339–355. [CrossRef]
16. Themens, D.R.; Jayachandran, P.T.; Galkin, I.; Hall, C. The Empirical Canadian High Arctic Ionospheric Model (E-CHAIM): NmF and hmF2. *J. Geophys. Res. Space Phys.* **2017**, *122*, 9015–9031. [CrossRef]
17. Sai, G.V.; Tulasi, R.S. An Artificial Neural Network based Ionospheric Model to predict NmF2 and hmF2 using long-tern data set of FORMOSAT-3/COSMIC radio occultation observations: Preliminary results. *J. Geophys. Res. Space Phys.* **2017**, *12* 11743–11755.
18. Tulasi, R.S.; Sai, G.V.; Mitra, A.; Reinisch, B. The improved two-dimensional artificial neural network-based ionospheric mode (ANNIM). *J. Geophys. Res. Space Phys.* **2018**, *123*, 5807–5820. [CrossRef]
19. Li, W.; Zhao, D.; He, C.; Hu, A.; Zhang, K. Advanced Machine Learning Optimized by The Genetic Algorithm in Ionospheri Models Using Long-Term Multi-Instrument Observations. *Remote Sens.* **2020**, *12*, 866. [CrossRef]
20. Fokoué, E. Model Selection for Optimal Prediction in Statistical Machine Learning. *Not. Am. Math. Soc.* **2020**, *67*, 2. [CrossRef]
21. Wang, J.; Yang, C.; An, W. Regional Refined Long-term Predictions Method of Usable Frequency for HF Communication Base on Machine Learning over Asia. *IEEE Trans. Antennas Propag.* **2022**, *70*, 4040–4055. [CrossRef]
22. Wang, J.; Yu, Q.; Shi, Y.; Liu, Y.; Yang, C. An Explainable Dynamic Prediction Method for Ionospheric foF2 Based on Machin Learning. *Remote Sens.* **2023**, *15*, 1256. [CrossRef]
23. Qian, L.; Burns, A.G.; Solomon, S.C. Annual/semiannual variation of the ionosphere. *Geophys. Res. Lett.* **2013**, *40*, 1928–1933.
24. Zhou, Z.H. *Machine Learning*, 2nd ed.; Tsinghua University Press: Beijing, China, 2016.
25. National Oceanic and Atmospheric Administration (NOAA). Available online: https://www.ngdc.noaa.gov/stp/space-weather solar-data/ (accessed on 28 October 2022).
26. Afraimovich, E.L.; Astafyeva, E.I.; Oinats, A.V. Global electron content: A new conception to track solar activity. *Ann. Geophys* **2008**, *26*, 335–344. [CrossRef]
27. Sunspot Number. Available online: https://www.sidc.be/silso/datafiles (accessed on 28 October 2022).
28. Data of Hydrogen Emission at 121.6 nm. Available online: https://lasp.colorado.edu/lisird/composite_timeseries.html (accesse on 27 April 2022).
29. Tapping, K.F. The 10.7cm solar radio flux (F10.7). *Space Weather* **2013**, *11*, 394–406. [CrossRef]
30. Solomon, S.C.; Qian, L.; Burns, A.G. The anomalous ionosphere between solar cycles 23 and 24. *J. Geophys. Res. Space Phys.* **2013** *118*, 6524–6535. [CrossRef]
31. Bai, H.M. *Ionospheric Model Research Based on Intelligent Information Processing Technology*; Tianjin University: Tianjin, China, 2022
32. Sun, W. *Study on Regional Ionospheric Characteristics Based on Ground-Based GPS and Occultation Technology*; Wuhan University Wuhan, China, 2015.
33. Zeng, W. Comparison of Different Detection Scenarios of Lyman-α. In *Highlights in Science, Engineering and Technology* Darcy & Roy Press: Portland, OR, USA, 2023; Volume 38, pp. 850–855.
34. Perna, L.; Pezzopane, M. foF2 vs solar indices for the Rome station: Looking for the best general relation which is able to describe the anomalous minimum between cycles 23 and 24. *J. Atmos. Sol.-Terr. Phys.* **2016**, *148*, 13–21. [CrossRef]
35. Wang, J.; Feng, F.; Bai, H.M.; Cao, Y.B.; Chen, Q.; Ma, J.G. A regional model for the prediction of M(3000)F2 over East Asia. *Adv Space Res.* **2020**, *65*, 2036–2051. [CrossRef]
36. International Reference Ionosphere. Available online: http://IRImodel.org/IRI-2016 (accessed on 18 April 2022).

remote sensing

MDPI

Article

An Integrated Framework for Spatiotemporally Merging Multi-Sources Precipitation Based on F-SVD and ConvLSTM

Sheng Sheng [1], Hua Chen [1,*], Kangling Lin [1], Nie Zhou [1], Bingru Tian [1] and Chong-Yu Xu [2]

1 State Key Laboratory of Water Resources Engineering and Management, Wuhan University, Wuhan 430072, China; shengsheng@whu.edu.cn (S.S.); linkangling@whu.edu.cn (K.L.); niezhou@whu.edu.cn (N.Z.); tianbingru@whu.edu.cn (B.T.)

2 Department of Geosciences, University of Oslo, P.O. Box 1047 Blindern, N-0316 Oslo, Norway; c.y.xu@geo.uio.no

* Correspondence: chua@whu.edu.cn

Abstract: To improve the accuracy and reliability of precipitation estimation, numerous models based on machine learning technology have been developed for integrating data from multiple sources. However, little attention has been paid to extracting the spatiotemporal correlation patterns between satellite products and rain gauge observations during the merging process. This paper focuses on this issue by proposing an integrated framework to generate an accurate and reliable spatiotemporal estimation of precipitation. The proposed framework integrates Funk-Singular Value Decomposition (F-SVD) in the recommender system to achieve the accurate spatial distribution of precipitation based on the spatiotemporal interpolation of rain gauge observations and Convolutional Long Short-Term Memory (ConvLSTM) to merge precipitation data from interpolation results and satellite observation through exploiting the spatiotemporal correlation pattern between them. The framework (FS-ConvLSTM) is utilized to obtain hourly precipitation merging data with a resolution of 0.1° in Jianxi Basin, southeast of China, from both rain gauge data and Global Precipitation Measurement (GPM) from 2006 to 2018. The LSTM and Inverse Distance Weighting (IDW) are constructed for comparison purposes. The results demonstrate that the framework could not only provide more accurate precipitation distribution but also achieve better stability and reliability. Compared with other models, it performs better in variation process description and rainfall capture capability, and the root mean square error (RSME) and probability of detection (POD) are improved by 63.6% and 22.9% from the original GPM, respectively. In addition, the merged precipitation combines the strength of different data while mitigating their weaknesses and has good agreement with observed precipitation in terms of magnitude and spatial distribution. Consequently, the proposed framework provides a valuable tool to improve the accuracy of precipitation estimation, which can have important implications for water resource management and natural disaster preparedness.

Keywords: spatiotemporal fusion; machine learning; multi-source precipitation; ConvLSTM; F-SVD

Citation: Sheng, S.; Chen, H.; Lin, K.; Zhou, N.; Tian, B.; Xu, C.-Y. An Integrated Framework for Spatiotemporally Merging Multi-Sources Precipitation Based on F-SVD and ConvLSTM. *Remote Sens.* **2023**, *15*, 3135. https://doi.org/10.3390/rs15123135

Academic Editor: Gwanggil Jeon

Received: 28 April 2023
Revised: 10 June 2023
Accepted: 13 June 2023
Published: 15 June 2023

1. Introduction

Precipitation is a critical meteorological variable with significant implications for many applications, including flood forecasting [1], agriculture [2], water resource management [3], and climate change studies [4]. Due to its large spatiotemporal variability, the accurate estimation of precipitation remains challenging, especially in areas with complex terrains and limited observational networks [5,6].

Traditional approaches to precipitation estimation rely on observations from rain gauges. These are simple devices that collect and measure the amount of precipitation. The observation data are usually considered reliable and accurate but have limited spatial coverage and are prone to errors due to gauge under-catch and exposure issues [7]. To overcome these limitations, many alternative approaches have been developed to estimate precipitation, including remote sensing technologies such as radar and satellite [8]. Radar waves provide information on precipitation location, intensity, and distribution with a high temporal and spatial resolution. In contrast, satellite observations provide a large-scale view of precipitation with a relatively high spatial resolution but lower temporal resolution than radar waves [9]. Both offer improved spatial and temporal coverage, but the data may not be as accurate as that obtained from rain gauges. Therefore, combining data from multiple sources, also known as multi-source precipitation fusion, has become a promising way to overcome these limitations and provide more accurate estimates of precipitation.

In recent years, numerous statistical methods have been introduced for integrating precipitation data from rain gauges and satellites, including bias correction [10], Kriging-based methods [11], linear regression model [12], geographical difference analysis method [13], Bayesian combination method [14], Kalman filter calibration method [15], and geographically weighted regression method [16]. Duan et al. [17] merged the precipitation data observed at ground stations with the TRMM 3B42 satellite precipitation data by separately employing linear regression, geographically weighted regression, Kalman filer fusion, and the optimal interpolation method. The comparison results show that linear regression shows the best merging effect across the daily scale, while at the monthly scale, the precipitation data processed using the Kalman filter presented the highest accuracy. However, the aforementioned methodologies rely heavily on mathematical equations and strong assumptions, which can result in various limitations. Given the rapid advancement of machine learning (ML) technology, it possesses the potential to surmount certain limitations intrinsic to the aforementioned methods [18]. Unlike traditional approaches, ML exhibits more robust learning and generalization abilities, allowing it to effectively manage complex nonlinear relationships without requiring explicit statistical models. Additionally, ML demonstrates superior efficiency in processing vast amounts of data, thus enhancing its computational performance. Zhang et al. [19] used five ML algorithms (extreme gradient boosting, gradient boosting decision tree, random forest, LightGBM, and multiple linear regression) together with auxiliary geographic parameters to merge hourly data from meteorological stations, Radar, and satellites. The results show that the random forest-based hourly precipitation merging model is suitable for analyzing monsoon rainstorm events, while the extreme gradient boosting-based hourly precipitation merging model is suitable for analyzing typhoon events. Zhang et al. [20] applied four ML approaches, including a support vector machine, a random forest algorithm, an artificial neural network, and extreme gradient boosting, to construct the estimation models in which cloud properties are taken as additional predictors to improve the early run of the Integrated Multi-satellite Retrievals for GPM (IMERG).

Despite the progress in the development of multi-source precipitation merging models based on ML, most studies have primarily focused on describing the relationship be-tween the precipitation data and complex environment variables, while the spatiotemporal correlation patterns between satellite products and rain gauge observations have received relatively limited attention. The long short-term memory network (LSTM) is a type of recurrent neural network (RNN) specifically designed to handle long-term dependencies by utilizing a gating mechanism to regulate the flow of information. Shen et al. [21] proposed an integrated framework to merge multi-satellite and gauge precipitation data, which integrates the geographically weighted regression to improve the spatial resolution of precipitation estimations and the LSTM to improve the precipitation estimation accuracy by exploiting the temporal correlation pattern between multi-satellite precipitation products and rain gauges. Wu et al. [22] proposed a spatiotemporal deep fusion model by combining the convolutional neural networks (CNN) and the LSTM to merge the TRMM 3B42 V7 satellite data, rain gauge data, and thermal infrared images. The CNN was used to extract spatial features from the radar and satellite data, while the LSTM was used to capture the temporal features of the precipitation data. The superiority of LSTM in merging precipitation data has been verified. On the other hand, Convolutional LSTM (ConvLSTM) is an extension of LSTM that incorporates convolutional layers into its network architecture, which allows it to perform spatiotemporal modeling of sequential data and capture spatiotemporal correlations more effectively due to its inherent convolutional structure. It has been successfully applied to precipitation nowcasting and verified to consistently outperform the fully connected LSTM [23]. To the best of our knowledge, there are no related studies merging satellite and gauge precipitation using a ConvLSTM network by formulating multi-source precipitation merging as a spatiotemporal sequence processing problem.

The purpose of our study is to address the challenge of spatiotemporally merging precipitation data from satellite and rain gauges by introducing an integrated framework. This framework aims to effectively combine spatiotemporal information from different data sources to enhance the accuracy and reliability of precipitation estimation. It integrates F-SVD in the recommender system to achieve the accurate spatial distribution of precipitation based on the spatiotemporal interpolation of rain gauge observations and ConvLSTM by exploiting the spatiotemporal correlation pattern between them. The framework (FS-ConvLSTM) is applied to the Jianxi Basin of China to generate hourly precipitation estimates with a resolution of 0.1° from the data of both rain gauges and GPM (IMERG V06) from 2006 to 2018.

2. Study Area and Materials

2.1. Study Area

The study area is located in the Jianxi basin in southeast China, between 117°31′–119°00′ east longitude and 26°31′–28°31′ north latitude (Figure 1). This basin is the largest tributary in the upper reaches of the Minjiang River, with its mainstream originating in Wuyishan and spanning 635.6 km. The entire drainage area of the Jianxi River basin covers 14,787 km^2, accounting for 27% of the total area of the Minjiang River basin. The basin is situated in a subtropical monsoon climate zone, characterized by humid air and abundant rainfall, with annual average rainfall ranging between 1800 and 2200 mm. Most rainfall occurs during the plum rain season from April to June and the typhoon rain season from July to September.

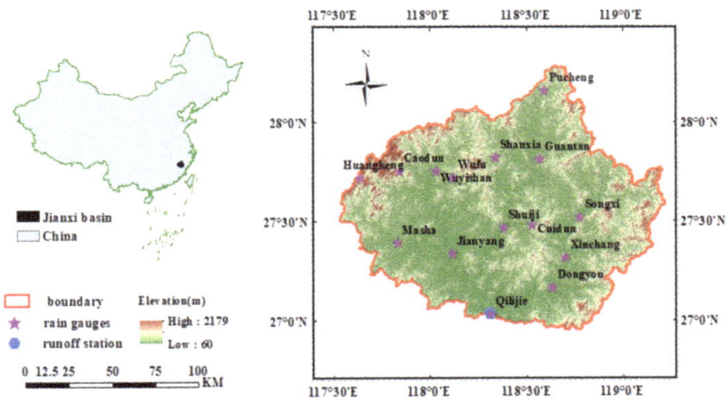

Figure 1. Location of the study area and spatial distribution of rain gauges.

2.2. Data

The network of rainfall gauges in the study area is characterized by its dense and evenly distributed nature, ensuring reliable data quality and high accuracy of the observations. The observation data used in this study were obtained from gauges operated by the Fujian Provincial Bureau of Hydrology, which are not classified as international exchange stations. The spatial distribution of meteorological stations can be seen in Figure 1, with a total of 15 rain gauges in the entire basin. Hourly data from 425 rainfall events from 2006 to 2018 were selected as the research data. The training period was from 2006 to 2014, and the testing period was from 2015 to 2018. The dataset was prepared through sliding windows, resulting in a training set with a sample size of 272,916 and a test set with a sample size of 109,858.

To evaluate the accuracy improvement of the fusion model for different magnitudes of rainfall data, the maximum 12 h cumulative rainfall was calculated based on the selected 425 rainfall events. Rainfall was classified into four categories: light, moderate, heavy, and torrential, based on their magnitude as per the classification criteria listed in Table 1.

Table 1. Rainfall type classification criteria and results.

Type	Light	Moderate	Heavy	Torrential
Maximum 12 h rainfall (mm)	<5	5~15	15~30	>30
Number of events	6	157	160	102
Number of events in the training stage	4	102	112	79
Number of events in the testing stage	2	52	48	23

The satellite precipitation data were obtained from the Integrated Multi-Satellite Retrievals for Global Precipitation Measurement Mission (IMERG). Global Precipitation Measurement (GPM) is an international satellite mission conducted by the National Aeronautics and Space Administration (NASA) and Japan Aerospace Exploration Agency (JAXA), which uses multi-sensors, multi-satellites, and multi-algorithms combined with satellite networks and rain gauge inversion to obtain more accurate precipitation data. IMERG algorithm is designed to calibrate, combine, and interpolate satellite microwave precipitation estimates from the TRMM and GPM, as well as microwave-calibrated infrared satellite estimates, precipitation measurement analyses, and potentially other precipitation estimates. It has now been updated to product version V06B. This study uses GPM (IMERG V06) Final Run data with a spatial resolution of 0.1° and a temporal resolution of half an hour from January 2006 to December 2018 (https://earthdata.nasa.gov/; accessed on 15 October 2022).

3. Methodology

This study proposes an integrated framework to spatiotemporally merge precipitation data from rain gauge and GPM observations. Figure 2 presents the main structure of the framework: the spatiotemporal interpolation method based on F-SVD (Figure 2a) and the fusion model based on ConvLSTM (Figure 2b). The relevant methods are briefly described as follows.

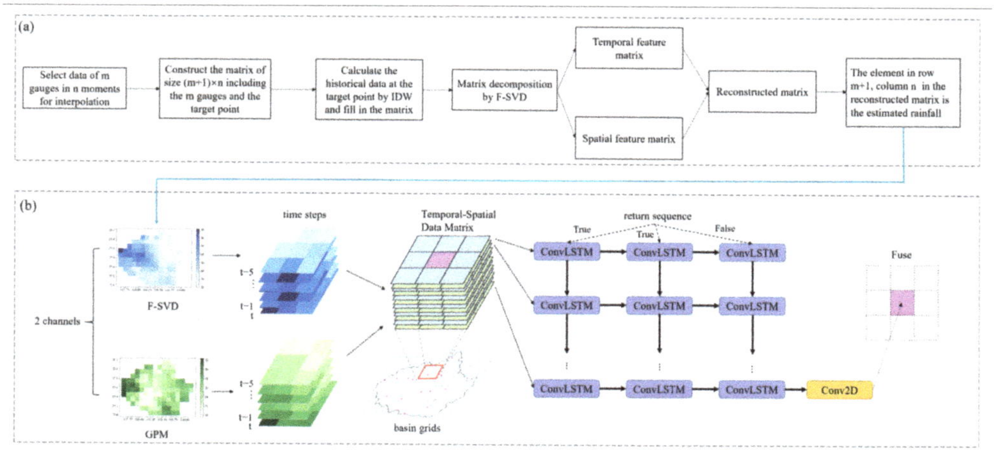

Figure 2. The integrated framework (FS-ConvLSTM) for the merging of precipitation data from rain gauge and GPM observations. (**a**) Spatiotemporal interpolation method based on F-SVD; (**b**) Fusion model based on ConvLSTM.

3.1. F-SVD

The interpolation method adopted in this study is a spatiotemporal precipitation method based on matrix decomposition (hereafter referred to as F-SVD) proposed by Chen et al. [24], which has been proven to outperform the traditional interpolation methods (inverse distance weight, ordinary kriging) through cross-validation and offer a better spatial estimation. The calculation process is presented in Figure 2a and contains the following steps:

(1) The historical precipitation information from m surrounding gauges and n past moments needs to be prepared for the interpolation at the target point at a certain moment.
(2) The precipitation data from the surrounding gauges and the target point from adjacent moments can form a spatiotemporal data matrix with dimensions of $(m+1) \times n$, where the rows represent time and columns represent space.
(3) If any precipitation values in the matrix representing the historical precipitation of the target point are unknown, traditional interpolation methods such as IDW should be used to calculate these values until only one null value representing the precipitation to be estimated remains.
(4) The F-SVD method is utilized to decompose the matrix into a temporal feature matrix X and a spatial feature matrix Y. The stochastic gradient descent algorithm is used for optimization.
(5) Then, the two optimal feature matrices are multiplied to reconstruct a matrix P, the element at the $m + 1$ row and n column in the reconstructed matrix represents the estimated precipitation, which is calculated as follows:

$$P_{\mathrm{i,j}} = \sum_{q=1}^{q} X_{\mathrm{i,q}} Y_{\mathrm{q,j}} \tag{1}$$

where q is the number of latent features.

3.2. ConvLSTM

ConvLSTM is an extension of LSTM, which uses convolutional layers to process spatiotemporal data [23]. It applies convolutions on the input data before passing it through the LSTM cells, allowing it to capture spatial and temporal dependencies within the input sequence [25]. As shown in Figure 3, the structure of ConvLSTM is similar to LSTM that contains forget gate (f_t), input gate (i_t), and output gate (O_t), but with convolutional layers instead of fully connected layers, which enables ConvLSTM to capture underlying spatial features [26]. The transmission relationship between the gates is expressed using the following equation:

$$i_t = \sigma(W_{xi} * \chi_t + W_{hi} * H_{t-1} + W_{ci} \circ C_{t-1} + b_i) \tag{2}$$

$$f_t = \sigma\left(W_{xf} * \chi_t + W_{hf} * H_{t-1} + W_{cf} \circ C_{t-1} + b_f\right) \tag{3}$$

$$C_t = f_t \circ C_{t-1} + i_t \circ \tanh(W_{xc} * \chi_t + W_{hc} * H_{t-1} + b_c) \tag{4}$$

$$o_t = \sigma(W_{xo} * \chi_t + W_{ho} * H_{t-1} + W_{co} \circ C_t + b_o) \tag{5}$$

$$H_t = o_t \circ \tanh(C_t) \tag{6}$$

where $\circ$ denotes the Hadamard product; $*$ denotes the convolution operator; σ is the sigmoid activation function, which is given by:

$$\sigma(x) = \frac{1}{1+e^{-x}} \tag{7}$$

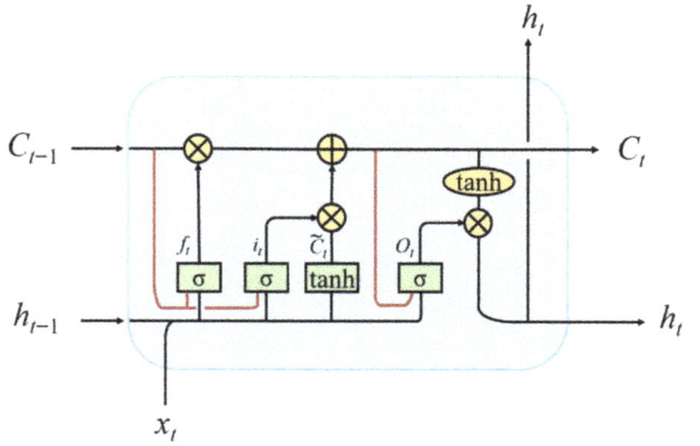

Figure 3. Structure of ConvLSTM cell.

3.3. The Spatiotemporal Fusion Model

The model diagram of the spatiotemporal fusion of precipitation data using ConvLSTM based on GPM satellite observation data and ground-based spatiotemporal interpolation data calculated by the F-SVD method is shown in Figure 2b.

At a point s in space, nine surrounding grid points, including itself, are selected to form the GPM satellite-observed precipitation matrix G^t and the ground spatiotemporal interpolated precipitation matrix I^t at time t. These matrices are defined as follows:

$$G^t = \begin{bmatrix} g_{s-4,t} & g_{s-3,t} & g_{s-2,t} \\ g_{s-1,t} & g_{s,t} & g_{s+1,t} \\ g_{s+2,t} & g_{s+3,t} & g_{s+4,t} \end{bmatrix} \quad I^t = \begin{bmatrix} i_{s-4,t} & i_{s-3,t} & i_{s-2,t} \\ i_{s-1,t} & i_{s,t} & i_{s+1,t} \\ i_{s+2,t} & i_{s+3,t} & i_{s+4,t} \end{bmatrix} \tag{8}$$

where $g_{s,t}$ and $i_{s,t}$ are the GPM satellite observation data and ground-based spatiotemporal interpolation data at time t, point s, respectively.

For the spatiotemporal precipitation fusion method, the GPM satellite observed precipitation matrix and the ground spatiotemporal interpolated precipitation matrix for the past six time periods need to be inputted into the constructed ConvLSTM to obtain the spatiotemporal fused precipitation $p_{s,t}$ at point s on space at time t in the following form:

$$p_{s,t} = f_{convlstm}\left[\left(G^{t-5}, G^{t-4}, \ldots, G^t\right), \left(I^{t-5}, I^{t-4}, \ldots, I^t\right)\right] \tag{9}$$

The structure of the ConvLSTM network model constructed for spatiotemporal precipitation fusion is presented in Figure 4. The figure illustrates the precipitation fusion process from input to output at a specific time point, with rectangles representing different neural network levels and rounded rectangles indicating the format of input, output, and intermediate variables. The input data have a tensor of size 6 × 3 × 3 × 2, where the dimensions represent time, row, column, and the number of channels, respectively. This tensor stores the data of GPM and ground-based spatiotemporal interpolation. The output data have a three-dimensional tensor of size 1 × 1 × 1, with dimensions of time, rows, and columns, respectively, which holds the fused precipitation.

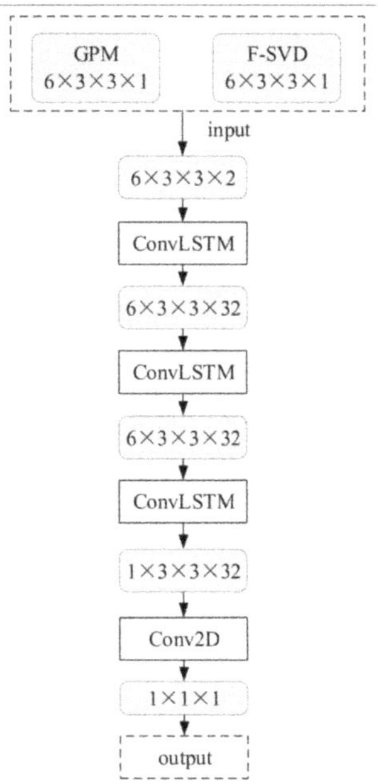

Figure 4. Structure of the three-layer ConvLSTM.

With the rain gauge as the central point, the nine surrounding grid points were selected. The input data for the ConvLSTM model comprised the GPM data and ground spatiotemporal interpolation data (F-SVD) of the past six moments, while the output was the precipitation observed at the station. The hourly precipitation data from 2006 to 2014 were used as the training set, and the hourly precipitation data from 2014 to 2018 were used as the testing set to evaluate the model's accuracy.

The main steps of the multi-source precipitation spatiotemporal fusion algorithm (FS-ConvLSTM) are as follows:

(1) Download GPM satellite observation data with a spatial resolution of 0.1° and a temporal resolution of 0.5 h. Process the data to obtain precipitation data with a time interval of 1 h.
(2) Collect and organize the rain gauge observation data of the study watershed and interpolate them into $0.1° \times 0.1°$ spatial grid point data using the F-SVD method.
(3) Construct input sample sets based on GPM and interpolation results of F-SVD, normalize the data, and use the data from 2006 to 2014 as the training set and the data from 2014 to 2018 as the testing set.
(4) Train the model at 15 rain gauges with precipitation observations as the true value and mean squared error as the loss function to minimize the training loss.
(5) Apply the trained model to each grid point within the study watershed for precipitation fusion.

3.4. Evaluation Indicators

The quality of precipitation data is evaluated using two types of indices: quantitative and categorical. The quantitative evaluation index assesses factors such as rainfall magnitude and temporal distribution. It employs several measures, including relative deviation (*BIAS*), root mean square error (*RSME*), correlation coefficient (*CC*), and rainfall ratio (*RATIO*). Meanwhile, the categorical evaluation index focuses on the spatial distribution of precipitation and uses probability of detection (*POD*), false alarm rate (*FAR*), threat score (*TS*), and missed alarm rate (*MAR*).

$$BIAS = \frac{\sum_{i=1}^{n} SAT_i - \sum_{i=1}^{n} OBS_i}{\sum_{i=1}^{n} OBS_i} \tag{10}$$

$$RSME = \sqrt{\frac{1}{n}\sum_{i=1}^{n} (SAT_i - OBS_i)^2} \tag{11}$$

$$CC = \frac{\sum_{i=1}^{n} (SAT_i - \overline{SAT})(OBS_i - \overline{OBS})}{\sqrt{\sum_{i=1}^{n} (SAT_i - \overline{SAT})^2 (OBS_i - \overline{OBS})^2}} \tag{12}$$

$$RATIO = \frac{\sum_{i=1}^{n} SAT_i}{\sum_{i=1}^{n} OBS_i} \tag{13}$$

$$POD = \frac{TP}{TP + FN} \tag{14}$$

$$FAR = \frac{FP}{TP + FP} \tag{15}$$

$$TS = \frac{TP}{TP + FN + FP} \tag{16}$$

$$MAR = \frac{FN}{FN + TP} \tag{17}$$

where n denotes the number of observation data; *SAT* and *OBS* refer to GPM observation and rain gauge observation, respectively; *TP* represents the number of accurately forecasted precipitation data; *FN* is the number of missed reports; and *FP* denotes the number of false reports. These variables are presented in Table 2. Specifically, *TP* indicates the number of precipitation data observed by both the satellite and the rain gauge, whereas *FP* refers to the number of precipitation data observed by the satellite but not by the rain gauge. On the other hand, *FN* represents the number of rainfall data observed by the rain gauge but not by the satellite.

Table 2. Description of TP, FP, FN, and TN.

Rain Gauge Observation	Satellite Product	
	Yes	**No**
Yes	TP	FN
No	FP	TN

4. Results

The study proposes an integrated framework (FS-ConvLSTM) and applies it to generate hourly spatial precipitation estimation through spatiotemporally merging data from rain gauge and GPM (IMERG V06) observations in the Jainxi basin of China from 2006 to 2018. The relevant results and findings are displayed in the following sub-sections: quality assessment on GPM (Section 4.1), accuracy evaluation of different models (Section 4.2), and model performance in typical rainfall events (Section 4.3).

4.1. Quality Assessment of GPM

The satellite rainfall observation dataset for the Jianxi watershed is constructed by collating GPM data at a 0.1° scale and 0.5 h interval from 2006 to 2018 into a precipitation dataset with a 1 h time scale. Based on this dataset, the eight evaluation indicators are applied to assess GPM at the grid points of the 15 rainfall stations, and the results are presented in Figure 5.

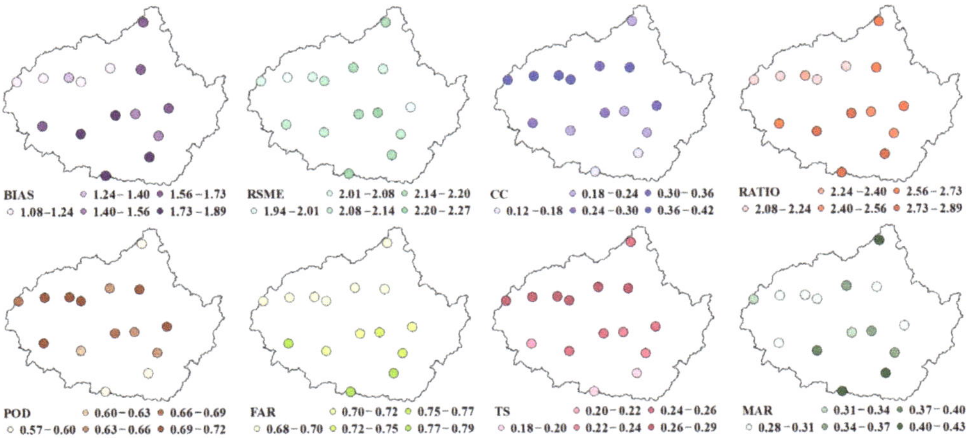

Figure 5. Evaluation of GPM with reference to rain gauge observations.

According to the quantitative evaluation results, on the whole, GPM overestimates rainfall events by approximately twice as much as the ground observations. The *BIAS* between GPM and gauge observation is greater than 1, and the *RATIO* is between 2 and 3. The *RSME* value does not vary significantly among gauges—mostly between 2 and 2.2 mm. In contrast, the *CC* varies widely among different gauges, with values ranging from 0.1 to 0.45. Qilijie station has the lowest *CC* value of 0.117, while Cao Dun station has the highest *CC* value of 0.422.

Regarding the categorical evaluation index results, the POD of GPM varies from 0.55 to 0.73 at each rainfall gauge, with an average value of 0.661. This indicates that approximately 66% of actual rainfall events can be captured and reflected by GPM. The *FAR* of GPM is around 0.7, which means that about 70% of the rainfall events observed by GPM are false alarms, i.e., no actual rainfall occurs. The *MAR* of GPM varies between 0.25 and 0.45 at each station, suggesting that approximately 30% of actual rainfall events correspond to GPM data of 0. It is important to note that individual classification index evaluation can be one-sided, and high *POD* may coincide with high *FAR*. *TS* integrates the evaluation of GPM on both hit and missed rainfall events, with an average value of 0.247 across all stations.

Overall, GPM data can partially reflect the actual rainfall patterns. However, there is an overall tendency to overestimate rainfall events, as well as omissions and the misreporting of some events, which indicates that the GPM data lacks stability.

4.2. Accuracy Evaluation of Different Models

4.2.1. Accuracy Evaluation at Rain Gauges

The accuracy evaluations of different precipitation data at the location of rain gauges during the training and testing stages are presented in Table 3. As can be seen from the table, for the training stage, FS-ConvLSTM exhibits superiority over other models in most indicators, except for two evaluation indicators related to total rainfall, including *BIAS* and *RATIO*, where F-SVD demonstrates better accuracy. For the testing stage, the advantage of FS-ConvLSTM is maintained, but LSTM shows better accuracy in *BIAS* and *RATIO*.

The evaluation results of different precipitation data show consistent performance in the training and testing stages.

Table 3. Performance of the FS-ConvLSTM fusion model in the training and testing stages.

Period	Data	*BIAS*	*RSME*	*CC*	*RATIO*	*POD*	*FAR*	*TS*	*MAR*
Training stage	GPM	1.029	3.874	0.295	2.029	0.524	0.496	0.346	0.476
	IDW	−0.036	1.998	0.489	0.964	0.552	0.326	0.436	0.448
	F-SVD	−0.027	1.481	0.753	0.973	0.634	0.186	0.554	0.366
	LSTM	−0.049	2.070	0.393	0.951	0.507	0.455	0.356	0.493
	FS-ConvLSTM	0.208	1.410	0.782	1.208	0.644	0.183	0.563	0.356
Testing stage	GPM	1.088	3.656	0.290	2.088	0.519	0.533	0.326	0.481
	IDW	−0.022	1.957	0.437	0.978	0.508	0.370	0.391	0.492
	F-SVD	−0.024	1.487	0.714	0.976	0.602	0.221	0.514	0.398
	LSTM	0.020	2.007	0.330	1.020	0.482	0.499	0.326	0.518
	FS-ConvLSTM	0.256	1.404	0.754	1.256	0.612	0.219	0.522	0.388

For the evaluation results of the quantitative indicators, the GPM data show an apparent overestimation. The ground-interpolated data slightly underestimate the rainfall, and the accuracy of F-SVD is better than that of IDW in each indicator value since the interpolation accounts for both spatial relationships and the trend of temporal changes. LSTM has lower *BIAS*, and *RATIO* is closer to 1, while FS-ConvLSTM has lower *RSME* and higher *CC*. Through fusing the data from rain gauges, the bias of GPM is improved, and the variation process fits better with the measured values. Compared with GPM, FS-ConvLSTM and LSTM reduce by 63.6% and 46.6% in *RSME* and improve by 165% and 33.2% in *CC*, respectively. Regarding the evaluation results of the categorical indicators, F-SVD shows obvious advantages over IDW, with higher *POD* and *TS* and lower *FAR* and *MAR*. The performance of LSTM is similar to that of GPM, while FS-ConvLSTM shows a great improvement in each indicator compared to GPM and LSTM. FS-ConvLSTM has the most accurate description of rainfall events, capturing and reflecting more than 60% of the rainfall events while reducing the cases of rainfall misreporting and omission. Compared with GPM, FS-ConvLSTM improves by 22.9% and 62.7% in *POD* and *TS* and is reduced by 63.1% and 25.2% in *FAR* and *MAR*. In summary, FS-ConvLSTM performs optimally in the description of rainfall variation process and event capture capability via the fusion of variation characteristics in the time and space of GPM and ground-interpolated rainfall data.

To compare the accuracy of rainfall data at different magnitudes more precisely, an accuracy evaluation was performed separately for light, medium, heavy, and torrential rainfall events during the testing stage, and the results are shown in Table 4. In light rainfall events, F-SVD performs best in the evaluation results of quantitative indicators, showing obvious advantages in *BIAS*, *RSME*, and *RATIO*. GPM performs best in the classification evaluation indicators, with the highest *POD* and *TS* and the lowest *MAR*. However, there is an apparent overestimation of rainfall in GPM, with values close to 3.8 times the measured values. In addition, FS-ConvLSTM has the highest *CC* and the lowest *FAR* among the models, which also indicates superiority. For the medium, heavy, and torrential rainfall events, FS-ConvLSTM has the lowest *RSME*, *FAR*, *MAR*, and the highest *CC*, *POD*, and *TS*, demonstrating that the spatiotemporal fusion data are closest to the variation process of the measured rainfall series and capture the rainfall events most accurately. Meanwhile, the ground-interpolated data performed best on *BIAS* and *RATIO* and are closest to the actual in terms of total rainfall.

Table 4. Evaluation results of the precipitation data in different types of rainfall events.

Type	Data	*BIAS*	*RSME*	*CC*	*RATIO*	*POD*	*FAR*	*TS*	*MAR*
Small rain	GPM	2.829	1.461	0.251	3.829	0.423	0.704	0.211	0.577
	IDW	−0.053	0.378	0.483	0.947	0.088	0.226	0.085	0.912
	F-SVD	−0.016	0.278	0.751	0.984	0.215	0.078	0.210	0.785
	LSTM	1.509	0.651	0.207	2.509	0.383	0.687	0.208	0.617
	FS-ConvLSTM	0.764	0.333	0.753	1.764	0.157	0.044	0.156	0.843
Moderate rain	GPM	0.970	2.317	0.151	1.970	0.381	0.629	0.232	0.619
	IDW	−0.020	1.469	0.247	0.980	0.332	0.470	0.256	0.668
	F-SVD	−0.011	1.102	0.649	0.989	0.463	0.277	0.394	0.537
	LSTM	0.323	1.447	0.214	1.323	0.329	0.595	0.222	0.671
	FS-ConvLSTM	0.463	1.018	0.724	1.463	0.464	0.262	0.398	0.536
Heavy rain	GPM	1.176	3.641	0.273	2.176	0.557	0.515	0.350	0.443
	IDW	−0.021	1.949	0.409	0.979	0.544	0.350	0.420	0.456
	F-SVD	−0.023	1.535	0.674	0.977	0.618	0.212	0.530	0.382
	LSTM	0.260	1.983	0.306	1.260	0.517	0.485	0.348	0.483
	FS-ConvLSTM	0.263	1.456	0.718	1.263	0.634	0.208	0.543	0.366
Torrential rain	GPM	1.063	4.903	0.314	2.063	0.594	0.475	0.386	0.406
	IDW	−0.024	2.477	0.485	0.976	0.623	0.334	0.475	0.377
	F-SVD	−0.027	1.846	0.748	0.973	0.708	0.209	0.602	0.292
	LSTM	−0.161	2.594	0.364	0.839	0.571	0.445	0.391	0.429
	FS-ConvLSTM	0.153	1.753	0.778	1.153	0.722	0.197	0.608	0.278

As the rainfall magnitude becomes larger, LSTM and FS-ConvLSTM are closer to the measured value in total rainfall; *BIAS* keeps decreasing, with *RATIO* becoming closer to 1. Since there is a direct relationship between *RSME* and rainfall magnitude, the *RSME* of the fused data also keeps increasing. The *CC* of FS-ConvLSTM fluctuates around 0.75, with the highest value of 0.778 for torrential rainfall events, while the *CC* of LSTM fluctuates between 0.2 and 0.4, becoming larger as the rainfall magnitude increases. In terms of classification indicators, as the rainfall event changes from light to heavy, the *POD* and *TS* of the fused data show an overall increasing trend, and the *MAR* and *FAR* also show an overall decreasing trend, suggesting that the rainfall capture capability keeps improving, and the FS-ConvLSTM performs better than the LSTM.

In general, FS-ConvLSTM exhibits advantages over other methods in rainfall events of different magnitudes, and its accuracy improves with the increase in rainfall magnitude.

4.2.2. Uncertainty Analysis

The estimation uncertainty of 15 rain gauges in each event during the testing stage was calculated, and box-line plots were generated to depict the distribution, as illustrated in Figure 6.

Based on the evaluation results of the quantitative indicators, the *BIAS* and *RATIO* of most precipitation data have a narrow error distribution interval, with a value around 1. In the case of light rain, the deviation of GPM is relatively larger, and the uncertainty interval wider compared to other data. The accuracy of the ground-interpolated data is slightly higher than that of the fused data, but this advantage decreases with increasing rainfall magnitude. FS-ConvLSTM has a smaller interval width than LSTM, which denotes higher stability. Regarding *RSME*, GPM has the highest median error in different magnitudes of rainfall, while F-SVD has a lower median value than IDW but with a larger interval width. The same difference exists between FS-ConvLSTM and LSTM, with the gap increasing as the magnitude of rainfall becomes larger. In terms of *CC*, the median values of GPM and LSTM are lower, while the median value of FS-ConvLSTM is the highest. F-SVD has a higher median value than IDW.

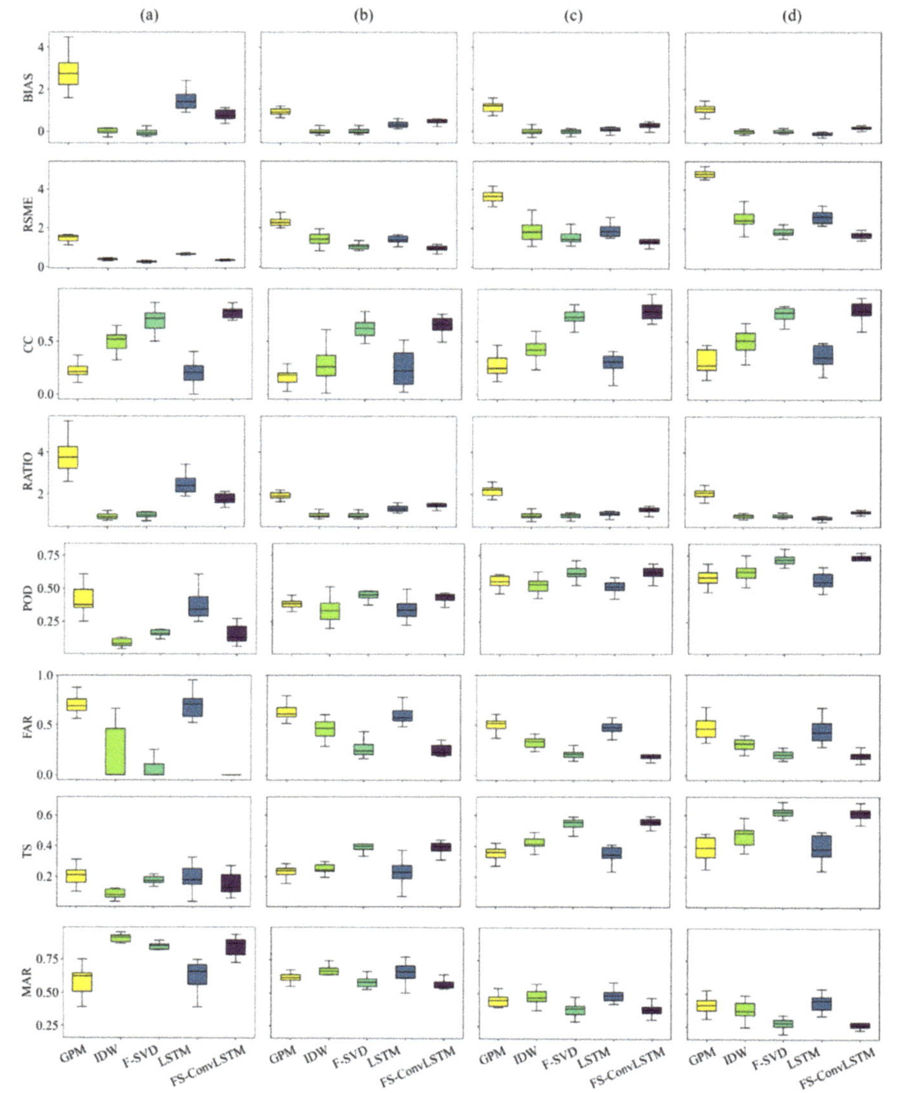

Figure 6. Boxplot of the evaluation indicators for different rainfall events. (**a**) Small rain; (**b**) Moderate rain; (**c**) Heavy rain; (**d**) Torrential rain.

Among the classification indicators, the values of *POD* and *TS* increase, and the values of *FAR* and *MAR* decrease as the rainfall magnitude increases. In light rainfall events, *POD*, *FAR*, and *MAR* exhibit large uncertainties. For ground-interpolated data, F-SVD has a better median and interval width of indicators than IDW. For the fused data, the median indicator of FS-ConvLSTM is generally better than LSTM, except in the light rainfall events where the median values of *POD*, *TS*, and *MAR* are worse for FS-ConvLSTM. As the rainfall magnitude changes from moderate to heavy rainfall, the difference between FS-ConvLSTM and LSTM in *POD* and *TS* increases while the difference in *FAR* decreases. FS-ConvLSTM also has a shorter interval width, indicating higher stability.

Based on the above analysis, FS-ConvLSTM shows better median accuracy than GPM and outperforms ground interpolation data and LSTM in most cases. The distribution width of FS-ConvLSTM is narrower than that of F-SVD and, in most cases, is narrower than that of the GPM. Overall, FS-ConvLSTM improves both the accuracy and the stability of the rainfall data.

4.3. Model Performance in Typical Rainfall Events

4.3.1. Comparison of Temporal Variation Processes

Four typical rainfall events were chosen during the testing stage, and the variation patterns of rainfall observed by rain gauges, GPM, spatiotemporal interpolation by F-SVD, and the spatiotemporal fused data by FS-ConvLSTM were compared. The results are displayed in Figure 7.

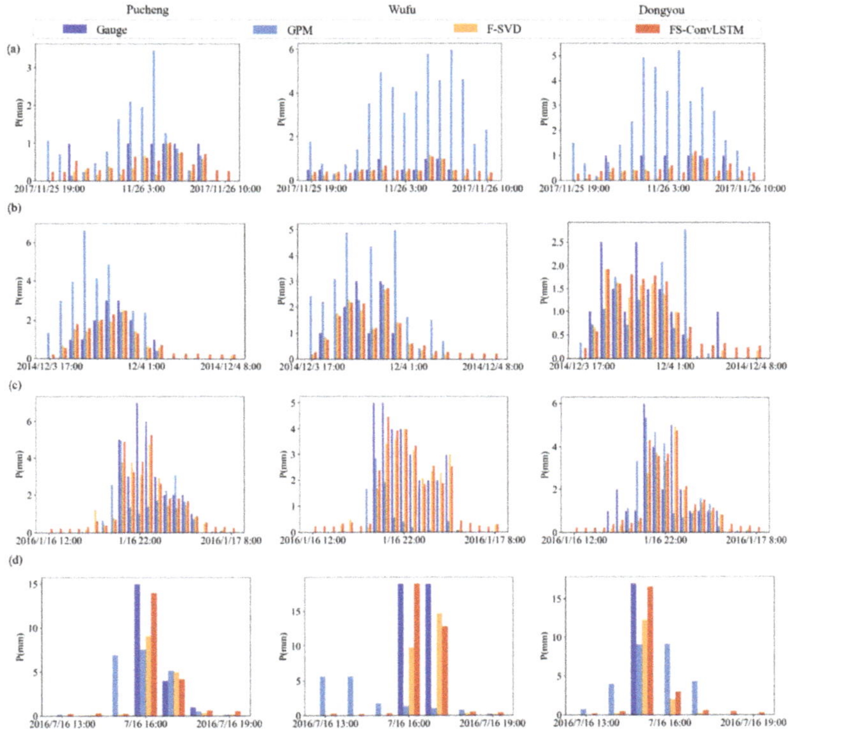

Figure 7. Comparison of the temporal variation process at three selected rain gauges. (**a**) Small rain; (**b**) Moderate rain; (**c**) Heavy rain; (**d**) Torrential rain.

From Figure 7a, it is evident that, for the light rainfall events, the rainfall intensities at all moments are below 1 mm/h. All types of rainfall data accurately capture the rainfall events, but there are differences in rainfall magnitudes. The GPM overestimates the rainfall significantly at all three stations, and the data of F-SVD and FS-ConvLSTM are closer to the actual measured values, with the fused data slightly higher than the ground observations. Notably, at Pucheng station, the value of FS-ConvLSTM is larger than that of F-SVD at the time of approaching the rain peak. Figure 7b shows that for the medium rainfall event, the rainfall intensity is below 3 mm/h at all moments. All three types of data reflect the rainfall event accurately, but GPM poorly describes the rain peak present time, and the rain peak at Pucheng station is earlier than the actual measurement, while the rain peak at Dongyou station is later than the actual measurement, showing considerably more uncertainty. At most moments, GPM is higher than the actual measurement, and the data of F-SVD and FS-ConvLSTM are slightly lower than the actual measurement. For the peak of rainfall, the values of FS-ConvLSTM are closer to the measured values compared to F-SVD. Figure 7c shows that, for heavy rainfall events, the maximum rainfall intensity is around 6 mm/h.

At Pucheng and Wufu stations, GPM appears to underestimate the rainfall, and the rainfall variation trend reflected at Pucheng station lags behind the measured data. Both the F-SVD and FS-ConvLSTM data accurately reflect the magnitudes and trends of the rainfall, but overall, FS-ConvLSTM performs better than F-SVD. Finally, Figure 7d illustrates that, for torrential rainfall events, the maximum rainfall intensity is above 15 mm/h. GPM underestimates the rainfall events, and the capture of rainfall trends at the Wufu station is also biased. There is also a slight underestimation in F-SVD for the rain peaks, and the rain peaks of FS-ConvLSTM are closer to the measured values than F-SVD.

After analyzing the four different magnitudes of rainfall events, it can be concluded that GPM is less reliable in describing the rainfall magnitude. The data tend to overestimate rainfall for light and moderate rainfall events while underestimating rainfall for heavy and torrential rain. On the other hand, F-SVD provides a more accurate description of rainfall magnitudes and rainfall trends, despite underestimating the rainfall peaks. FS-ConvLSTM outperforms F-SVD in accurately reflecting the rainfall events, particularly in rainfall peaks, which is more evident in heavy and torrential rainfall events.

4.3.2. Comparison of the Spatial Distribution

During the testing stage, two rainfall events of different magnitudes were selected, and their cumulative rainfall spatial distributions were plotted. The resulting figures (Figures 8 and 9) show the distribution of GPM, F-SVD, and FS-ConvLSTM from left to right.

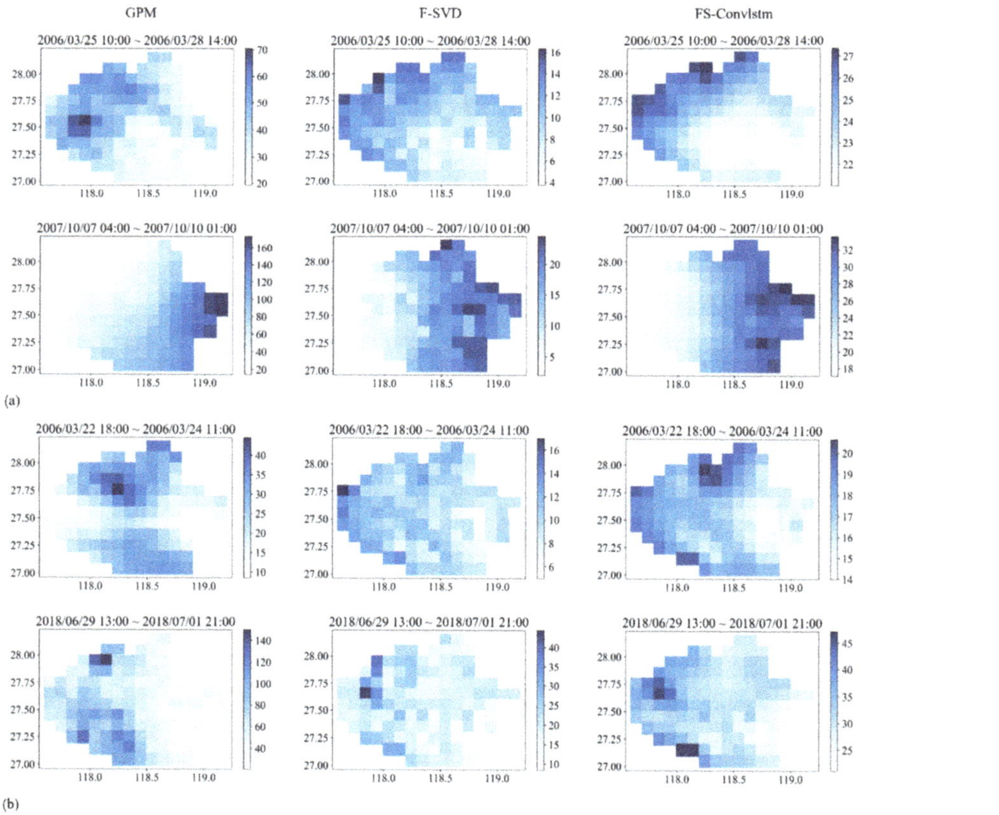

Figure 8. Spatial distribution of the accumulated precipitation for small and moderate rainfall events. (**a**) Small rain; (**b**) Moderate rain.

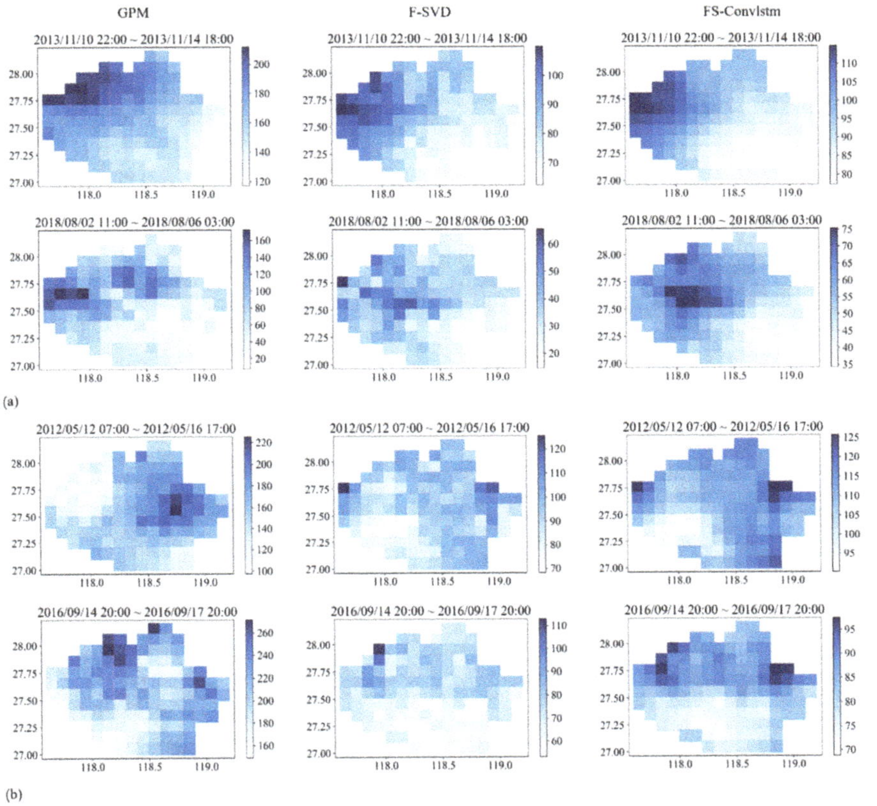

Figure 9. Spatial distribution of the accumulated precipitation for heavy and torrential rainfall events. (**a**) Heavy rain; (**b**) Torrential rain.

As seen in Figure 8a, the rainfall epoch for light rainfall events is around 3 days. The values of F-SVD and FS-ConvLSTM are closer, while the magnitude of the GPM data is significantly larger. For spatial extreme values of accumulated rainfall, the minimal values of FS-ConvLSTM and the minimal values of GPM are closer, and the maximal values of FS-ConvLSTM are slightly larger than those of F-SVD. GPM and FS-ConvLSTM are more continuous in spatial variation, while F-SVD has multiple isolated points in space that can produce abrupt changes in rainfall amounts. The rainfall centers of GPM and F-SVD are slightly different in space, and the rainfall centers of the two data and the rainfall distribution of the surrounding points are reflected in the fusion data. Similar results are found for the medium rainfall event, as seen in Figure 8b, where the duration of the medium rainfall event is around two days. The magnitude of GPM estimation is higher than the other two data, and the spatial extremes of F-SVD and FS-ConvLSTM are very close. The spatial distributions of the three data are relatively similar, but the values for F-SVD are more discrete in terms of spatial variation compared to the continuous spatial variation of GPM and FS-ConvLSTM. The fused data combine the distribution characteristics of GPM and F-SVD, with the spatial aggregation of rainfall of both reflected in the fusion.

For heavy rainfall events, as depicted in Figure 9a, the rainfall duration lasts for approximately four days. GPM continues to overestimate the rainfall magnitude, while the magnitudes of F-SVD and FS-ConvLSTM are closer and the spatial extremes are similar. The spatial distribution of rainfall reflected by GPM and F-SVD are similar, and the location of the rainfall center is also close, corresponding to a similar distribution in the fused data. FS-ConvLSTM shows better spatial continuity. As for the torrential rainfall events, as shown in Figure 9b, the rainfall duration is about 3–4 days. Regarding rainfall magnitude, GPM is relatively large compared to the other two types of data. In terms of spatial distribution F-SVD shows abrupt changes, and the rainfall at a few points is significantly higher than at other surrounding points, resulting in poor spatial continuity. Moreover, GPM and F-SVD have single and multiple rainfall centers, respectively, and the distribution of both rainfall centers is reflected in the fusion.

In general, GPM exhibits continuous spatial distribution but overestimates rainfall events. F-SVD shows abrupt spatial variations but is more accurate in its response to magnitude. FS-ConvLSTM combines the advantages of GPM in spatial distribution and the advantages of F-SVD in magnitude, which display continuous spatial distribution and closely approximate the actual rainfall values.

5. Discussion

The proposed FS-ConvLSTM framework effectively merges hourly precipitation data from rain gauge and GPM observations, resulting in improved accuracy and reduced uncertainty when estimating spatial precipitation. The framework's ability to capture the precipitation variation process and its superior performance compared to alternative models highlight its potential for practical application in precipitation estimation.

The comparison results presented in Table 3 and Figure 6 demonstrate that the proposed FS-ConvLSTM outperforms LSTM in terms of accuracy and stability. While traditional LSTM models only consider time-series dependencies, ConvLSTM combines the strengths of LSTM and convolutional neural networks (CNN), allowing it to effectively capture spatiotemporal dependencies and model both temporal and spatial dimensions [27]. Additionally, traditional LSTM primarily relies on matrix multiplication and element-wise operations [28], which limits its capacity to model data nonlinearly. In contrast, by utilizing multiple filters and the local connectivity of the convolution kernel, ConvLSTM can learn richer feature representations and extract more complex spatiotemporal patterns in the data. These characteristics contribute to the improved accuracy and stability of FS-ConvLSTM observed in the comparison results.

Comparing the fusion results of FS-ConvLSTM for different magnitudes of rainfall data (Table 4, Figures 6 and 7), it can be observed that the fusion results for heavy rainfall are better. Heavy rains exhibit stronger spatial and temporal correlation and spatial expansion, whereas small rains tend to be more localized and discrete. The convolution operation in ConvLSTM utilizes shared weights across different spatial locations [29], enabling it to effectively capture the spatial characteristics of heavy rainfall. Furthermore, heavy rainfall events have longer durations, and past rainfall conditions can influence future rainfall. The gating mechanism and memory units in ConvLSTM aid the model in retaining and updating key spatiotemporal information [30], facilitating the capture of long-term dependencies associated with heavy rainfall. These capabilities of ConvLSTM contribute to its superior performance in understanding and fusing heavy rainfall data.

From the spatial distribution of the rainfall data before and after fusion presented in Figures 8 and 9, it is evident that the fusion process retains the distribution features of both data sources while achieving higher accuracy. The introduction of multiple input channels in the ConvLSTM model enables the simultaneous processing of inputs from multiple data sources [31]. This allows the model to leverage the strengths of each data source, capture their spatial features, and retain the expression of these features in the fusion results. Moreover, the ConvLSTM model exhibits a large capacity and nonlinear modeling capability, enabling it to effectively handle the heterogeneity and nonlinear correlation between different data sources during the fusion process [32]. The model learns adaptively and adjusts the weights based on the importance and contribution of each data source. This ability enables the model to identify and reduce the impact of data with large deviations on the accuracy of fusion results, resulting in more accurate outcomes.

While the proposed framework successfully merges precipitation data from satellites and rain gauges, it has certain limitations. The ConvLSTM method was chosen for this study due to its ability to capture spatial and temporal dependencies in rainfall data, but alternative deep learning models such as Transformer-based models may also have advantages in processing spatiotemporal structured data. Exploring and comparing the performance of different architectures for precipitation data fusion would be a valuable direction for future research. Additionally, it is important to note that the framework proposed in this paper was tested in a single study region with a dense network of rain gauges in southeast China, so expanding the study area and incorporating data from additional stations would provide a more comprehensive understanding of the framework's performance across a larger spatial extent.

6. Conclusions

This study proposes an integrated framework (FS-ConvLSTM) to spatiotemporally merge precipitation data from rain gauge observations and GPM (IMERG V06). The proposed framework integrates F-SVD in the recommender system to improve the accuracy of spatiotemporal interpolation based on rain gauge observations, and ConvLSTM merges precipitation data from satellites and rain gauge interpolations by exploiting the spatiotemporal correlation pattern between them. The FS-ConvLSTM framework was applied to estimate the hourly spatial precipitation in the Jianxi Basin of China from 2006 to 2018. The findings are summarized as follows:

(1) The proposed FS-ConvLSTM framework outperforms the comparative models (IDW, F-SVD, and LSTM) in terms of precipitation variation process description and rainfall capture capability, reducing the *RSME* and *FAR* of the original GPM data by 63.6% and 63.1%, respectively, and increasing the *CC* and *POD* by 165% and 22.9%, respectively.
(2) The merged data not only improve the accuracy of the precipitation but also reduce the uncertainty in precipitation estimation. As the intensity of precipitation increases, the precipitation capture ability substantially improves, and the estimation more closely matches the measured data in terms of total rainfall.
(3) Due to the powerful feature extraction capability of ConvLSTM, the merged precipitation data combines the advantages of GPM and ground interpolation data with the continuous spatial distribution data and values close to the actual one, and the spatial aggregation of both data is reflected in the fusion.

The two-step merging framework proposed in this study demonstrates satisfactory performance in merging hourly spatial precipitation data in the Jianxi basin of China by exploring the spatiotemporal dependence between rain gauge and GPM. Different types of precipitation data, such as Tropical Rainfall Measuring Mission (TRMM), the Precipitation Estimation from Remotely Sensed Information using Artificial Neural Networks (PERSIANN), and the Climate Hazards Group InfraRed Precipitation with Station (CHIRPS), possess their unique characteristics and advantages. In this study, we only considered merging data from rain gauges and GPM. To maximize the benefits of each satellite product's data and enhance the accuracy of spatial precipitation estimations, future work should consider incorporating more multi-source precipitation observation data.

Author Contributions: Conceptualization, S.S. and H.C.; methodology, S.S., H.C. and K.L.; software, S.S. and K.L.; validation, S.S., N.Z. and B.T.; formal analysis, S.S.; investigation, N.Z. and B.T.; resources, S.S. and K.L.; data curation, N.Z., B.T. and K.L.; writing—original draft preparation, S.S.; writing—review and editing, H.C. and C.-Y.X.; visualization, S.S.; supervision, H.C. and C.-Y.X.; project administration, K.L.; funding acquisition, H.C. All authors have read and agreed to the published version of the manuscript.

Funding: This research was funded by the National Key Research and Development Program (2022YFC3002701) and Water Science and Technology Project in Fujian Province, China.

Data Availability Statement: The rain gauge observation data used in this study are confidential and the GPM (IMERG V06) Final Run data were downloaded from NASA (https://earthdata.nasa.gov/; accessed on 15 October 2022).

Conflicts of Interest: The authors declare no conflict of interest.

References

1. Hinge, G.; Hamouda, M.A.; Long, D.; Mohamed, M.M. Hydrologic utility of satellite precipitation products in flood prediction: A meta-data analysis and lessons learnt. *J. Hydrol.* **2022**, *612*, 128103. [CrossRef]
2. Estébanez-Camarena, M.; Taormina, R.; van de Giesen, N.; ten Veldhuis, M.-C. The Potential of Deep Learning for Satellite Rainfall Detection over Data-Scarce Regions, the West African Savanna. *Remote Sens.* **2023**, *15*, 1922. [CrossRef]
3. Song, J.-H.; Her, Y.; Kang, M.-S. Estimating Reservoir Inflow and Outflow From Water Level Observations Using Expert Knowledge: Dealing With an Ill-Posed Water Balance Equation in Reservoir Management. *Water Resour. Res.* **2022**, *58*, e2020WR028183. [CrossRef]
4. Talchabhadel, R.; Shah, S.; Aryal, B. Evaluation of the Spatiotemporal Distribution of Precipitation Using 28 Precipitation Indices and 4 IMERG Datasets over Nepal. *Remote Sens.* **2022**, *14*, 5954. [CrossRef]
5. Ramanathan, A.; Versini, P.A.; Schertzer, D.; Perrin, R.; Sindt, L.; Tchiguirinskaia, I. Stochastic simulation of reference rainfall scenarios for hydrological applications using a universal multi-fractal approach. *Hydrol. Earth Syst. Sci.* **2022**, *26*, 6477–6491. [CrossRef]
6. Gofa, F.; Flocas, H.; Louka, P.; Samos, I. A Coherent Approach to Evaluating Precipitation Forecasts over Complex Terrain. *Atmosphere* **2022**, *13*, 1164. [CrossRef]
7. Gyasi-Agyei, Y. A framework for comparing two rainfields based on spatial structure: A case of radar against selected satellite precipitation products over southeast Queensland, Australia. *J. Hydrol.* **2022**, *613*, 128356. [CrossRef]
8. Sreeparvathy, V.; Srinivas, V.V. A Bayesian Fuzzy Clustering Approach for Design of Precipitation Gauge Network Using Merged Remote Sensing and Ground-Based Precipitation Products. *Water Resour. Res.* **2022**, *58*, e2021WR030612. [CrossRef]
9. Noor, R.; Arshad, A.; Shafeeque, M.; Liu, J.; Baig, A.; Ali, S.; Maqsood, A.; Pham, Q.B.; Dilawar, A.; Khan, S.N.; et al. Combining APHRODITE Rain Gauges-Based Precipitation with Downscaled-TRMM Data to Translate High-Resolution Precipitation Estimates in the Indus Basin. *Remote Sens.* **2023**, *15*, 318. [CrossRef]
10. Iqbal, Z.; Shahid, S.; Ahmed, K.; Wang, X.; Ismail, T.; Gabriel, H.F. Bias correction method of high-resolution satellite-based precipitation product for Peninsular Malaysia. *Theor. Appl. Clim.* **2022**, *148*, 1429–1446. [CrossRef]
11. Varouchakis, E.A.; Kamińska-Chuchmała, A.; Kowalik, G.; Spanoudaki, K.; Graña, M. Combining Geostatistics and Remote Sensing Data to Improve Spatiotemporal Analysis of Precipitation. *Sensors* **2021**, *21*, 3132. [CrossRef]
12. Papacharalampous, G.; Tyralis, H.; Doulamis, A.; Doulamis, N. Comparison of Tree-Based Ensemble Algorithms for Merging Satellite and Earth-Observed Precipitation Data at the Daily Time Scale. *Hydrology* **2023**, *10*, 50. [CrossRef]
13. Chen, S.; Li, Q.; Zhong, W.; Wang, R.; Chen, D.; Pan, S. Improved Monitoring and Assessment of Meteorological Drought Based on Multi-Source Fused Precipitation Data. *Int. J. Environ. Res. Public Health* **2022**, *19*, 1542. [CrossRef]
14. Yumnam, K.; Kumar Guntu, R.; Rathinasamy, M.; Agarwal, A. Quantile-based Bayesian Model Averaging approach towards merging of precipitation products. *J. Hydrol.* **2022**, *604*, 127206. [CrossRef]
15. Shao, Y.; Fu, A.; Zhao, J.; Xu, J.; Wu, J. Improving quantitative precipitation estimates by radar-rain gauge merging and an integration algorithm in the Yishu River catchment, China. *Theor. Appl. Clim.* **2021**, *144*, 611–623. [CrossRef]
16. Pan, Y.; Yuan, Q.; Ma, J.; Wang, L. Improved Daily Spatial Precipitation Estimation by Merging Multi-Source Precipitation Data Based on the Geographically Weighted Regression Method: A Case Study of Taihu Lake Basin, China. *Int. J. Environ. Res. Public Health* **2022**, *19*, 13866. [CrossRef]
17. Duan, Z.; Ren, Y.; Liu, X.; Lei, H.; Hua, X.; Shu, X.; Zhou, L. A comprehensive comparison of data fusion approaches to multi-source precipitation observations: A case study in Sichuan province, China. *Environ. Monit. Assess.* **2022**, *194*, 422. [CrossRef]
18. Lei, H.; Zhao, H.; Ao, T. A two-step merging strategy for incorporating multi-source precipitation products and gauge observations using machine learning classification and regression over China. *Hydrol. Earth Syst. Sci.* **2022**, *26*, 2969–2995. [CrossRef]
19. Zhang, J.; Xu, J.; Dai, X.; Ruan, H.; Liu, X.; Jing, W. Multi-Source Precipitation Data Merging for Heavy Rainfall Events Based on Cokriging and Machine Learning Methods. *Remote Sens.* **2022**, *14*, 1750. [CrossRef]
20. Zhang, Z.; Wang, D.; Qiu, J.; Zhu, J.; Wang, T. Machine Learning Approaches for Improving Near-Real-Time IMERG Rainfall Estimates by Integrating Cloud Properties from NOAA CDR PATMOS-x. *J. Hydrometeorol.* **2021**, *22*, 2767–2781. [CrossRef]
21. Shen, J.; Liu, P.; Xia, J.; Zhao, Y.; Dong, Y. Merging Multisatellite and Gauge Precipitation Based on Geographically Weighted Regression and Long Short-Term Memory Network. *Remote Sens.* **2022**, *14*, 3939. [CrossRef]
22. Wu, H.; Yang, Q.; Liu, J.; Wang, G. A spatiotemporal deep fusion model for merging satellite and gauge precipitation in China. *J. Hydrol.* **2020**, *584*, 124664. [CrossRef]
23. Shi, X.; Chen, Z.; Wang, H.; Yeung, D.-Y.; Wong, W.-K.; Woo, W.-C. Convolutional LSTM network: A machine learning approach for precipitation nowcasting. *Adv. Neural Inf. Process. Syst.* **2015**, *28*, 802–810.
24. Chen, H.; Sheng, S.; Xu, C.-Y.; Li, Z.; Zhang, W.; Wang, S.; Guo, S. A spatiotemporal estimation method for hourly rainfall based on F-SVD in the recommender system. *Environ. Modell. Softw.* **2021**, *144*, 105148. [CrossRef]
25. Durrani, A.u.R.; Minallah, N.; Aziz, N.; Frnda, J.; Khan, W.; Nedoma, J. Effect of hyper-parameters on the performance of ConvLSTM based deep neural network in crop classification. *PLoS ONE* **2023**, *18*, e0275653. [CrossRef]

6. Hu, W.S.; Li, H.C.; Pan, L.; Li, W.; Tao, R.; Du, Q. Spatial–Spectral Feature Extraction via Deep ConvLSTM Neural Networks for Hyperspectral Image Classification. *IEEE Trans. Geosci. Remote Sens.* **2020**, *58*, 4237–4250. [CrossRef]
7. Dizaji, M.S.; Mao, Z.; Haile, M. A hybrid-attention-ConvLSTM-based deep learning architecture to extract modal frequencies from limited data using transfer learning. *Mech. Syst. Signal Process.* **2023**, *187*, 109949. [CrossRef]
8. Zhang, W.; Ge, F.; Cui, C.; Yang, Y.; Zhou, F.; Wu, N. Design and Implementation of LSTM Accelerator Based on FPGA. In Proceedings of the 2020 IEEE 20th International Conference on Communication Technology (ICCT), Nanning, China, 28–31 October 2020; pp. 1675–1679.
9. De Medrano, R.; Aznarte, J.L. On the inclusion of spatial information for spatio-temporal neural networks. *Neural Comput. Appl.* **2021**, *33*, 14723–14740. [CrossRef]
0. Li, Y.; Chai, S.; Wang, G.; Zhang, X.; Qiu, J. Quantifying the Uncertainty in Long-Term Traffic Prediction Based on PI-ConvLSTM Network. *IEEE Trans. Intell. Transp. Syst.* **2022**, *23*, 20429–20441. [CrossRef]
1. Eide, S.S.; Riegler, M.A.; Hammer, H.L.; Bremnes, J.B. Deep Tower Networks for Efficient Temperature Forecasting from Multiple Data Sources. *Sensors* **2022**, *22*, 2802. [CrossRef]
2. Zhang, X.; Zhou, Y.n.; Luo, J. Deep learning for processing and analysis of remote sensing big data: A technical review. *Big Earth Data* **2022**, *6*, 527–560. [CrossRef]

remote sensing

MDPI

Article

Estimation of the Two-Dimensional Direction of Arrival for Low-Elevation and Non-Low-Elevation Targets Based on Dilated Convolutional Networks

Guoping Hu [1], Fangzheng Zhao [1] and Bingqi Liu [2,*]

1 Air and Missile Defense College, Air Force Engineering University, Xi'an 710043, China
2 Graduate School, Air Force Engineering University, Xi'an 710043, China
* Correspondence: bingqi0828_liu@163.com

Abstract: This paper addresses the problem of the two-dimensional direction-of-arrival (2D DOA) estimation of low-elevation or non-low-elevation targets using L-shaped uniform and sparse arrays by analyzing the signal models' features and their mapping to 2D DOA. This paper proposes a 2D DOA estimation algorithm based on the dilated convolutional network model, which consists of two components: a dilated convolutional autoencoder and a dilated convolutional neural network. If there are targets at low elevation, the dilated convolutional autoencoder suppresses the multipath signal and outputs a new signal covariance matrix as the input of the dilated convolutional neural network to directly perform 2D DOA estimation in the absence of a low-elevation target. The algorithm employs 3D convolution to fully retain and extract features. The simulation experiments and the analysis of their results revealed that for both L-shaped uniform and L-shaped sparse arrays, the dilated convolutional autoencoder could effectively suppress the multipath signals without affecting the direct wave and non-low-elevation targets, whereas the dilated convolutional neural network could effectively achieve 2D DOA estimation with a matching rate and an effective ratio of pitch and azimuth angles close to 100% without the need for additional parameter matching. Under the condition of a low signal-to-noise ratio, the estimation accuracy of the proposed algorithm was significantly higher than that of the traditional DOA estimation.

Keywords: 2D DOA estimation; low-elevation-angle targets; L-shaped uniform array; L-shaped sparse array; dilated convolutional autoencoder; dilated convolutional neural network; 3D convolution

Citation: Hu, G.; Zhao, F.; Liu, B. Estimation of the Two-Dimensional Direction of Arrival for Low-Elevation and Non-Low-Elevation Targets Based on Dilated Convolutional Networks. *Remote Sens.* **2023**, *15*, 3117. https://doi.org/10.3390/rs15123117

Academic Editor: Gwanggil Jeon

Received: 6 March 2023
Revised: 15 May 2023
Accepted: 22 May 2023
Published: 14 June 2023

1. Introduction

Array signal processing, which has a wide range of applications in communications, remote sensing, detection, and radar, involves the use of sensor arrays to achieve signal parameter estimation, signal enhancement [1], etc. Accordingly, direction-of-arrival (DOA) estimation is an important branch of research. This involves estimating the direction of arrival of one or more signals in a region of space using theoretical or technical methods. One-dimensional (1D) DOA estimation is the estimation of the elevation angle of targets. Two-dimensional (2D) DOA estimation, as an extension of 1D DOA estimation, enables the estimation of both the elevation and the azimuth angles [2]. Two-dimensional DOA is of greater importance for spatial localization and is, therefore, one of the main focuses of current research in the field. Two-dimensional DOA estimation requires arrays to be arranged in a 2D plane, generally using L-shaped arrays, surface arrays, parallel arrays, or vector sensors [3,4]. Most 2D DOA estimation algorithms extend the 1D DOA estimation algorithm to a 2D spatial spectrum, such as the 2D multiple-signal classification (MUSIC) [5] algorithm and 2D estimating signal parameter via rotational invariance techniques (ESPRIT) algorithm. The former can produce asymptotic unbiased estimates with high estimation accuracy without the need for parameter matching, but this algorithm requires a 2D spatial spectrum search and has a high computational demand. On the other hand, the latter does not require a spatial spectral search, and the elevation and azimuth angles can also be automatically matched, but the estimation accuracy of this method is low, especially when the signal-to-noise ratio (SNR) is low. Yin et al. proposed a DOA direction matrix method [6], where the elevation and azimuth angles could be directly obtained via eigendecomposition of the DOA direction matrix, with automatic parameter matching; however, this method is only applicable to specific arrays such as parallel linear arrays. To improve the estimation accuracy and spatial freedom, sparse arrays are often used in practice instead of uniform arrays [7]. In a study on the 2D DOA estimation of sparse arrays, Liu et al. proposed a 2D DOA estimation method based on singular value decomposition [8], taking advantage of the structural characteristics of T-shaped arrays and co-prime array arrays to obtain three signal subspaces without using virtual elements before using the signal subspaces to perform 2D DOA estimation. Wang et al. designed a generalized coprime parallel linear array instead of the traditional parallel uniform linear array, then improved the differential virtual array to obtain greater degrees of freedom, and finally simplified the 2D search to two 1D searches to reduce the number of operations [9]. However, the algorithm led to an increase in the influence of the mutual coupling between array elements, and the compression factor needed to be artificially chosen, restricting the performance of the algorithm. In addition, when the elevation angle of the target incident array is low, multipath effects can occur, which can result in the received signal including reflected waves that are coherent with the direct wave, thereby complicating 2D DOA estimation. For the 2D DOA estimation problem of low-elevation-angle targets, Ma et al. proposed a 2D DOA estimation algorithm based on the alternating direction method of multipliers [10], which transforms the 2D DOA estimation into two 1D DOA estimation problems and avoids the problem of the high computational demand caused by 2D joint estimation; however, the algorithm could only solve the 2D DOA estimation of a single target. Su et al. and Park et al. proposed 2D DOA estimation algorithms for coherent signals based on sparse reconstruction [11,12], which could be used for the decoherence of low-elevation targets; however, the algorithms had a complex arithmetic process. Liang et al. proposed a 2D DOA estimation algorithm for coherent sources based on Toeplitz matrix reconstruction [13], which could estimate the elevation and azimuth angles without loss of array aperture through a 1D search only; however, the algorithm was only applicable to uniform arrays. Molaei et al. proposed a k-medoids clustering signal separation method that could realize the 2D DOA estimation of multipath signals and effectively separate coherent and noncoherent signals [14]; however, the method was only applicable to rectangular arrays.

Usually, physical model algorithms suffer from limited applicability and complex computational processes, whereas data-driven deep-learning-based algorithms have greater applicability. Compared with traditional signal processing algorithms, deep-learning-like algorithms convert the DOA estimation problem into a high-dimensional nonlinear mapping relationship, i.e., realizing mapping between the covariance matrix of the received signal or other variables and the DOA, which provides a new way of thinking for the study of 2D DOA estimation methods. Marija et al. implemented the fast estimation of spatial single-target 2D DOA using a multilayer perceptron [15]; however, the artificial neural network (ANN) model needed to expand the signal covariance matrix into 1D data as input, thereby losing the spatial characteristics of the covariance matrix. Zhu proposed a 2D DOA estimation algorithm based on deep ensemble learning [16], using multiple convolutional neural networks to output the elevation and azimuth angles. This approach was not limited by the deployment method; however, there was a matching problem of elevation and azimuth angles.

To address the practical problems of the above algorithms, this paper proposes a 2D DOA estimation model based on the combination of a dilated convolutional autoencoder and a dilated convolutional neural network, whereby the former solves the coherence problem of direct and reflected waves by suppressing the multipath signal, i.e., filtering out the reflected wave components of the signal covariance matrix, while the latter is used to implement 2D DOA estimation. Both the dilated convolutional autoencoder and the dilated convolutional neural network are convolved in three dimensions to fully extract the spatial features of the data; accordingly, the model is able to achieve the 2D DOA estimation of non-low-elevation targets and hybrid targets in L-shaped uniform and L-shaped sparse arrays without the need for parameter matching.

2. Signal Model

2.1. L-Shaped Array Signal Model

When the array arrangement is in one dimension, only 1D DOA estimation can be realized. If 2D DOA estimation is required for the source, i.e., elevation and azimuth, the array arrangement needs to be at least 2D. In this study, an L-shaped array was designed to perform 2D DOA estimation. When the L-shaped array consists of two mutually perpendicular uniform line arrays, its arrangement is as shown in Figure 1.

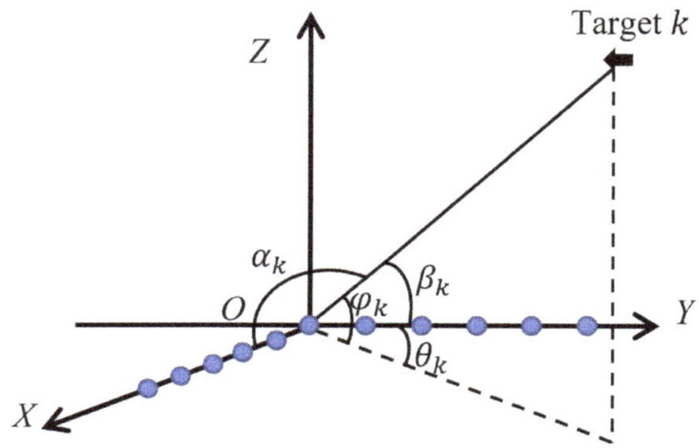

Figure 1. L-shaped array arrangement.

The array elements are uniformly arranged along the X and Y axes, with an array element spacing of d and less than half a wavelength λ; and the number of X- and Y-axis array elements are M and N, respectively, with overlapping array elements at the origin of the coordinate axis, so the total number of array elements is $(M+N-1)$. Consider $K(K<M+N-1)$ far-field uncorrelated narrowband signals incident to the L-shaped array in the directions (α_1,β_1), (α_2,β_2) ... (α_K,β_K) $(k=1,2,\ldots,K)$, where α_k and β_k are the angles of the target to the X and Y axis, respectively, also known as the spatial phase factor; φ_k and θ_k denote the elevation and azimuth angles of the target, respectively; and the correspondence between α_k, β_k and φ_k, θ_k is shown below:

$$cos\alpha_k = sin\varphi_k cos\theta_k, \tag{1}$$

$$cos\beta_k = sin\varphi_k sin\theta_k, \tag{2}$$

The received signals for a uniform line array along the X- and Y-axis directions are

$$\boldsymbol{x}_1(t) = \boldsymbol{A}_1\boldsymbol{s}(t) + \boldsymbol{n}_1(t), \tag{3}$$

$$\boldsymbol{x}_2(t) = \boldsymbol{A}_2\boldsymbol{s}(t) + \boldsymbol{n}_2(t), \tag{4}$$

where $\boldsymbol{s}(t) = [s_1(t), s_2(t), \ldots, s_K(t)]^T$ denotes the signal vector; $\boldsymbol{n}_1(t)$ and $\boldsymbol{n}_2(t)$ denote gaussian white noise with noise power σ^2 and uncorrelated with the signal, respectively;

and A_1 and A_2 are the direction vectors of uniform line arrays in the X- and Y-axis directions, respectively.

$$A_1 = [a(\alpha_1), a(\alpha_2), \ldots, a(\alpha_K)], \tag{5}$$

$$A_2 = [a(\beta_1), a(\beta_2), \ldots, a(\beta_K)], \tag{6}$$

$$X : a(\alpha_k) = \left[1, e^{-\frac{j2\pi d cos\alpha_k}{\lambda}}, \ldots, e^{-\frac{j2\pi(M-1)d cos\alpha_k}{\lambda}}\right]^T, \tag{7}$$

$$Y : a(\beta_k) = \left[e^{-\frac{j2\pi d cos\beta_k}{\lambda}}, e^{-\frac{j2\pi 2d cos\beta_k}{\lambda}} \ldots, e^{-\frac{j2\pi(N-1)d cos\beta_k}{\lambda}}\right]^T, \tag{8}$$

Combining Equations (3) and (4) yields

$$x(t) = \left[x_1{}^H(t) x_2{}^H(t)\right]^T = B(\varphi, \theta)s(t) + n(t), \tag{9}$$

where $B(\varphi, \theta) = \left[A_1{}^H, A_2{}^H\right]^T$ and $n(t) = \left[n_1{}^H(t), n_2{}^H(t)\right]^T$, calculate the received signal covariance matrix according to Equation (9), i.e.,

$$R_x = E\left[x(t)x^H(t)\right] = B(\varphi, \theta)R_s B^H(\varphi, \theta) + \sigma^2 I_{M+N-1}, \tag{10}$$

where $R_s = E\left[s(t)s^H(t)\right]$ denotes the incident signal covariance matrix, and I_{M+N-1} denotes the unit matrix of dimension $M + N - 1$. The eigendecomposition of the received signal covariance matrix R_x can be divided into a signal subspace and a noise subspace,

$$R_x = U\Sigma U^H = U_s \Sigma_s U_s^H + U_n \Sigma_n U_n^H, \tag{11}$$

where Σ denotes the diagonal matrix constructed from all the eigenvalues obtained from the eigen decomposition; U denotes the eigenvector matrix; Σ_s denotes the diagonal matrix constructed from the K largest eigenvalues in Σ equal to the number of signals; U_s denotes the eigenvector corresponding to the K largest eigenvalues, considered as the signal subspace; Σ_n denotes the diagonal matrix constructed from the remaining $(M + N - 1 - K)$ eigenvalues; U_n the eigenvectors corresponding to the remaining eigenvalues, which are regarded as the noise subspace. According to the theory of the MUSIC algorithm, the signal subspace and the noise subspace have orthogonal properties, and U_n is orthogonal to $b(\varphi, \theta)$ column vector in $B(\varphi, \theta)$, and the spatial spectrum $P(\varphi, \theta)$ is calculated according to the 2D MUSIC algorithm, as follows

$$b(\varphi, \theta) = \left[a^H(\alpha), a^H(\beta)\right]^T, \tag{12}$$

$$P(\varphi, \theta) == \frac{1}{b^H(\varphi, \theta)U_n U_n^H b(\varphi, \theta)}, \tag{13}$$

The elevation and azimuth angles (φ, θ) can be obtained by searching for the peak points of the 2D spatial spectrum within the target airspace. Taking an L-shaped uniform array with the number of elements in the X and Y axis being 8 and 9, respectively, as an example, when three targets with elevation and azimuth angles of $(10°, 30°)$, $(20°, 10°)$, and $(40°, 20°)$ are incident on the array with SNR = 10 dB and snapshots = 100, the spatial spectrum and its top view were obtained after a 2D search, as shown in Figure 2.

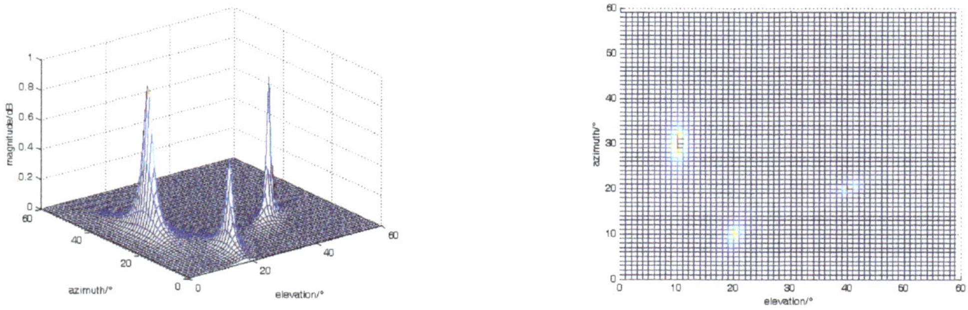

(a) (b)

Figure 2. Spatial spectrum and top view of L-shaped uniform array. (**a**) Spatial spectrum of L-shaped uniform array. (**b**) Top view of L-shaped uniform array.

Figure 2 shows that the 2D MUSIC algorithm can effectively estimate elevation and azimuth with a high degree of accuracy. In practice, sparse arrays are often used instead of uniform line arrays to reduce the effect of the mutual coupling between array elements on the accuracy of DOA estimation and to increase the number of measurable sources [17]. Although the arrangement of sparse arrays can largely reduce the actual number of array elements, they often produce ambiguous angles, i.e., spurious spectral peaks, which interfere with the judgement. Taking an L-shaped uniform sparse array with an array element spacing of 2λ as an example, the spatial spectrum and its top view under the same conditions as above are shown in Figure 3.

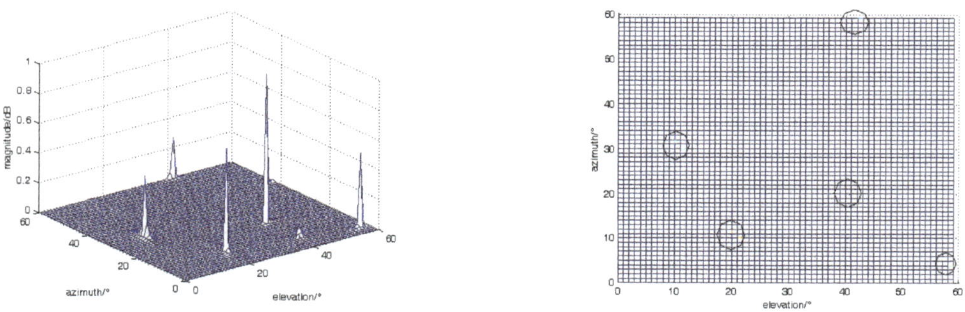

(a) (b)

Figure 3. Spatial spectrum and top view of L-shaped uniform sparse array. (**a**) Spatial spectrum of L-shaped uniform sparse array. (**b**) Top view of L-shaped uniform sparse array.

Figure 3a shows that the spatial spectrum contains five distinct spectral peaks, the corresponding coordinates of which coincide with the center of the circle in Figure 3b, which are sharper than the spectral peaks in Figure 2. However, there are two blurred angles in it. To address the problem of the blurring generated by sparse arrays, the use of coprime arrays can avoid the generation of blurred angles, so coprime arrays are widely used in practice. The array element arrangements of the X and Y axes are changed to the mutual prime number (4, 5) and (3, 7), respectively; and the array element arrangement of the X and Y axis are

$$X : (0, 4, 5, 8, 10, 12, 15, 16)\lambda/2, \tag{14}$$

$$Y : (0, 3, 6, 7, 9, 12, 14, 15, 18)\lambda/2, \tag{15}$$

The number of arrays on the X and Y axis is 8 and 9, respectively; and, under the same conditions as above, the spatial spectrum and its top view were calculated as shown in Figure 4.

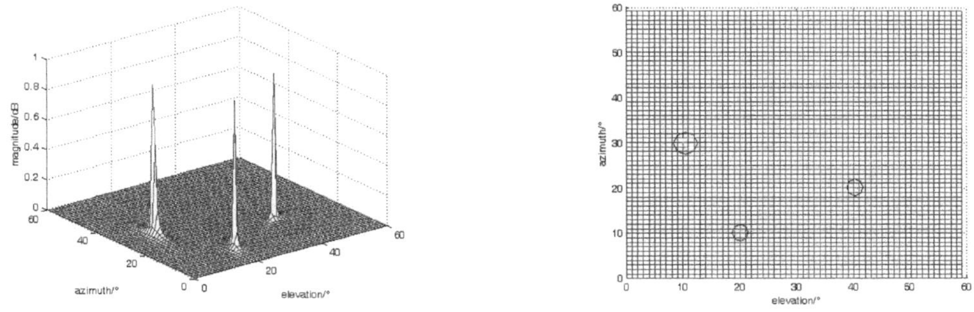

(a) (b)

Figure 4. Spatial spectrum and top view of L-shaped coprime array. (**a**) Spatial spectrum of L-shaped coprime array. (**b**) Top view of L-shaped coprime array.

As can be seen in Figure 4, the spatial spectrum contains three spectral peaks, which correspond to the center of the circle in the top view. The sharpness of the spectral peaks is similar to that in Figure 3 and better than that in Figure 2, but there is no blurring of the angles, and the resulting elevation and azimuth angles of the target are both highly accurate. When replacing only the uniform line array in the X or Y axis with a coprime array, but not both, the spatial spectrum is obtained as follows.

The blurred spectral peaks are also avoided when the array with only one axis is replaced with a coprime array, as shown in Figure 5, which is slightly less sharp compared with those in Figures 3a and 4a. A comparison of Figure 5a,b shows that the spectral peaks are narrower in elevation when the X axis is a coprime array and narrower in azimuth when the Y axis is a coprime array but still better than that in Figure 2a, overall.

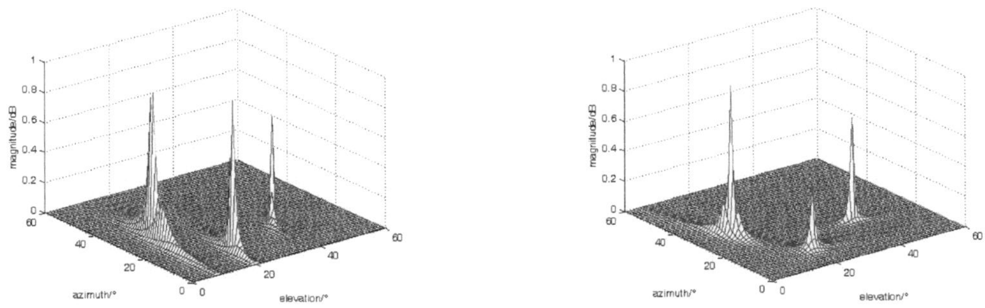

(a) (b)

Figure 5. Spatial spectrum when the X or Y axis is a coprime array. (**a**) Coprime array in X axis. (**b**) Coprime array in Y axis.

2.2. Low-Elevation-Target Signal Model

The multipath effect occurs when the elevation angle of the incident to the array is low, producing a reflected wave that is coherent with the direct wave [18] in its elevation angle dimension. The multipath effect is shown in Figure 6.

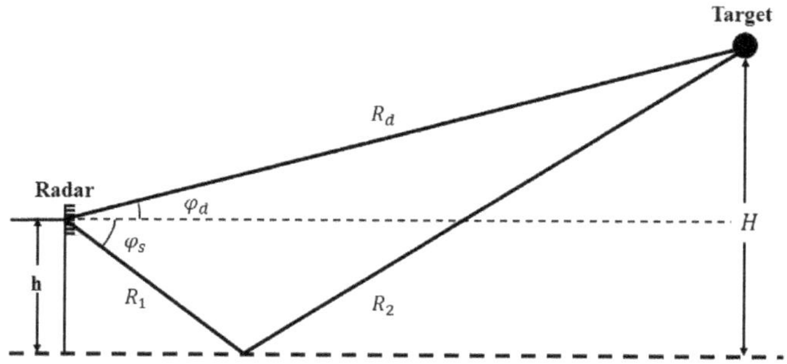

Figure 6. Schematic representation of the multipath effect in the elevation angle dimension.

To simplify the model, only multipath effects at the receiver side are considered. The target direct wave signal is incident on the array at an elevation angle φ_d, reflected by a smooth surface, and the reflected wave signal is incident on the array at an angle $\varphi_s(\varphi_s < 0)$. Let the height of the target be H and the height of the center of the array be h. The difference in the wave range between the direct and reflected waves [19] is approximated as

$$\Delta R = R_d - R_s = R_d - (R_1 + R_2) \approx 2hsin\varphi_d, \tag{16}$$

According to the position relationship in Figure 6, the relationship between the direct and reflected angles satisfies Equation (17),

$$\varphi_s = \arctan(\frac{H+h}{H-h}tan\varphi_d), \tag{17}$$

When there are K $(K < M + N - 1)$ incoherent sources in a space that includes $Q(Q \leq K)$ low-elevation targets, the number of signals received by the array is $(K + Q)$, which includes Q direct wave signals from low-elevation targets, Q reflected wave signals from low-elevation targets, and $(K - Q)$ non-low-elevation signals. The direction vectors $\boldsymbol{a}(\alpha_k)$ and $\boldsymbol{a}(\beta_k)$ for non-low-elevation targets have the same Equation (5) to (7), while the direction vectors $\boldsymbol{a}(\alpha_q)$ and $\boldsymbol{a}(\beta_q)$ for low-elevation targets can be synthesized to include both direct and reflected angles and are expressed as follows

$$\boldsymbol{a}(\alpha_q) = \boldsymbol{a}\left(\varphi_{qd}\right) + \rho\boldsymbol{a}(\varphi_{qs}), \tag{18}$$

$$X: \begin{cases} a\left(\varphi_{qd}\right) = \left[1, e^{-\frac{j2\pi dsin\varphi_{qd}cos\theta_{qd}}{\lambda}}, \ldots, e^{-\frac{j2\pi(M-1)dsin\varphi_{qd}cos\theta_{qd}}{\lambda}}\right]^T \\ a(\varphi_{qs}) = \left[1, e^{-\frac{j2\pi dsin\varphi_{qs}cos\theta_{qs}}{\lambda}}, \ldots, e^{-\frac{j2\pi(M-1)dsin\varphi_{qs}cos\theta_{qs}}{\lambda}}\right]^T \end{cases} \tag{19}$$

$$\boldsymbol{a}(\beta_q) = \boldsymbol{a}\left(\theta_{qd}\right) + \rho\boldsymbol{a}(\theta_{qs}) \tag{20}$$

$$Y: \begin{cases} a\left(\theta_{qd}\right) = \left[e^{-\frac{j2\pi dsin\varphi_{qd}sin\theta_{qd}}{\lambda}}, e^{-\frac{j2\pi 2dsin\varphi_{qd}sin\theta_{qd}}{\lambda}} \ldots, e^{-\frac{j2\pi(N-1)dsin\varphi_{qd}sin\theta_{qd}}{\lambda}}\right]^T \\ a(\theta_{qs}) = \left[e^{-\frac{j2\pi dsin\varphi_{qs}sin\theta_{qs}}{\lambda}}, e^{-\frac{j2\pi 2dsin\varphi_{qs}sin\theta_{qs}}{\lambda}} \ldots, e^{-\frac{j2\pi(N-1)dsin\varphi_{qs}sin\theta_{qs}}{\lambda}}\right]^T \end{cases} \tag{21}$$

where $\boldsymbol{a}\left(\varphi_{qd}\right)$ and $\boldsymbol{a}(\varphi_{qs})$ denote the direction vectors of the direct and reflected angles in the X axis, respectively; $\boldsymbol{a}\left(\theta_{qd}\right)$ and $\boldsymbol{a}(\theta_{qs})$ denote the direction vectors of the direct and reflected angles in the Y axis,; $\rho = \rho_0\exp(-j2\pi\Delta R/\lambda)$ denotes the multipath attenuation coefficient; and ρ_0 denotes the specular reflection coefficient. In the spatial model, the azimuthal angles of the direct and reflected waves are equal, i.e.,

$$\theta_q = \theta_{qd} = \theta_{qs} \tag{22}$$

Substituting Equations (17) and (22) into $\boldsymbol{a}\left(\varphi_{qd}\right)$ and $\boldsymbol{a}(\varphi_{qs})$ as well as $\boldsymbol{a}\left(\theta_{qd}\right)$ and $\boldsymbol{a}(\theta_{qs})$, i.e.,

$$X:\begin{cases} \boldsymbol{a}\left(\varphi_{qd}\right) = \left(\exp\left(-j2\pi(i)dsin\varphi_{qd}cos\theta_q/\lambda\right)\right)_{1\times M}, i=0,1,M-1 \\ \boldsymbol{a}(\varphi_{qs}) = \left(\exp\left(-j2\pi(i)dsin\left(\arctan\left(\frac{H+h}{H-h}tan\varphi_{qd}\right)\right)cos\theta_q/\lambda\right)\right)_{1\times M}, i=0,1,M-1 \end{cases} \tag{23}$$

$$Y:\begin{cases} \boldsymbol{a}\left(\theta_{qd}\right) = \left(\exp\left(-j2\pi idsin\varphi_{qd}sin\theta_q/\lambda\right)\right)_{1\times(N-1)}, i=1,2,\ldots,N-1 \\ \boldsymbol{a}(\theta_{qs}) = \left(\exp\left(-j2\pi(i)dsin\left(\arctan\left(\frac{H+h}{H-h}tan\varphi_{qd}\right)\right)sin\theta_q/\lambda\right)\right)_{1\times(N-1)}, i=1,2,\ldots,N-1 \end{cases} \tag{24}$$

The direction vectors in the X and Y axis are

$$\boldsymbol{A}_1 = \left[\boldsymbol{a}(\varphi_{1d}) + \rho\boldsymbol{a}(\varphi_{1s}),\ldots,\boldsymbol{a}(\varphi_{Qd}) + \rho\boldsymbol{a}(\varphi_{Qs}),\boldsymbol{a}(\alpha_{Q+1}),\ldots,\boldsymbol{a}(\alpha_K)\right], \tag{25}$$

$$\boldsymbol{A}_2 = \left[\boldsymbol{a}(\theta_{1d}) + \rho\boldsymbol{a}(\theta_{1s}),\ldots,\boldsymbol{a}(\theta_{Qd}) + \rho\boldsymbol{a}(\theta_{Qs}),\boldsymbol{a}(\beta_{Q+1}),\ldots,\boldsymbol{a}(\beta_K)\right] \tag{26}$$

The received signal and its covariance can be calculated according to Equations (9) and (10). When the array is an L-shaped sparse array, the spacing of the array elements in the signal direction vector will change, corresponding to the sparse array element spacing, and the received signal and signal covariance matrix will change accordingly. The 2D DOA estimation becomes more complex when there is a low-elevation signal in the received signal, and the existing algorithms, whether for L-shaped uniform arrays or L-shaped sparse arrays, are not easy and accurate to implement 2D DOA estimation, and most of them can only be used for a specific array structure or a single low-elevation signal [20]. In contrast, from the signal model, there is a correspondence between the array signal covariance matrix and the elevation and azimuth angles of the targets (including low-elevation targets), i.e., in the absence of low-elevation targets, the 2D DOA relationship between the received signal covariance matrix and the target can be regarded as

$$\boldsymbol{R_x} \to f((\varphi_1,\theta_1),(\varphi_2,\theta_2),\ldots,(\varphi_K,\theta_K)) \tag{27}$$

When low-elevation targets are present,

$$\boldsymbol{R_x} \to f((\varphi_{1d},\theta_1),(\varphi_{1s},\theta_1),\ldots,(\varphi_{Qd},\theta_Q),(\varphi_{Qs},\theta_Q),(\varphi_{Q+1},\theta_{Q+1}),(\varphi_K,\theta_K)) \tag{28}$$

which includes Q low-elevation targets, combined with Equation (17) above, Equation (23) can be further rewritten as

$$\boldsymbol{R_x} \to f'((\varphi_{1d},\theta_1),\ldots,(\varphi_{Qd},\theta_Q),,(\varphi_{Q+1},\theta_{Q+1}),(\varphi_K,\theta_K)) \tag{29}$$

On this basis, the above mapping relations can be obtained with the help of deep learning, providing new ideas and methods to solve the problem of 2D DOA estimation for L-shaped uniform arrays or sparse arrays in the presence of low-elevation-angle signals.

3. Dilated Convolution Network Model

Due to the significant difference in the direction vector generation process between low-elevation signals and non-low-elevation signals, when there are low-elevation targets in space, conventional algorithms will first decoherence and then implement DOA estimation. The flow of the algorithm proposed in this paper is shown in Figure 7 below.

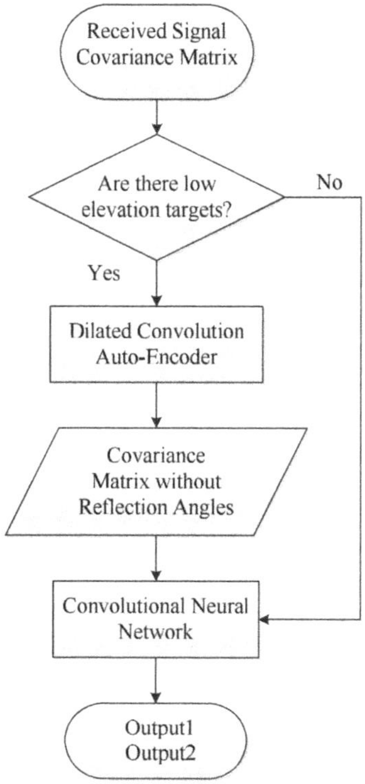

Figure 7. The flow of dilated convolution networks model.

As can be seen in Figure 7, when solving the DOA estimation problem, it firstly determines whether there are low-elevation targets. When there are low-elevation targets the reflected wave components are filtered out by the dilated convolutional autoencoder (DCAE) to achieve multipath suppression, and then the 2D DOA estimation is achieved by the dilated convolutional neural network (DCNN). When there is no low-elevation target signal in space, no multipath suppression is required, so the 2D DOA estimation can be directly achieved by the DCNN. Output1 and Output2 in Figure 7 are two output branches, which are the elevation angle sequence and azimuth angle sequence, respectively corresponding to the same position in two sequences that belongs to the same target, which can be automatically matched.

It should be added that when the covariance matrix of the received signal is used as the input to a neural network model for model training (decoherence or angle estimation), the real part of the covariance matrix is usually retained or the real and imaginary parts are stitched together to form an $N \times 2N$ (N denotes the total number of array elements) real matrix, which may make the data information incomplete or affect the extraction of spatial features. In the model design process, the real and imaginary parts of the covariance matrix are expanded into a 3D matrix to form an $N \times N \times 2$ 3D matrix as the input and for training, so that the spatial features can be more fully and comprehensively extracted.

3.1. Dilated Convolutional Autoencoder Mode

The convolutional autoencoder is a type of autoencoder, which is a self-supervised learning algorithm that encodes and decodes data through convolutional operations so that the output data can reproduce the input data, and has a wide range of applications in data compression, data denoising, and anomaly detection [21,22]. The traditional convolutional autoencoder consists of an encoding process and a decoding process, in which the former consists of alternating convolutional and pooling layers, with the convolutional layer used to extract features and the pooling layer used to reduce the dimensionality of the data; the latter consists of alternating deconvolution and upsampling layers, with the deconvolution being essentially the same as the convolutional layer [23], and the upsampling layer mainly achieving the recovery of data dimensionality. However, for the array received signal covariance matrix, the number of array elements is limited and the size of the covariance matrix is limited and often not very large, so the pooling layer is likely to cause insufficient feature extraction and loss of relevant features. Therefore, we discarded the pooling layer on the basis of the traditional convolutional autoencoder, and we introduced the dilation convolution to achieve data compression without data loss, DCAE model is as shown in Figure 8.

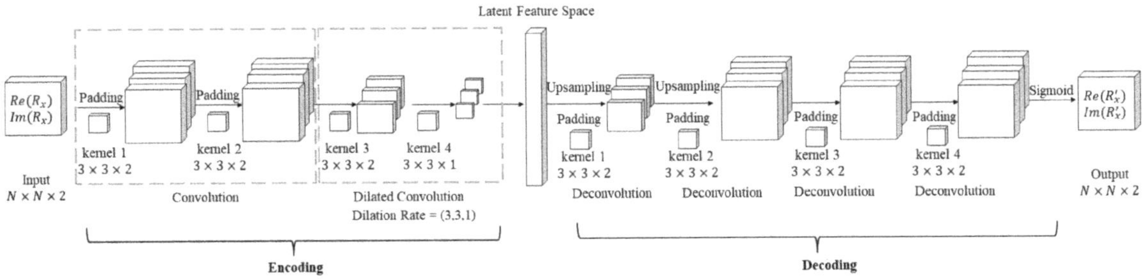

Figure 8. The model of DCAE.

In Figure 8, R_x denotes the received signal covariance matrix containing the reflected angle, i.e., the original signal covariance matrix; R_x' denotes the received signal covariance matrix without the reflected angle, containing only the direct and azimuth angles of the low-elevation target and the elevation and azimuth angles of the non-low-elevation target; and $Re(*)$ and $Im(*)$ denote the real and imaginary parts of the signal, respectively. The encoding and decoding processes are abstracted into the following mapping relationships, respectively,

$$Encoding : \boldsymbol{y} = f_e(\boldsymbol{R_x}), \tag{30}$$

$$Decoding : \boldsymbol{R_x^{'}} = f_d(\boldsymbol{y}), \tag{31}$$

Then, the dilated convolutional autoencoder action proposed in this paper can be further described as

$$\boldsymbol{R_x^{'}} = f_d(f_e(\boldsymbol{R_x})), \tag{32}$$

This means that a mapping between a covariance matrix with reflection angles and a covariance matrix without reflection angles is achieved. In the encoding process, the convolution operation proceeds as

$$\boldsymbol{h}^n = f(\boldsymbol{R} * \boldsymbol{w}^n + \boldsymbol{b}^n), \tag{33}$$

where $\boldsymbol{R}$ denotes the input 3D matrix; w denotes the 3D convolution kernel, whose number is n; $\boldsymbol{b}^n$ denotes the bias; and $f(*)$ denotes the activation function. The decoding process performs the deconvolution operation, which is essentially the same as the convolution operation. The two convolutional layers in the blue box in Figure 8 are regular convolutional operations with padding, which aims to preserve the boundary features. The two convolutional layers in the green box are dilated convolutional operations, the sizes of the convolutional kernels are $3 \times 3 \times 2$ and $3 \times 3 \times 1$, and the dilation rate is

(3,3,1). The loss function of the DCAE model is a binary cross-entropy function, whose expression is

$$bce = -\sum_{i=1}^{N}\sum_{j=1}^{N}(r_{ij}\log(r'_{ij}) + (1 - r_{ij})\log(1 - r'_{ij})), \tag{34}$$

where N denotes the total number of samples; and r_{ij} and r'_{ij} denote the predicted and true values, respectively.

Dilated Convolution

Dilated convolution is a kind of convolution idea to address the problem of information loss caused by connecting pooling layers after the standard convolution process [24]. The principle involves adding holes to the standard convolution map, using the holes to make the original convolution kernel have a larger reception field without increasing the number of parameters and operations [25]. Taking 2D convolution as an example, the dilation rate contains two values, which represent the magnitude of the distance between the value in the convolution kernel in the horizontal and vertical directions and its adjacent value position; when the convolution kernel size is 3×3, the reception field at different dilation rates is as shown in Figure 9.

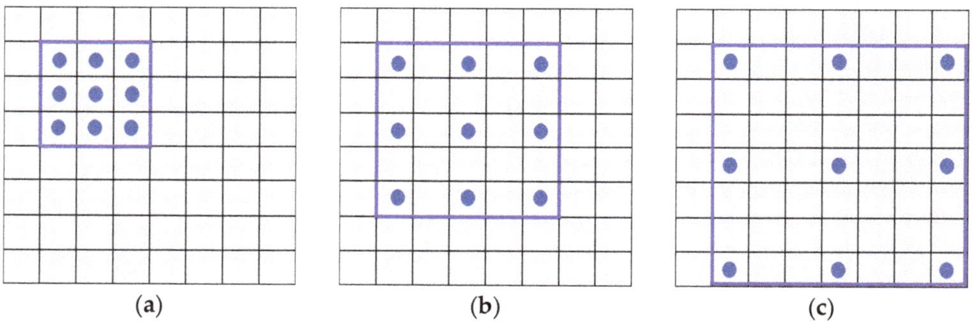

Figure 9. Reception fields at different dilation rates: (**a**) dilation rate = (1,1), (**b**) dilation rate = (2,2), and (**c**) dilation rate = (3,3).

The blue dots in Figure 9 represent the values of the convolution kernel, and the blue boxes represent the receptive fields under the convolution kernel; the positions in the receptive field area not filled with dots are hole., When the convolution operation is performed, the empty positions are filled with a value of 0. As shown in Figure 9a, the dilation rate is (1, 1), which is the standard convolution; the values in the convolution kernel are adjacent to each other; and the sizes of the receptive field and the convolution kernel are the same. The dilation rate in Figure 9b is (2, 2), i.e., the difference in the position between adjacent values in the convolution kernel is 2, so when the dilation rate is (2, 2), the size of the receptive field is the same as when the convolution kernel is 5×5. In Figure 9c, the dilation rate is (3, 3), the difference in the position of the values is 3, and the size of the receptive field is 7×7. The dilated convolution achieves an increase in the receptive field with the same convolution kernel and avoids an increase in computational effort.

3.2. Dilated Convolutional Neural Network Model

When there is no low-elevation target in the space target or the signal covariance matrix containing the low-elevation target has been suppressed by the DCAE model, the elevation and azimuth angles of the signal are obtained by the DCNN model. The structure of the model is shown in Figure 10.

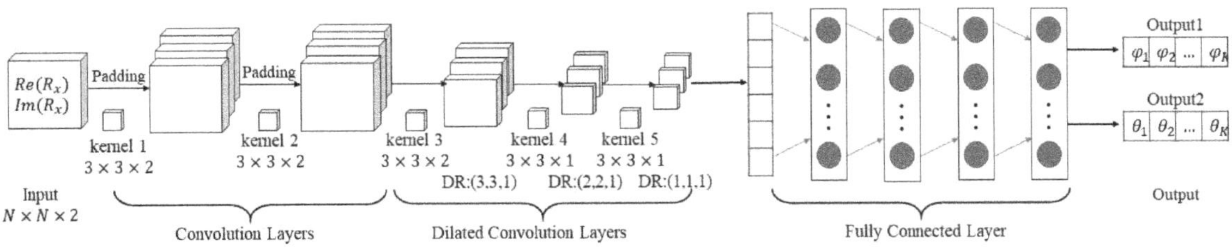

Figure 10. The model of DCNN. DR = dilation rate.

As can be seen in Figure 10, the model consists of five convolutional layers and four fully connected layers. The first two of the convolutional layers are standard convolutional layers with convolutional kernel sizes of $3 \times 3 \times 2$, and the last three are dilated convolutional layers with convolutional kernel sizes of $3 \times 3 \times 2$, $3 \times 3 \times 1$, and $3 \times 3 \times 1$, separately. The model contains two separate output branches for elevation and azimuth angles. Because the output of a convolutional neural network is sensitive to order, the two output branches for the elevation and azimuth angles correspond in order and no matching is required. Similar to DCAE, the input to the model, which consists of the real and imaginary parts of the covariance matrix in three dimensions, and the convolutional layers in the model are all 3D convolutional operations; the pooling layer is also removed from the convolutional neural network. The activation function for both the convolutional and fully connected layers is the ReLU function [26], which is characterized by fast convergence and no saturation of gradients, so is widely used in the training of convolutional neural network models [27]. Both output branches of the model are estimated angular values, which are regression problems, so the loss function of the model is the mean squared loss function, i.e.,

$$mse = \frac{\sum_i^N \left((\varphi_i - \varphi_i')^2 + (\theta_i - \theta_i')^2 \right)}{2N}, \quad (35)$$

where φ_i and φ_i' denote the real and estimated values of the elevation angle, respectively; θ_i and θ_i' denote the real and estimated values of the azimuth angle, respectively; and N denotes the number of targets.

4. Simulation Experiments and Analysis of Results

In the simulation experiment, we used 16 array elements; 8 and 9 uniform arrays in the X and Y axis, respectively; and the sparse arrays were two coprime arrays with coprime numbers (4, 5) and (3, 7). The total number of arrays was 16 due to the existence of a common element at the origin. The range of low elevation angles in the spatial signal where multipath effects occur was $(0°, 10°]$, the range of non-low elevation angles was $(10°, 60°]$, and the range of azimuth angles was $[-90°, 90°]$. The DCAE and DCNN models are shown in Figures 8 and 10 above. The number of convolutional kernels for each layer of the encoding process in the DCAE model was 200, 200, 150, and 150 in order in the decoding process, i.e., 150, 150, 200, and 200. The size of the kernels and the dilation rate were set as in Figures 8 and 10. The number of neurons in each layer of the fully connected layer was 1500, 1500, 1000, and 1000 in that order. The capacity of the training set for different formations was 50,000, the size of the test set was 2000, the number of iterations was 5000, and the batch size was 100.

4.1. *Verification of Dilated Convolutional Autoencoder Mode Validity*

Test case 1: The formation is an L-shaped uniform array. There are two targets in space, one of which is a low-elevation target with direct and reflected angles of 3.216° and −5.957° for elevation, respectively, and 38.472° for the azimuth; the other is a non-low-elevation target with a 38.293° elevation and 48.506° azimuth; SNR is 10 dB; and snapshot is 100. After multipath suppression, the angle was estimated by the 2D MUSIC algorithm (the 2D search angle interval was 1°), and the spatial spectrum is shown in Figures 11 and 12.

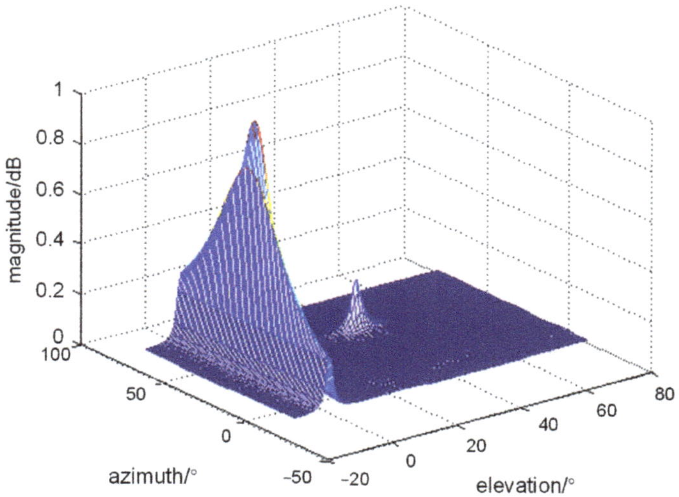

Figure 11. Spatial spectrum of test case 1 obtained by DCAE MUSIC.

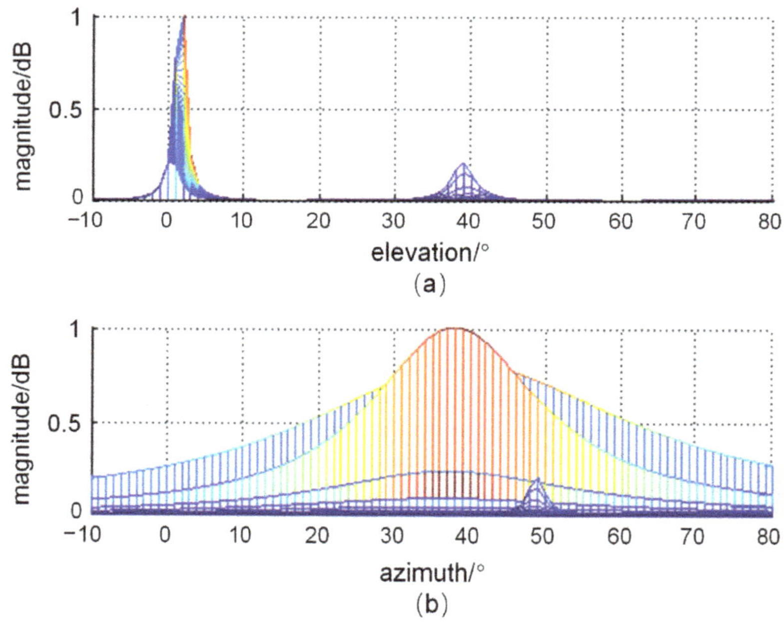

Figure 12. Spatial spectrum of (**a**) elevation and (**b**) azimuth angles of test case 1 obtained by DCAE MUSIC.

As can be seen in Figures 11 and 12, there are two distinct spectral peaks in the spatial spectrum, corresponding to the low-elevation target and the non-low-elevation target in the signal; there is no spectral peak for the reflected angle in Figure 12a. Comparing the azimuth of the low-altitude target, the spectral peaks of its elevation angle are sharper, while the difference between the sharpness of the azimuth and elevation angles of the non-low-altitude target is not significant. From the angle estimation accuracy and Figure 12a,b, we concluded that the elevation and azimuth angles of the low-altitude target are about 3° and 38°, respectively; and the elevation and azimuth angles of the non-low-altitude target are about 38° and 49°, respectively, which are close to the target angle in test case 1, for the existence of a low-elevation target in space. The DCAE model can effectively filter the reflected angular component of a low-elevation target and a non-low-elevation target in space, without interfering with the direct angle of arrival and the non-low-elevation target.

Test case 2: The array is L-shaped sparse array. There are two low-elevation targets and one non-low-elevation target in space, where the direct and reflected angles of the low-elevation targets are 1.999° and −3.708°, and 8.000° and −18.937°, respectively; the azimuth angles of the two low-elevation targets are 1.904° and 1.478°; the elevation and azimuth angles of the non-low-elevation targets are 33.743° and 30.509°, respectively; SNR is 10 dB; and the snapshot is 100. After filtering the reflected angle, the angle was estimated by the 2D MUSIC algorithm (the 2D search angle interval was 1°), and the spatial spectrum is shown in Figure 13.

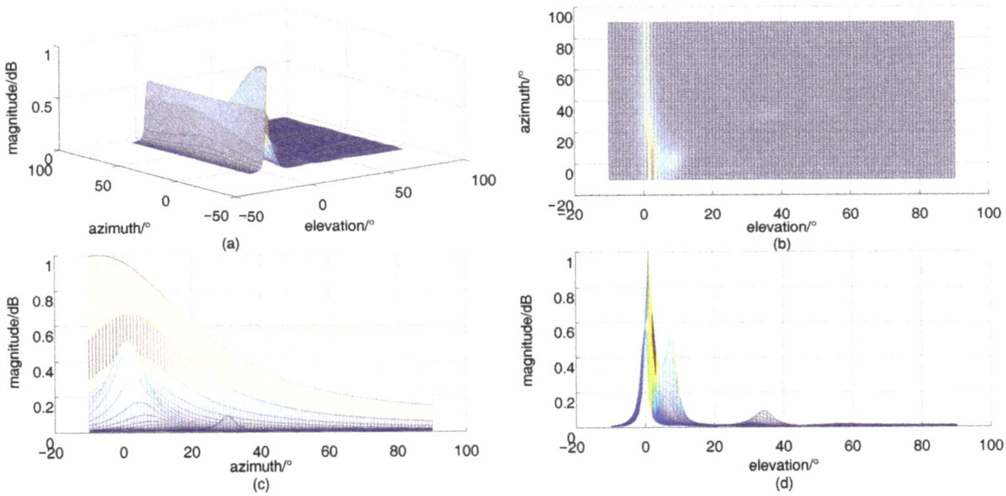

Figure 13. Spatial spectrum of test case 2 obtained by DCAE MUSIC: (**a**) 3D view of the 2D spatial spectrum, (**b**) top view of the 2D spatial spectrum, (**c**) spatial spectrum of azimuth angles, and (**d**) spatial spectrum of elevation angles.

Figure 13a shows the spatial spectrum of the 2D search, and Figure 13b shows the top view of the spatial spectrum, from which it can be seen that there are three spectral peaks in the spatial spectrum, with the non-low-elevation target having the lowest peak value. Figure 13c shows the azimuth angle, containing three spectral peaks corresponding to 1°, 2°, and 31°; Figure 12d shows the elevation angle, also containing three spectral peaks corresponding to 1°, 8° and 33°. The elevation and azimuth angles of the three targets obtained from Figure 13 are essentially the same as the actual angles in test case 2 and are not affected by the formation. When the three targets in test case 2 were estimated with 2D MUSIC (without the reflected angles), the spatial spectrum was obtained as shown in Figure 14.

Comparing Figures 13 and 14, the two spatial spectral distributions are basically the same. We verified that when there are multiple low-elevation targets in space, the DCAE algorithm can effectively suppress multipath without interfering with the estimation of direct-angle and non-low-elevation targets. Test cases 1 and 2 verify that the DCAE model can effectively achieve "de-multipathing" and that the DCAE model is valid.

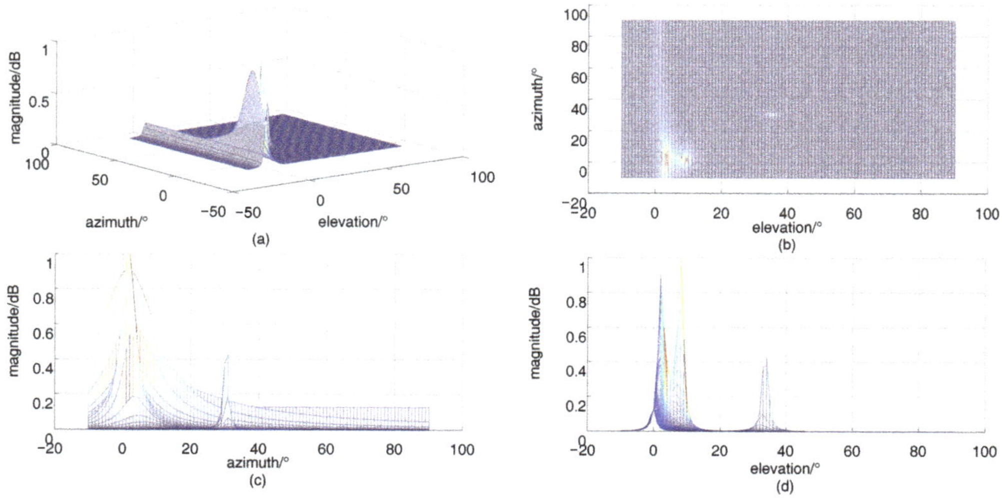

Figure 14. Spatial spectrum of test case 2 (without the reflected angles) obtained with 2D MUSIC: (**a**) 3D view of the 2D spatial spectrum. (**b**) Top view of the 2D spatial spectrum. (**c**) Spatial spectrum of azimuth angles. (**d**) Spatial spectrum of elevation angles.

4.2. Verification of Dilated Convolutional Neural Network Model Validity

The arrays were L-shaped uniform (LUA) and L-shaped sparse array (LSA), as described above; SNR was 10 dB; snapshot was 100; the numbers of targets were 2 and 3, respectively; and all were non-low-elevation targets. The efficiency rate, matching rate, and root mean square error of the angle estimation were used as the performance evaluation metrics of DCNN for 2D DOA estimation. When the angular error in the output is not greater than 5°, the angle is regarded as a valid angle. The proportion of valid angles to all output angles is the efficiency rate P_E; the matching rate P_M indicates the proportion of azimuth and elevation angles that are accurately matched according to their positions, and the root mean square error (RMSE) is a common measure in DOA angle estimation; its expression is

$$RMSE = \sqrt{\frac{1}{N}\sum_{i=1}^{N}(\theta_i' - \theta_i)^2}, \tag{36}$$

where N denotes the total number of test sets, θ_i' denotes the angle estimate output by the model, and θ_i denotes the actual angle value. After 200 Monte Carlo experiments, P_E, P_M, and RMSE were statistically obtained as shown in Table 1,

Table 1. P_E, P_M, and RMSE for non-low-elevation targets obtained with DCNN.

Type	Target Number	P_E/%	P_M/%	RMSE/°	RMSE$_e$/°	RMSE$_a$/°
LUA	2	99.98	100	0.3697	0.2901	0.4352
	3	99.97	100	0.3083	0.2710	0.3412
LSA	2	99.99	100	0.3507	0.2548	0.4256
	3	100	100	0.2768	0.2602	0.2759

In Table 1, RMSE/° indicates the RMSE for all outputs; RMSE$_e$ and RMSE$_a$ denote the RMSE for elevation and azimuth angles respectively. From Table 1, it can be seen that all P_E values are close to 100%, P_M reaches 100%, and the elevation and azimuth angles of the targets in the two output branches can achieve one-to-one correspondence without parameter matching. From the RMSE results, 2D DOA estimation accuracy is better than that of the L-shaped uniform array when the array type is L-shaped sparse array, and the estimation effect is better than that when the number of sources is three than when the number of sources is two. The estimation accuracy for elevation angle is slightly higher than that for azimuth angle in all conditions above.

When there were three targets in a space that contains two low-elevation-angle targets, P_E, P_M, and RMSE were calculated after 200 Monte Carlo experiments using the DCAE-DCNN and DCNN models (without the reflected angle in the model output); the results are shown in Table 2.

Table 2. P_E, P_M, and RMSE for low-elevation targets obtained with DCNN and DCAE-DCNN. "Low" and "Non-Low" denote the elevation angle of low-elevation target and non-low-elevation target in all targets, respectively.

Type	Model	P_E/%	P_E/%	RMSE/°	RMSE$_e$/°		RMSE$_a$/°
					Low	Non-Low	
LUA	DCNN	85.74	97.35	1.6782	1.5757	1.5302	1.7946
	DCAE-DCNN	99.98	100	0.3441	0.2913	0.2893	0.3898
LSA	DCNN	87.36	96.77	1.6813	1.6710	1.6276	1.7125
	DCAE-DCNN	99.99	100	0.2975	0.2851	0.2720	0.3124

Table 2 shows that when only DCNN was applied for the 2D DOA estimation for multiple targets including low-elevation targets, it was not effective. Despite the high P_E, P_M is low, and the RMSEs of the elevation and azimuth angles are much higher than those of DCAE-DCNN method, which indicates that the 2D DOA estimation problem could not be directly solved when the signal contained low-elevation targets using the DCNN method alone. As such, decoherence or de-multipathing of the received signal is necessary. Additionally, when the DCAE-DCNN algorithm was used, P_E and P_M were close to 100%, RMSEs were lower, and P_E and P_M were higher when the array type was LSA than LUA. The RMSE of the elevation angle was slightly lower than that of the azimuth angle; the RMSEs of the non-low-elevation targets were slightly lower than those of the low-elevation targets. Comparing Tables 1 and 2, the results using the DCAE-DCNN algorithm when low-elevation angle targets are present in the signal are similar to those when only the DCNN algorithm is used when low-elevation angle targets are not present. The RMSE of the former is slightly higher, which proves that the DCNN algorithm is effective and stable, and the DCAE-DCNN algorithm has a better estimation effect for the presence of low-elevation targets.

4.3. RMSE of 2D DOA Estimation at Different SNRs with Non-Low-Elevation Targets

In general, the variation in the SNR has a significant effect on the accuracy of DOA estimation. In this set of simulation experiments, 3 non-low-altitude targets were in space, array types were LUA and LSA, the number of snapshots was 200, and SNR was −10 dB, −5 dB, 0 dB, 5 dB, 10 dB, 15 dB, or 20 dB. The proposed DCNN model was used for angle estimation, and its results were compared with those of the 2D MUSIC algorithm to calculate the RMSE, as shown in Figure 15.

As can be seen from Figure 15a,b, the RMSE of both the elevation and azimuth angles decrease as SNR increases, and the higher SNR, the higher the estimation accuracy. From Figure 15a,b, it can be seen that the estimation accuracy of LSA is higher than that of LUA for the same algorithm. For the same array type, when the SNR was less than 10 dB, the estimation accuracy of both the elevation and azimuth angles significantly improved as the SNR increased, and the estimation performance of the DCNN algorithm was significantly better than that of 2D MUSIC. When the SNR was greater than 10dB, the decreasing trend of the RMSE became slower; for LUA, the DCNN algorithm's estimation accuracy for the azimuth angles was slightly lower than that of 2D MUSIC, and for elevation, it is slightly higher than that of 2D MUSIC. For LSA, DCNN algorithm's estimation accuracy for elevation was better than that of 2D MUSIC, while the estimation accuracy for the azimuth angles was approximately equal between the two. By comparing Figure 15a,b, it can be seen that for either array type, DCNN algorithm's estimation accuracy for elevation angles is higher than that for the azimuth angles under each SNR condition, while the difference in the estimation performance of the 2D MUSIC algorithm for elevation and azimuth angles is not significant.

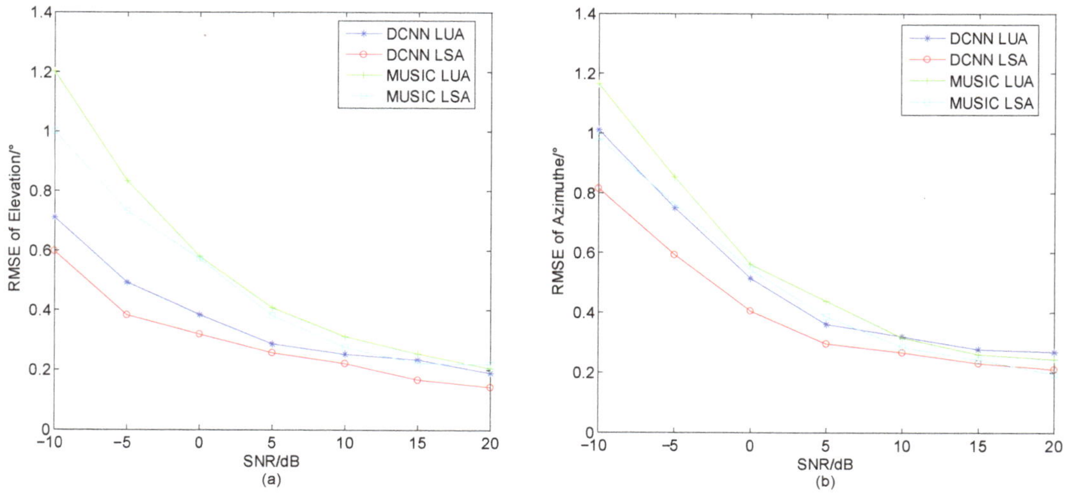

Figure 15. RMSE comparison at different SNRs with non-low-elevation targets: (**a**) elevation angles and (**b**) azimuth angles.

4.4. RMSE of 2D DOA Estimation at Different SNRs with Low-Elevation Targets

In this set of simulations, the number of spatial targets was 3, including 2 low-elevation targets; the array type was the above LUA and LSA; the number of snapshots was 200; and the SNR was −10 dB, −5 dB, 0 dB, 5 dB, 10 dB, 15 dB, or 20 dB. The DCAE-DCNN model proposed in this paper was used to perform 2D DOA estimation according to the process in Figure 7. Because most of the sparse array decoherence before 2D DOA estimation is for uniform surface arrays or other specific arrays, the Toeplitz matrix reconstruction algorithm proposed in the literature [13] and MSSP-MUSIC for 2D DOA estimation of LUA were used as comparison experiments in this set of simulations. The calculated RMSEs for the elevation and azimuth angles under each algorithm are shown in Figure 16.

In Figure 16, with the increase in the SNR, the RMSE of each algorithm for the estimation of elevation and azimuth angles shows a decreasing trend, and the estimation performance increases accordingly. When the SNR is less than 10 dB, RMSE significantly decreases as SNR increases, and the estimation accuracy of the proposed algorithm is significantly higher than that of the other two algorithms. When the SNR is greater than 10 dB, the decreasing trend in RMSE decreases, and the estimation accuracy of the proposed algorithm for LUA is close to multiple Toeplitz matrices reconstruction (MTOEP) method in the literature [13]. Comparing Figures 15 and 16, when there is a low-elevation target in the signal, the estimation accuracy of both the elevation and azimuth angles is degraded after de-multipathing by the proposed DCAE model, because when the signal is de-multipathed by the DCAE model, it may lead to new errors in the signal covariance matrix, which affects the estimation accuracy to a certain extent.

The estimation accuracy for the elevation angle is slightly higher than that for the azimuth angle for both algorithms proposed in this paper and MTOEP method proposed in the literature [13] in Figures 15 and 16. For the proposed algorithm, the reason for this is that when designing the DCNN model, the output sequence of the elevation angle is arranged in the order from smallest to largest, and the angles with the same position number in both outputs correspond to the same target, while the order of the azimuth angle is affected by the elevation angle, resulting in the output of branch 2 being affected by branch 1. The MTOEP method in the literature [13] estimates the azimuthal angle based on the elevation angle first, so the RMSEs for the azimuthal angle of the above two algorithms are slightly larger than those for the elevation angle. However, when the MUSIC algorithm performs a two-dimensional search, it traverses the entire two-dimensional space, and the priority traversal order of the elevation and azimuthal angles does not affect the results, which are equivalent, so the difference between the RMSE of the elevation angle and azimuthal angles is not significant.

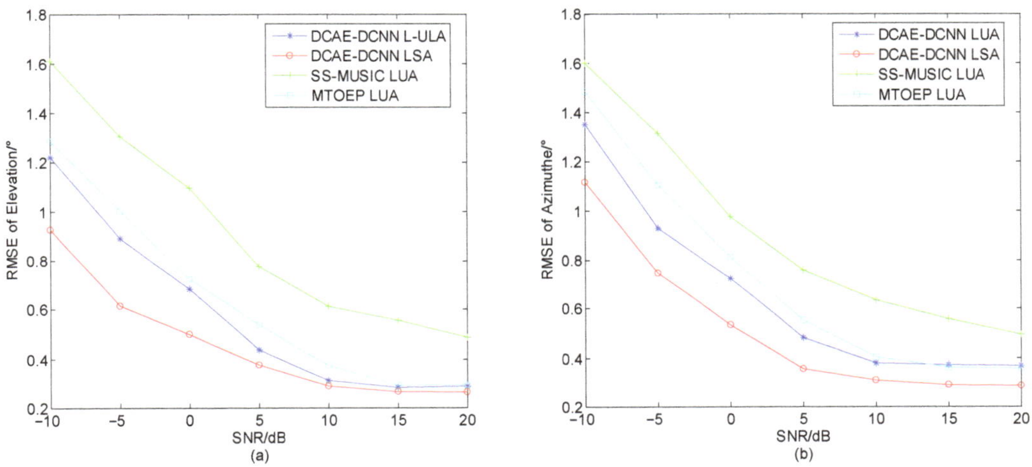

Figure 16. RMSE comparison at different SNRs with low-elevation targets: (**a**) elevation angles and (**b**) azimuth angles.

5. Discussion

For a long time, 2D DOA estimation has been of great importance in the field of array signal processing. The 2D DOA estimation of low-elevation targets, especially when the array elements are sparsely arranged, is a key and difficult problem for research in this field. The development of deep learning has provided new ideas to solve such problems. To address this problem, we developed a 2D DOA estimation algorithm based on a dilated convolutional autoencoder and s dilated convolutional neural network, which requires the total number of targets in space and the presence of low-elevation angles to be known quantities. When low-elevation targets are present in space, multipath suppression is applied to the received signal covariance matrix with DCAE, and then DCNN is used for 2D DOA estimation. Additionally, when there is no low-elevation target in space, 2D DOA estimation can be directly achieved using DCNN. The simulation experiments showed that when there are low-elevation targets in space, DCAE can effectively achieve multipath suppression and filter out the reflected angle components in the covariance matrix; when there is no low-elevation target in space or after multipath suppression is completed, DCNN can effectively achieve 2D DOA estimation with high estimation accuracy and without the need for further parameter matching.

In the proposed algorithm, the choice of hyperparameters for the model is not strict and needs to be optimized and adjusted according to the output results. In addition, we used simulation data for validation and comparison experiments, and there are certain differences between the simulation and measured data. The next study will focus on analyzing and comparing the similarities and differences between the simulation data and the measured data, so that the simulation data and the experimental scenarios can be set closer to the actual situation, thus increasing the applicability of the.

Author Contributions: This article was coauthored by G.H., F.Z. and B.L.; the major individual contributions are as follows: conceptualization, F.Z. and G.H.; methodology, G.H., F.Z. and B.L.; software, F.Z. and B.L.; validation, F.Z. and B.L.; formal analysis, G.H.; investigation, F.Z. and B.L.; resources, G.H.; data curation, F.Z. and B.L.; writing—original draft preparation, G.H. and F.Z.; writing—review and editing, F.Z.; supervision, G.H. and B.L.; project administration, G.H.; funding acquisition, G.H. and B.L. All authors have read and agreed to the published version of the manuscript.

Funding: This study was funded by the National Natural Science Foundation of China, grant number 61871395.

Data Availability Statement: The data presented in this study are available on request from the corresponding author. The data are not publicly available, due to the data in this paper not being from publicly available datasets but obtained from the simulation of the signal models listed in the article.

Acknowledgments: We thank the college for providing us with an efficient simulation platform so that we could complete the experimental simulation as scheduled. Funding from the National Natural Science Foundation of China (No. 6207011332) is gratefully acknowledged.

Conflicts of Interest: The authors declare no conflict of interest.

References

1. Richards, M. *Fundamentals of Radar Signal Processing*, 2nd ed.; IET: London, UK, 2014; pp. 1–16.
2. Ning, G.; Zhang, S.; Zhang, J. Velocity-independent two-dimensional direction-of-arrival estimation algorithm with three parallel linear arrays. *IET Signal Process.* **2022**, *1*, 106–116. [CrossRef]
3. Zhang, H.; Gao, K.; Xing, J. 2D Direction-of-arrival Estimation for Sparse L-shaped Array based on Recursive Gridding. In Proceedings of the 9th Asia-Pacific Conference on Antennas and Propagation (APCAP), Xiamen, China, 4–7 August 2020.
4. Xiong, Y.; Li, Z.; Wen, F. 2D DOA Estimation for Uniform Rectangular Array with One-bit Measurement. In Proceedings of the IEEE 11th Sensor Array and Multichannel Signal Processing Workshop (SAM), Hangzhou, China, 8–11 June 2020.
5. Zhao, X.; Zhou, J.; Fan, H. Improved 2D-MUSIC estimation for low intercept coprime MIMO radar. *J. Phys. Conf. Ser.* **2021**, *1971*, 012007. [CrossRef]
6. Yin, Q.; Zou, L. A Robert High Resolution Approach to 2D Signal Parameters Estimation-DOA Matrix Method. *J. China Inst. Commun.* **1991**, *4*, 1–7, 44.
7. Liang, L.; Shi, Y.; Shi, Y. Two-dimensional DOA estimation method of acoustic vector sensor array based on sparse recovery. *Digit. Signal Process.* **2022**, *120*, 103294. [CrossRef]
8. Liu, S.; Zhao, J.; Wu, D. 2D DOA estimation by a large-space T-shaped array. *Digit. Signal Process.* **2022**, *130*, 103699. [CrossRef]
9. Wang, H.; He, P.; Yu, W. Two-dimensional DOA Estimation Based on Generalized Coprime Double Parallel Arrays. *J. Signal Process.* **2022**, *38*, 223–231.
10. Ma, J.; Wei, S.; Ma, H. Two-dimensional DOA Estimation for Low-angle Target Based on ADMM. *J. Electron. Inf. Technol.* **2022**, *44*, 2859–2866.
11. Su, X.; Liu, Z.; Peng, B. A Sparse Representation Method for Coherent Sources Angle Estimation with Uniform Circular Array. *Int. J. Antennas Propag.* **2019**, *2019*, 3849791. [CrossRef]
12. Park, H.; Li, J. Efficient sparse parameter estimation-based methods for two-dimensional DOA estimation of coherent signals. *IET Signal Process.* **2020**, *14*, 643–651. [CrossRef]
13. Liang, H.; Li, X. Two-Dimensional DOA Estimation of Coherent Signals Based on the Toeplitz Matrix Reconstruction. *Electron. Inf. Warf. Technol.* **2012**, *27*, 23–27.
14. Molaei, A.; Zakeri, B.; Andargoli, S. Two-dimensional DOA estimation for multi-path environments by accurate separation of signals using k-medoids clustering. *IET Commun.* **2019**, *13*, 1141–1147. [CrossRef]
15. Agatonovic, M.; Stanković, Z.; Milovanovic, I. Efficient Neural Network Approach for 2d DOA Estimation based on Antenna Array Measurements. *Prog. Electromagn. Res.* **2013**, *137*, 741–758. [CrossRef]
16. Zhu, W.; Zhang, M.; Li, P. Two-Dimensional DOA Estimation via Deep Ensemble Learning. *IEEE Access* **2020**, *8*, 124544–124552. [CrossRef]
17. Yang, X.; Liu, L.; Li, L. A Method for Estimating 2D Direction of Arrival Based on Coprime Array with L-shaped Structure. *J. Xi'an Jiaotong Univ.* **2020**, *54*, 144–149.
18. Sinha, A.; Bar-Shalom, Y.; Blair, W. Radar measurement extraction in the presence of sea-surface multipath. *IEEE Trans. Aerosp. Electron. Syst.* **2003**, *39*, 550–567. [CrossRef]
19. Zhang, L.; Zhang, Y.; Yun, Y. Direct signal DOA estimation algorithm in radar low angle bearing environment. *Aerosp. Electron. Warf.* **2009**, *25*, 29–31.
20. Gu, J.; Wei, P.; Tai, H. 2-D Direction-of-Arrival Estimation of Coherent Signals using Cross-Correlation Matrix. *Signal Process.* **2008**, *88*, 75–85. [CrossRef]

1. Zhao, F.; Hu, G.; Zhou, H.; Zhan, C. CAE-CNN-Based DOA Estimation Method for Low-Elevation-Angle Target. *Remote Sens.* **2023**, *15*, 185. [CrossRef]
2. Vu, T.; Yang, H.; Nguyen, V. Multimodal Learning using Convolution Neural Network and Sparse Autoencoder. In Proceedings of the IEEE International Conference on Big Data and Smart Computing, Jeju Island, Republic of Korea, 13–16 February 2017.
3. Firat, O.; Vural, F. Representation Learning with Convolutional Sparse Autoencoders for Remote Sensing. In Proceedings of the 21st Signal Processing and Communications Applications Conference (SIU), Haspolat, Turkey, 24–26 April 2013.
4. Yao, J.; Wang, D.; Hu, H. ADCNN: Towards Learning Adaptive Dilation for Convolutional Neural Networks. *Pattern Recognit.* **2022**, *123*, 108369. [CrossRef]
5. Chalavadi, V.; Jeripothula, P.; Datla, R. mSODANet: A Network for Multi-Scale Object Detection in Aerial Images using Hierarchical Dilated Convolutions. *Pattern Recognit.* **2022**, *126*, 108548. [CrossRef]
6. Nair, V.; Hinton, G.E. Rectified Linear Units Improve Restricted Boltzmann Machines. In Proceedings of the International Conference on Machine Learning, Haifa, Israel, 21–24 June 2010.
7. Su, X.; Hu, P.; Liu, Z. Mixed Near-Field and Far-Field Source Localization Based on Convolution Neural Networks via Symmetric Nested Array. *IEEE Trans. Veh. Technol.* **2021**, *70*, 7908–7920. [CrossRef]

remote sensing

MDPI

Article

Network Collaborative Pruning Method for Hyperspectral Image Classification Based on Evolutionary Multi-Task Optimization

Yu Lei, Dayu Wang, Shenghui Yang, Jiao Shi, Dayong Tian and Lingtong Min *

School of Electronics and Information, Northwestern Polytechnical University, 127 West Youyi Road, Xi'an 710072, China; leiy@nwpu.edu.cn (Y.L.); wang_day@mail.nwpu.edu.cn (D.W.); sh_yang@mail.nwpu.edu.cn (S.Y.); jiaoshi@nwpu.edu.cn (J.S.); dayong.tian@nwpu.edu.cn (D.T.)

* Correspondence: minlingtong@nwpu.edu.cn

Abstract: Neural network models for hyperspectral images classification are complex and therefore difficult to deploy directly onto mobile platforms. Neural network model compression methods can effectively optimize the storage space and inference time of the model while maintaining the accuracy. Although automated pruning methods can avoid designing pruning rules, they face the problem of search efficiency when optimizing complex networks. In this paper, a network collaborative pruning method is proposed for hyperspectral image classification based on evolutionary multi-task optimization. The proposed method allows classification networks to perform the model pruning task on multiple hyperspectral images simultaneously. Knowledge (the important local sparse structure of the network) is automatically searched and updated by using knowledge transfer between different tasks. The self-adaptive knowledge transfer strategy based on historical information and dormancy mechanism is designed to avoid possible negative transfer and unnecessary consumption of computing resources. The pruned networks can achieve high classification accuracy on hyperspectral data with limited labeled samples. Experiments on multiple hyperspectral images show that the proposed method can effectively realize the compression of the network model and the classification of hyperspectral images.

Keywords: hyperspectral images classification; network pruning; multi-task optimization; knowledge transfer; multi-objective optimization

Citation: Lei, Y.; Wang, D.; Yang, S.; Shi, J.; Tian, D.; Min, L. Network Collaborative Pruning Method for Hyperspectral Image Classification Based on Evolutionary Multi-Task Optimization. *Remote Sens.* **2023**, *15*, 3084. https://doi.org/10.3390/rs15123084

Academic Editor: Saeid Homayouni

Received: 30 April 2023
Revised: 26 May 2023
Accepted: 9 June 2023
Published: 13 June 2023

1. Introduction

Hyperspectral images (HSIs) have become an important tool for resource exploration and environmental monitoring because they contain a lot of spectral segments and extensive spatial information. By using a convolutional neural network (CNN) [1–4], features of HSIs were extracted [5] and classified, which greatly improved the classification performance. Therefore, deep network methods have been widely applied in HSI classification.

However, the powerful feature representation ability of CNN relies on the complex structure of the model and a large number of parameters. With the development of remote sensing technology, the resolution is improved, which makes the size of the image larger and such data size significantly influences the computational and storage requirements [6,7]. This hinders the application of networks to satellites, aircraft, or other mobile platforms, which greatly reduces the practical efficiency of remote sensing images. Therefore, reducing the complexity of deep network models is an enduring problem for deploying on limited resource devices [8]. Neural network model compression can be used to solve the problem.

Neural network pruning is regarded as a simple yet efficient technique to compress model while maintaining their performance [9], which makes it possible to deploy the remote sensing lightweight analysis model on hardware. Generally speaking, network pruning methods can be classified as manual and automatic pruning methods. Pruning rules and selection of solutions in traditional manual methods are designed by domain experts. LeCun [10] first proposed optimal brain damage (OBD), which removed the low-value parameters by calculating the second derivative of parameters and sorting them. Han et al. [11] used an iterative pruning method to prune the weights that were less than a manually preset layer threshold. Lee et al. [12] proposed an importance score for global pruning; the score was a rescaling of weight magnitude that incorporates the model-level distortion incurred by pruning, and did not require any hyperparameter tuning. Recent advances in neural tangent kernel (NTK) theory have suggested that the training dynamics of sufficiently large neural networks was closely related to the spectrum of the NTK. Motivated by this finding, Wang et al. [13] pruned the connections that had the least influence on the spectrum of the NTK. The pruning method was applied to remote sensing images. Qi et al. [14] used the original network as a teacher model and guided the model to pruning through loss. Wang et al. [15] pruned according to the scaling factor of the BatchNorm layer. Guo et al. [16] designed a sensitivity function to evaluate the pruning effect of channels in each layer. Furthermore, the pruning rate of each layer was adaptively corrected. It is important to note that the criteria of manual pruning methods are not uniform, such as the absolute value of the network weights, the activation value of the neurons, and so on. As a result, a lot of time and labor costs are required to design and select appropriate pruning criteria for different networks. Furthermore, the sparse network obtained by manual pruning is generally not optimal due to the limited exploration space [17].

Different from the traditional manual pruning methods, automatic pruning methods can reduce the design cost [18]. As an automatic pruning method, evolution-based pruning methods constructed the pruning of the network as an optimization task, which can find and retain better sparse network structure in discrete space. Zhou et al. [19] implemented pruning of medical image segmentation CNNS by encoding filter and skipping some sensitive layers. By considering the sensitivity of each layer, our previous work proposed a differential evolutionary pruning method based on layer-wise weight pruning (DENNC) [20]. In addition, a multi-objective pruning method (MONNP) [21] was proposed, which can balance the network accuracy and network complexity at the same time. Furthermore, MONNP generated different sparse networks to meet various hardware constraints and requirements more efficiently. Zhou et al. [22] searched sparse networks at the knee point on Pareto-optimal front, and the networks create a trade-off between accuracy and sparsity. Zhao et al. [23] compressed the model with a pruning filter and applied the multi-objective optimization of CNN model compression to remote sensing images. Wei et al. [24] proposed a channel pruning method based on differentiable neural architecture search to automatically prune CNN models. The importance of each channel was measured by a trainable score. In conclusion, evolutionary pruning methods reduce the cost of manually designing pruning rules; however, network structures designed for hyperspectral data are becoming more and more complex, which also causes certain difficulties in evolutionary pruning methods.

For cases where the task is difficult to optimize, introducing additional knowledge to facilitate the search process of the target task provides feasible ideas. Ma et al. [25] proposed a multi-task model ESMM, which contains a main task CVR (post-click conversion rate) prediction, and an auxiliary task CTCVR (post-view click-through conversion rate) prediction. The CTCVR task was used to help the learning of CVR to avoid problems such as over-fitting and poor generalization of CVR prediction due to small samples. Ruder [26] pointed out that in multi-task learning, by constructing additional tasks, the prompts of these tasks can promote the learning of the main task. Feng et al. [27] considered the random embedding space as additional task for the target problem, which ensured the effectiveness of the search on the target problem by simultaneously optimizing the original task and the embedding task. Evolutionary multitasking can be used to optimize multiple tasks simultaneously to achieve the promotion of their respective tasks. In evolutionary multi-task optimization, effective facilitation between tasks relies on task similarity.

In HSI classification, if there exists different HSIs from the same sensor, the spectral information has a similar physical meaning (radiance or reflectivity) [28,29], and the similarity between two images is high. As shown in Figure 1, the HSIs obtained by the same sensor had the same spectral range. The comparison of spectral curves of the Indian Pines and Salinas reflected the similarity between HSIs. If the ground features of different HSIs are close, there is an underlying similarity between them. When the same network is trained on similar data, the distribution of network parameters is close. Thus, there are also similarities between structural sparsification tasks on different datasets. When dealing with HSI, deep neural networks mainly learn the spectral characteristics of the data through the convolution layer, and the parameters of the convolution layers realize the feature extraction of the data. Therefore, the structural information of the neural network is regarded as the transferred knowledge, which can be used as prior knowledge for other parallel tasks. In addition, the labels of hyperspectral data are limited, and CNN need enough data to learn features, which affects the training process of neural networks. When distribution of network parameter is close, knowledge transfer can obtain useful representation information from other image to alleviate the problem of limited labeled samples.

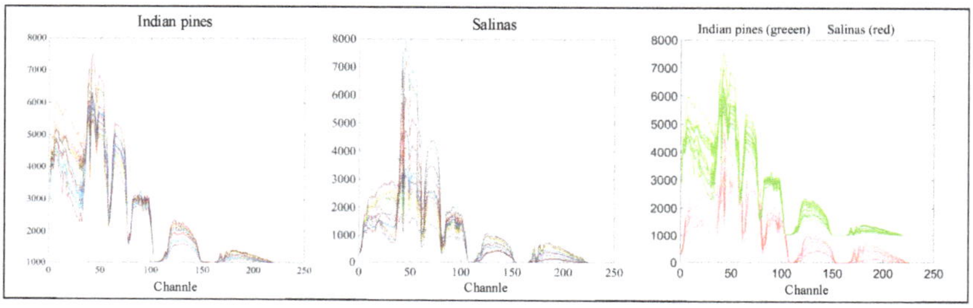

Figure 1. Spectral curves of Indian Pines and Salinas under AVIRIS.

In this paper, a network collaborative pruning method is proposed for HSI classification based on evolutionary multi-task optimization. The main contributions of this paper are as follows:

- A multi-task pruning algorithm: by exploiting the similarity between HSIs, different HSI classification networks can be pruned simultaneously. Through parallel optimization, the optimization efficiency of each task can be improved. The pruned networks can be applied to the classification of limited labeled sample HSIs.
- Model pruning based on evolutionary multi-objective optimization: the potential excellent sparse networks are searched by an evolutionary algorithm. Multi-objective optimization optimizes the sparsity and accuracy of the networks at the same time, and can obtain a set of sparse networks to meet different requirements.
- To ensure effective knowledge transfer, the network sparse structure is the transfer of knowledge, using knowledge transfer between multiple tasks to achieve the knowledge of the search and update. A self-adaptive knowledge transfer strategy based on the historical information of task and dormancy mechanism is proposed to effectively prevent negative transfer.

The rest of this paper is organized as follows. Section 2 reviews the background. The motivation of the proposed method is also introduced. Section 3 describes the model compression methods for HSI classification in detail. Section 4 presents the experimental study. Section 5 presents the conclusions of this paper.

2. Background and Motivation

2.1. HSI Classification Methods

Classification methods based on deep neural networks utilize its strong representation learning ability in the image field to automatically construct a representation structure that extracts spectral and spatial features and realize the classification of pixels. The HSI classification methods based on deep learning require data preprocessing and construction of the neural network structure before finally classifying the data [30], as shown in Figure 2. In recent years, the commonly used deep learning network models have included stacked autoencoder (SAE) [31], recurrent neural network (RNN) [32], convolutional neural network (CNN) [5], and graph convolutional network (GCN) [33–35]. Hamida [36] proposed a 3D-DL approach that enables joint spectral and spatial information processing. The 3D-DL method combines the traditional CNN network with the application of 3D convolution operations instead of using 1D convolution operators that only inspect the spectral content of the data. The deep CNN with a large parameter scale has stronger nonlinearity, which leads to high complexity and calculation of the neural network. If trained on limited labeled samples, a neural network is overparameterized with respect to the limited training samples, which causes the CNN to tend to overfit, so a large number of training samples was needed to improve the generalization ability of the model and alleviate overfitting in the case of limited samples.

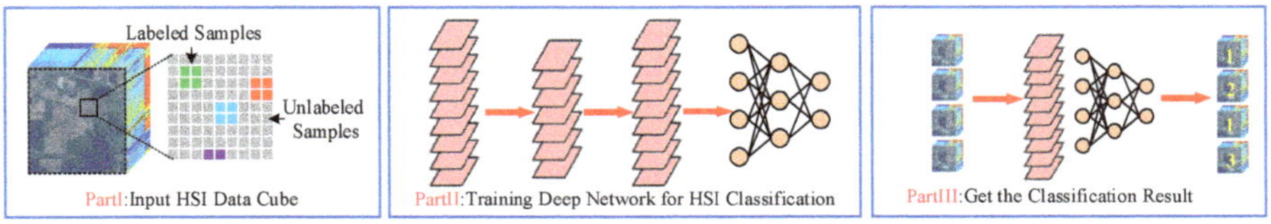

Figure 2. HSI classification based on neural networks.

A lightweight model can alleviate the requirement for the number of labeled samples. Simplification methods of the model are mainly divided into model compression and lightweight model design. Li et al. [37] proposed a compression network considering the high dimensionality of HSI. A fast and compact 3-D-CNN with few parameters was developed in [38]. Some efficient convolution operations have been explored to reduce the number of network parameters. Lightweight model design still requires prior knowledge to design the network structure. In the model compression method, this mainly includes network parameter quantization, neural network pruning, knowledge distillation, and tensor decomposition methods. Cao et al. [39] proposed a compressed neural network-based HSI classification method that uses a large teacher network to guide the training of a small student network, thereby achieving similar performance to the teacher network under the premise of low complexity. Compared with other model compression methods, neural network pruning is efficient and simple and has strong generalization. It can compress the network model and prevent the network from overfitting.

2.2. Neural Network Pruning

Neural network pruning is a classic technique in the model compression filed. As shown in Figure 3, network pruning requires a trained network, which is usually over-parameterized. For a network N of depth L, the overall parameters contained can be obtained by $W = \{w^1, \ldots, w^L\}$, where w^i denotes the parameter matrix of the i-th layer of the network.

Neural network pruning is usually achieved by pruning mask $M = \{m^1, \ldots, m^L\}$[40]. m_i represents the pruning mask of each layer of the network, which is usually represented by a binary matrix with the same dimension as w_i. Specifically, 0 means that the parameter is pruned and 1 means that the parameter is preserved. The pruned weight w^i_{prun} is obtained by performing a Hadamard product on m^i and w^i, and it can be expressed as $w^i_{prun} = w^i \odot m^i$. The process of neural network pruning is also shown in Figure 3.

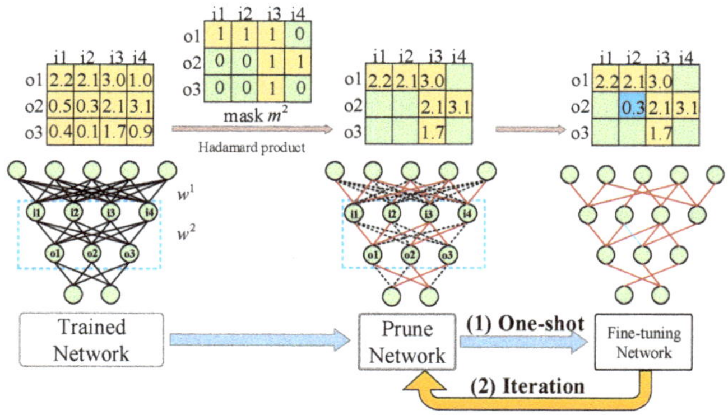

Figure 3. The procedure of neural network pruning.

Finally, the pruned network is fine-tuned. According to the pruning process, it can be divided into iterative pruning and one-shot pruning, the difference between the two pruning process is represented in Figure 3. Iterative pruning is a cyclical process of pruning and retraining, and many successful pruning methods [11,41,42] in the past have been based on iterative pruning. However, recent research [43,44] has suggested that such heavy consumption and the selection of design undermine their utility. One-shot pruning is trained after a one-time pruning process, and it can avoid the problem of iterative pruning.

2.3. Evolutionary Multi-Task Optimization

Evolutionary multi-task optimization (EMTO) [45–49] is an emerging paradigm in the field of evolutionary computation. By sharing searched knowledge in similar tasks, EMTO can improve the convergence characteristics and searching efficiency for each task [50]. As shown in Figure 4, EMTO randomly marks the individuals with different task cultures and maps them to the corresponding task space for evolving. Furthermore, the knowledge in each task is transferred by genetic material among individuals in a unified space. Furthermore, EMTO has been studied to solve similar tasks parallelly [51] and handling optimization problems efficiently by building module tasks [52–55]. In avoiding the possible negative transfer of knowledge, Gao et al. [56] reduced the divergence between subpopulations belonging to different tasks by aligning the distributions in the subspaces.

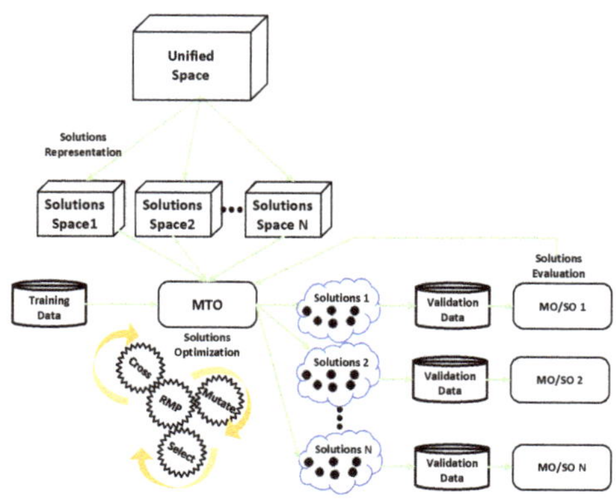

Figure 4. The overview of evolutionary multi-task optimization.

A minimization EMTO problem with K optimization tasks have a unified space Ω. The j task, denoted as T_j, is considered to have a search space Ω_j on which the objective function $F_j : \Omega_j \rightarrow \Omega$ implements a mapping from subsearch space Ω_j to uniform space Ω. In addition, each task may be constrained by several equality and/or inequality conditions that must be satisfied for a solution to be considered feasible. EMTO aims to optimize all tasks:

$$\text{minimize}\{F_1(\boldsymbol{x}_1), \cdots, F_t(\boldsymbol{x}_t), \cdots, F_k(\boldsymbol{x}_k)\} \tag{1}$$

In evolutionary multi-task optimization, each individual is assigned a skill factor indicating the cultural trait of the associated task [51]. Then, the individuals are encoded in a unified search space and the genetic operators are applied to produce offspring in this space. The offspring also inherit the parents' skill factors through the vertical cultural transmission.

2.4. Motivation

Deep neural networks achieve good classification results based on large-scale parameters. The complex nonlinear structure leads to complex calculation, which affects the application of neural network for HSI classification on mobile platforms. Therefore, it is necessary to compress the model of the existing large-scale network. Moreover, the training of neural networks relies on a large number of training samples. HSIs need to be manually labeled, so the labeled samples of HSIs are limited, which will lead to overfitting and classification difficulties during complex neural network training.

Traditional network pruning methods based on deep neural networks only deal with one image at a time, which has limited learning knowledge and does not make full use of the common features between similar images. The multi-task framework can be used to simultaneously prune the classification networks of multiple different images.Taking advantage of the potential similarities between optimization tasks, the multi-task framework can be used to simultaneously prune the classification networks of multiple different images. Using existing HSI with high similarity, when the same network architecture is trained on different datasets, its parameters characterize different datasets, so interaction between tasks can alleviate the limited sample problem on a single dataset and help the classification of the respective task. Although the existing evolutionary pruning methods can avoid the cost and prior knowledge requirements of designing pruning rules, they are difficult to optimize when facing more complex network structures. The proposed multi-task optimization framework, using knowledge transfer between tasks, can also effectively facilitate the respective optimization tasks.

3. Methodology

This section provides a comprehensive description of the proposed network collaborative pruning method for HSI classification. Firstly, the overall framework of the method is introduced. Secondly, compression of the model is achieved by an evolutionary multi-task pruning algorithm, the algorithm is introduced, and the initialization of individual and population, genetic operators, and self-adaptive knowledge transfer strategy are described in detail. Finally, the complexity of the proposed method is calculated.

3.1. The Framework of the Proposed Network Collaborative Pruning Method for HSI Classification

The overall framework of the proposed method is shown in Figure 5. First, different optimization tasks are constructed for two similar HSIs, i.e., there is a similarity between the two sparsification tasks. The evolutionary algorithm is used to search the potential excellent sparse network structure on the respective HSI. Genetic operators are designed according to the representation of the network structure. In the process of the parallel optimization of two tasks, interaction between tasks is needed to transfer the local sparse network structure. At the same time, in order to avoid the possible negative transfer, the self-adaptive knowledge transfer strategy is used to control the interaction strength between tasks. After completing the pruning search in different tasks and fine-tuning on the respective HSI, a set of sparse networks is obtained.

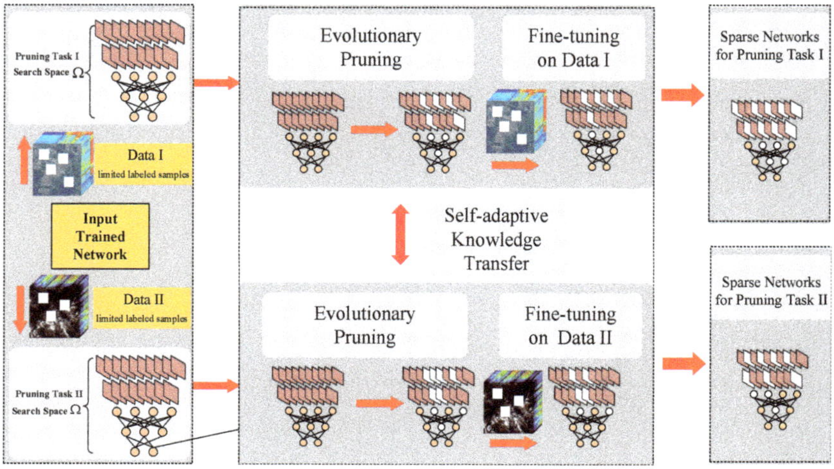

Figure 5. Overall framework of proposed network collaborative pruning method for HSI classification.

3.2. Evolutionary Multi-Task Pruning Algorithm

3.2.1. Mathematical Models of Multi-Tasks

In the evolutionary pruning algorithm, modeling is performed on different HSIs and the similarity between images is high. Therefore, the models of multi-tasks are given in (2).

$$\begin{cases} T_I = max(f_{acc}(W_{taskI}), f_{spar}(W_{taskI})) & W_{taskI} \in \Omega \\ T_{II} = max(f_{acc}(W_{taskII}), f_{spar}(W_{taskII})) & W_{taskII} \in \Omega \end{cases} \quad (2)$$

$$\begin{cases} f_{acc}(W_{prun}) = 1 - eval(D_{test}, W_{prun}) \\ f_{spar}(W, W_{prun}) = \frac{\left\| \sum_{i=1}^{L} w_{prun}^{i} \right\|_{l0}}{\left| \sum_{i=1}^{L} w^{i} \right|} \end{cases} \quad (3)$$

where T_I represents the classification and structure sparsification task on a certain HSI and the search space of T_I is Ω. Furthermore, the optimization of the task is achieved by searching the result pruned network weights W_{taskI}. Similarly, T_{II} represents the classification and structure sparsification task on a different HSI, the search space of T_{II} is also Ω, and the pruned network weights obtained by searching is W_{taskII}.

Each task is a multi-objective optimization model which can be expressed by (3). Generally speaking, in the search process, when the network sparsity is reduced, the accuracy of the network will reduce; sparsity and accuracy are two conflicting goals. One objective function f_{acc} represents the accuracy of the neural network on the test dataset D_{test}, and another objective function f_{spar} represents the sparsity of the network, which can be represented by the pruning rate of the network. Specifically, sparsity can be expressed as the ratio of the number of all elements that are not zero to the number of all elements.

3.2.2. Overall Framework of Proposed Evolutionary Multi-Task Pruning Algorithm

The evolutionary pruning algorithm is shown in Figure 6. One-dimensional vectors are designed for different tasks to represent different pruning schemes, which can also be regarded as a set of sparse networks. In these two optimization tasks, the stepwise optimization of the network structure within the task is achieved. Through the knowledge transfer between different tasks, the optimization efficiency of the two tasks is further improved. After the evolution is completed, a set of network pruning schemes that can balance accuracy and sparsity are obtained. The specific implementation of the evolutionary pruning algorithm based on multi-task parallel optimization is shown in Algorithm 1.

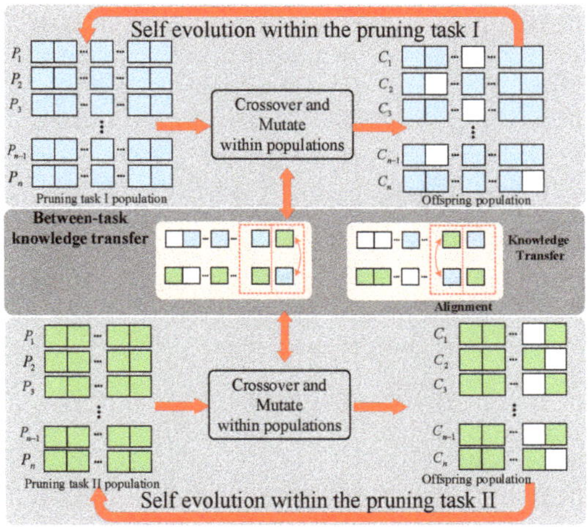

Figure 6. The proposed evolutionary multi-task pruning algorithm.

Algorithm 1 The proposed evolutionary multi-task pruning algorithm

Input: *pop*: task population size, *t*: number of evolutionary iterations, *P*: parent population, *rmp*: random mating probability, *gen*: maximum number of generation
Output: a set of trade-off sparse networks for multiple HSIs
Step (1) Train a state-of-the-art network N
Step (2) Construct task T_I and task T_{II} in Ω
Step (3) Pruning
  Set $t = 1$ then initialize the population P_t
  while $(t < gen)$ **do**
    $P_t \leftarrow$ Binary Tournament Selection (P_t)
    Generate offspring $C_t \rightarrow$ Refer Algorithm 2
    $R_t = C_t \cup P_t$
    Update scalar fitness in R_t
    Select *pop* fittest members from R_t to form P_{t+1} by NSGA-II
    Self-adaptively update *rmp* $\rightarrow$ Refer Algorithm 3
    $t = t + 1$
  end while
Step (4) Fine-tuning the optimized results in T_I and task T_{II}

3.2.3. Representation and Initialization

In this paper, we adopt a one-dimensional vector to represent a layer-by-layer differentiated pruning scheme, which can also represent a unique sparse network. This can more comprehensively reflect the sensitivity differences of different layers in the neural network, so as to achieve more refined and differentiated pruning. This encoding method can be well extended to a variety of networks, only needing to determine the depth of the network to achieve encoding and pruning. On the other hand, the use of one-dimensional vector encoding makes the design of genetic operators more convenient. Each element in the vector represents the weight pruning ratio of each layer of the network, which is the proportion of 0 elements in the w_i matrix. Thus, the encoding vector of layer i can be represented by the w^i as:

$$vector[i] = \frac{||w^i||_{l0}}{|w^i|} \tag{4}$$

Similar to (3), $||w^i||_{l0}$ represents the number of nonzero elements in the layer i, and $|w^i|$ represents the number of elements in this layer. In the pruning process, the weights are sorted from small to large according to the element value of the i-th bit of the one-dimensional vector, and the weight of the former $vector[i]\%$ is pruned. The upper and lowe bounds of $vector[i]$ are 0 and 1, respectively. In this way, the network weights are prune layer by layer, and the sparse network structure corresponding to the one-dimensiona vector can be finally obtained. The search process tries to approach the real Pareto-optima front. The decoding operation is the reverse process of the encoding operation.

Specifically, as shown in Figure 7, for a pruning scheme, its i-th element is a and its j-t element is b. Firstly, the weights of layers i and j are arranged from small to large. Suppos that pruning $a \times 100\%$ of the weights in the i-th convolution layer, the total parameter $|w^i$ of this layer is $k_w^i \times k_h^i \times f^i$, where k_h^i represents the height of the convolution kernel, k_w^i represents the width of the convolution kernel, f^i represents the number of convolutio filters in this layer. Suppose that pruning $b \times 100\%$ of the weights in the j-th fully connecte layer, the total parameter $|w^j|$ is the product of the input neurons n_{in}^j and output neuron n_{out}^j. After determining the pruned parameter,the corresponding bit is set to zero to indicat that the parameter is pruned.

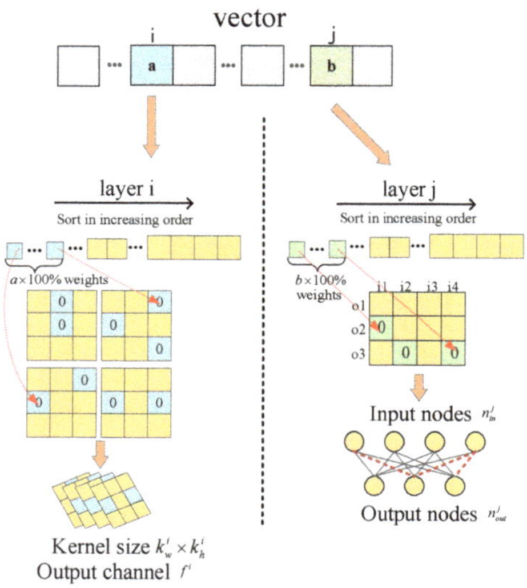

Figure 7. The representation of individual initialization.

According to the depth L of the network and the population size pop, pop one dimensional vectors of length L are randomly generated to form the initial population o task. This represents pop pruning schemes, which can also be regarded as pop differen sparse networks. The population is initialized in the same way for different tasks.

3.2.4. Genetic Operator

The genetic operators used in proposed algorithm include crossover and mutatio operators. It is necessary to judge the skill factor of the individual when two individual crossover. This is similar to MFEA [45]. If two randomly selected parent pruning scheme have the same skill factor, they come from the same task and crossover directly. Otherwise, i comes from different tasks, and rmp is needed to determine whether to carry out knowledg transfer between tasks. After completing the crossover operation, the individual perform the mutation operation. The generated offspring individuals inherit the skill factor of the parent individual. If within-task crossover is performed, the skill factor of the offspring is the same as that of the parents, otherwise, the offspring randomly inherits the skill facto of one parent. The details are shown in Algorithm 2.

Algorithm 2 Genetic operations

Input: p_1, p_2: candidate parent individuals, τ_i: the skill factor of the parent, rmp: random mating probability, $rand$: a random number between 0 and 1
Output: offspring individual c_1, c_2

if $\tau_1 == \tau_2$ **then** or $rand < rmp$
    $c_1, c_2 \leftarrow$ Crossover(p_1, p_2)
    for i $select\ from\{1,2\}$ **do**
        $c_i \leftarrow$ Mutate(p_i)
    end for
    if $\tau_1 == \tau_2$ **then**
        c_i inherits the skill factor from p_i
    else
        if $rand < 0.5$ **then**
            c_1, c_2 inherits τ_1 from p_1
        else
            c_1, c_2 inherits τ_2 from p_2
        end if
    end if
else
    for i $select\ from\{1,2\}$ **do**
        $c_i \leftarrow$ Mutate(p_i)
        c_i inherits the skill factor from p_i
    end for
end if

Both between-task and within-task crossover operators are designed in the same single-point crossover. The i-th value in *vector* of parents p_1 and p_2 are swapped to generate two new individuals c_1 and c_2. As shown in Figure 8, when individuals crossover at a certain bit, the bit on different individual vectors is swapped directly. Because pruning rate and sparse structure correspond one-to-one, it is also directly exchanged at the weight matrix of the network.

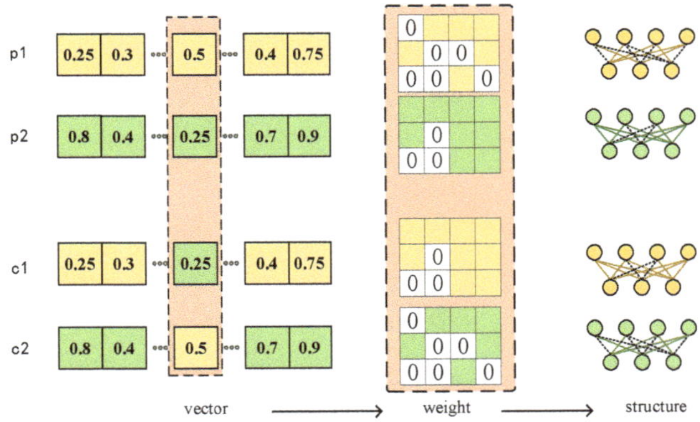

Figure 8. The illustration of crossover operator.

A polynomial-mutation [57] is designed when the crossover operation is complete. Figure 9 depicts the mechanism of the designed mutation operator. Taking individual $p1$ for example, the i-th value changes as preset mutate probability from 0 to 0.25, which can be calculated from the polynomial mutation in Figure 9. The change quantity β_i in layer i is related to the $u_i \in [0, 1)$ and the non-negative real number η_u. η_u is the distribution exponent. The larger this value is, the more similar the offspring and the parent are, so $\eta_u = 10$ is set as the mutation probability. There are four input neurons and three output neurons in this layer for a total of 12 weight parameters. During pruning, the weights are sorted, then select the weight from small to large for pruning, and the sparse structure obtained after mutation operation is unique. Therefore, a total of three bits in the matrix need to be changed.

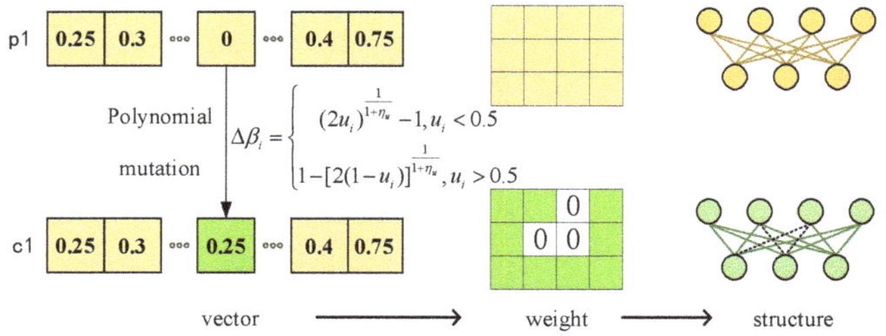

Figure 9. Illustration of mutation operator.

The crossover and mutation operators adopted in this paper not only realize the self-evolution within tasks but also transfer the effective sparse structure so as to promote the search efficiency of two tasks.

3.2.5. Self-adaptive Knowledge Transfer Strategy

Although there is a high similarity between the two tasks [58], negative transfer is still inevitable; this affects the search efficiency and solution quality. So, a self-adaptive knowledge transfer strategy based on historical information and a dormancy mechanism is designed. The intensity of transfer can be adjusted adaptively by taking advantage of individual contributions. The dormancy mechanism is used to suppress irrelevant knowledge transfer, reduce the interference of useless knowledge to task search, and save computing resources.

Algorithm 3 introduces the self-adaptive knowledge transfer strategy. New individuals generated by knowledge transfer between tasks are labeled as $\{p_{tki}|i = 1, 2, \ldots, n\}$. After the fitness evaluation of the generated offspring, the Pareto rank of the offspring individual in the non-dominated ranking is obtained. The knowledge transfer contribution $TKCR$ can then be represented by the rank of the individual with the best non-dominated rank result among these newly generated individuals. Then, $TKCR$ controls the value of rmp. Notice that when comparing the Pareto rank of the offspring, the task to which the offspring belongs is not distinguished.

Algorithm 3 Self-adaptive knowledge transfer strategy

Input: N_{p1}, N_{p2}: the population size in multi tasks, $rank_{min}$: minimum rank of non-dominated sort, p_{tki}: new individuals generated by knowledge transfer, ϵ: preset threshold
Output: random mating probability rmp
$rank_{min} \leftarrow \min_{rank}(p_{tk1}, p_{tk2}, \ldots, p_{tkm})$
$\delta \leftarrow rank_{min}/(N_{p1} + N_{p2})$
Transfer knowledge contribution $TKCR \leftarrow 1 - \delta$
if $TKCR > \epsilon$ **then**
    $rmp \leftarrow TKCR$
else
    $rmp \leftarrow 0.1$
end if

When the value of $TKCR$ is less than the set threshold ϵ of population interaction, the dormancy condition of the population is reached, and rmp is set to a small fixed value. When the value of $TKCR$ is greater than ϵ, the transfer of useful knowledge is detected at this time, the self-adaptive update is resumed, and then, the value of rmp is the value of $TKCR$. Through the self-adaptive strategy to control the frequency of knowledge transfer in the evolution process and the dormancy mechanism, the impact of negative transfer between tasks on task performance can be effectively avoided.

3.3. Fine-Tune Pruned Neural Networks

After pruning, a set of sparse networks is obtained. Then, they are retrained, as studied in [59]. In detail, these networks are trained with the Adam optimizer, and the initial learning rate, weight decay, and training epochs are set differently according to different data. The learning rate is adjusted by cosine annealing with the default setting.

3.4. Computational Complexity of Proposed Method

An analysis of the computational complexity of the proposed method is calculated in two parts: the computational cost of evolutionary computation and the computational cost of fine-tuning. In the pruning parts, the computational complexity is $O(GPC)$, where G is the number of generations, P is the number of individuals, and C is the cost of given function. Assuming the computational cost of training for each epoch is $O(T)$, the fine-tuning computational complexity is $O(ET)$, E denotes the number of training epochs. Therefore, the computational complexity of the proposed approach is $O(GPC + PTE)$. Because the proposed method is multi-task optimization and is able to handle two HSIs pruning tasks simultaneously, it is twice the computational complexity of a single evolution and fine-tuning process.

4. Experiments

In this part, experiments that are carried out on HSIs to verify the effectiveness of the proposed method are described. Firstly, it is verified that the pruned network has better classification accuracy with limited labeled samples on multiple HSIs. The proposed method is compared with other neural network pruning methods, and the relevant parameters of the pruned network are compared with other methods. After that, the sparse networks obtained on the Pareto-optimal front are compared to prove the effectiveness of the multi-objective optimization. The effectiveness of the proposed self-adaptive knowledge transfer strategy is proven by quantifying the knowledge transfer between tasks. Finally, the proposed method is validated on more complex networks and larger HSI.

4.1. Experimental Setting

A 3DCNN [36] trained on the HSI was used to validate proposed method. The structure of network is composed of convolutional layers of different stride. The convolutional layer with stride 1 is called Conv, and the convolutional layer with stride 2 is called ConvPool. Excluding the classification layer, the number in the network structure is the number of the filter of the convolutional layer, and the network structure can be expressed as: 3DConv(20)-1DConvPool(2)- 3DConv(35)- 1DConvPool(2)-3DConv(35)- 1DConvPool(2)-3DConv(35)-1DConvPool(35)-1DConv(35)-1DconvPool(35).

HSIs use Indian Pines, Salinas, and University of Pavia datasets. Data in the real world not only have the problem of limited labeled samples, but also the labeled samples often cannot reflect the real distribution of the data. For example, only part of the HSI in a certain area of the ground are sampled in the detection, and these data are continuous but may not be comprehensive. In order to simulate limited sample data, 10% labeled samples were set for each dataset, and the sample of the corresponding comparison methods was also 10%.

The Indian Pines (IP) dataset is collected by the sensor AVIRIS [60] from a pine forest test site in northwest India. Its wavelength range is 400–2500 nm. After removing the water absorption area, there are 200 spectral segments in total, and the spatial image size of each spectral segment is 145×145, with a total of 16 types of labels. The spatial resolution of this dataset is only 20 m. Figure 10 shows the pseudo-color plots and labels of Indian Pines.

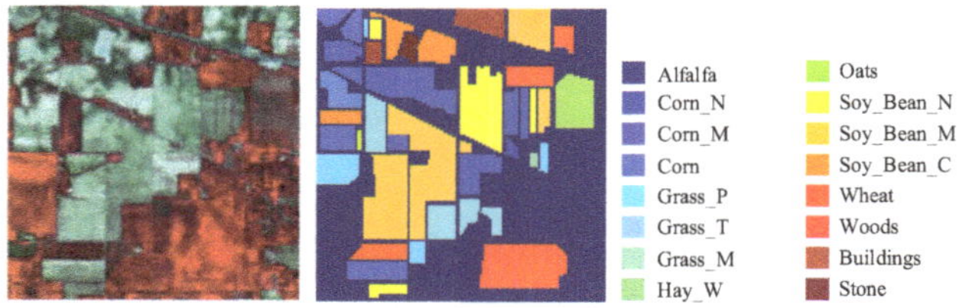

Figure 10. The false-color image and reference image on Indian Pines dataset.

The Salinas (SA) dataset is collected from the Salinas Valley in California by the sensor AVIRIS. After removing the water absorption area, there are a total of 200 spectral segments and the spatial image size of each spectral segment is 521 × 217, with a total of 16 labels. The spatial resolution of this dataset is 3.7 m. Figure 11 show the pseudo-color plots and labels of Salinas.

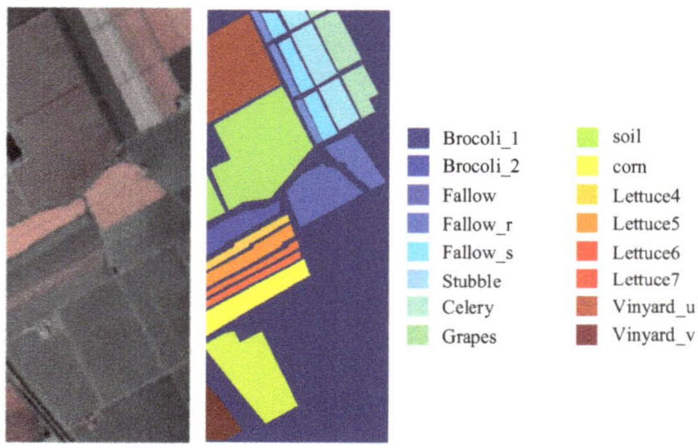

Figure 11. The false-color image and reference image on Salinas dataset.

The University of Pavia (PU) dataset is collected by the sensor ROSIS near the University of Pavia, Italy. After removing the water absorption area, there are a total of 103 spectral segments, and the spatial image size of each spectral segment is 610 × 340, with a total of nine categories of labels. The spatial resolution of this dataset is 1.3 m. Figure 12 shows the pseudo-color plot and labels of the University of Pavia.

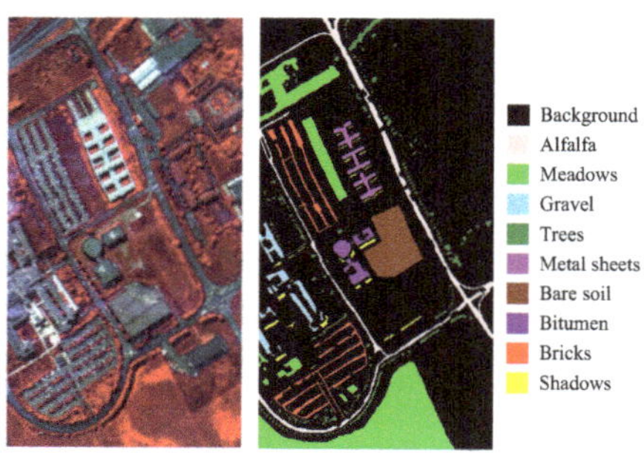

Figure 12. The false-color image and reference image on University of Pavia dataset.

The proposed method was compared with five deep learning methods, including 1DCNN [61], 3DCNN [62], M3DCNN [63], DCCN [64],HybridSN [65], ResNet [66], and DPRN [67]. In the experiment, three evaluation metrics—overall accuracy (OA), average accuracy (AA), and Kappa coefficient (κ)—were used to evaluate the classification effect of the proposed method. The parameters of our proposed method are shown in Table 1.

Table 1. Parameters used in proposed method.

	HSI Datasets
Offspring size in pruning task I	50
Offspring size in pruning task II	50
Maximum number of generation	50
Mutation probability	10
Crossover probability	10
The initial value of transfer	0.5
The dormancy condition	0.1

The experimental server included four Intel(R) Xeon(R) Silver 4214R cpus @ 2.40 GHz, 192 GB DDR4 RAM, Two NVIDIA Tesla K40 12 GB Gpus and eight NVIDIA Tesla v1000s Gpus were used. The software environment used the Ubuntu operating system with Pytorch framework and Python 3.6 as the programming language. The optimizer of the convolutional neural network was set to Adam optimizer, the weight decay was 0, betas = (0.9, 0.999), and eps = 1×10^{-8}. The initial learning rate was 1×10^{-4}, the learning rate decay was adopted by cosine annealing, the number of training epochs of the network was 200, and the batch size was 100.

4.2. Results on HSIs

4.2.1. Classification Results

In the experiment, two groups of experiments were constructed to analyze the influence on the performance of the proposed method. The first group uses the Indian Pines dataset and the Salinas dataset, and the second group uses the University of Pavia dataset and the Salinas dataset. The Indian Pines dataset and Salinas dataset are from the same sensor, and the University of Pavia dataset and Salinas dataset are from different sensors.

The classification result of the Indian Pines dataset is shown in Figure 13, and the specific classification result table is shown in Table 2. Although the pruned network do not obtain the best results on the three evaluation metrics, it obtain the highest classification accuracy on the seven categories, all of which are 100%. The network for Indian pines dataset is able to prune 91.2% of the parameters.

From the overall evaluation metrics, it can be seen that when the Indian Pines dataset from the same sensor is used as an another task, it obtains relatively better results, and pruning 87.2% of the network weights. By transferring the existing knowledge, the method successfully improves the classification accuracy of the network and greatly reduces the complexity of the network model. It is basically superior to other deep learning methods in the OA and AA. Although the number of samples in each category of data is not balanced, the knowledge transfer can improve the overall performance of the sparse network, so that the network still achieves a high κ, that is, the distribution of classification accuracy on each category is balanced.

The classification result of the University of Pavia dataset is shown in Figure 13, and the specific classification results are shown in Table 3. It can be seen that although 83.1% of the parameters are pruned, the pruned network still obtains high OA, AA, and κ values, which are 97.57%, 97.84%, and 96.79%, respectively. In addition to this, the best results are achieved in three categories. This proves that leveraging the knowledge transferred from other images can facilitate the training of the network on the current image.

Table 2. Classification accuracy (%) for the collaborative pruning task (Indian Pines and Salinas). Best results are reported in bold.

Category	1DCNN	3DDL	M3DCNN	DCCN	HybridSN	ResNet	DPRN	Pruned 87.15%
OA (%)	91.78 ± 1.45	92.05 ± 1.37	90.51 ± 0.98	95.66 ± 2.06	91.68 ± 1.71	93.68 ± 1.03	**97.14 ± 0.77**	95.70 ± 1.31
AA (%)	96.13 ± 2.33	95.50 ± 2.67	95.41 ± 2.56	98.05 ± 0.42	96.10 ± 2.11	97.46 ± 1.68	**98.59 ± 1.09**	98.14 ± 0.69
Kappa (%)	90.87 ± 2.06	91.13 ± 2.01	89.45 ± 2.79	95.17 ± 1.78	90.77 ± 2.21	92.99 ± 1.51	**96.10 ± 0.68**	95.14 ± 0.74
1	**99.95**	99.90	99.70	98.45	98.35	99.75	99.10	99.70
2	99.59	99.81	99.27	99.78	99.81	**100.00**	99.88	**100.00**
3	98.93	86.33	97.36	99.84	98.27	99.39	**100.00**	99.24
4	99.78	**99.92**	99.28	98.78	99.71	99.85	99.03	99.07
5	98.39	98.99	99.62	**100.00**	96.34	98.80	99.49	99.03
6	99.99	99.99	99.98	99.99	99.99	**100.00**	**100.00**	99.97
7	99.52	99.30	99.46	99.94	99.49	**99.97**	99.81	99.47
8	80.09	88.59	77.82	85.04	76.69	78.79	**93.17**	88.31
9	99.06	99.64	98.37	**99.91**	98.59	99.48	99.82	99.77
10	90.69	95.21	91.03	96.06	93.47	97.98	**98.42**	98.26
11	99.06	99.90	99.25	99.06	98.68	99.06	**100.00**	99.34
12	99.01	97.76	99.01	**100.00**	98.96	99.89	99.71	99.95
13	99.34	99.45	99.23	99.01	98.79	**100.00**	**100.00**	**100.00**
14	99.06	97.38	97.75	99.34	98.69	98.31	99.55	**100.00**
15	77.06	66.82	72.45	**94.04**	81.89	88.44	89.32	88.37
16	98.61	99.00	96.90	99.50	99.88	99.61	99.74	**99.89**
Category	**1DCNN**	**3DDL**	**M3DCNN**	**DCNN**	**HybridSN**	**ResNet**	**DPRN**	**Pruned 91.27%**
OA (%)	80.93 ± 4.37	91.45 ± 3.62	95.18 ± 3.74	92.17 ± 3.79	95.38 ± 2.91	93.24 ± 2.86	**97.46 ± 1.50**	88.90 ± 1.27
AA (%)	90.15 ± 3.77	96.84 ± 2.81	98.07 ± 1.72	93.45 ± 2.26	**98.12 ± 0.58**	97.89 ± 1.43	98.05 ± 0.49	95.38 ± 0.81
Kappa (%)	78.38 ± 4.69	90.33 ± 2.84	94.52 ± 2.94	91.11 ± 3.19	94.75 ± 1.67	93.11 ± 2.48	**95.97 ± 1.39**	87.46 ± 0.93
1	**100.00**	**100.00**	**100.00**	**100.00**	**100.00**	**100.00**	**100.00**	**100.00**
2	65.68	96.42	93.20	82.56	89.28	91.75	**96.68**	76.96
3	73.97	96.14	97.34	91.20	97.22	96.26	**98.23**	95.66
4	**100.00**	**100.00**	**100.00**	95.78	**100.00**	**100.00**	**100.00**	**100.00**
5	96.48	**100.00**	**100.00**	96.27	99.17	**100.00**	**100.00**	99.37
6	99.31	99.86	99.45	98.63	99.86	**100.00**	**100.00**	98.35
7	**100.00**	**100.00**	**100.00**	92.86	**100.00**	**100.00**	**100.00**	**100.00**
8	**100.00**	**100.00**	**100.00**	**100.00**	**100.00**	**100.00**	**100.00**	**100.00**
9	**100.00**	**100.00**	**100.00**	90.00	**100.00**	**100.00**	**100.00**	**100.00**
10	64.19	88.58	93.10	91.67	**100.00**	93.18	97.48	91.15
11	72.66	73.28	88.55	91.57	**97.42**	81.92	93.74	76.65
12	75.71	97.97	98.65	86.34	90.17	97.95	**99.03**	92.91
13	**100.00**	**100.00**	**100.00**	**100.00**	98.14	**100.00**	**100.00**	**100.00**
14	95.25	99.84	98.81	97.00	**100.00**	98.37	99.28	95.81
15	99.22	97.40	**100.00**	95.34	98.89	**100.00**	**100.00**	99.22
16	**100.00**	**100.00**	**100.00**	86.02	99.74	**100.00**	**100.00**	**100.00**

Sub-optimal results were obtained on the different sensor University of Pavia dataset, which still has certain advantages compared with other deep learning methods. Using the University of Pavia dataset as another task, 84.3% of network parameters were pruned. Compared with the results on the Indian Pines dataset, the number of retained parameters is greater, and the classification performance and consistency are lower.

These two groups of experiments show that the search efficiency of task can be promoted by transferring the important sparse structure of the SOTA network from the another task. In view of the differences between the two groups of experiments due to the same physical imaging logic under the same sensor device the similarity between the datasets is higher, and the spectral features are more common, so the better results can be achieved. Due to the lack of labeled training samples and the high complexity of the network model, the parameters are too large, so the evaluation metrics of the unpruned neural network is low, which reflects the limitation of the lack of labeled samples on the network training.

Table 3. Classification accuracy (%) for the collaborative pruning task (University of Pavia and Salinas). Best result are reported in bold.

Category	1DCNN	3DDL	M3DCNN	DCCN	HybridSN	ResNet	DPRN	Pruned 84.30%
OA (%)	91.78 ± 1.45	92.05 ± 1.37	90.51 ± 0.98	95.66 ± 2.06	91.68 ± 1.71	93.68 ± 1.03	**97.14 ± 0.77**	95.02 ± 0.98
AA (%)	96.13 ± 2.33	95.50 ± 2.67	95.41 ± 2.56	98.05 ± 0.42	96.10 ± 2.11	97.46 ± 1.68	**98.59 ± 1.09**	98.03 ± 0.30
Kappa (%)	90.87 ± 2.06	91.13 ± 2.01	89.45 ± 2.79	95.17 ± 1.78	90.77 ± 2.21	92.99 ± 1.51	**96.10 ± 0.68**	94.03 ± 1.12
1	**99.95**	99.90	99.70	98.45	98.35	99.75	99.10	99.60
2	99.59	99.81	99.27	99.78	99.81	**100.00**	99.88	99.97
3	98.93	86.33	97.36	99.84	98.27	99.39	**100.00**	99.60
4	99.78	**99.92**	99.28	98.78	99.71	99.85	99.03	99.28
5	98.39	98.99	99.62	**100.00**	96.34	98.80	99.49	99.44
6	99.99	99.99	99.98	99.99	99.99	**100.00**	**100.00**	99.97
7	99.52	99.30	99.46	99.94	99.49	**99.97**	99.81	99.66
8	80.09	88.59	77.82	85.04	76.69	78.79	**93.17**	84.86
9	99.06	99.64	98.37	99.91	98.59	99.48	99.82	**99.97**
10	90.69	95.21	91.03	96.06	93.47	97.98	**98.42**	98.14
11	99.06	99.90	99.25	99.06	98.68	99.06	**100.00**	**100.00**
12	99.01	97.76	99.01	**100.00**	98.96	99.89	99.71	99.95
13	99.34	99.45	99.23	99.01	98.79	**100.00**	**100.00**	**100.00**
14	99.06	97.38	97.75	99.34	98.69	98.31	99.55	**99.91**
15	77.06	66.82	72.45	**94.04**	81.89	88.44	89.32	88.06
16	98.61	99.00	96.90	99.50	99.88	99.61	99.74	**100.00**
Category	**1DCNN**	**3DDL**	**M3DCNN**	**DCCN**	**HybridSN**	**ResNet**	**DPRN**	**Pruned 83.14%**
OA (%)	88.32 ± 3.76	81.67 ± 3.17	94.36 ± 1.43	97.43 ± 1.12	93.47 ± 1.69	97.72 ± 1.19	**98.48 ± 0.86**	97.57 ± 1.40
AA (%)	91.29 ± 2.86	85.11 ± 3.84	94.87 ± 2.77	96.12 ± 2.01	94.81 ± 2.17	97.14 ± 1.28	**98.36 ± 0.92**	97.84 ± 0.95
Kappa (%)	84.85 ± 3.21	76.24 ± 3.65	92.59 ± 1.79	96.60 ± 2.24	91.46 ± 2.60	96.91 ± 1.28	**97.19 ± 1.06**	96.79 ± 0.69
1	83.47	69.91	85.03	95.53	86.98	92.35	94.12	**95.58**
2	87.08	82.99	96.24	**99.52**	93.71	98.92	99.48	98.14
3	88.42	74.08	89.09	88.61	88.58	95.41	96.86	**96.95**
4	96.57	94.48	96.34	96.01	96.96	96.99	**97.94**	97.74
5	99.99	99.95	99.99	**100.00**	99.99	**100.00**	**100.00**	99.77
6	91.05	72.51	98.03	98.01	97.43	**99.97**	99.63	98.60
7	91.42	83.75	95.78	97.66	96.76	98.84	99.60	**99.92**
8	84.46	90.82	94.16	95.54	93.18	**97.48**	94.41	95.05
9	99.15	97.57	99.15	94.19	99.47	**99.62**	99.52	98.83

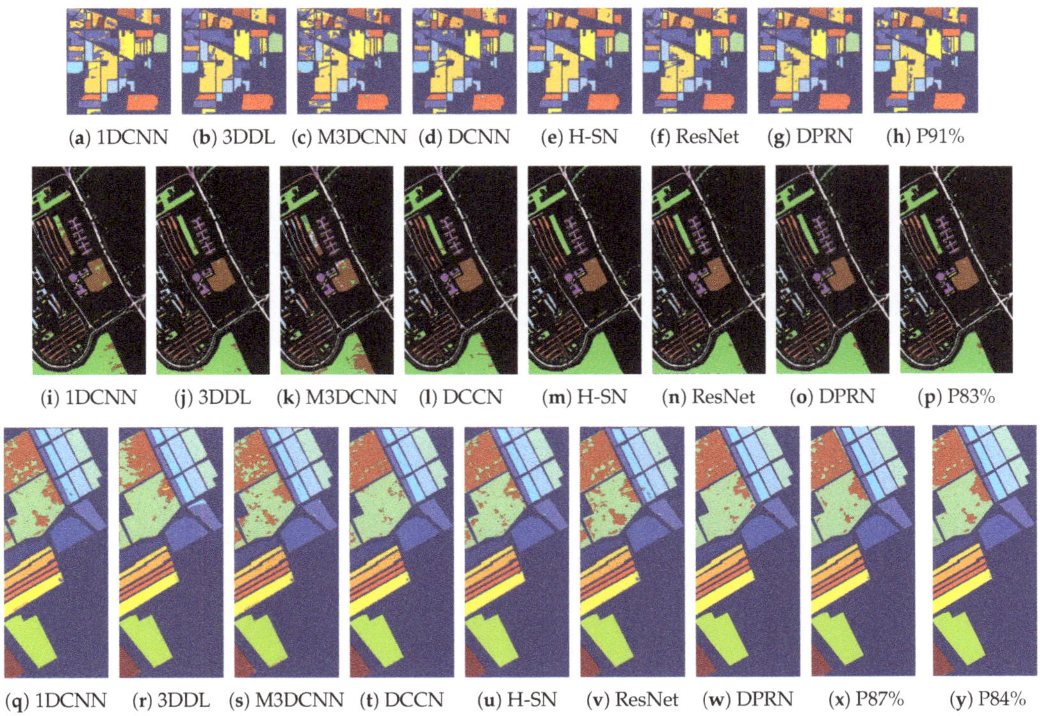

Figure 13. Classification maps on Indian Pines, Salinas and University of Pavia. Where P represents Pruned Network.

4.2.2. Comparison with other Neural Network Pruning Methods

The proposed method was compared to three neural network pruning methods in Table 4. NCPM is the network collaborative pruning method proposed in this paper. Because NCPM is a multi-objective optimization method, it selects a sparse network on the Pareto-optimal front.

The first pruning method L2Norm [68] is based on L2 norm, which sets a threshold for pruning for each layer by comparing the weight value of network parameters in each layer. In addition, NCPM is compared with MOPSO [21], a method based on particle swarm optimization. LAMP [12] is an iterative pruning method. LAMP utilizes a layer-adaptive global pruning importance score for pruning.

The three comparison methods and the proposed method all use the 3D-DL network. The original three pruning methods are all proposed based on 2DCNN and are suitable for image classification datasets, such as MNIST and CIFAR10. Therefore, the original pruning method needs to be changed to the pruning of 3DCNN. When training the network model, the same experimental settings such as the optimizer and learning rate are used as in NCPM.

Table 4. Classification results of the networks obtained by different pruning methods on the three HSIs. Best result are reported in bold.

HSI	Method	L2Norm	MOPSO	LAMP	NCPM
Salinas	Pruned (%)	87.00	85.24	87.00	**87.15**
	OA (%)	86.66	90.40	94.28	**95.02**
	AA (%)	91.48	94.65	97.68	**98.03**
	Kappa (%)	85.24	89.31	93.64	**94.03**
Indian Pines	Pruned (%)	91.00	90.23	**91.00**	**91.27**
	OA (%)	66.49	72.68	**89.31**	**88.90**
	AA (%)	81.44	84.61	**94.90**	**95.38**
	Kappa (%)	62.52	69.23	**87.90**	**87.46**
University of Pavia	Pruned (%)	83.00	84.11	83.00	**83.14**
	OA (%)	87.03	90.67	96.86	**97.57**
	AA (%)	87.4	87.70	97.54	**97.84**
	Kappa (%)	83.1	87.51	95.87	**96.79**

NCPM obtains the best pruning results on Salinas and University of Pavia, and the OA of the pruned network is much better than that of L2Norm and MOPSO with the same pruning rate. The pruned network on Indian Pines is highly similar to the LAMP method, but both are better than L2Norm and MOPSO.

From the three HSIs, it can be clearly seen that the sparse network searched by the L2Norm is sub-optimal due to the single redundancy evaluation criterion, and the evolutionary pruning method can search a better sparse network structure. Due to the lack of diversity in selecting solutions, the sparse network searched by MOPSO is inferior to the NCPM method. The LAMP method is an iterative pruning method, and it will be retrained in an iteration process, which will cause additional computational complexity.

Compared with other pruning methods, NCPM can simultaneously prune two hyperspectral data classification networks, which improves the search efficiency. At the same time, the multi-objective optimization of the sparsity and accuracy of the network structure can obtain a set of sparse networks after one run.

4.2.3. Complexity Results of the Pruned Network

Table 5 shows the comparison results between the pruned network and the original network, as well as other neural networks, where the training time refers to a training time of 200 epochs. Our method is able to prune the 3D-DL network, and when compared with the original network, 3D-DL, the pruned network can cut most of the parameters and can also accelerate the test time of the network in a certain range. On the Univeristy of Pavia dataset, the training time was reduced by 18.23%, on the Salinas dataset, the training time was reduced by 4.18%, and on the Indian Pines, the time was almost unchanged

The pruned network achieves the best results when compared to other methods the Indian Pines and University of Pavia datasets. The comparison experiment proves the significance and necessity of neural network pruning.

Table 5. Comparison results of the complexity of the pruned network.

HSIs	Methods	1DCNN	M3DCNN	HybridSN	ResNet	3DDL	Pruned
Indian pines	EpochTrainTime/s	40.5771	49.8241	73.0636	67.3195	60.1175	60.4814
	Parameter	246,409	263,584	534,656	414,333	259,864	22,868
	OA (%)	80.93	95.18	95.38	93.24	91.45	88.90
Pavia University	EpochTrainTime/s	40.3595	43.3423	79.7415	56.5454	41.6149	34.0278
	Parameter	246,409	263,584	534,656	534,656	259,864	43,918
	OA (%)	88.32	94.36	93.47	97.50	81.67	95.02
Salinas	EpochTrainTime/s	64.9641	85.6064	173.6447	134.5664	68.5989	65.7300
	Parameter	246,409	263,584	534,656	534,656	259,864	33,262
	OA (%)	91.78	90.51	91.68	93.68	92.05	95.70

4.2.4. The Result of the Sparse Networks Obtained by Multi-Objective Optimization

Figure 14 represents the Pareto-optimal front without fine-tuning in both two experiments. The Pareto-optimal front obtained for the Indian Pines dataset is uniformly distributed, whereas the Pareto-optimal front obtained for the University of Pavia is sparsely distributed. For the comparison of the Salinas dataset Pareto-optimal front in different experiments, the diversity of solutions is better in the multi-task optimization experiment of the Indian Pines dataset with the same sensor.

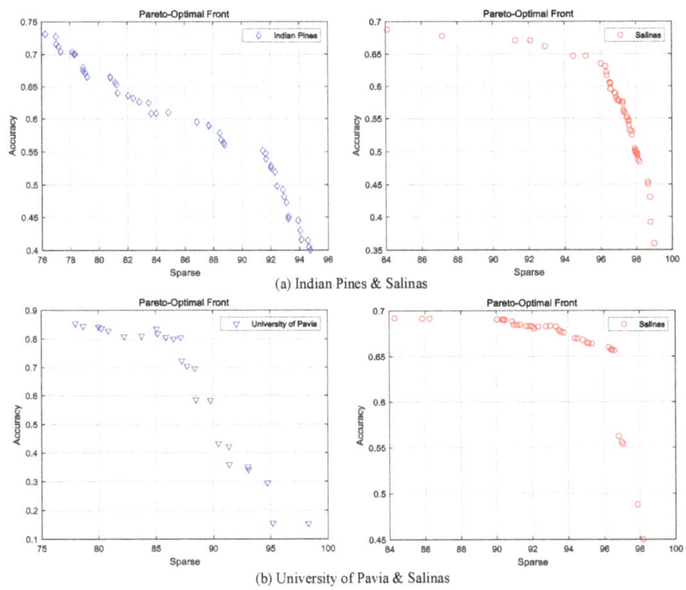

Figure 14. The Pareto-optimal front without fine-tuning after completing evolutionary search on two groups experiments.

The hypervolume curve Figure 15 is used to represent the convergence of the evolutionary search process. The hypervolume of each generation is determined by the sparse network on the Pareto-optimal front, and the diversity and quality of the sparse network affect the hypervolume. The initialization of the two experiments is random, so the initial *hv* is different. By comparing the results on the Salinas dataset in different experiments, it can be seen that the Salinas hypervolume curve optimized by the Indian Pines multi-task optimization converges faster and improves more, which again verifies the influence of the similarity between tasks on the results of multi-task optimization. In addition, the growth trend of the *hvscore* is the same in the two sets of experiments, and the period of faster growth of *hvscore* coincides, which can be understood as the promotion effect of knowledge transfer between the two tasks for their respective tasks.

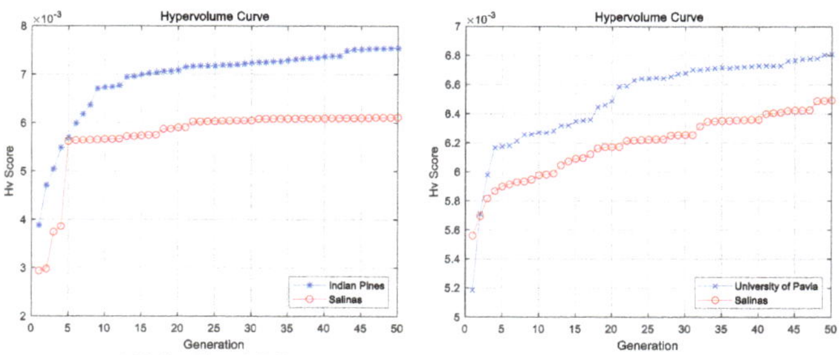

(a) Indian Pines & Salinas (b) University of Pavia & Salinas

Figure 15. Hypervolume curves of the evolutionary process on two groups experiments.

Four networks on the Indian Pines dataset were selected for comparison with the original unpruned network in Table 6. We can see that although about 80–90% of the parameters were pruned, after fine-tuning, the total accuracy was about 3% different from the original network. In some categories, such as classes 1, 4, and 7, the classification accuracy can be basically guaranteed to be 100%. Through multi-objective optimization, a set of sparse network structures can be obtained after one run, which have different sparsity and accuracy, and are suitable for different application conditions and application scenarios.

Table 6. Results after fine-tuning sparse networks on the Pareto-optimal front on collaborative pruning task (Indian Pines and Salinas). Best results are reported in bold.

Category	ORG	Pruned Networks in Salinas					Category	ORG	Pruned Networks in Indian Pines			
Pruned (%)	0.00	84.09	87.15	92.93	96.49	97.21	Pruned (%)	0.00	83.66	84.00	84.86	91.27
OA (%)	92.05	95.25	**95.70**	95.42	95.51	95.42	OA (%)	**91.45**	89.75	89.87	89.64	88.90
KAPPA (%)	91.13	94.72	**95.22**	94.90	95.01	94.90	KAPPA (%)	**90.33**	88.39	88.52	88.29	87.46
AA (%)	95.50	97.97	**98.14**	97.96	97.98	97.83	AA (%)	**96.84**	95.02	94.85	95.02	95.38
1	99.90	99.55	99.70	**100.00**	99.65	99.65	1	**100.00**	**100.00**	**100.00**	**100.00**	**100.00**
2	99.81	99.75	**100.00**	**100.00**	**100.00**	99.86	2	**96.42**	88.16	79.20	83.89	76.96
3	86.33	99.24	99.24	98.83	99.03	98.07	3	**96.14**	91.20	95.54	89.75	95.66
4	**99.92**	99.56	99.06	98.42	99.42	98.63	4	**100.00**	**100.00**	97.46	97.46	**100.00**
5	98.99	98.84	99.02	98.73	99.25	98.31	5	**100.00**	96.48	94.61	93.78	99.37
6	99.99	99.94	99.97	99.97	99.97	**100.00**	6	**99.86**	98.63	96.71	98.08	98.35
7	99.30	98.60	99.46	99.66	99.86	99.46	7	**100.00**	**100.00**	**100.00**	**100.00**	**100.00**
8	88.59	86.65	88.30	87.96	**89.73**	87.75	8	**100.00**	**100.00**	**100.00**	**100.00**	**100.00**
9	99.64	99.59	99.77	99.48	99.96	99.14	9	**100.00**	**100.00**	**100.00**	**100.00**	**100.00**
10	95.21	**98.35**	98.26	97.13	97.22	97.31	10	88.58	83.12	86.41	**93.10**	91.15
11	99.90	99.90	99.34	100.00	99.90	99.81	11	73.28	80.61	**85.41**	78.28	76.65
12	97.76	100.00	99.94	99.89	99.89	99.74	12	**97.97**	87.52	91.23	88.36	92.91
13	99.45	**100.00**	**100.00**	**100.00**	99.78	99.89	13	**100.00**	**100.00**	**100.00**	**100.00**	**100.00**
14	97.38	99.43	**100.00**	99.81	99.53	98.87	14	**99.84**	94.70	91.85	97.94	95.81
15	66.82	88.23	88.37	87.65	84.86	**88.96**	15	97.40	**100.00**	99.22	99.74	99.22
16	99.00	99.94	99.88	99.88	99.66	99.88	16	**100.00**	**100.00**	**100.00**	**100.00**	**100.00**

Four networks on the University of Pavia dataset were selected for comparison with the original unpruned network in Table 7. Compared with the original network, the OA of the pruned network was improved, and the OA reached 97.58% when the pruning rate was 92.93%. With the improvement of pruning rate, the obtained sparse network can still maintain the optimal classification accuracy on many categories.

Five of the sparse networks on the Salinas dataset obtained from each of the two experiments were selected for comparison with the original unpruned networks in Tables 6 and 7. Implementing multi-task pruning with the Indian Pines dataset pruned 87.15% of the networks, and obtained the best results. Each class in the original network did not reach 100%, but the network after pruning can be completely classified correctly in multiple classes, which indicates that the training of the network is limited in the case of limited samples, and the problem of limited samples can be alleviated after knowledge transfer between tasks. Different sparse networks obtain the best classification accuracy on different categories, which provides a choice for different classification requirements.

Table 7. Results after fine-tuning of sparse networks on the Pareto-optimal front on collaborative pruning task (University of Pavia and Salinas). Best results are reported in bold.

Category	ORG	Pruned Networks in Salinas					Category	ORG	Pruned Networks in University of Pavia			
Pruned (%)	0.00	84.30	90.33	93.33	95.97	98.18	Pruned (%)	0.00	83.14	89.98	92.08	92.93
OA (%)	92.05	95.02	95.26	95.46	94.26	93.83	OA (%)	91.67	97.57	97.55	97.18	**97.58**
KAPPA (%)	91.13	94.46	94.73	94.95	93.61	93.13	KAPPA (%)	76.24	96.79	96.76	96.28	**96.80**
AA (%)	95.50	98.02	98.08	98.07	97.17	96.95	AA (%)	85.11	**97.84**	97.77	97.48	97.66
1	99.90	99.60	99.95	99.90	99.95	99.80	1	69.91	**95.58**	95.11	94.78	94.85
2	99.81	99.97	**100.00**	99.91	**100.00**	99.43	2	82.99	98.14	98.25	97.73	**98.72**
3	86.33	99.59	**99.74**	99.24	95.95	95.34	3	74.08	96.95	**97.33**	94.94	95.66
4	**99.92**	99.28	99.06	99.28	98.70	98.85	4	94.48	97.74	96.86	96.96	**98.95**
5	98.99	99.43	99.47	**99.66**	97.90	97.34	5	99.95	99.77	**100.00**	99.62	**100.00**
6	99.99	99.97	99.97	99.97	**100.00**	**100.00**	6	72.51	98.60	**99.18**	98.52	97.81
7	99.30	99.66	**100.00**	99.91	99.63	98.99	7	83.75	**99.92**	99.24	99.17	99.62
8	88.59	84.86	87.08	87.88	87.68	86.40	8	90.82	95.05	94.94	**96.19**	94.32
9	99.64	**99.96**	99.51	99.06	98.98	99.16	9	97.57	98.83	99.04	**99.36**	99.04
10	95.21	98.13	98.23	97.62	95.72	95.85						
11	99.90	**100.00**	**100.00**	**100.00**	99.53	99.62						
12	97.76	99.94	**100.00**	**100.00**	**100.00**	99.94						
13	99.45	**100.00**	99.89	99.89	99.89	**100.00**						
14	97.38	99.90	**100.00**	99.25	99.53	99.71						
15	66.82	88.05	86.46	87.60	81.70	81.24						
16	99.00	**100.00**	99.94	**100.00**	99.66	99.55						

The proposed method uses the evolutionary multi-objective optimization model to realize the simultaneous optimization of network performance and network complexity, and automatically obtains multiple sparse networks. Some points on the Pareto-optimal front are selected for comparison, the classification results of the pruned network obtained on the Pareto-optimal front on different HSIs are shown in Figure 16. With the increase in the sparsity, the OA and AA of the network gradually decrease, but they are better than the neural network method directly trained on limited labeled sample data. In general, the proposed method can obtain a set of non-dominated sparse network solution at the same time, and the quality of sparse network is high, which can provide reference for practical datasets without labeled, and the method can be applied to different datasets.

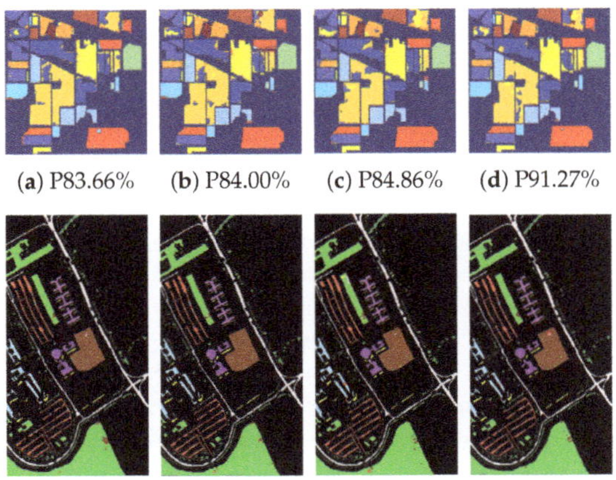

(**a**) P83.66% (**b**) P84.00% (**c**) P84.86% (**d**) P91.27%

(**e**) P83.14% (**f**) P89.98% (**g**) P92.08% (**h**) P92.93%

Figure 16. *Cont.*

(**i**) P84.09% (**j**) P87.15% (**k**) P92.93% (**l**) P96.49% (**m**) P97.21% (**n**) P84.30% (**o**) P90.33% (**p**) P93.33% (**q**) P95.97% (**r**) P98.18%

Figure 16. Classification maps on Indian Pines, Salinas, and University of Pavia datasets. GT represents ground truth and P represents pruned network.

4.2.5. Effectiveness Analysis of Self-Adaptive Knowledge Transfer strategy

For the quality of knowledge transfer between tasks, three metrics are given:

- Proportion of migrated individuals: After the elite retention operation of NSGA-II, the proportion of individuals who survived through knowledge transfer in the new population is calculated in the whole population, and the overall quality of the transfer is evaluated. The higher the ratio is, the better the quality of knowledge transfer is, which can greatly promote the population optimization.
- Transfer knowledge contribution degree: the minimum non-dominated rank of all transfer individuals after non-dominated sorting of the main task. The smaller the rank is, the more excellent the transfer individual is in the population, which indicates the greater contribution of the population optimization.
- Self-adaptive knowledge transfer probability (rmp): the variable used to control the degree of knowledge transfer in the self-adaptive transfer strategy. A larger value of rmp represents a stronger degree of interaction.

As shown in Figure 17, there are more individuals with transfer knowledge in the early stage of evolution, with the proportion distribution ranging from 50% to 10%. Although the rmp curve shows that the strength of knowledge transfer is almost the same, which indicates that the knowledge transfer in the early evolution can greatly help the search, but with the continuous optimization and convergence of the population, the effect of knowledge transfer is declining. Because of the contribution degree of transfer knowledge—although fewer individuals survive through knowledge transfer—part of the knowledge is still of high quality, which is still very effective for promoting the optimization of tasks.

Because the search of the task has not converged in the early stage, knowledge can provide a general network structure to guide the search. However, with the continuous optimization of the task, it is necessary to transfer very high-quality knowledge to promote search. At this time, although the knowledge transfer is heavy, only the part of individuals containing high quality can survive. Therefore, the self-adaptive knowledge transfer strategy based on the historical information is necessary.

During the evolution of the University of Pavia dataset as another task, as shown in Figure 17, a long dormancy mechanism is triggered, which indicates that the self-adaptive transfer strategy during this period considers the knowledge as invalid and intrusive. This may be due to the fact that there are differences between the datasets collected by different detection devices and there are few spectral features in common. Therefore, it is more useful to build multi-task optimization with datasets collected by the same sensor.

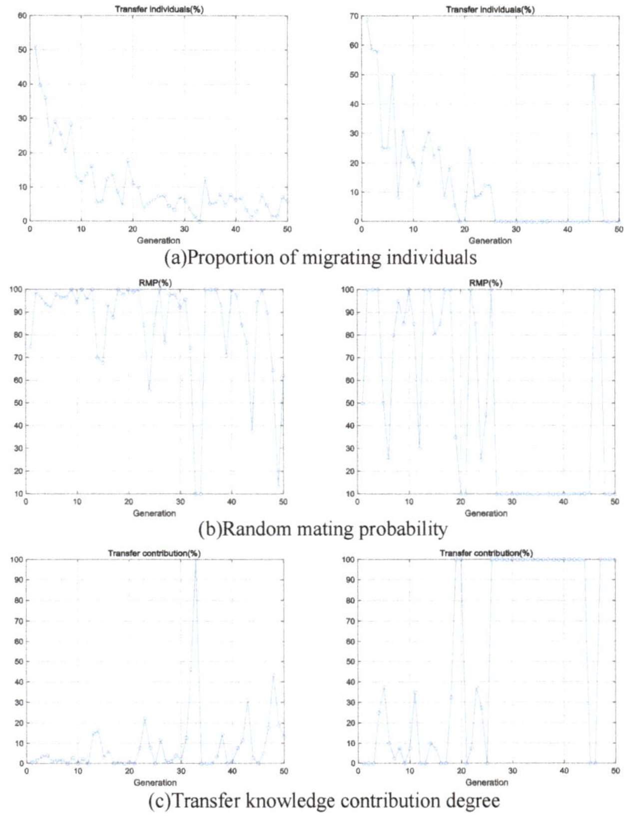

Figure 17. Knowledge transfer between tasks: the left column uses the Indian Pines dataset as another task of collaborative pruning, and the right column uses the University of Pavia datatset as another task of collaborative pruning.

4.2.6. Discussion

In this part, the proposed method is validated on more complex networks and larger HSI dataset. The proposed method is used to prune the complex network CMR-CNN [69] for HSI classification, the number of parameters is 28,779,784. A new cross-mixing residual network denoted by CMR-CNN is developed, wherein one three-dimensional (3D) residual structure responsible for extracting the spectral characteristics, one two-dimensional (2D) residual structure responsible for extracting the spatial characteristics, and one assisted feature extraction (AFE) structure responsible for linking the first two structures are designed.

Table 8 shows the pruning results of CMR-CNN on different HSIs. For this network, there is almost no decrease in the OA of the network after pruning nearly 75% of the parameters, and the OA of the network on Indian Pines is improved by 0.46%, which proves that our method can be applied to complex networks and can alleviate the overfitting problem of training on complex networks. Compared with the original network, the pruned network can cut most of the parameters, and can also accelerate the test time of the network in a certain range. On the University of Pavia dataset, the training time is reduced by 9.58% and on the Salinas dataset, the training time is reduced by 14.8%. The above comparison experiment proves the significance and necessity of neural network pruning.

Table 8. Pruning results of CMR-CNN.

HSIs	Salinas		Indian Pines		University of Pavia	
Method	CMR-CNN	NCPM	CMR-CNN	NCPM	CMR-CNN	NCPM
Pruned (%)	0.00	73.44	0.00	76.85	0.00	75.2
TrainTime (s)	9283	7909	2088	2058	7832	7082
Parameter	28,779,784	7,643,640	28,779,784	6,662,135	28,779,784	7,137,390
OA (%)	99.97	99.97	98.69	99.15	99.65	99.63
AA (%)	99.94	99.93	98.6	98.52	99.32	99.05
Kappa (%)	99.97	99.97	98.51	99.03	99.54	99.5

In addition, AlexNet [6] and VGG-16 [7] are pruned on image classification dataset CIFAR10, The Naive-Cut [70] method is a manual pruning method that uses the weight size as the redundancy.

The comparison results after fine-tuning are shown in Table 9. As the complexity of the network and the number of parameters increase, the gap between the proposed method and other neural network pruning methods becomes larger. Compared with the traditional single-objective pruning methods Naive-Cut and L2-pruning, the proposed method can obtain a set of networks with different sparsity and accuracy values in one run. At close accuracy, the solution obtains more sparse results. This is because the proposed evolution-based method has strong local search capability and is able to obtain sparse network structures in the search space. Due to the higher search efficiency and better diversity maintenance strategy, the proposed method can better ensure the population diversity in the evolution process than MOPSO.

Table 9. Pruning results of AlexNet and VGG-16.Best results are reported in bold.

Models	Methods	Accuracy	Parameter	Pruned (%)	CR
AlexNet	Naive-Cut	80.33	564,791	85.00	6.7×
	L2-pruning	80.90	338,874	91.00	11.1×
	MOPSO	80.97	364,854	90.31	10.3×
	NCPM	**95.18**	**304,610**	**91.91**	**12.4×**
VGG-16	Naive-Cut	87.47	6,772,112	53.98	2.17×
	MOPSO	83.69	1,358,248	90.77	10.83×
	NCPM	**95.91**	**2,096,970**	**85.75**	**7.017×**

The Pavia Center is captured by the ROSE-3 satellite, and the photographed terrain is the urban space of the University of Pavia, Italy. This dataset has a spatial resolution of 1.3 m and an image size of 1096 × 715 pixels. The dataset contains 114 spectral bands with spectral wavelengths ranging from 430 to 860 nm. After removing the noise bands, the number of bands used for classification is 104. Figure 18 show the pseudo-color plots and labels of Pavia Center.

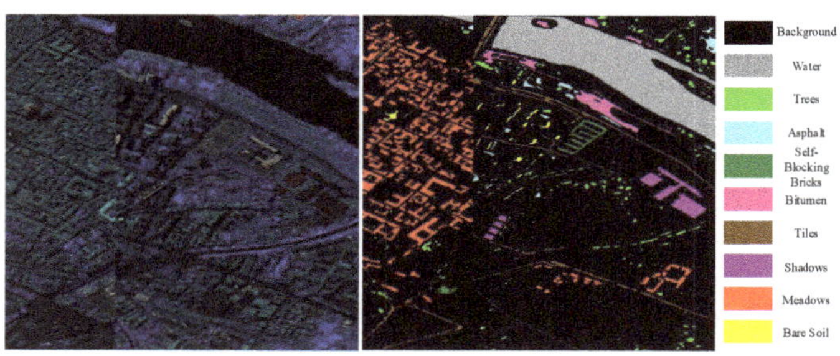

Figure 18. The false-color image and reference image on Pavia Center dataset.

Table 10 compares the classification results of the pruned network with the results of other neural network methods. Figure 19 shows the classification maps of different methods on Pavia Center. In the collaborative pruning task in the University of Pavia and Pavia Center datasets, a sparser network structure is obtained on the Pavia Center. OA is still maintained at 97.45%. On the University of Pavia dataset, a 97.39% OA is obtained in Pavia Center, which is better than the original network 3DDL, as well as the results on 1DCNN and M3DCNN. This also proves that proposed method can be applied to larger HSIs.

Table 10. Classification accuracy (%) for collaborative pruning task (University of Pavia and Pavia Center datasets). Best results are reported in bold.

Category	1DCNN	3DDL	M3DCNN	DCCN	HybridSN	ResNet	DPRN	Pruned 90.88%
OA (%)	88.32	81.67	94.36	97.43	93.47	97.72	**98.48**	97.45
AA (%)	91.29	85.11	94.87	96.12	94.81	97.14	**98.36**	96.25
Kappa (%)	84.85	76.24	92.59	96.60	91.46	96.91	**97.19**	96.62
1	83.47	69.91	85.03	95.53	86.98	92.35	94.12	**97.78**
2	87.08	82.99	96.24	**99.52**	93.71	98.92	99.48	99.43
3	88.42	74.08	89.09	88.61	88.58	95.41	**96.86**	89.37
4	96.57	94.48	96.34	96.01	96.96	96.99	**97.94**	96.02
5	99.99	99.95	99.99	**100.00**	99.99	**100.00**	**100.00**	99.85
6	91.05	72.51	98.03	98.01	97.43	**99.97**	99.63	95.13
7	91.42	83.75	95.78	97.66	96.76	98.84	**99.60**	93.08
8	84.46	90.82	94.16	95.54	93.18	**97.48**	94.41	96.03
9	99.15	97.57	99.15	94.19	99.47	**99.62**	99.52	99.57
Category	**1DCNN**	**3DDL**	**M3DCNN**	**DCCN**	**HybridSN**	**ResNet**	**DPRN**	**Pruned 91.70%**
OA (%)	96.55	97.71	97.90	**99.55**	99.20	99.06	99.10	97.39
AA (%)	89.57	92.57	92.50	**98.71**	96.92	96.78	96.75	91.32
Kappa (%)	95.11	96.76	97.03	**99.37**	98.87	98.68	99.16	96.91
1	99.63	99.93	**99.99**	**99.99**	**99.99**	99.94	**99.99**	99.96
2	95.65	95.64	96.51	96.77	97.06	98.31	**99.43**	95.76
3	89.51	94.43	89.44	98.83	96.18	90.45	**99.31**	91.13
4	67.37	81.48	79.32	97.24	88.97	96.01	**99.53**	70.73
5	83.38	92.64	96.47	99.72	98.73	**99.72**	99.17	92.95
6	97.05	96.30	96.75	98.36	99.18	**99.59**	99.19	98.14
7	84.67	85.22	87.42	99.17	98.94	94.82	**99.86**	83.47
8	98.67	99.80	99.70	**99.93**	99.71	99.59	99.18	99.13
9	90.18	87.67	86.90	98.39	93.53	92.59	**99.01**	90.63

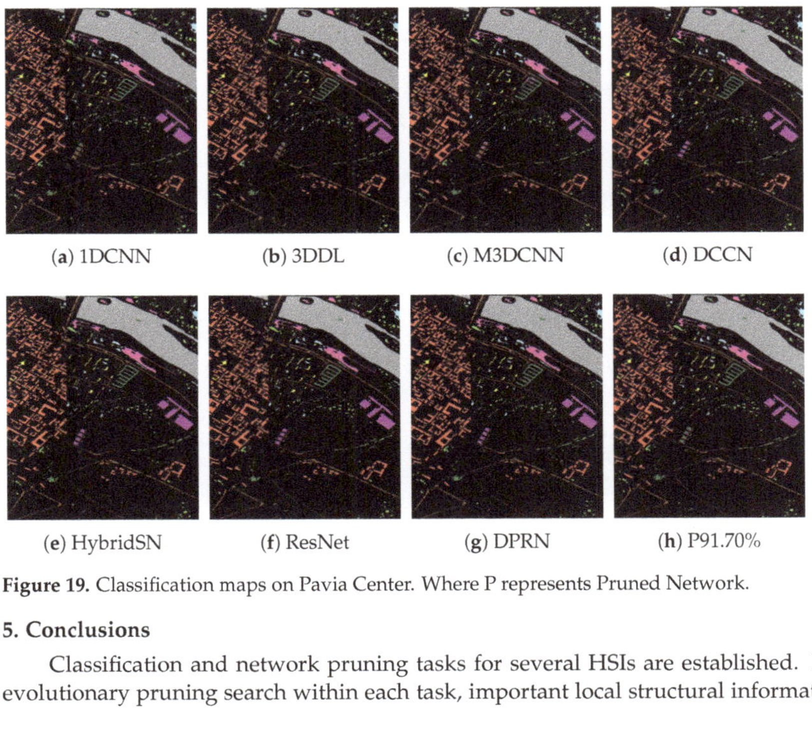

(**a**) 1DCNN (**b**) 3DDL (**c**) M3DCNN (**d**) DCCN

(**e**) HybridSN (**f**) ResNet (**g**) DPRN (**h**) P91.70%

Figure 19. Classification maps on Pavia Center. Where P represents Pruned Network.

5. Conclusions

Classification and network pruning tasks for several HSIs are established. In the evolutionary pruning search within each task, important local structural information is acquired and learned. Knowledge transfer between tasks is used to transfer important structures for representation in other tasks to the current task, which guides the learning and optimization of the network on limited labeled samples. It effectively improves the problem of network model overfitting and difficult training caused by limited labeled samples in each task. The self-adaptive transfer strategy based on historical information and dormancy mechanism achieves the original design goal: transferring as much good knowledge as possible and avoiding as much negative knowledge as possible.

Experiments on HSIs show that the proposed method can simultaneously realize classification and structure sparsification on multiple images. By comparing with other pruning methods on image classification data, the proposed method can search for sparse networks while maintaining accuracy. For structured pruning, which is currently more popular, the computation of sparse weight matrices can be avoided, so our future work will consider applying the proposed framework to structured pruning. Therefore, it is necessary to consider knowledge and knowledge transfer strategy in structured pruning. This will further expand our work in the area of neural network architecture optimization. Finally, the proposed method needs to be tested on hardware devices to verify the feasibility and practicability of the method.

Author Contributions: Conceptualization, Y.L. and D.W.; methodology, Y.L.; software, Y.L.; validation, Y.L. and S.Y.; formal analysis, Y.L.; investigation, Y.L.; resources, J.S. and S.Y.; data curation, J.S. and D.W.; writing—original draft preparation, Y.L.; writing—review and editing, D.W.; visualization, J.S.; supervision, D.T.; project administration, L.M.; funding acquisition, J.S. All authors have read and agreed to the published version of the manuscript.

Funding: This work was supported by the National Natural Science Foundation of China (Grant 62076204 and Grant 62206221), the National Natural Science Foundation of Shaanxi Province (Grant 2020JQ-197) and the Fundamental Research Funds for the Central Universities.

Institutional Review Board Statement: Not applicable.

Informed Consent Statement: Not applicable.

Data Availability Statement: Not applicable.

Acknowledgments: Thanks to NASA-JPL for providing AVIRIS data. The University of Pavia is collected by the sensor ROSIS near the University of Pavia, Italy. Thanks to Pavia university for providing the Pavia data.

Conflicts of Interest: The authors declare no conflict of interest.

References

1. LeCun, Y.; Bengio, Y.; Hinton, G. Deep learning. *Nature* **2015**, *521*, 436–444. [CrossRef] [PubMed]
2. Ding, Y.; Zhao, X.; Zhang, Z.; Cai, W.; Yang, N.; Zhan, Y. Semi-supervised locality preserving dense graph neural network with ARMA filters and context-aware learning for hyperspectral image classification. *IEEE Trans. Geosci. Remote Sens.* **2021**, *60*, 1–12. [CrossRef]
3. Ding, Y.; Zhang, Z.; Zhao, X.; Cai, W.; Yang, N.; Hu, H.; Huang, X.; Cao, Y.; Cai, W. Unsupervised self-correlated learning smoothy enhanced locality preserving graph convolution embedding clustering for hyperspectral images. *IEEE Trans. Geosci. Remote Sens.* **2022**, *60*, 5536716. [CrossRef]
4. Ding, Y.; Zhang, Z.; Zhao, X.; Cai, Y.; Li, S.; Deng, B.; Cai, W. Self-supervised locality preserving low-pass graph convolutional embedding for large-scale hyperspectral image clustering. *IEEE Trans. Geosci. Remote Sens.* **2022**, *60*, 5536016. [CrossRef]
5. Zhang, M.; Li, W.; Du, Q. Diverse region-based CNN for hyperspectral image classification. *IEEE Trans. Image Process.* **2018**, *27*, 2623–2634. [CrossRef]
6. Krizhevsky, A.; Sutskever, I.; Hinton, G.E. ImageNet classification with deep convolutional neural networks. *Commun. ACM* **2017**, *60*, 84–90. [CrossRef]
7. Simonyan, K.; Zisserman, A. Very Deep Convolutional Networks for Large-Scale Image Recognition. In Proceedings of the ICLR 2015: International Conference on Learning Representations 2015, San Diego, CA, USA, 7–9 May 2015.
8. Wang, H.; Wu, Z.; Liu, Z.; Cai, H.; Zhu, L.; Gan, C.; Han, S. HAT: Hardware-Aware Transformers for Efficient Natural Language Processing. In Proceedings of the 58th Annual Meeting of the Association for Computational Linguistics, Online, 5–10 July 2020; pp. 7675–7688.

9. Liu, Z.; Sun, M.; Zhou, T.; Huang, G.; Darrell, T. Rethinking the Value of Network Pruning. In Proceedings of the International Conference on Learning Representations, Vancouver, BC, Canada, 30 April–3 May 2018.
10. LeCun, Y.; Denker, J.S.; Solla, S.A. Optimal Brain Damage. *Adv. Neural Inf. Process. Syst.* **1989**, *2*, 598–605.
11. Han, S.; Pool, J.; Tran, J.; Dally, W.J. Learning Both Weights and Connections for Efficient Neural Networks. In Proceedings of the 28th International Conference on Neural Information Processing Systems, NIPS'15, Montreal, QC, Canada, 7–12 December 2015; MIT Press: Cambridge, MA, USA, 2015; Volume 1, pp. 1135–1143.
12. Lee, J.; Park, S.; Mo, S.; Ahn, S.; Shin, J. Layer-adaptive sparsity for the magnitude-based pruning. *arXiv* **2020**, arXiv:2010.07611.
13. Wang, Y.; Li, D.; Sun, R. NTK-SAP: Improving neural network pruning by aligning training dynamics. *arXiv* **2023**, arXiv:2304.02840.
14. Qi, B.; Chen, H.; Zhuang, Y.; Liu, S.; Chen, L. A Network Pruning Method for Remote Sensing Image Scene Classification. In Proceedings of the 2019 IEEE International Conference on Signal, Information and Data Processing (ICSIDP), Chongqing, China, 11–13 December 2019; pp. 1–4.
15. Wang, Z.; Xue, W.; Chen, K.; Ma, S. Remote Sensing Image Classification Based on Lightweight Network and Pruning. In Proceedings of the 2022 China Automation Congress (CAC), Xiamen, China, 25–27 November 2022; pp. 3186–3191.
16. Guo, X.; Hou, B.; Ren, B.; Ren, Z.; Jiao, L. Network pruning for remote sensing images classification based on interpretable CNNs. *IEEE Trans. Geosci. Remote Sens.* **2021**, *60*, 1–15. [CrossRef]
17. Jung, I.; You, K.; Noh, H.; Cho, M.; Han, B. Real-time object tracking via meta-learning: Efficient model adaptation and one-shot channel pruning. In Proceedings of the AAAI Conference on Artificial Intelligence, New York, NY, USA, 7–12 February 2020; Volume 34, pp. 11205–11212.
18. He, Y.; Lin, J.; Liu, Z.; Wang, H.; Li, L.J.; Han, S. Amc: Automl for model compression and acceleration on mobile devices. In Proceedings of the European conference on computer vision (ECCV), Munich, Germany, 8–14 September 2018; pp. 784–800.
19. Zhou, Y.; Yen, G.G.; Yi, Z. Evolutionary compression of deep neural networks for biomedical image segmentation. *IEEE Trans. Neural Netw. Learn. Syst.* **2019**, *31*, 2916–2929. [CrossRef]
20. Wu, T.; Li, X.; Zhou, D.; Li, N.; Shi, J. Differential Evolution Based Layer-Wise Weight Pruning for Compressing Deep Neural Networks. *Sensors* **2021**, *21*, 880. [CrossRef]
21. Wu, T.; Shi, J.; Zhou, D.; Lei, Y.; Gong, M. A Multi-objective Particle Swarm Optimization for Neural Networks Pruning. In Proceedings of the 2019 IEEE Congress on Evolutionary Computation (CEC), Wellington, New Zealand, 10–13 June 2019; pp. 570–577.
22. Zhou, Y.; Yen, G.G.; Yi, Z. A Knee-Guided Evolutionary Algorithm for Compressing Deep Neural Networks. *IEEE Trans. Syst. Man Cybern.* **2021**, *51*, 1626–1638. [CrossRef]
23. Zhao, J.; Yang, C.; Zhou, Y.; Zhou, Y.; Jiang, Z.; Chen, Y. Multi-Objective Net Architecture Pruning for Remote Sensing Classification. In Proceedings of the 2021 IEEE International Geoscience and Remote Sensing Symposium IGARSS, Brussels, Belgium, 11–16 July 2021; pp. 4940–4943.
24. Wei, X.; Zhang, N.; Liu, W.; Chen, H. NAS-Based CNN Channel Pruning for Remote Sensing Scene Classification. *IEEE Geosci. Remote Sens. Lett.* **2022**, *19*, 1–5. [CrossRef]
25. Ma, X.; Zhao, L.; Huang, G.; Wang, Z.; Hu, Z.; Zhu, X.; Gai, K. Entire space multi-task model: An effective approach for estimating post-click conversion rate. In Proceedings of the 41st International ACM SIGIR Conference on Research & Development in Information Retrieval, Ann Arbor, MI, USA, 8–12 July 2018; pp. 1137–1140.
26. Ruder, S. An overview of multi-task learning in deep neural networks. *arXiv* **2017**, arXiv:1706.05098.
27. Hou, Y.; Jiang, N.; Ge, H.; Zhang, Q.; Qu, X.; Feng, L.; Gupta, A. Memetic Multi-agent Optimization in High Dimensions using Random Embeddings. In Proceedings of the 2019 IEEE Congress on Evolutionary Computation (CEC), Wellington, New Zealand, 10–13 June 2019; pp. 135–141. [CrossRef]
28. Shi, J.; Zhang, X.; Tan, C.; Lei, Y.; Li, N.; Zhou, D. Multiple datasets collaborative analysis for hyperspectral band selection. *IEEE Geosci. Remote Sens. Lett.* **2021**, *19*, 1–5. [CrossRef]
29. Liu, S.; Shi, Q. Multitask deep learning with spectral knowledge for hyperspectral image classification. *IEEE Geosci. Remote Sens. Lett.* **2020**, *17*, 2110–2114. [CrossRef]
30. Zhang, L.; Zhang, L.; Du, B. Deep learning for remote sensing data: A technical tutorial on the state of the art. *IEEE Geosci. Remote Sens. Mag.* **2016**, *4*, 22–40. [CrossRef]
31. Chen, Y.; Lin, Z.; Zhao, X.; Wang, G.; Gu, Y. Deep learning-based classification of hyperspectral data. *IEEE J. Sel. Top. Appl. Earth Obs. Remote Sens.* **2014**, *7*, 2094–2107. [CrossRef]
32. Mou, L.; Ghamisi, P.; Zhu, X.X. Deep recurrent neural networks for hyperspectral image classification. *IEEE Trans. Geosci. Remote Sens.* **2017**, *55*, 3639–3655. [CrossRef]
33. Ding, Y.; Zhang, Z.; Zhao, X.; Hong, D.; Cai, W.; Yang, N.; Wang, B. Multi-scale receptive fields: Graph attention neural network for hyperspectral image classification. *Expert Syst. Appl.* **2023**, *223*, 119858. [CrossRef]
34. Ding, Y.; Zhang, Z.; Zhao, X.; Hong, D.; Cai, W.; Yu, C.; Yang, N.; Cai, W. Multi-feature fusion: Graph neural network and CNN combining for hyperspectral image classification. *Neurocomputing* **2022**, *501*, 246–257. [CrossRef]
35. Zhang, Z.; Ding, Y.; Zhao, X.; Siye, L.; Yang, N.; Cai, Y.; Zhan, Y. Multireceptive field: An adaptive path aggregation graph neural framework for hyperspectral image classification. *Expert Syst. Appl.* **2023**, *217*, 119508. [CrossRef]

36. Hamida, A.B.; Benoit, A.; Lambert, P.; Amar, C.B. 3-D deep learning approach for remote sensing image classification. *IEEE Trans. Geosci. Remote Sens.* **2018**, *56*, 4420–4434. [CrossRef]
37. Li, H.C.; Lin, Z.X.; Ma, T.Y.; Zhao, X.L.; Plaza, A.; Emery, W.J. Hybrid Fully Connected Tensorized Compression Network for Hyperspectral Image Classification. *IEEE Trans. Geosci. Remote Sens.* **2023**, *61*, 1–16. [CrossRef]
38. Ahmad, M.; Khan, A.M.; Mazzara, M.; Distefano, S.; Ali, M.; Sarfraz, M.S. A fast and compact 3-D CNN for hyperspectral image classification. *IEEE Geosci. Remote Sens. Lett.* **2020**, *19*, 1–5. [CrossRef]
39. Cao, X.; Ren, M.; Zhao, J.; Li, H.; Jiao, L. Hyperspectral imagery classification based on compressed convolutional neural network. *IEEE Geosci. Remote Sens. Lett.* **2019**, *17*, 1583–1587. [CrossRef]
40. Verma, V.K.; Singh, P.; Namboodri, V.; Rai, P. A"Network Pruning Network"Approach to Deep Model Compression. In Proceedings of the IEEE/CVF Winter Conference on Applications of Computer Vision, Snowmass Village, CO, USA, 1–5 March 2020; pp. 3009–3018.
41. Castellano, G.; Fanelli, A.M.; Pelillo, M. An iterative pruning algorithm for feedforward neural networks. *IEEE Trans. Neural Netw.* **1997**, *8*, 519–531. [CrossRef]
42. Li, H.; Kadav, A.; Durdanovic, I.; Samet, H.; Graf, H.P. Pruning filters for efficient convnets. *arXiv* **2016**, arXiv:1608.08710.
43. Zhang, S.; Stadie, B.C. One-Shot Pruning of Recurrent Neural Networks by Jacobian Spectrum Evaluation. In Proceedings of the ICLR 2020: Eighth International Conference on Learning Representations, Addis Ababa, Ethiopia, 26–30 April 2020.
44. Chen, T.; Ji, B.; Ding, T.; Fang, B.; Wang, G.; Zhu, Z.; Liang, L.; Shi, Y.; Yi, S.; Tu, X. Only train once: A one-shot neural network training and pruning framework. *Adv. Neural Inf. Process. Syst.* **2021**, *34*, 19637–19651.
45. Gupta, A.; Ong, Y.S.; Feng, L. Multifactorial Evolution: Toward Evolutionary Multitasking. *IEEE Trans. Evol. Comput.* **2016**, *20*, 343–357. [CrossRef]
46. Gupta, A.; Ong, Y.S.; Feng, L.; Tan, K.C. Multiobjective Multifactorial Optimization in Evolutionary Multitasking. *IEEE Trans. Syst. Man Cybern.* **2017**, *47*, 1652–1665. [CrossRef]
47. Tan, K.C.; Feng, L.; Jiang, M. Evolutionary transfer optimization-a new frontier in evolutionary computation research. *IEEE Comput. Intell. Mag.* **2021**, *16*, 22–33. [CrossRef]
48. Thang, T.B.; Dao, T.C.; Long, N.H.; Binh, H.T.T. Parameter adaptation in multifactorial evolutionary algorithm for many-task optimization. *Memetic Comput.* **2021**, *13*, 433–446. [CrossRef]
49. Shen, F.; Liu, J.; Wu, K. Evolutionary multitasking network reconstruction from time series with online parameter estimation. *Knowl.-Based Syst.* **2021**, *222*, 107019. [CrossRef]
50. Tang, Z.; Gong, M.; Xie, Y.; Li, H.; Qin, A.K. Multi-task particle swarm optimization with dynamic neighbor and level-based inter-task learning. *IEEE Trans. Emerg. Top. Comput. Intell.* **2021**, *6*, 300–314. [CrossRef]
51. Li, H.; Ong, Y.S.; Gong, M.; Wang, Z. Evolutionary Multitasking Sparse Reconstruction: Framework and Case Study. *IEEE Trans. Evol. Comput.* **2019**, *23*, 733–747. [CrossRef]
52. Chandra, R.; Gupta, A.; Ong, Y.S.; Goh, C.K. Evolutionary Multi-task Learning for Modular Training of Feedforward Neural Networks. In Proceedings of the 23rd International Conference on Neural Information Processing, Kyoto, Japan, 16–21 October 2016; Volume 9948, pp. 37–46.
53. Chandra, R.; Gupta, A.; Ong, Y.S.; Goh, C.K. Evolutionary Multi-task Learning for Modular Knowledge Representation in Neural Networks. *Neural Process. Lett.* **2018**, *47*, 993–1009. [CrossRef]
54. Chandra, R. Co-evolutionary Multi-task Learning for Modular Pattern Classification. In Proceedings of the International Conference on Neural Information Processing, Guangzhou, China, 14–18 November 2017; pp. 692–701.
55. Tang, Z.; Gong, M.; Zhang, M. Evolutionary multi-task learning for modular extremal learning machine. In Proceedings of the 2017 IEEE Congress on Evolutionary Computation (CEC), Donostia, Spain, 5–8 June 2017.
56. Gao, W.; Cheng, J.; Gong, M.; Li, H.; Xie, J. Multiobjective Multitasking Optimization With Subspace Distribution Alignment and Decision Variable Transfer. *IEEE Trans. Emerg. Top. Comput. Intell.* **2021**, *6*, 818–827. [CrossRef]
57. Deb, K.; Goyal, M. A combined genetic adaptive search (GeneAS) for engineering design. *Comput. Sci. Inform.* **1996**, *26*, 30–45.
58. Caruana, R. Multitask learning. *Mach. Learn.* **1997**, *28*, 41–75. [CrossRef]
59. Frankle, J.; Carbin, M. The lottery ticket hypothesis: Finding sparse, trainable neural networks. *arXiv* **2018**, arXiv:1803.03635.
60. Green, R.O.; Eastwood, M.L.; Sarture, C.M.; Chrien, T.G.; Aronsson, M.; Chippendale, B.J.; Faust, J.A.; Pavri, B.E.; Chovit, C.J.; Solis, M.; et al. Imaging spectroscopy and the airborne visible/infrared imaging spectrometer (AVIRIS). *Remote Sens. Environ.* **1998**, *65*, 227–248. [CrossRef]
61. Li, Y.; Zhang, H.; Shen, Q. Spectral–spatial classification of hyperspectral imagery with 3D convolutional neural network. *Remote Sens.* **2017**, *9*, 67. [CrossRef]
62. He, M.; Li, B.; Chen, H. Multi-scale 3D deep convolutional neural network for hyperspectral image classification. In Proceedings of the 2017 IEEE International Conference on Image Processing (ICIP), Beijing, China, 17–20 September 2017; pp. 3904–3908.
63. Hu, W.; Huang, Y.; Wei, L.; Zhang, F.; Li, H. Deep convolutional neural networks for hyperspectral image classification. *J. Sens.* **2015**, *2015*, 258619. [CrossRef]
64. Yu, H.; Zhang, H.; Liu, Y.; Zheng, K.; Xu, Z.; Xiao, C. Dual-channel convolution network with image-based global learning framework for hyperspectral image classification. *IEEE Geosci. Remote Sens. Lett.* **2021**, *19*, 1–5. [CrossRef]
65. Roy, S.K.; Krishna, G.; Dubey, S.R.; Chaudhuri, B.B. HybridSN: Exploring 3-D–2-D CNN feature hierarchy for hyperspectral image classification. *IEEE Geosci. Remote Sens. Lett.* **2019**, *17*, 277–281. [CrossRef]

66. Roy, S.K.; Manna, S.; Song, T.; Bruzzone, L. Attention-based adaptive spectral–spatial kernel ResNet for hyperspectral image classification. *IEEE Trans. Geosci. Remote Sens.* **2020**, *59*, 7831–7843. [CrossRef]
67. Paoletti, M.E.; Haut, J.M.; Fernandez-Beltran, R.; Plaza, J.; Plaza, A.J.; Pla, F. Deep pyramidal residual networks for spectral–spatial hyperspectral image classification. *IEEE Trans. Geosci. Remote Sens.* **2018**, *57*, 740–754. [CrossRef]
68. Han, S.; Mao, H.; Dally, W.J. Deep Compression: Compressing Deep Neural Networks with Pruning, Trained Quantization and Huffman Coding. In Proceedings of the ICLR 2016: International Conference on Learning Representations 2016, San Juan, Puerto Rico, 2–4 May 2016.
69. Yang, Z.; Xi, Z.; Zhang, T.; Guo, W.; Zhang, Z.; Li, H.C. CMR-CNN: Cross-mixing residual network for hyperspectral image classification. *IEEE J. Sel. Top. Appl. Earth Obs. Remote Sens.* **2022**, *15*, 8974–8989. [CrossRef]
70. Srinivas, S.; Babu, R.V. Data-free parameter pruning for deep neural networks. *arXiv* **2015**, arXiv:1507.06149.

remote sensing

Article

Adversarial Robustness Enhancement of UAV-Oriented Automatic Image Recognition Based on Deep Ensemble Models

Zihao Lu, Hao Sun * and Yanjie Xu

College of Electronic Science, National University of Defense Technology, Changsha 410073, China; luzihao21@nudt.edu.cn (Z.L.); xuyanjie@nudt.edu.cn (Y.X.)
* Correspondence: sunhao@nudt.edu.cn

Abstract: Deep neural networks (DNNs) have been widely utilized in automatic visual navigation and recognition on modern unmanned aerial vehicles (UAVs), achieving state-of-the-art performances. However, DNN-based visual recognition systems on UAVs show serious vulnerability to adversarial camouflage patterns on targets and well-designed imperceptible perturbations in real-time images, which poses a threat to safety-related applications. Considering a scenario in which a UAV is suffering from adversarial attack, in this paper, we investigate and construct two ensemble approaches with CNN and transformer for both proactive (i.e., generate robust models) and reactive (i.e., adversarial detection) adversarial defense. They are expected to be secure under attack and adapt to the resource-limited environment on UAVs. Specifically, the probability distributions of output layers from base DNN models in the ensemble are combined in the proactive defense, which mainly exploits the weak adversarial transferability between the CNN and transformer. For the reactive defense, we integrate the scoring functions of several adversarial detectors with the hidden features and average the output confidence scores from ResNets and ViTs as a second integration. To verify their effectiveness in the recognition task of remote sensing images, we conduct experiments on both optical and synthetic aperture radar (SAR) datasets. We find that the ensemble model in proactive defense performs as well as three popular counterparts, and both of the ensemble approaches can achieve much more satisfactory results than a single base model/detector, which effectively alleviates adversarial vulnerability without extra re-training. In addition, we establish a one-stop platform for conveniently evaluating adversarial robustness and performing defense on recognition models called AREP-RSIs, which is beneficial for the future research of the remote sensing field.

Keywords: deep neural network; adversarial defense; deep ensemble model; unmanned aerial vehicle; remote sensing; image recognition

Citation: Lu, Z.; Sun, H.; Xu, Y. Adversarial Robustness Enhancement of UAV-Oriented Automatic Image Recognition Based on Deep Ensemble Models. *Remote Sens.* **2023**, *15*, 3007. https://doi.org/10.3390/rs15123007

Academic Editor: Gwanggil Jeon

Received: 27 April 2023
Revised: 31 May 2023
Accepted: 7 June 2023
Published: 8 June 2023

1. Introduction

Over the past several decades, an abundance of remote sensing images (RSIs) have been continuously collected from UAVs with massive and detailed information that allows researchers to observe the Earth more precisely. Nevertheless, the mode of image interpretation, which relies only on expert knowledge and handcrafted features, can no longer meet the requirements of higher accuracy and efficiency. Fortunately, the substantial progress of DNNs [1] in computer vision has achieved the state-of-the-art performances in the various tasks of remote sensing field and supported on-device inference for the real-time demands. Well-trained DNNs can be deployed on UAVs for the tasks including image recognition, object detection, image matching and so on, which enables quick feedback with useful analysis for both military (e.g., target acquisition [2–5], battlefield reconnaissance [6], communications [7–9]) and civilian (e.g., land surveys [10], delivery service [11,12], medical rescue [13,14]) use.

However, hidden dangers lurk in the working process of UAV, and a great diversity of counter-UAV attacks have been extensively developed that are targeted at its vulnerability.

which mainly exists in the cyber, sensing, and kinetic domains [15]. Distribution drifts [16–18] and common corruptions such as blur, weather and noise [19] also interfere with the automatic interpretations of RSIs in the image domain. Meanwhile, a new kind of threat has emerged due to the security and reliability issues with DNN models [20–22], which is known as *adversarial vulnerability* and potentially has devastating effects on the UAVs with autonomous visual navigation and recognition systems. For example, when such a UAV carries out a target recognition task, particularly for the non-cooperative vehicles on military missions, the suspicious vehicles with carefully designed camouflage patterns (i.e., physical adversarial attacks) or a leakage of real-time images with malicious perturbations (i.e., digital adversarial attacks) can mislead the DNNs on UAVs to make wrong predictions and violate the integrity of the outputs. In this way, the enemy's targets are likely to evade the automatic recognition, causing a severe disadvantage to the battlefield reconnaissance. Thus, the harmful effects of adversarial vulnerability in DNN models need to be taken more seriously for modern UAVs. Moreover, compared with the natural images such as ImageNet [23], not as many RSIs are labeled in a dataset. Therefore, the trained DNNs in the remote sensing field tend to be sensitive to adversarial attacks [24], which puts forward a higher requirement on the adversarial robustness.

Under threat from the adversarial attacks, researchers are motivated to propose effective defense methods mainly in the context of natural images. The defense strategies can be divided into two categories. The first is *proactive defense* to generate robust DNNs aimed at correctly classifying all the attacked images. Adversarial training (AT) [25] is a commonly used approach belonging to this category, which minimizes the training loss with online-generated adversarial examples. However, standard AT counts on prior knowledge with no awareness of new attacks and can decrease the accuracy of benign data. So, many improved versions such as TRADES [26], FAT [27], and LAS-AT [28] are developed. In addition, an attack designed for one DNN model may not confuse another DNN, which makes ensemble methods [29–32] an attractive defense strategy while bridging the gap between benign and adversarial accuracy. Ensemble methods against adversarial attacks often combine the output predictions or fuse the features extracted from the intermediate layers of several DNNs.

However, given the fact that obtaining a sufficiently robust DNN against any kind of attack is not realistic, some research efforts have been turned to *reactive defense*, namely detecting the input image whether it has been attacked or not. The detection strategy can be classified into three categories, including statistical [33–38], prediction inconsistency-based [39,40] and auxiliary model [41–44] strategies. In reactive defense, we do not modify the original victim models during the detection and train a detector with a certain strategy as a 3rd-party entity. Moreover, the reactive defense is valuable when the output of a baseline DNN does not agree with the one from a robust DNN strengthened by a proactive defense method [45].

In this article, we consider the case that the DNN-based visual navigation and recognition systems on UAVs are suffering from adversarial attacks when performing an important task after take-off. Aimed at this intractable scenario and several analyzed motives, we propose to investigate the ensemble strategy to address the problem for both proactive and reactive defense **only using base DNN models**:

- In proactive defense, standard AT and its variants need re-training and model updates if UAVs meet unknown attacks, which does not suit the environment of edge devices with limited resources (e.g., latency, memory, energy); thus, an ensemble of base DNN models can be an alternative strategy. Intuitively, an ensemble is expected to be more robust than an individual model, as the adversary needs to fool the majority of the sub-models. As the representative models of CNNs and transformers, ResNet [46] and Vision Transformer (ViT) [47] have different network architectures and mechanisms in extracting discriminative features. We also verify that the adversarial examples of RSIs show weak transferability between CNNs and transformers. Therefore, we combine the probability distributions of output layers from CNNs and transformers with standard supervised training for a better performance under adversarial attacks in the recognition of RSIs.

- In terms of reactive defense, we consider a case study with the framework of ENsemble Adversarial Detector (ENAD) [48], which combines scoring functions computed by multiple adversarial detection algorithms with intermediate activation values in a well-trained ResNet. Based on the original framework, we further integrate the scoring functions from ViT with the ones from ResNet, forming a connection with the ensemble method in proactive defense. Therefore, the ensemble has two levels of meaning: one is combining layer-specific values from multiple adversarial detection algorithms, and the other is integrating the results from CNNs and transformers. Different detection algorithms with different network architectures can exploit distinct statistical features of the images, so this ensemble strategy is highly suitable for RSIs with rich information.

Both of the defenses in the form of an ensemble will be activated when the controller realizes that the outputs from the system on UAVs are obviously manipulated. The supposed scenarios and the role of ensemble defense are illustrated as Figure 1. To verify their effectiveness, we conduct a series of experiments with the datasets including optical and SAR RSIs. For proactive defense, we compare the performances regarding the *Attack Success Rate* of an ensemble of base ResNets and ViTs for different adversarial attack algorithms with three other proactive defense to improve the robustness of base DNN models. In terms of reactive defense, we compare the ensemble framework with three stand-alone adversarial detectors, which are also the components in the ensemble framework. The metrics of detection are the Area Under the Receiver Operating Curve (AUROC) and the Area Under Precision Recall (AUPR).

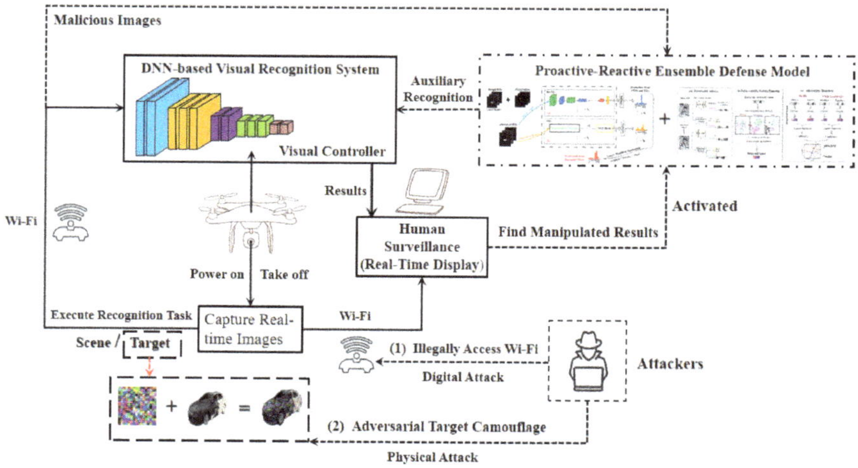

Figure 1. The threat scenarios caused by adversarial vulnerability in modern UAVs and the role of our adversarial ensemble defense (blue lines: general working mode of UAVs; red lines: confrontation with adversarial attacks).

From the experimental results, we find that an ensemble of base ResNets and ViTs demonstrates good defensive capability in most experimental configurations of proactive defense. It does not need a re-training but can be on a par with the methods based on AT. Moreover, an ensemble framework modified from ENAD can yield AUROC and AUPR of over 90 in gradient-based attacks of optical datasets. The performances of the ensemble method slightly decrease on Deepfool, C&W and adversarial examples of SAR RSIs, but it is still generally better than the stand-alone adversarial detectors.

Based on the above work, we establish a one-stop integrated platform for evaluating the adversarial robustness of DNNs trained with optical or SAR RSIs and conducting adversarial defenses on the models called *Adversarial Robustness Evaluation Platform for*

Remote Sensing Images (AREP-RSIs). Users can operate just on AREP-RSIs to perform a complete robustness evaluation with all necessary procedures, including training, adversarial attacks, tests for recognition accuracy, proactive defense and reactive defense. AREP-RSIs can be deployed on the edge devices such as UAVs and connected with cameras for real-time recognition as well. Equipped with various network architectures, several training paradigms, and classical defense methods, to the best of our knowledge, AREP-RSIs is the first platform for adversarial robustness improvements and evaluations in the remote sensing field. More importantly, the framework of AREP-RSIs is flexibly extendable. Users can add the model architecture files, load their own weight configurations, and register new attack and defense methods for a customized DNN, which greatly facilities designing robust DNN-based recognition models in the remote sensing field for the future research. The AREP-RSIs can be available at Github (https://github.com/ZeoLuuuuuu/AREP-RSIs, accessed on 26 April 2023).

In summary, the main contributions of this paper are as follows.

- We innovatively analyze the adversarial vulnerability from a scenario in which the edge-deployed DNN-based system for visual navigation and recognition on a modern UAV is suffering from adversarial attacks produced by the physical camouflage patterns or digital imperceptible perturbations.
- To cope with the intractable condition, we investigate the ensemble of ResNets and ViTs for both proactive and reactive defense for the first time in the remote sensing field. We conduct experiments with optical and SAR remote sensing datasets to verify that the ensemble strategies have good efficacy and show a favorable prospect against adversarial vulnerability in the DNN-based visual recognition task.
- We finally integrate all the procedures of performing adversarial defenses and evaluating adversarial robustness into a platform called AREP-RSIs. Equipped with various network architectures, several training paradigms, and defense methods, users can verify if a specific model has good adversarial robustness or not just through this one-stop platform AREP-RSIs.

The rest of this paper is organized as follows. Section 2 introduces the background knowledge, related works and threat model utilized in this article. Section 3 tells why we use the ensemble strategy, specific methods and our developed platform in detail. Section 4 reports on the experimental results and provides an analysis. Finally, the conclusions are given in Section 5.

2. Background and Related Works

This section briefly reviews the causes of adversarial vulnerability in image recognition tasks and existing research of the adversarial vulnerability in the remote senisng field and DNN-based UAVs. Finally, we provide a threat model including the potential approaches of attacking the automatic recognition systems of UAVs with adversarial examples, some possible goals and the access level of models for attackers.

2.1. Causes of Adversarial Vulnerability in Image Recognition

To better learn the adversarial vulnerability in an image recognition system, its possible causes are discussed theoretically. Sun et al. [49] give a comprehensive analysis, and based on their work, we briefly review the reasons why adversarial vulnerability is a common problem for image recognition.

- **Dependency on Training Data**: The accuracy and robustness of an image recognition model are highly dependent on the quantity and quality of training data. During the training process, DNN models only learn the correlations from data, which tend to vary with data distribution. In many security-sensitive fields, the severe scarcity of large-scale high-quality training data and the problem of category imbalance in the training datasets can exacerbate the risk of adversarial vulnerability of DNN models.
- **High-Dimensionality of Input Space**: The training dataset only covers a very small part of the input space portion, and a large amount of potential input data are not

utilized. Moreover, hundreds of parameters are optimized during the training process and the space formed by parameters is also huge. Therefore, the generalized decision boundaries in the input space are just roughly approximated by DNNs, which cannot completely overlap with the ground-truth decision boundaries. The adversarial examples may exist in the gap between them.

- **Black-box property of DNNs**: Due to the complex network architectures and optimization process, it is hard to directly translate the internal representation of a DNN into a tool for understanding its wrong outputs under an adversarial attack. So, this black-box property of DNNs makes it more difficult to design a universal defense technique against adversarial perturbations from the perspective of the model itself.

2.2. Adversarial Vulnerability in DNN-based UAVs

In recent years, as DNNs are increasingly applied to the visual navigation and recognition systems on UAVs, the security threat produced by adversarial attacks has been a formidable problem, which can be utilized by the attackers with motives for maliciously permeating into the working process of these DNN-based UAVs.

Previous research has indicated that this security problem exists in DNN models for RSI recognition, which poses a threat to the modern UAVs. Most of them still focus on the digital attacks, which directly manipulate the pixel values in RSIs and suppose full access to the images for attackers. In terms of scene recognition, Li et al. [50] and Xu et al. [51] both used various adversarial attacks to fool multiple high-accuracy models trained on different scene datasets. In another article, Xu et al. also provided a black-box universal dataset with adversarial examples called UAE-RS [52], which serve as a benchmark to design DNNs with higher robustness. Even further, Li et al. [53] proposed a soft threshold defense for scene recognition to judge whether an input RSI is adversarial or not. Focused on SAR target recognition, Li et al. [54] mounted white-box attacks on SAR images and proposed a new metric to successfully explain the phenomenon of attack selectivity. Du et al. [55] proposed a fast C&W algorithm for DNN-based SAR classifiers, using a deep coded network to replace the search process in the original C&W algorithm. Zhou et al. [56] focused on the sparsity of SAR images and applied the sparse attack methods on the MSTAR dataset to verify their effectiveness in SAR target recognition.

In addition, there are also explorations into physical adversarial attacks applied to RSIs. Czaja et al. [57] conducted attacks through adversarial patches to confuse the victim DNN among four scene classes, and den Hollander et al. [58] generated the patches for the task of object detection. However, they only restricted their patches to the digital domain and did not print them. The most relevant to our assumed scenario is the work of Du et al. [59], in which they optimized, fabricated and installed their designed patches on or around a car to significantly reduce the efficacy of a DNN-based detector on a UAV. They also experimented under different atmospheric factors (lighting, weathers, seasons) and distance between the camera and target. Their results indicated the realistic threat of adversarial vulnerability on DNN-based intelligent systems on UAVs.

Moreover, some research has discussed the adversarial vulnerability in the context of UAVs. Doyle et al. [15] considered two common operations for a navigation system of UAVs: follow and waypoint missions to develop a threat model from the perspective of attackers. They sketched state diagrams and analyzed the potential attacks for each state transition. Torens et al. [60] give a comprehensive review for the verification and safety of machine learning in UAVs. Tian et al. [61] proposed two adversarial attacks for the regression problems of predicting steering angles and collision probabilities from real-time images in UAVs. They also investigated standard AT and defensive distillation against the two designed attacks.

2.3. Threat Model

We denote a real-time image captured and processed by the sensors as $x \in \mathbb{R}^{h \times w \times c}$ with h, w, c representing height, width and channel (c = 3 for optical images and c = 1 for SAR images), which is also the input of a DNN-based visual recognition system $\mathcal{M}(\cdot)$ deployed on UAVs. In addition, each image has a potential groundtruth label $y \in \mathcal{Y} = \{0, 1, \ldots, K-1\}$ where K is the number of recognizable categories for the system. A well-trained system $\mathcal{M}(\cdot)$ can correctly recognize the scene or targets for most of x, namely $\mathcal{M}(x) = y$.

We suppose two possible approaches that attackers can exploit to attack the DNN-based visual recognition system on UAVs.

(1) The first approach is to illegally access the Wi-Fi communication between the sensors (i.e., cameras) and the controller for UAVs. The attackers can spoof imperceptible perturbations ρ to the images provided by the sensors to craft adversarial examples $\hat{x} = x + \rho$ through the communication link. The wrong predictions $y' = \mathcal{M}(\hat{x}) \neq y$ for most of $\hat{x}$ can influence the next commands and actions for UAVs.

(2) The second approach is physically realizing the perturbations as "ground camouflage" based on adversarial patches [62], especially for the task of target recognition. An adversarial patch is generally optimized in the form of sub-images by modifying the pixel values within a confined area, and the attacker then prints the patch as a sticker or poster. Ref. [59] gives a real-world experiment for this approach by pasting designed patches on top of or around vehicles to highly reduce the probabilities of detection and recognition rates. Even if the patterns are noticeable to our human eyes, they can effectively confuse the recognition system.

There are several reasons why attackers hope to do harm to the visual navigation and recognition system on a UAV. For scene recognition, attackers can mislead UAVs to incorrect situational awareness for military use. In addition, the misclassification of the scene may make the navigation system confuse the current environment, become lost, and hover in the air. For target recognition, once non-cooperative targets of high military value are camouflaged, UAVs will not be able to accurately detect and recognize them, which aims at evading aerial reconnaissance or targeted strikes in the battlefield.

The access level of the victim DNN models for attackers is an important factor. White-box attackers are the strongest in all conditions. They can obtain the network structures, weights and even the training data. In contrast, black-box attackers only query the outputs at each attempt, craft adversarial examples against a substitute model or search randomly. Moreover, whether they mislead DNNs to a specified class distinguishes an attack as a targeted or untargeted one. In our threat model, we consider both white-box and black-box settings during our experiments with the more general untargeted condition.

3. Methodology

This section will briefly analyze the motives of exploiting the ensemble strategy in Section 3.1. Then, it will present the proactive–reactive defensive ensemble framework in detail in Section 3.2 and finally introduce our edge-deployed platform AREP-RSIs for adversarial robustness improvements and evaluations in Section 3.3.

3.1. Motives of Ensemble

As the most representative models of CNNs and transformers, ResNet and ViT are mainly discussed within the defensive ensemble framework. Before a detailed description of the defense method, we start with the reasons why the ensemble strategy should be selected and attempted in the supposed scenario of this article.

3.1.1. Different Mechanisms for Feature Representations

Recently, ViTs have drawn great attention as a fundamentally new model structure offering impressive performances in image recognition and robustness benefits as well [63]. Compared with CNNs, ViTs have striking differences in their feature representations [64].

Specifically, CNNs share kernels in each convolution layer (Conv) that locally perceive a small part of the input image (i.e., receptive field) to extract features. The powerful inductive bias of translation equivariance and locality correlation within the convolutional layers make CNNs excellent in learning general-purpose visual representations. However, the receptive fields are limited with a fixed size, which is not conducive to obtaining global information. In contrast, ViT processes an image as a sequence of image patches, and each patch is linearly projected into a representation vector with a positional embedding. Moreover, a learnable class token is also attached for the image. As the main component in ViT, multi-head self-attention modules (MSAs) are then connected for an aggregation of the information from all patches to have an entire view of the image.

More importantly, [65] revealed that the MSAs in ViT exhibit opposite behaviors with the Convs in ResNet by performing the Fourier analysis of feature maps from both models. The Convs act like a low-pass filter that tends to reduce low-frequency signals, while MSAs are high-pass filters that are robust against high-frequency noise in images. In addition, [64] found that ViT incorporates more global information and has more uniform representations with greater similarity throughout the layers. There have been many hybrid architectures that combine CNNs and transformers to inherit both of their advantages [66–69]. Therefore, to some extent, ViT can be complementary to ResNet, which intuitively enlightens us about the selection of network architectures in the ensemble.

3.1.2. Weak Adversarial Transferability

Reducing the adversarial transferability among base models in an ensemble can achieve good robustness without sacrificing benign accuracy [70–72]. To further verify the differences between ResNet and ViT in the context of remote sensing, we found empirical evidence that the adversarial examples of RSIs tend to have weak transferability between CNNs and ViTs, which facilities constructing ensemble classifiers to generate a more robust model. For the details of transferability experiments, we trained a set of various CNNs (including ResNet-18, ResNet-50, DenseNet-121, DenseNet-201, MobileNet-V2 and ShuffleNet-V2) and two ViT variants (ViT-Base/16 and ViT-Large/16) with the same training setting on the MSTAR dataset. A white-box attack, PGD-l_∞ [25], is applied on the test set of MSTAR with a different attack radius against the victim models of ResNet-18 and ViT-Base/16, respectively. Then, both sets of generated adversarial examples are recognized under each well-trained DNN model. Similarly, we conducted the experiments on the UC Merced LandUse, which is an optical scene RSI dataset again. The results are illustrated in Figures 2 and 3. From the results, for both datasets, we observe that the adversarial examples crafted against ResNet-18 generally have much better performance of recognition accuracy in ViTs and vice versa.

3.1.3. Defects in AT and Our solution for Edge Environment

One of the most commonly used methods for improving adversarial robustness is still AT, which trained DNNs with both natural data and its corresponding adversarial variants. Even though previous research indicated that AT can force DNNs to learn robust features and gained better performance on adversarial robustness, the absence of non-robust features can lead to a drop in generalization and the accuracy on the benign data [73]. This trade-off between adversarial robustness and natural accuracy still needs to be considered when using AT. Moreover, AT sometimes heavily counts on such prior knowledge and cannot achieve a sufficient robustness against an unknown attack.

Generally, modern UAVs are equipped with different base DNN models instead of the DNN models trained with AT for standard automatic visual recognition. When the UAVs suffer from adversarial attacks in performing a recognition task, it is time-consuming to make an extra re-training to obtain a new robust model and replace the base models on the ground. Training on edge devices is also impractical because of the resource-limited environment. Therefore, our proposed solution for this problem is attempting an ensemble of base DNN models, especially DNNs with different network architectures and feature extraction mechanisms. Based on the analysis of CNNs and transformers above, we decide to use ResNet and ViT, which are two standard popular DNN architectures in the ensemble.

They will be trained solely with benign data to improve adversarial robustness while guaranteeing natural accuracy in our supposed scenario.

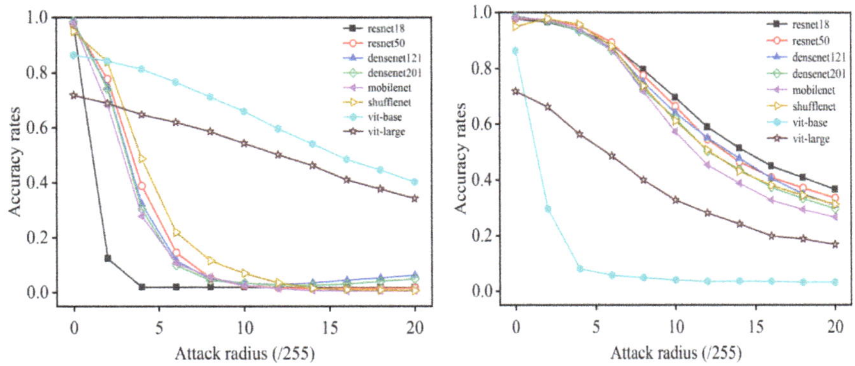

Figure 2. The transferability test of PGD on MSTAR against ResNet-18 (**left**) and ViT-Base/16 (**right**).

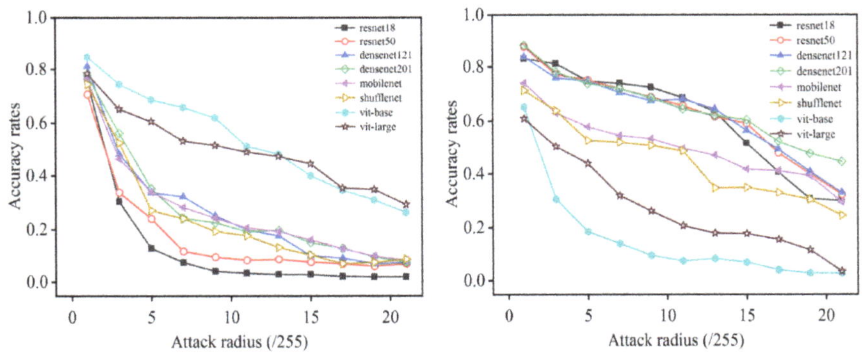

Figure 3. The transferability test of PGD on UC Merced LandUse against ResNet-18 (**left**) and ViT-Base/16 (**right**).

In addition, we report the computation and memory footprints of base DNN models used in our ensemble as shown in Table 1, including the number of parameters within network architecture (Params), floating point operations (FLOPs) and parameter memory footprint (Param. Mem). The specific network architectures consist of ResNet-18, ResNet-50, and ResNet-101 for CNN and ViT-Base/16, ViT-Large/16, and ViT-Base/32 for transformer.

Table 1. The computation and memory footprints of base DNN models used in our ensemble.

	Params (M)	**FLOPs (GFLOPs)**	**Param. Mem (MB)**
ResNet-18	11.69	2	45
ResNet-50	25.56	4	98
ResNet-101	44.55	8	170
ViT-Base/16	86.86	17.6	327
ViT-Base/32	88.30	8.56	336
ViT-Large/16	304.72	61.6	1053

As shown in Table 1, several of the network architectures we used such as ResNet-101 and ViT-Large/16 seem to be less suitable for the edge environment; however, our intention of this attempt is to first verify that the ensemble of CNNs and transformers can resist adversarial data of RSIs in both proactive and reactive defense. So, the two most commonly used DNNs are selected for the paradigm in our article. In practice, we can replace them with more light-weight DNNs such as MobileNet, ShuffleNet, and Inception-V3 for CNN and ViT-Tiny/16, EfficientFormer for transformer.

3.2. Proactive–Reactive Defensive Ensemble Method

3.2.1. Proactive Defense

In the non-ensemble schemes, a single base model is provided to attackers, which can be attacked with the worst perturbations. However, based on the analyzed motives above, an ensemble of CNNs and transformers suits our supposed scenario better. Our defensive ensemble model includes both proactive and reactive defense. For proactive defense, an ensemble model is a weighted average of N random base ResNets with different depths denoted as $\Omega_1 = \{\mathcal{R}_1, \mathcal{R}_2, ..., \mathcal{R}_N\}$ and N base ViT variants denoted as $\Omega_2 = \{\mathcal{R}_{N+1}, \mathcal{R}_{N+2}, ..., \mathcal{R}_{2N}\}$. To confuse the whole ensemble model, an attacker has to design an attack against both types of DNN with more difficulty [74].

Specifically, we can train two sets of base DNNs including ResNets and ViTs with $N = 3$ (i.e., $\Omega_1 = \{$ResNet-18, ResNet-50, ResNet-101$\}$, $\Omega_2 = \{$ViT-Base/16, ViT-Large/16, ViT-Base/32$\}$). Ω_1 and Ω_2 form the overall set of base models Ω. We denote $\{\mathcal{D}_j\}_{j=1}^{2N}$ as a large set including the probability distributions $\{d_{jk}\}_{k=1}^{K}$ predicted by each base DNN model, where d_{jk} is the confidence score of category k predicted by the jth base model and K denotes the number of recognizable categories. Therefore, the probability distribution for each base model can be expressed as (1).

$$\mathcal{D}_j = \{d_{j1}, d_{j2}, \dots d_{jK}\},\ j = 1, 2, \dots, 2N \tag{1}$$

Then, we can weight the $2N$ models with non-negative values $(\omega_1, \omega_2, ..., \omega_{2N})$ that add up to 1. Let a vector $\mathcal{W}$ denote these weights, and we can obtain a new probability distribution $\mathcal{D}'$ of the deep ensemble model with new confidence scores $\{d'_k\}_{k=1}^{K}$ by taking a linearly weighted summation as (2).

$$\mathcal{D}' = \mathcal{W} \cdot (\mathcal{D}_1, \mathcal{D}_2, \dots, \mathcal{D}_{2N})^T = (d'_1, d'_2, \dots, d'_K) \tag{2}$$

In fact, the new probability distribution $\mathcal{D}'$ is a fusion on the decision level, integrating the opinions from CNNs and transformers. In addition, from the perspective of base models, we can also express the DNN-based ensemble model $\mathcal{M}(x, \mathcal{W})$ as (3).

$$\mathcal{M}(x, \mathcal{W}) = \mathcal{W} \cdot \Omega = \sum_{j=1}^{2N} \omega_j \cdot \mathcal{R}_j \tag{3}$$

The framework of this ensemble model for proactive defense is illustrated as Figure 4. As shown in Figure 4, if a modern UAV captures real-time RSIs with BMP2 vehicles but suffers from adversarial perturbations crafted for CNN architecture, these RSIs can be sent to the proposed deep ensemble model and inferred by all of the base models simultaneously. Even though the adversarial RSIs can mislead the predictions from CNNs, the outputs from transformers are still correct. The model will fuse the opinions of CNNs and transformers on the decision level, namely making a linear weighted summation as mentioned above, to obtain the final correct prediction.

In terms of the weights of base models $(\omega_1, \omega_2, \dots, \omega_{2N})$ in the ensemble, one solution is to weight them with fixed values, and we can search for the better set of values manually. The other solution of deciding the models' weights is to make them learnable, so the weights can be adjusted automatically during the training time. In our following experiments, we choose the former for simplicity.

This deep ensemble model for proactive defense only exploits standard base models and does not need to require extra re-training such as AT, which constitutes a practical attempt for improving the adversarial robustness of automatic recognition systems on edge devices such as UAVs. It is also the first DNN-based ensemble model against adversarial attacks in the remote sensing field. In this way, more adversarial examples are expected to be correctly recognized when confronting adversarial attacks. We will compare the ensemble of ResNets and ViTs with a victim model without any defense and trained with standard AT [25], Trades [26] and GAL [75] against malicious RSIs crafted by different adversarial attack methods. The experimental results will be collected in the next section. The Attack Success Rate (ASR) (i.e., the number of wrongly recognized RSIs divided by the number of RSIs in the whole test set) will be the metric for the proactive defense.

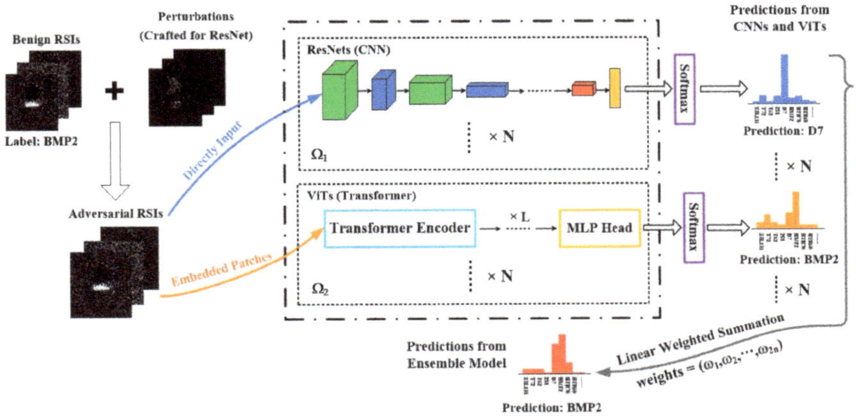

Figure 4. The process of proactive ensemble method with MSTAR dataset and victim model ResNet-18.

3.2.2. Reactive Defense

Considering the fact that it is not possible for proactive defense to classify all types of adversarial examples, detection-based methods (i.e., reactive defense) also deserve an exploration to indirectly enhance the adversarial robustness. To pursue better performance than individual adversarial detectors, we selected and modified an excellent deep ensemble framework called ENAD [48], which integrates the scoring functions from different adversarial detection algorithms on the hidden features of intermediate layers of CNNs. Our modified version repeats these procedures on ViTs with the features extracted from a transformer encoder and averages the detection outputs from both types of DNN at the end of the framework as the second integration. The structure of an ensemble including CNNs and transformers also matches our proposed model in the proactive defense.

The specific procedures of the ensemble model in reactive defense are illustrated in detail in Figure 5. A real-time RSI captured by UAVs, which can be either benign or maliciously attacked, is input to a well-trained ResNet and ViT. For ResNet, the activation values from several selected hidden layers are then extracted. Next, the model will compute layer-specific scores through three commonly used adversarial detection algorithms: Local Intrinsic Dimensionality (LID) [34], Kernel Density Estimation (KDE) [35] and Mahalanobis Distance (MD) [36]. Each detection algorithm measures the "distance" as the score based on each activation value of the real-time RSI with respect to training examples and the paradigm learned during the training time. The layer-specific scores for each detection algorithm are fused to obtain the detector-specific scores, namely three final scores, which are input to a logistic regression to compute a probability c_1 of classifying the test RSI as benign or adversarial. In the meantime, the above procedures are also performed in parallel on ViT with the activation values extracted from multi-head self-attention in several transformer encoders. The predicted probability from ViT is denoted as c_2. c_1 and c_2 from ResNet and ViT are averaged to obtain a final result p. The ensemble model will decide an RSI image as the adversarial one if c is greater than 0.5, and it is benign otherwise.

In terms of the individual detectors (i.e., LID, KDE and MD) in the ensemble model for reactive defense, there is one trick that needs to be considered. Apart from benign and adversarial examples, we also craft noisy examples with Gaussian noise that are treated as benign examples during the training time for better generalization.

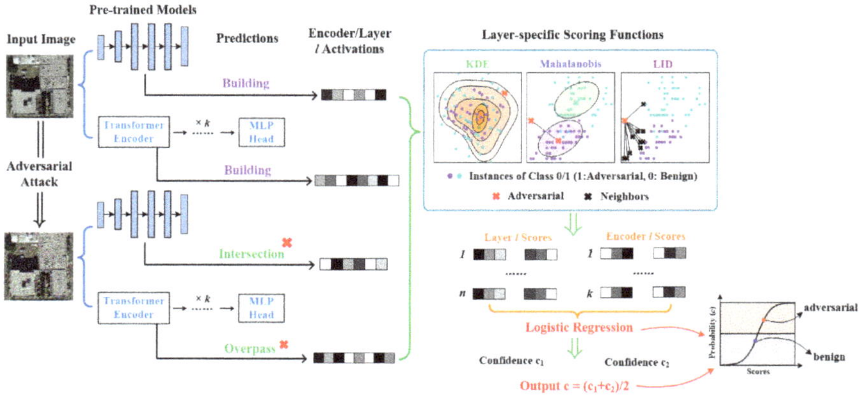

Figure 5. The procedures of reactive defense ensemble model modified from ENAD [48] (assuming the number of layers in ResNet is n and the number of transformer encoders in ViT is k).

The extracted activation values are high-dimensional in both types of DNN model, so different detection algorithms can use distinct statistical features of input images. The ensemble idea integrates the features and is expected to be perfectly suited for the adversarial detection of RSIs, because RSIs have rich information such as color, texture, spatial and spectral features. Moreover, the first integration of multiple adversarial detectors can better alleviate the problems of overfitting and generalization than just using one detector. The second integration benefits from different feature representations in ResNet and ViT. To evaluate the performances of detection, we take two standard metrics, Area Under the Receiver Operating Characteristic (AUROC) and Area Under Precision Recall (AUPR). The correctly detected adversarial and benign RSIs are true positives (TP) and true negatives (TN), respectively. On the contrary, the wrongly detected adversarial and benign RSIs are false negatives (FN) and false positives (FP), respectively.

3.3. *Adversarial Robustness Evaluation Platform for Remote Sensing Images (AREP-RSIs)*

Based on the above work, we further developed a one-stop platform for conducting adversarial defense and conveniently evaluating the adversarial robustness of a DNN-based visual recognition system on UAVs called *Adversarial Robustness Evaluation Platform for Remote Sensing Images* (AREP-RSIs). AREP-RSIs are multi-functional, and users can readily operate on this platform to evaluate the defensive performance for a DNN model trained with RSIs. In addition, if we load a well-trained DNN model, AREP-RSIs connected with cameras can predict the category of a scene or target for a real-time image and output the confidence scores in the main interface.

As shown in Figure 6, AREP-RSIs is built as a modular framework with 6 sub-modules including datasets, models, training, adversarial attack, test for recognition accuracy and adversarial defense. For example, the module of a test for recognition accuracy has two sub-models: single image test and batch images test. In the single image test, users can load an RSI and an arbitrary DNN model file to obtain the predicted category and the maximum confidence scores. If the selected RSI is detected as an adversarial example, the activated feature maps of this adversarial image and its corresponding original image are displayed. In the batch images test, a batch of RSIs is input to the selected DNN model, and the interface will show a confusion matrix to visualize the recognition performance. We can also know the recognition accuracy of this batch of RSIs.

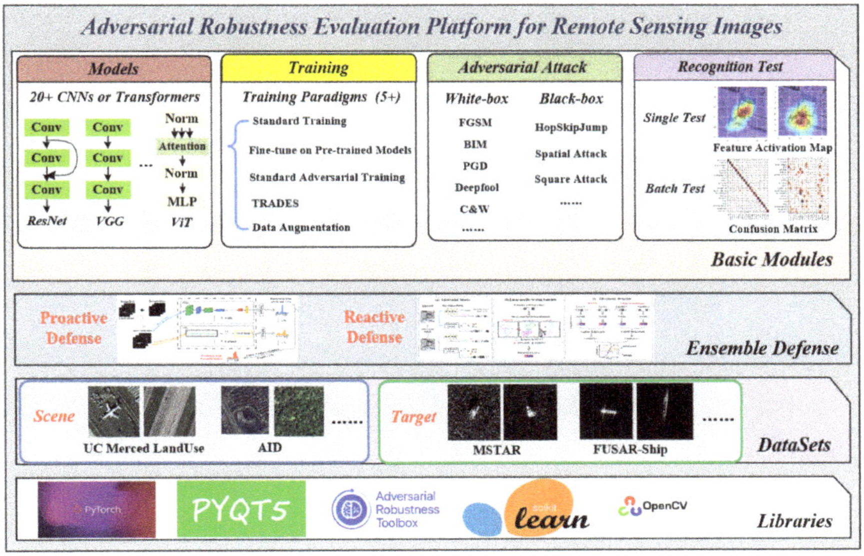

Figure 6. The overall framework of AREP-RSIs with 6 modules.

The graphic interface of this platform is designed with PyQt [76] and built upon necessary libraries such as Pytorch [77], Adversarial Robustness Toolbox (ART) [78], OpenCV [79] and Scikit-learn [80].AREP-RSIs includes several popular optical and SAR RSI datasets for scene/target recognition. We show the screenshots of a graphic interface of two modules in use, recognition test (single image test) and performing the proactive defense as shown in Figures 7 and 8.

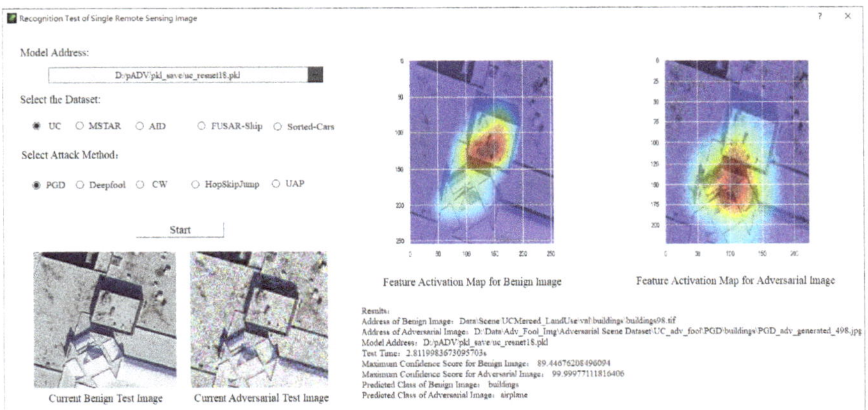

Figure 7. The interface of a recognition test of a single RSI in AREP-RSIs.

Moreover, all of these modules are highly extendable, which greatly facilities designing robust DNN-based recognition models in the remote sensing field for future research. For instance, we can include other adversarial defense methods for RSI recognition, such as TRADES, GAL [75], and DVERGE [81] in proactive defense and FS [40] and DNR [42] in reactive defense into AREP-RSIs. New DNN model architecture, training paradigms and adversarial attacks can also be flexibly registered in AREP-RSIs for users to compare the adversarial robustness before and after performing a specific adversarial defense scheme to a base DNN model. In the current AREP-RSIs, we have embedded more than 20 types of DNNs with different training schemes and various mainstream adversarial attacks. We will make the AREP-RSIs open source at Github (https://github.com/ZeoLuuuuuu/AREP-RSIs, accessed on 26 April 2023).

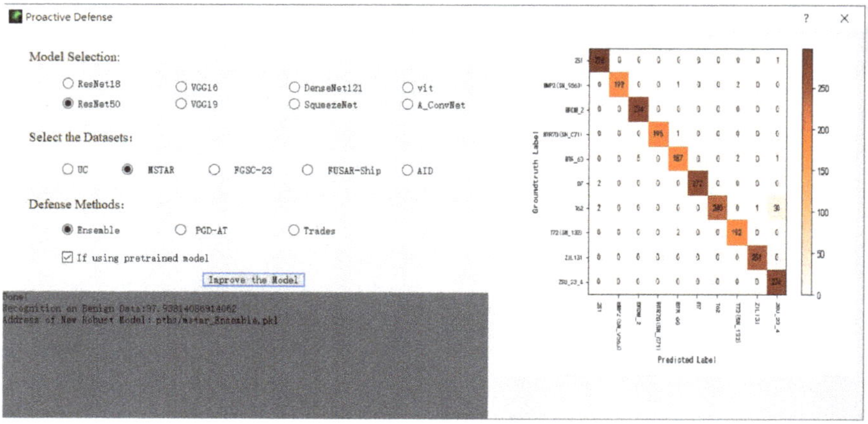

Figure 8. The interface of performing proactive defense to generate robust models in AREP-RSIs.

4. Experiments

4.1. Datasets

(1) Scene Recognition: Two high-quality datasets for scene classification, UCM [82] and AID [83], are selected for our experiments. Both of them include optical RSIs with scene only. The RSI examples for each dataset are illustrated in Figures 9 and 10.

Figure 9. RSI examples randomly selected for each class from UCM.

UCM: The UC Merced LandUse Dataset contains 2100 RSIs from 21 different land-use classes, each of which contains 100 256 $\times$ 256 images with a spatial resolution of 0.3 m per pixel in the RGB color space. The dataset is derived from the National Map Urban Area Imagery collection, which captures the scenes of nationwide towns across the United States.

AID: AID is a large RSI dataset that collects scene images from Google Earth. The dataset comprises 10,000 labeled RSIs containing 30 categories of scenes, approximately 200–420 images per category with an image size of 600 $\times$ 600 pixels. Even if the Google Earth images are post-processed using RGB renderings of the original aerial images, this does not affect its use in evaluating scene classification algorithms.

(2) Target Recognition: Two benchmark datasets for target recognition, MSTAR [84] and FUSAR-Ship [85], are also utilized in the experiments. The RSI examples for each dataset are illustrated in Figures 11 and 12.

Figure 10. RSI examples randomly selected for each class from AID.

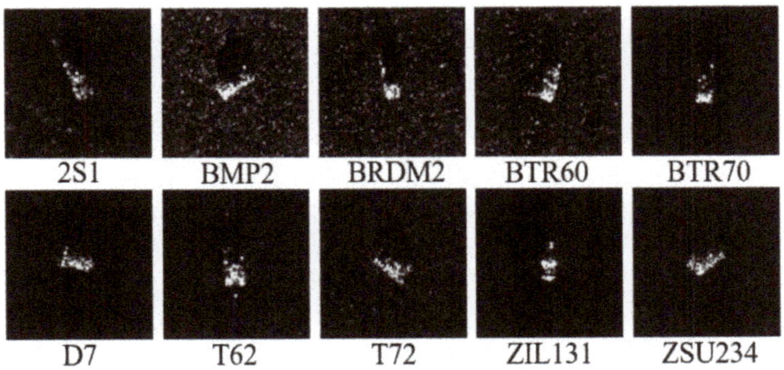

Figure 11. RSI examples randomly selected for each class from MSTAR.

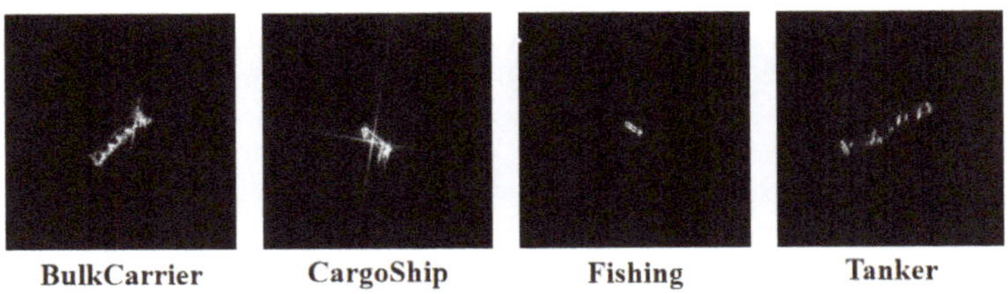

Figure 12. RSI examples randomly selected for each class from FUSAR-Ship.

MSTAR: MSTAR is from the publicly available Moving and Stationary Target Acquisition and Recognition (MSTAR) dataset produced by the US Defense Advanced Research Projects Agency. This dataset contains 5172 SAR sliced images of stationary vehicles with 10 categories acquired at various azimuths. The sensor is a high-resolution cluster SAR with a resolution of 0.3 m $\times$ 0.3 m, operating in the X-band.

FUSAR-Ship: FUSAR-Ship is a high-resolution SAR dataset obtained from GF-3 for ship detection and recognition. The maritime targets are divided into two branches, ship and non-ship. Here, we selected four sub-classes, bulk carrier, cargo ship, fishing and tanker from ship targets, collecting 420 images in total.

4.2. Experimental Setup and Results

We designed our experiment in a systematic manner to verify the adversarial robustness improvement of DNNs for RSI recognition after performing an ensemble strategy. In fact, our experiments include four procedures, which are training and testing base DNNs for recognition in RSIs, performing adversarial attacks with RSIs against the base models, improving adversarial robustness with the proactive ensemble model and detecting adversarial examples with the reactive ensemble model. All the experiments are implemented on a server equipped with an Intel Core i9-12900KF 3.19 GHz CPU, 32 GB of RAM and one NVIDIA GeForce RTX 3090 Ti GPU (24 GB Video RAM). The deep learning framework is Pytorch 1.8. All of the above experiments can be performed on the one-stop integrated platform AREP-RSIs, which makes it greatly convenient for users to evaluate the defensive effectiveness and adversarial robustness.

In this part, we collected all the quantitative results presented in the form of a graph or table, and in the following part, we analyzed the results adequately to verify if the ensemble models for both proactive and reactive defense are effective for the RSI recognition task.

In the first part, the training sets are randomly selected with 80% labeled images in each dataset, and the remaining images make up the test set. The trained base models are also the components in the following proactive ensemble model including ResNet-18, ResNet-50, ResNet-101, ViT-Base/16, ViT-Base/32 and ViT-Large/16. We train all models for 100 epochs with batch size = 32, and the optimizer as Adam [86]. We collected the recognition accuracy of the test set for these base models, as shown in Table 2.

Table 2. Recognition accuracy of base DNN models for test set of RSI datasets (the values below are averaged from 10 repeated experiments).

	UCM	AID	MSTAR	FUSAR-Ship
ResNet-18	96.19%	95.65%	94.80%	**81.40**%
ResNet-50	**96.67**%	96.05%	**97.73**%	80.95%
ResNet-101	92.38%	**97.90**%	93.32%	77.91%
ViT-Base/16	94.80%	92.70%	88.21%	79.76%
ViT-Base/32	91.80%	91.20%	82.64%	76.19%
ViT-Large/16	87.38%	93.34%	88.08%	78.57%

In terms of adversarial attacks, both white-box and black-box conditions are considered. Specifically, we choose 4 white-box and 2 black-box attack algorithms including the Fast Gradient Sign Method (FGSM) [25], Basic Iterative Method (BIM) [87], Carlini and Warger Attack (C&W Attack) [88], Deepfool [89], Square Attack (SA) [90] and HopSkipJump Attack (HSJA) [91]. The settings for attacks in our experiment are shown in Table 3. The victim model is ResNet-18 and ViT-Base/16.

Table 3. Important parameters of attack algorithms utilized in the experiments.

	Batch Size	Norm of Perturbation	Maximum Perturbation	Number of Iterations
FGSM	32	L_2	0.25	–
BIM	32	L_2	0.125	25
C&W	32	L_∞	–	20
Deepfool	8	–	–	50
SA	16	L_2	0.3	50
HSJA	16	L_2	–	50

In the part of proactive defense, we will recognize the generated adversarial data with the victim base models (i.e., ResNet-18 and ViT-Base/16). We set the weight of each base model in the ensemble as the same value, namely $1/2N$. The results of ASR from the victim model will be viewed as the performances before the defense. To evaluate the effectiveness of the ensemble model, we also conduct three counterparts in proactive defense of PGD-AT (adversarial training with PGD-perturbed RSIs), TRADES and GAL on the victim base models. The results for proactive defense are graphed as shown in Figures 13 and 14, and the victim model is labeled as *Without Defense* in both graphs.

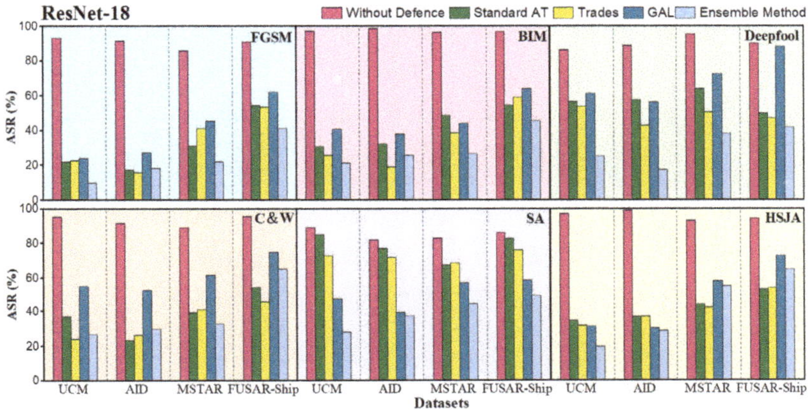

Figure 13. Comparisons in ASR of ensemble model in proactive defense with that of base model ResNet-18 and its three counterparts (the results are averaged from 10 repeated experiments).

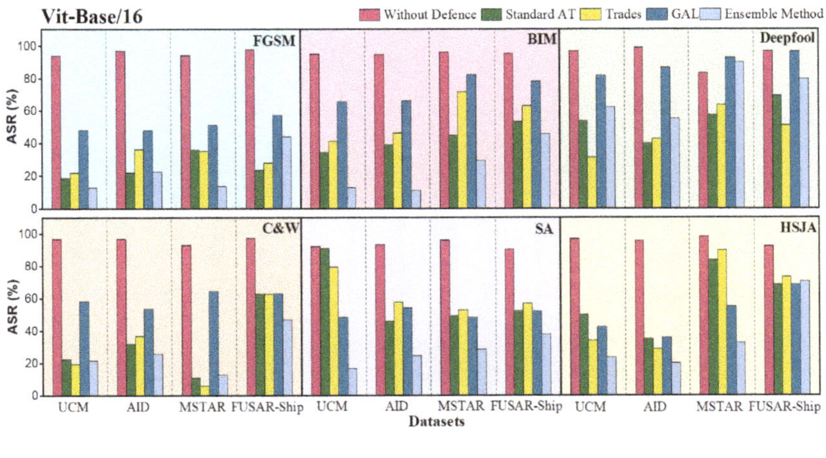

Figure 14. Comparisons in ASR of ensemble model in proactive defense with that of base model ViT-Base/16 and its three counterparts (the results are averaged from 10 repeated experiments).

In the last part of reactive defense, we compare the performances of the ensemble model with stand-alone detectors (i.e., KDE, LID and MD) in the ensemble framework. All four detectors exploit layer-specific scores from several intermediate layers of ResNet-18 and transformer encoders of ViT-Base/16 through logistic regression, and they detect if the input RSI is adversarial or benign. We only selected white-box attacks on UCM, AID and MSTAR for the experiments of this part because the RSIs in the test set of FUSAR-Ship are too inadequate to obtain stable data and analyze meaningful conclusions. The results of reactive defense are shown in Tables 4–6.

Table 4. Performances of ensemble method on UCM with three individual detection algorithms (the results below are averaged from 10 repeated experiments).

Dataset		FGSM		BIM		DeepFool		C & W	
		AUPR	**AUROC**	**AUPR**	**AUROC**	**AUPR**	**AUROC**	**AUPR**	**AUROC**
UCM	LID	88.89	90.30	88.52	89.53	57.91	65.22	64.27	72.35
	MD	92.95	88.51	90.24	84.83	67.45	74.28	76.44	83.42
	KDE	88.67	89.56	83.13	84.63	66.26	75.21	61.75	75.27
	Ensemble	**93.26**	**94.15**	**91.35**	**94.10**	**75.73**	**82.29**	**80.26**	**85.18**

Table 5. Performances of ensemble method on AID with three individual detection algorithms (the results below are averaged from 10 repeated experiments).

Dataset		FGSM		BIM		DeepFool		C & W	
		AUPR	**AUROC**	**AUPR**	**AUROC**	**AUPR**	**AUROC**	**AUPR**	**AUROC**
AID	LID	89.38	90.25	89.47	84.35	60.12	74.39	73.72	71.43
	MD	92.67	93.33	89.46	92.23	71.15	77.23	77.41	**85.63**
	KDE	87.59	83.84	80.93	83.30	68.18	78.32	61.51	73.77
	Ensemble	**95.73**	**95.93**	**93.37**	**95.10**	**74.05**	**81.08**	**80.40**	84.15

Table 6. Performances of ensemble method on MSTAR with three individual detection algorithms (the results below are averaged from 10 repeated experiments).

Dataset		FGSM		BIM		DeepFool		C & W	
		AUPR	**AUROC**	**AUPR**	**AUROC**	**AUPR**	**AUROC**	**AUPR**	**AUROC**
MSTAR	LID	81.58	83.17	81.13	82.79	71.80	74.83	68.47	71.61
	MD	**83.45**	84.85	85.73	77.67	67.25	72.23	75.41	78.24
	KDE	74.51	73.84	77.93	82.17	**73.60**	**78.74**	69.54	68.50
	Ensemble	82.43	**86.91**	**86.57**	**87.04**	72.13	77.37	**75.67**	**78.59**

4.3. Discussion

4.3.1. Recognition Performance of Base Models

First, for the base models in an ensemble of proactive defense, we trained them with the same setting and reported the recognition accuracy on the test sets. It can be observed that most of the 24 models yield very good performances with an accuracy of more than 85% except for the models on FUSAR-Ship. The reason for a drop in FUSAR-Ship is probably that the number of RSIs in FUSAR-Ship is scarce (only 420 RSIs in total) and the appearances of targets in four categories are similar, which makes it hard for the DNN model to learn the discriminative features to correctly distinguish them. The highest accuracy comes from ResNet-101 on AID, which can reach 97.86%. Models with deeper layers and more complex architectures perform a little bit worse such as ResNet-101 and ViT-Large/16 on UCM, which may be caused by a slight overfitting problem as the train data are not that sufficient. Nevertheless, all of these base models are well-trained and will be utilized in the later experiments of ensemble strategy for adversarial defense.

4.3.2. Analysis on Proactive Defense

We crafted adversarial examples against the ResNet-18 and ViT-Base/16, respectively, for each dataset with adversarial attack methods. The adversarial data are then recognized by the corresponding victim base model, our proposed ensemble model, and the victim base model is strengthened by three popular proactive defense methods. It can be noticed that in Figures 11 and 12, the height of all pink columns indicates that the ASR of these attacks reaches a very high level for the victim base model, which exhibits serious adversarial vulnerability and needs to be reduced urgently.

For adversarial examples generated against ResNet-18, we find that the ensemble of ResNets and ViTs performs well in optical datasets, especially with FGSM, BIM, Deepfool and HSJA attacks. In an optical setting, the ensemble can perform more consistently than other proactive defense methods. For example, ResNet-18 with Trades can correctly recognize more adversarial examples in BIM, but it has unsatisfactory performance in Deepfool. For the ensemble model, the best result is from the FGSM of UCM, with only 9.52% ASR. For SAR configurations, the ensemble of base models obtains better results in MSTAR than FUSAR-Ship, while it is worse than those from UCM and AID. In general, if we say an ASR below 30% is qualified, the ensemble has a good result in 15 out of 24 scenarios.

For adversarial examples generated against ViT-Base/16, the ensemble of ResNets and ViTs also maintains relatively low ASRs for most adversarial attack methods in optical RSI datasets. It is interesting to find that the ensemble model performs even worse than the base model without defense in Deepfool of MSTAR, but in C&W, another attack with very imperceptible noise, it yields decent values for MSTAR. Still, if we say an ASR below 30% is qualified, the ensemble has an acceptable result in 14 out of 24 scenarios.

Overall, compared with the models without defense under an adversarial attack, the ensemble strategy effectively improves the adversarial robustness and can rival or even perform better than the three other popular adversarial proactive defense methods.

4.3.3. Analysis on Reactive Defense

Last but not least, for reactive defense, we first discuss the results in optical RSI datasets. It can be observed that the ensemble method obtains the best AUPR or AUROC in 15 out of 16 scenarios. For gradient-based attacks of FGSM and BIM, the ensemble model can yield AUPR and AUROC values of more than 90%, which are obviously better than those from Deepfool and C&W. That is because Deepfool finds the shortest path to guide original RSIs across a decision boundary to generate adversarial examples, and C&W is an optimized-based attack with very small perturbations added to the original RSIs. The best result comes from the ensemble model in detecting FGSM on AID, with AUPR and AUROC values of 95.73 and 95.93, respectively. In addition, the results of FGSM are slightly better than those of the BIM attack, which is probably because the maximum perturbation in FGSM-perturbed RSIs is a little larger; thus, it leads to more obvious changes in feature representation. With respect to two harder situations, Deepfool and C&W, the ensemble model still shows better ability than stand-alone adversarial detection algorithms, especially with obvious improvements in Deepfool and C&W on UCM. MD only yields AUPR and AUROC values of Deepfool on UCM as 67.25 and 74.28, while our modified ENAD framework improve the metrics to 75.73 and 82.29. The results are not as good as those in gradient-based attacks, but compared with stand-alone detectors, these improvements show that the ensemble of detection algorithms and base DNN models has brought substantial benefits. In general, the ensemble framework has the potential to perform very well in RSI recognition for optical configuration.

In terms of results in MSTAR, the SAR dataset of target recognition, the values of output are generally lower than those of UCM and AID. The performances of the ensemble model are decreasing with the five best out of eight results. One possible reason for this phenomenon may lie in that the channel of SAR RSIs is 1 and most of an RSI in MSTAR is background without useful information, which inhibits the detector from extracting representative features except the target itself. Nevertheless, the detection of gradient-based attacks remains at a high level, with the AUPR and AUROC at around 85. The highest value comes from the BIM attack with 87.04 and the lowest is from the Deepfool attack with 73.60. The Deepfool and C&W attacks are still challenging situations with more imperceptible perturbations. In Deepfool, the results from the ensemble model are even lower than the stand-alone detector KDE, and in C&W, it performs at almost the same level as MD. Therefore, in such a case, an ensemble framework is not recommended, and it is worthwhile to further modify the ensemble model for a better detection in the SAR recognition dataset, especially for very imperceptible noise in the digital domain.

5. Conclusions

Stability and reliability are significant factors in the working process of modern UAVs with DNN-based visual navigation and recognition systems. However, there exists severe adversarial vulnerability when performing scene and target recognition tasks. We build a threat model when attackers maliciously access the communication link or place physical adversarial camouflage on targets. In the scenario, considering that AT is not adaptive for the resource-limited edge environment like UAVs and single adversarial detectors not performing well in reactive defense, we exploit the different mechanisms of feature extractions and weak adversarial transferability between the two mainstrean DNN models, CNN and transformer, to build deep ensemble models for both proactive and reactive adversarial defense only with base DNN models for the RSI recognition task. In addition, a one-stop platform for conducting adversarial defenses and evaluating adversarial robustness for DNN-based RSI recognition models called AREP-RSIs is developed, which can be edge-deployed to achieve real-time recognition and greatly facilitate designing more robust defense strategies in the remote sensing field for future research.

To evaluate the effectiveness of the two ensemble strategies, a series of experiments are conducted with both optical and SAR RSI datasets. We find that an ensemble of ResNets and ViTs can yield very satisfactory results in recognizing and detecting adversarial examples generated by gradient-based attacks such as FGSM and BIM. In proactive defense, compared with the three other popular defense methods, the ensemble can be more stable in different configurations. In reactive defense, our ensemble model integrates the scoring values from multiple detection algorithms and confidence scores from different base models, performing much better than stand-alone detectors in most experimental settings. Even though the proposed model does not perform as well on some attacks of SAR datasets, this ensemble strategy has shown the favorable potential to improve detection rates with the DNN models trained for RSI recognition.

In our future work, we will further optimize both of the deep ensemble frameworks, including exploring the defensive effectiveness against other types of adversarial attack in the RSI recognition task, replacing the current DNNs in the ensemble with more lightweight network architectures to suit the edge environment better and making the models' weights learnable during the training time to find the best combination. Therefore, as the first exploration of a deep ensemble method against adversarial RSIs in resource-limited environments, we need to conduct more experiments and report them in our next article. Finally, we will deploy the two deep ensemble models and AREP-RSIs on the edge devices to truly achieve a practical application.

Author Contributions: Methodology, Z.L.; software, Z.L. and Y.X.; validation, Z.L., H.S. and Y.X.; original draft preparation, Y.X.; writing—review and editing, Z.L.; supervision, H.S. and Y.X. All authors have read and agreed to the published version of the manuscript.

Funding: This work was supported by the National Natural Science Foundation of China under Grant 61971426.

Data Availability Statement: The UCM, AID, MSTAR and FUSAR-Ship dataset are available in the references of this paper.

Conflicts of Interest: The authors declare no conflict of interest.

References

1. LeCun, Y.; Bengio, Y.; Hinton, G. Deep learning. *Nature* **2015**, *521*, 436–444. [CrossRef] [PubMed]
2. He, H.; Wang, S.; Yang, D.; Wang, S. SAR target recognition and unsupervised detection based on convolutional neural network. In Proceedings of the 2017 Chinese Automation Congress (CAC), Jinan, China, 20–22 October 2017; pp. 435–438. [CrossRef]
3. Cho, J.H.; Park, C.G. Multiple Feature Aggregation Using Convolutional Neural Networks for SAR Image-Based Automatic Target Recognition. *IEEE Geosci. Remote Sens. Lett.* **2018**, *15*, 1882–1886. [CrossRef]
4. Wang, L.; Yang, X.; Tan, H.; Bai, X.; Zhou, F. Few-Shot Class-Incremental SAR Target Recognition Based on Hierarchical Embedding and Incremental Evolutionary Network. *IEEE Trans. Geosci. Remote Sens.* **2023**, *61*, 5204111. [CrossRef]

5. Ding, B.; Wen, G.; Ma, C.; Yang, X. An Efficient and Robust Framework for SAR Target Recognition by Hierarchically Fusing Global and Local Features. *IEEE Trans. Image Process.* **2018**, *27*, 5983–5995. [CrossRef] [PubMed]
6. Deng, H.; Huang, J.; Liu, Q.; Zhao, T.; Zhou, C.; Gao, J. A Distributed Collaborative Allocation Method of Reconnaissance and Strike Tasks for Heterogeneous UAVs. *Drones* **2023**, *7*, 138. [CrossRef]
7. Li; Bin; Fei, Z.; Zhang, Y. UAV communications for 5G and beyond: Recent advances and future trends. *IEEE Internet Things J.* **2018**, *6*, 2241–2263. [CrossRef]
8. Khuwaja, A.A.; Chen, Y.; Zhao, N.; Alouini, M.S.; Dobbins, P. A survey of channel modeling for UAV communications. *IEEE Commun. Surv. Tutorials* **2018**, *20*, 2804–2821. [CrossRef]
9. Azari, M.M.; Geraci, G.; Garcia-Rodriguez, A.; Pollin, S. UAV-to-UAV communications in cellular networks. *IEEE Trans. Wirel. Commun.* **2020**, *19*, 6130–6144. [CrossRef]
10. El Meouche, R.; Hijazi, I.; Poncet, P.A.; Abunemeh, M.; Rezoug, M. Uav Photogrammetry Implementation to Enhance Land Surveying, Comparisons and Possibilities. *Int. Arch. Photogramm. Remote Sens. Spat. Inf. Sci.* **2016**, *42*, 107–114. [CrossRef]
11. Jung, S.; Kim, H. Analysis of amazon prime air uav delivery service. *J. Knowl. Inf. Technol. Syst.* **2017**, *12*, 253–266.
12. She, R.; Ouyang, Y. Efficiency of UAV-based last-mile delivery under congestion in low-altitude air. *Transp. Res. Part C Emerg. Technol.* **2021**, *122*, 102878. [CrossRef]
13. Thiels, C.A.; Aho, J.M.; Zietlow, S.P.; Jenkins, D.H. Use of unmanned aerial vehicles for medical product transport. *Air Med. J.* **2015**, *34*, 104–108. [CrossRef] [PubMed]
14. Konert, A.; Smereka, J.; Szarpak, L. The use of drones in emergency medicine: Practical and legal aspects. *Emerg. Med. Int.* **2019**, *2019*, 3589792. [CrossRef]
15. Michael, D.; Josh, H.; Keith, M.; Mikel, R. The vulnerability of UAVs: An adversarial machine learning perspective. In Proceedings of the Geospatial Informatics XI, SPIE, Online, 22 April 2021; Volume 11733, pp. 81–92. [CrossRef]
16. Barbu, A.; Mayo, D.; Alverio, J.; Luo, W.; Wang, C.; Gutfreund, D.; Tenenbaum, J.; Katz, B. Objectnet: A large-scale bias-controlled dataset for pushing the limits of object recognition models. In Proceedings of the 33rd International Conference on Neural Information Processing Systems, Vancouver, BC, Canada, 8–14 December 2019; pp. 9453–9463.
17. Hendrycks, D.; Dietterich, T. Benchmarking neural network robustness to common corruptions and perturbations. International Conference on Learning Representation. *arXiv* **2019**, arXiv:1903.12261.
18. Dong, Y.; Ruan, S.; Su, H.; Kang, C.; Wei, X.; Zhu, J. Viewfool: Evaluating the robustness of visual recognition to adversarial viewpoints. Advances in Neural Information Processing Systems. *arXiv* **2022**, arXiv:2210.03895v1.
19. Hendrycks, D.; Lee, K.; Mazeika, M. Using pretraining can improve model robustness and uncertainty. *Int. Conf. Mach. Learn.* **2019**, *97*, 2712–2721.
20. Akhtar, N.; Mian, A. Threat of Adversarial Attacks on Deep Learning in Computer Vision: A Survey. *IEEE Access* **2018**, *6*, 14410–14430. [CrossRef]
21. Akhtar, N.; Mian, A.; Kardan, N.; Shah, M. Advances in Adversarial Attacks and Defenses in Computer Vision: A Survey. *IEEE Access* **2021**, *9*, 155161–155196. [CrossRef]
22. Khamaiseh, S.Y.; Bagagem, D.; Al-Alaj, A.; Mancino, M.; Alomari, H.W. Adversarial Deep Learning: A Survey on Adversarial Attacks and Defense Mechanisms on Image Classification. *IEEE Access* **2022**, *10*, 102266–102291. 2022.3208131. [CrossRef]
23. Deng, J.; Dong, W.; Socher, R.; Li, L.J.; Li, K.; Li, F.-F. ImageNet: A large-scale hierarchical image database. In Proceedings of the 2009 IEEE Conference on Computer Vision and Pattern Recognition, Miami, FL, USA, 20–25 June 2009; pp. 248–255. [CrossRef]
24. Zhang, L.; Zhang, L. Artificial Intelligence for Remote Sensing Data Analysis: A review of challenges and opportunities. *IEEE Geosci. Remote Sens. Mag.* **2022**, *10*, 270–294. [CrossRef]
25. Goodfellow, I.J.; Shlens, J.; Szegedy, C. Explaining and harnessing adversarial examples. *arXiv* **2014**, arXiv:1412.6572.
26. Zhang, H.Y.; Yu, Y.D.; Jiao, J.T.; Xing, E.P.; Ghaoui, L.E.; Jordan, M. Theoretically principled trade-off between robustness and accuracy. *arXiv* **2019**, arXiv:1901.08573.
27. Zhang, J.F.; Xu, X.L.; Han, B.; Niu, G.; Cui, L.Z.; Sugiyama, M.; Kankanhalli, M. Attacks which do not kill training make adversarial learning stronger. *Int. Conf. Mach. Learn.* **2020**, *119*, 11278–11287.
28. Jia, X.J.; Zhang, Y.; Wu, B.Y.; Ma, K.; Wang, J.; Cao, X.C. LAS-AT: Adversarial Training with Learnable Attack Strategy. *arXiv* **2022**, arXiv:2203.06616.
29. Saligrama, A.; Leclerc, G. Revisiting Ensembles in an Adversarial Context: Improving Natural Accuracy. *arXiv* **2020**, arXiv:2002.11572.
30. Li, N.; Yu, Y.; Zhou, Z.H. Diversity regularized ensemble pruning. In Proceedings of the Joint European Conference on Machine Learning and Knowledge Discovery in Databases, Bilbao, Spain, 13–17 September 2012; pp. 330–345.
31. Wang, X.; Xing, H.; Hua, Q.; Dong, C.R.; Pedrycz, W. A study on relationship between generalization abilities and fuzziness of base classifiers in ensemble learning. *IEEE Trans. Fuzzy Syst.* **2015**, *23*, 1638–1654. [CrossRef]
32. Sun, T.; Zhou, Z.H. Structural diversity for decision tree ensemble learning. *Front. Comput. Sci.* **2018**, *12*, 560–570. [CrossRef]
33. Cohen, G.; Sapiro, G.; Giryes, R. Detecting adversarial samples using influence functions and nearest neighbors. In Proceedings of the IEEE/CVF Conference on Computer Vision and Pattern Recognition, Seattle, WA, USA, 13–19 June 2020; pp. 14453–14462.
34. Ma, X.J.; Li, B.; Wang, Y.S.; Erfani, S.M.; Wijewickrema, S.N.R.; Schoenebeck, G.; Song, D.; Houle, M.E.; Bailey, J. Characterizing adversarial subspaces using local intrinsic dimensionality. In Proceedings of the 6th International Conference on Learning Representations, ICLR, Vancouver, BC, Canada, 30 April–3 May 2018.

35. Feinman, R.; Curtin, R.R.; Shintre, S.; Gardner, A.B. Detecting Adversarial Samples from Artifacts. *arXiv* **2017**, arXiv:1703.0041(
36. Lee, K.; Lee, K.; Lee, H.; Shin, J. A simple unified framework for detecting out-of distribution samples and adversarial attacks *Adv. Neural Inf. Process. Syst.* **2018**, *31*, 7167–7177.
37. Hendrycks, D.; Gimpel, K. Early methods for detecting adversarial images. In Proceedings of the 5th International Conference on Learning Representations, ICLR, Toulon, France, 24–26 April 2017.
38. Zheng, Z.H.; Hong, P.Y. Robust detection of adversarial attacks by modeling the intrinsic properties of deep neural networks *Adv. Neural Inf. Process. Syst.* **2018**, *31*, 7913–7922.
39. Liang, B.; Li, H.C.; Su, M.Q.; Li, X.R.; Shi, W.C.; Wang, X.F. Detecting adversarial image examples in deep neural networks with adaptive noise reduction. *IEEE Trans. Dependable Secur. Comput.* **2021**, *18*, 72–85. [CrossRef]
40. Xu, W.L.; Evans, D.; Qi, Y.J. Feature squeezing: Detecting adversarial examples in deep neural networks. In Proceedings of the 25th Annual Network and Distributed System Security Symposium, NDSS 2018, San Diego, CA, USA, 18–21 February 2018.
41. Kherchouche, A.; Fezza, S.A.; Hamidouche, W.; Deforges, O. Detection of adversarial examples in deep neural networks with natural scene statistics. In Proceedings of the 2020 International Joint Conference on Neural Networks (IJCNN), IEEE, Glasgow, UK, 19–24 July 2020; pp. 1–7.
42. Sotgiu, A.; Demontis, A.; Melis, M.; Biggio, B.; Fumera, G.; Feng, X.Y.; Roli, F. Deep neural rejection against adversarial examples *Eurasip J. Inf. Secur.* **2020**, *2020*, 5. [CrossRef]
43. Aldahdooh, A.; Hamidouche, W.; Deforges, O. Revisiting model's uncertainty and confidences for adversarial example detection *arXiv* **2021**, arXiv:2103.05354.
44. Carrara, F.; Falchi, F.; Caldelli, R.; Amato, G.; Fumarola, R.; Becarelli, R. Detecting adversarial example attacks to deep neural networks. In Proceedings of the 15th International Workshop on Content-Based Multimedia, Florence, Italy, 19–21 June 2017.
45. Aldahdooh, A.; Hamidouch, W.; Fezza, S.A.; Deforges, O. Adversarial Example Detection for DNN Models: A Review and Experimental Comparison. *arXiv* **2021**, arXiv:2105.00203.
46. He, K.; Zhang, X.; Ren, S.; Sun, J. Deep Residual Learning for Image Recognition. In Proceedings of the 2016 IEEE Conference on Computer Vision and Pattern Recognition (CVPR), Las Vegas, NV, USA, 27–30 June 2016; pp. 770–778. [CrossRef]
47. Dosovitskiy, A.; Beyer, L.; Kolesnikov, A.; Weissenborn, D.; Zhai, X.; Unterthiner, T.; Dehghani, M.; Minderer, M.; Heigold, G. Gelly, S.; et al. An Image is Worth 16 × 16 Words: Transformers for Image Recognition at Scale. *arXiv* **2020**, arXiv:2010.11929.
48. Craighero, F.; Angaroni, F.; Stella, F.; Damiani, C.; Antoniotti, M.; Graudenzi, A. Unity is strength: Improving the detection of adversarial examples with ensemble approaches. *arXiv* **2021**, arXiv:2111.12631.
49. Sun, H.; Chen, J.; Lei, L.; Ji, K.; Kuang, G. Adversarial robustness of deep convolutional neural network-based image recognition models: A review. *J. Radars* **2021**, *10*, 571–594. [CrossRef]
50. Chen, L.; Zhu, G.; Li, Q.; Li, H. Adversarial example in remote sensing image recognition. *arXiv* **2019**, arXiv:1910.13222.
51. Xu, Y.; Du, B.; Zhang, L. Assessing the threat of adversarial examples on deep neural networks for remote sensing scene classification: Attacks and defenses. *IEEE Trans. Geosci. Remote Sens.* **2021**, *59*, 1604–1617. [CrossRef]
52. Xu, Y.; Ghamisi, P. Universal Adversarial Examples in Remote Sensing: Methodology and Benchmark. *IEEE Trans. Geosci. Remote Sens.* **2022**, *60*, 1–15. [CrossRef]
53. Chen, L.; Xiao, J.; Zou, P.; Li, H. Lie to Me: A Soft Threshold Defense Method for Adversarial Examples of Remote Sensing Images *IEEE Geosci. Remote Sens. Lett.* **2022**, *19*, 8016905. [CrossRef]
54. Li, H.; Huang, H.; Chen, L.; Peng, J.; Huang, H.; Cui, Z.; Mei, X.; Wu, G. Adversarial Examples for CNN-Based SAR Image Classification: An Experience Study. *IEEE J. Sel. Top. Appl. Earth Obs. Remote Sens.* **2021**, *14*, 1333–1347. [CrossRef]
55. Du, C.; Huo, C.; Zhang, L.; Chen, B.; Yuan, Y. Fast C&W: A Fast Adversarial Attack Algorithm to Fool SAR Target Recognition with Deep Convolutional Neural Networks. *IEEE Geosci. Remote Sens. Lett.* **2022**, *19*, 4010005. [CrossRef]
56. Zhou, J.; Sun, H.; Lei, L.; Ke, J.; Kuang, G. Sparse Adversarial Attack of SAR Image. *J. Signal Process.* **2021**, *37*, 11.
57. Czaja, W.; Fendley, N.; Pekala, M.; Ratto, C.; Wang, I.J. Adversarial examples in remote sensing. In Proceedings of the 26th ACM SIGSPATIAL International Conference on Advances in Geographic Information Systems, Washington, DC, USA, 6–9 November 2018; pp. 408–411.
58. Hollander, R.; Adhikari, A.; Tolios, I.; van Bekkum, M.; Bal, A.; Hendriks, S.; Kruithof, M.; Gross, D.; Jansen, N.; Perez, G. et al. Adversarial patch camouflage against aerial detection. In Proceedings of the Artificial Intelligence and Machine Learning in Defense Applications II, International Society for Optics and Photonics, SPIE, Online, 20 September 2020; Volume 11543, pp. 77–86.
59. Du, A.; Chen, B.; Chin, T.-J.; Law, Y.W.; Sasdelli, M.; Rajasegaran, R.; Campbell, D. Physical adversarial attacks on an aerial imagery object detector. In Proceedings of the IEEE/CVF Winter Conference on Applications of Computer Vision, Waikoloa, HI, USA, 4–8 January 2022.
60. Torens, C.; Juenger, F.; Schirmer, S.; Schopferer, S.; Maienschein, T.D.; Dauer, J.C. Machine Learning Verification and Safety for Unmanned Aircraft-A Literature Study. In Proceedings of the AIAA Scitech 2022 Forum, San Diego, CA, USA, 3–7 January 2022
61. Tian, J.; Wang, B.; Guo, R.; Wang, Z.; Cao, K.; Wang, X. Adversarial Attacks and Defenses for Deep-Learning-Based Unmanned Aerial Vehicles. *IEEE Internet Things J.* **2021**, *9*, 22399–22409. [CrossRef]
62. Brown, T.B.; Mané, D.; Roy, A.; Abadi, M.; Gilmer, J. Adversarial patch. *arXiv* **2017**, arXiv:1712.09665.
63. Gu, J.; Tresp, V.; Qin, Y. Are vision transformers robust to patch perturbations? In Proceedings of the Computer Vision—ECCV 2022: 17th European Conference, Tel Aviv, Israel, 23–27 October 2022; Proceedings, Part XII; Springer Nature: Cham, Switzerland, 2022.

64. Raghu, M.; Unterthiner, T.; Kornblith, S.; Zhang, C.; Dosovitskiy, A. Do vision transformers see like convolutional neural networks? *Adv. Neural Inf. Process. Syst.* **2021**, *34*, 12116–12128.
65. Namuk, P.; Kim, S. How do vision transformers work? *arXiv* **2022**, arXiv:2202.06709.
66. Rao, Y.; Zhao, W.; Tang, Y.; Zhou, J.; Lim, S.N.; Lu, J. Hornet: Efficient high-order spatial interactions with recursive gated convolutions. *Adv. Neural Inf. Process. Syst.* **2022**, *35*, 10353–10366.
67. Si, C.; Yu, W.; Zhou, P.; Zhou, Y.; Wang, X.; Yan, S. Inception transformer. *arXiv* **2022**, arXiv:2205.12956.
68. Li, J.; Xia, X.; Li, W.; Li, H.; Wang, X.; Xiao, X.; Wang, R.; Zheng, M.; Pan, X. Next-vit: Next generation vision transformer for efficient deployment in realistic industrial scenarios. *arXiv* **2022**, arXiv:2207.05501.
69. Yang, T.; Zhang, H.; Hu, W.; Chen, C.; Wang, X. Fast-ParC: Position Aware Global Kernel for ConvNets and ViTs. *arXiv* **2022**, arXiv:2210.04020.
70. Cai, Y.; Ning, X.; Yang, H.; Wang, Y. Ensemble-in-One: Learning Ensemble within Random Gated Networks for Enhanced Adversarial Robustness. *arXiv* **2021**, arXiv:2103.14795.
71. Pang, T.; Xu, K.; Du, C.; Chen, N.; Zhu, J. Improving adversarial robustness via promoting ensemble diversity. In Proceedings of the 36th International Conference on Machine Learning, PMLR, Long Beach, CA, USA, 9–15 June 2019.
72. Teresa, Y.; Kar, O.F.; Zamir, A. Robustness via cross-domain ensembles. In Proceedings of the IEEE/CVF International Conference on Computer Vision, Montreal, QC, Canada, 10–17 October 2021.
73. Tsipras, D.; Santurkar, S.; Engstrom, L.; Turner, A.; Madry, A. Robustness may be at odds with accuracy. *arXiv* **2018**, arXiv:1805.12152.
74. Mahmood, K.; Mahmood, R.; van Dijk, M. On the Robustness of Vision Transformers to Adversarial Examples. In Proceedings of the 2021 IEEE/CVF International Conference on Computer Vision (ICCV), Montreal, QC, Canada, 10–17 October 2021; pp. 7818–7827. [CrossRef]
75. Sanjay, K.; Qureshi, M.K. Improving adversarial robustness of ensembles with diversity training. *arXiv* **2019**, arXiv:1901.09981.
76. Summerfield, M. *Rapid GUI Programming with Python and Qt: The Definitive Guide to PyQt Programming (Paperback)*; Pearson Education: London, UK, 2007.
77. Paszke, A.; Gross, S.; Massa, F.; Lerer, A.; Bradbury, J.; Chanan, G.; Killeen, T.; Lin, Z.; Gimelshein, N.; Antiga, L.; et al. Pytorch: An imperative style, high-performance deep learning library. *Adv. Neural Inf. Process. Syst.* **2019**, *32*, 4970–4979.
78. Nicolae, M.-M.; Sinn, M.; Tran, M.N.; Buesser, B.; Rawat, A.; Wistuba, M.; Zantedeschi, V.; Baracaldo, N.; Chen, B.; Ludwig, H.; et al. Adversarial Robustness Toolbox v1.0.0. *arXiv* **2018**, arXiv:1807.01069.
79. Bradski, G. The openCV library. *Dr. Dobb's J. Softw. Tools Prof. Program.* **2000**, *25*, 120–123.
80. Pedregosa, F.; Varoquaux, G.; Gramfort, A.; Michel, V.; Thirion, B.; Grisel, O.;Blondel, M.; Prettenhofer, P.; Weiss, R.; Dubourg, V.; et al. Scikit-learn: Machine learning in Python. *J. Mach. Learn. Res.* **2011**, *12*, 2825–2830.
81. Yang, H.; Zhang, J.; Dong, H.; Inkawhich, N.; Gardner, A.; Touchet, A.; Wilkes, W.; Berry, H.; Li, H. DVERGE: Diversifying vulnerabilities for enhanced robust generation of ensembles. *Adv. Neural Inf. Process. Syst.* **2020**, *33*, 5505–5515.
82. Yang, Y.; Newsam, S. Bag-of-visual-words and spatial extensions for land-use classification. In Proceedings of the 18th SIGSPATIAL International Conference on Advances in Geographic Information Systems, San Jose, CA, USA, 2–5 November 2010; pp. 270–279.
83. Xia, G.-S.; Hu, J.; Hu, F.; Shi, B.; Bai, X.; Zhong, Y.; Zhang, L. AID: A Benchmark Data Set for Performance Evaluation of Aerial Scene Classification. *IEEE Trans. Geosci. Remote Sens.* **2017**, *55*, 3965–3981. [CrossRef]
84. Ross, T.D.; Worrell, S.W.; Velten, V.J.; Mossing, J.C.; Bryant, M.L. Standard SAR ATR evaluation experiments using the MSTAR public release data set. *Proc. SPIE* **1998**, *3370*, 566–573.
85. Hou, X.; Ao, W.; Song, Q.; Lai, J.; Wang, H.; Xu, F. FUSAR-Ship: Building a high-resolution SAR-AIS matchup dataset of Gaofen-3 for ship detection and recognition. *Sci. China Inf. Sci.* **2020**, *63*, 140303. [CrossRef]
86. Kingma, D.P.; Ba, J. Adam: A method for stochastic optimization. *arXiv* **2014**, arXiv:1412.6980.
87. Kurakin, A.; Goodfellow, I.J.; Bengio, S. Adversarial examples in the physical world. In *Artificial Intelligence Safety and Security*; Chapman and Hall/CRC: Boca Raton, FL, USA, 2018; pp. 99–112.
88. Carlini, N.; Wagner, D. Adversarial examples are not easily detected: Bypassing ten detection methods. In Proceedings of the 10th ACM Workshop on Artificial Intelligence and Security, New York, NY, USA, 3 November 2017.
89. Dezfooli, M.; Mohsen, S.; Fawzi, A.; Frossard, P. Deepfool: A simple and accurate method to fool deep neural networks. In Proceedings of the IEEE Conference on Computer Vision and Pattern Recognition, New Orleans, LA, USA, 18–20 June 2016.
90. Andriushchenko, M.; Croce, F.; Flammarion, N.; Hein, M. Square attack: A query-efficient black-box adversarial attack via random search. In Proceedings of the Computer Vision—ECCV 2020: 16th European Conference, Glasgow, UK, 23–28 August 2020; Proceedings, Part XXIII; Springer International Publishing: Cham, Switzerland, 2020.
91. Chen, J.; Jordan, M.I.; Wainwright, M.J. Hopskipjumpattack: A query-efficient decision-based attack. In Proceedings of the 2020 IEEE Symposium on Security and Privacy (sp), IEEE, San Francisco, CA, USA, 18–21 May 2020.

remote sensing

MDPI

Article

Probabilistic Wildfire Segmentation Using Supervised Deep Generative Model from Satellite Imagery

Ata Akbari Asanjan [1,2,*], Milad Memarzadeh [1,2], Paul Aaron Lott [2,3], Eleanor Rieffel [3] and Shon Grabbe [3]

[1] Data Science Group (DSG), NASA Ames Research Center, Moffett Field, CA 94035, USA; milad.memarzadeh@nasa.gov
[2] USRA Research Institute for Advanced Computer Science (RIACS), Washington, DC 20024, USA; plott@usra.edu
[3] Quantum Artificial Intelligence Laboratory (QuAIL), NASA Ames Research Center, Moffett Field, CA 94035, USA; eleanor.rieffel@nasa.gov (E.R.); shon.grabbe@nasa.gov (S.G.)
* Correspondence: ata.akbariasanjan@nasa.gov

Abstract: Wildfires are one of the major disasters among many and are responsible for more than 6 million acres burned in the United States alone every year. Accurate, insightful, and timely wildfire detection is needed to help authorities mitigate and prevent further destruction. Uncertainty quantification is always a crucial part of the detection of natural disasters, such as wildfires, and modeling products can be misinterpreted without proper uncertainty quantification. In this study, we propose a supervised deep generative machine-learning model that generates stochastic wildfire detection, allowing fast and comprehensive uncertainty quantification for individual and collective events. In the proposed approach, we also aim to address the patchy and discontinuous Moderate Resolution Imaging Spectroradiometer (MODIS) wildfire product by training the proposed model with MODIS raw and combined bands to detect fire. This approach allows us to generate diverse but plausible segmentations to represent the disagreements regarding the delineation of wildfire boundaries by subject matter experts. The proposed approach generates stochastic segmentation via two model streams in which one learns meaningful stochastic latent distributions, and the other learns the visual features. Two model branches join eventually to become a supervised stochastic image-to-image wildfire detection model. The model is compared to two baseline stochastic machine-learning models: (1) with permanent dropout in training and test phases and (2) with Stochastic ReLU activations. The visual and statistical metrics demonstrate better agreements between the ground truth and the proposed model segmentations. Furthermore, we used multiple scenarios to evaluate the model comprehension, and the proposed Probabilistic U-Net model demonstrates a better understanding of the underlying physical dynamics of wildfires compared to the baselines.

Keywords: wildfire detection; generative machine-learning; stochastic modeling; remote sensing; segmentation; uncertainty analysis

Citation: Akbari Asanjan, A.; Memarzadeh, M.; Lott, P.A.; Rieffel E.; Grabbe S. Probabilistic Wildfire Segmentation Using Supervised Deep Generative Model from Satellite Imagery. *Remote Sens.* **2023**, *15*, 2718. https://doi.org/10.3390/rs15112718

Academic Editor: Gwanggil Jeon

Received: 4 April 2023
Revised: 4 May 2023
Accepted: 8 May 2023
Published: 24 May 2023

1. Introduction

Wildfires are one of the necessary dynamic components of terrestrial ecosystems, and they provide significant ecological benefits [1,2]. Natural wildfires offer significant ecological benefits through promoting forest rejuvenation, nutrient cycling, and habitat diversity, all of which contribute to the overall health and resilience of ecosystems [3].

However, it is important to acknowledge the growing trends in wildfire size, frequency, and intensity, which are largely influenced by human activities and interventions. Factors such as wildfire suppression efforts and urban/wildlife encroachment have contributed to increases in wildfire size, frequency, and intensity [4,5]. These anthropogenic influences have transformed wildfires into a global problem in recent decades [2,6]. Consequently, wildfires have emerged as one of the most destructive natural hazards, with severe consequences for both human and ecological systems.

The scale of devastation caused by wildfires is alarming. In the United States alone, wildfires have burned over 6 million acres of land [7], while globally, the figure exceeds 1 billion acres [8]. These staggering numbers underscore the urgent need for effective wildfire management strategies, enhanced understanding of fire behavior, and improved decision-making processes to mitigate the devastating impacts of wildfires on communities, ecosystems, and economies. By leveraging advanced technologies, such as remote sensing, data analytics, and predictive modeling, researchers and practitioners can gain valuable insights into wildfire dynamics and develop proactive measures to reduce risks and enhance resilience in fire-prone regions.

Preventive and mitigative decision-making and proper resource management are tied to the availability of insightful, accurate, and timely wildfire monitoring, along with a deep understanding of wildfire dynamics. For this reason, wildfire detection is an active research field focusing on understanding wildfire's complex processes and correlated factors (e.g., fuel load and structure, vegetation health and types) [9–11]. Many studies have been conducted to improve the accuracy and detection latency [12–15] for the benefit of acute and agile decision-making. Wildfire studies can be categorized into two main directions: (1) deterministic and (2) stochastic models. Deterministic models are a category of simulation that assumes the simulation process is fully resolved and the simulations can be conducted with negligible errors [16]. Many studies use such an assumption to address this problem using deterministic tools [17–19]. For instance, Toan et al., 2019 [18] developed a deterministic machine-learning model that uses geostationary satellites (GOES-16) as input to detect wildfires at the pixel level. The paper reports robustness against different wildfire types and adversarial conditions. In another study, Sayed et al., 2019 [19] created a wildfire dataset from satellite imagery and used a feed-forward neural network and Support Vector Machine (SVM) to detect wildfire events. Although deterministic models can solve complex highly non-linear scenarios, they are still insufficient in fully resolving the process behaviors [20]. Additionally, deterministic models are limited from the uncertainty quantification perspective [13,21,22].

Stochastic models, as opposed to deterministic models, acknowledge the presence of unresolved subprocesses and seek to incorporate them into the modeling framework. These stochastic approaches introduce randomness into the inference process, resulting in varying outcomes even under identical conditions. This variability gives rise to a distribution of possible outcomes, providing a comprehensive view of the wildfire segmentation. Such stochastic models prove to be well-suited for wildfire analysis due to their ability to capture the inherent uncertainty and complexity associated with these natural phenomena [23,24]. By considering multiple plausible generated outcomes, stochastic models offer informative insights into the characteristics of potential wildfire patterns and aid in assessing the uncertainty and variability associated with the segmentation results.

The modeling characteristics of stochastic models enable a more comprehensive understanding of the inherent uncertainties and complexities associated with wildfires [25,26]. By considering the variability and uncertainty in the data, stochastic models can capture the inherent stochasticity in wildfire processes and provide valuable insights into the spatial dynamics of wildfire events. This is important in wildfire analysis as it allows for the exploration of various scenarios and the assessment of the likelihood of different outcomes. Moreover, stochastic models facilitate probabilistic-based decision-making processes, enabling more informed and robust wildfire management strategies. The utilization of stochastic models in wildfire analysis has shown promising results in improving our understanding of fire behavior, predicting fire spread patterns [25,26]. Through the incorporation of stochastic modeling approaches, we can enhance our ability to effectively understand and mitigate the risks associated with wildfires, ultimately contributing to more resilient and sustainable fire management practices [25].

Additionally, wildfire ground-truth segmentations are arbitrary and/or noisy to some level, due to human labeling, instrument differences, and other artifacts, affecting the wildfire segmentation shapes. A good example of the wildfire discrepancies can be seen in the comparison of MODIS and Visible Infrared Imaging Radiometer Suite (VIIRS) fire radiative power products [27]. Even subject matter experts (SMEs), assigned to wildfire delineation tasks, often disagree on the active fire's spatial extent. These plausible but discrepant takes from the same events prompt a different look at wildfire detection where wildfire segmentations are considered a distribution of events instead of a single unified segmentation.

With the advent of terrestrial and atmospheric remote sensing, mainly supported by satellite and aviation platforms, the means to monitor and detect wildfires have been more accessible [28]. Advances in observation sensors, and specific enhancement of spatial, temporal, and spectral resolution, allow more in-depth studies and reveal some of the unknown dynamics of fires such as holdover fires [28,29]. However, with the increase in the number of satellites/aviation missions, and the increase of retrieved information, efficient and effective land management through remote sensing has been challenging [15].

Machine learning proposes an opportunity to extract useful information from a large volume of remote sensing datasets. Unsupervised methods are popular for such processes due to generally limited labels for remote sensing in Earth science. Methods such as Auto-Encoders (AEs) are widely used for such tasks where the network is an encoder–decoder architecture and the model aims to learn a compressed representation of the data with minimum information loss [30]. The main issue of these deterministic models in image-to-image translation is their loss of resolution problem. The encoder part of the model subsamples the spatial information to compress the data and due to such an operation the decoder is not able to recover the spatial information effectively [31,32]. To remedy this issue, [32] proposed U-NET which is encoder–decoder architecture with skip-connections in all spatial resolution levels from encoded activations to the corresponding decoding layers to preserve the spatial information. Despite the wide applications of AEs and U-NETs, they are not capable of learning distributions around events which limits their expressibility of data. Generative models such as variational inference methods enable characterizing stochastic behaviors in data [33], such as ones in wildfire processes.

Variational Auto-Encoders (VAEs) are among the most popular unsupervised variational inference techniques in machine learning. We propose a supervised version of VAEs developed by [34], where the model consists of four submodels: (1) prior network in charge of learning the latent prior distribution of input data, (2) posterior network in charge of learning the latent posterior distribution of input and target data, (3) U-NET network in charge of feature extraction of inputs, and (4) Combination network that uses the U-NET features and samples from latent distribution to generate stochastic wildfire segmentations.

The main contributions of this work are: (1) developing a stochastic machine-learning model with accurate and fast probabilistic inference on target wildfire segmentation, (2) conducting uncertainty quantification by drawing a significant number of samples, and (3) performing what-if scenarios to understand the impact of inputs variability.

The rest of the paper is structured as follows: Section 2 presents the methodology and proposed model, Section 3 shows the obtained results, uncertainty quantification, and comparison with baseline along with discussions, and Section 4 focuses on the conclusion and summary of findings.

2. Methodology

2.1. Variational Autoencoder

To gain a comprehensive understanding of the proposed methodology, it is imperative to establish a foundational understanding of variational autoencoders (VAEs). VAEs represent a key component in elucidating the intricacies of the proposed approach. Variational autoencoders (VAEs) are powerful *unsupervised* generative models that combine the concepts of autoencoders and variational inference. They are designed to learn a low-dimensional latent space representation of complex high-dimensional input data. The latent space is a continuous multivariate distribution that captures the underlying structure and variations within the data. VAEs consist of two main components: (1) an encoder and (2) a decoder. The encoder maps input data into the latent space, while the decoder reconstructs the data from the latent space back to the original input space.

In VAEs, instead of directly encoding input data into a single point in the latent representation, the data is encoded into probability distributions over the latent variables [35]. This probabilistic representation allows for more flexibility and uncertainty modeling. It enables VAEs to not only reconstruct the input data but also generate new samples by sampling from the learned probability distributions in the latent representation and decoding them using the decoder network.

The fundamental idea underlying VAEs is to approximate the input data distribution (i.e. marginal likelihood) noted by $P_\theta(x)$. VAEs achieve this goal by maximize the evidence lower bound (ELBO), which serves as the objective function during the training process (1). The ELBO consists of two main components: the reconstruction loss, which measures how well the VAE can reconstruct the input data, and the regularization term that encourages the latent space to adhere to a predefined prior distribution, often a multivariate Gaussian distribution. By maximizing the ELBO, VAEs achieve a delicate balance between accurately reconstructing the input data and regularizing the latent space to follow the prior distribution.

$$\log P_{\theta}(\mathbf{x}) \geq \mathcal{L}(\boldsymbol{\theta}, \boldsymbol{\phi}; \mathbf{x}) = \mathbb{E}_{Q_{\boldsymbol{\phi}}(\mathbf{z}|\mathbf{x})}\left[-\log Q_{\boldsymbol{\phi}}(\mathbf{z} \mid \mathbf{x}) + \log P_{\theta}(\mathbf{x}, \mathbf{z})\right] \tag{1}$$

Equation (1) can be re-written as equation below with the ELBO loss on left. Right hand side of the equation presents the regularization term, called Kullback-Leibler divergence, and reconstruction term.

$$\mathcal{L}(\boldsymbol{\theta}, \boldsymbol{\phi}; \mathbf{x}) = -D_{KL}\left(Q_{\boldsymbol{\phi}}(\mathbf{z} \mid \mathbf{x}) \| P_{\theta}(\mathbf{z})\right) + \mathbb{E}_{Q_{\boldsymbol{\phi}}(\mathbf{z}|\mathbf{x})}[\log P_{\theta}(\mathbf{x} \mid \mathbf{z})] \tag{2}$$

$D_{KL}\left(Q_{\boldsymbol{\phi}}(\mathbf{z} \mid \mathbf{x}) | P_{\theta}(\mathbf{z})\right)$ term represents the Kullback-Leibler (KL) divergence between the posterior distribution $Q_{\boldsymbol{\phi}}(\mathbf{z} \mid \mathbf{x})$ and the prior distribution $P_{\theta}(\mathbf{z})$. It measures the discrepancy or difference between these two distributions. $\mathbb{E}_{Q_{\boldsymbol{\phi}}(\mathbf{z}|\mathbf{x})}[\log P_{\theta}(\mathbf{x} \mid \mathbf{z})]$ represents the expected log-likelihood of the reconstruction, where $\mathbf{x}$ is the input data and $\mathbf{z}$ is a latent variable sampled from the posterior distribution $Q_{\boldsymbol{\phi}}(\mathbf{z} \mid \mathbf{x})$. It measures how well the VAE can reconstruct the input data given a sampled latent variable.

This regularization process encourages the latent space of the VAE to capture meaningful and continuous representations of the data. It facilitates various tasks, including data generation and interpolation, by ensuring that similar input data points are mapped to nearby regions in the latent space. As a result, VAEs provide a powerful framework for learning complex data distributions and exploring the latent space in a probabilistic manner.

VAEs focus on unsupervised learning and aim to learn meaningful representations of the input data by modeling the underlying probability distributions. Probabilistic U-Net extends the variational capabilities of VAEs to supervised learning and create possibilities to perform tasks such as segmentation in variational context.

2.2. Proposed Approach

Image segmentation is the process of identifying and isolating objects or features of interest in input images. One of the commonly used techniques for the segmentation of instances is the U-NET model, initially developed for biomedical image segmentation but also applicable to other fields such as Earth sciences and space exploration. U-NET is a deep convolutional neural network that performs image-to-image translation by taking an image as input and generating a segmentation map as output. The model is trained using supervised learning, which involves providing accurate segmented images to train the network to map input images to their corresponding segmentations. Despite the impressive performance of U-NET in image segmentation tasks, its deterministic nature poses a limitation. The mapping from input images to output segmentation maps is fully deterministic and fails to consider sources of uncertainty and stochasticity, which can lead to overfitting and poor generalization to new data. Moreover, the deterministic nature of U-NET limits its ability to perform "what-if" analysis and provide probabilistic segmentations.

Kohl et al. [34] proposed a novel Probabilistic U-Net model for image segmentation that combines the U-NET with a Conditional Variational Auto-Encoder (CVAE) [36,37]. The CVAE framework allows the model to generate plausible hypotheses and explore "what-if" scenarios. The architecture of the proposed model is depicted in Figure 1. Specifically, the U-NET generates segmentations that are conditioned on the samples drawn from the latent feature space of the VAE. This low-dimensional space captures the range of possible segmentation variations and can be used to evaluate "what-if" scenarios during the evaluation phase. By conditioning the segmentation generation on the latent space, the model can produce multiple segmentation maps for a single input image, corresponding to different regions of the latent feature space that are sampled. According to the authors, this capability enables the model to "*learn hypotheses that have a low probability and to predict them with the corresponding frequency*" (Kohl et al., 2018).

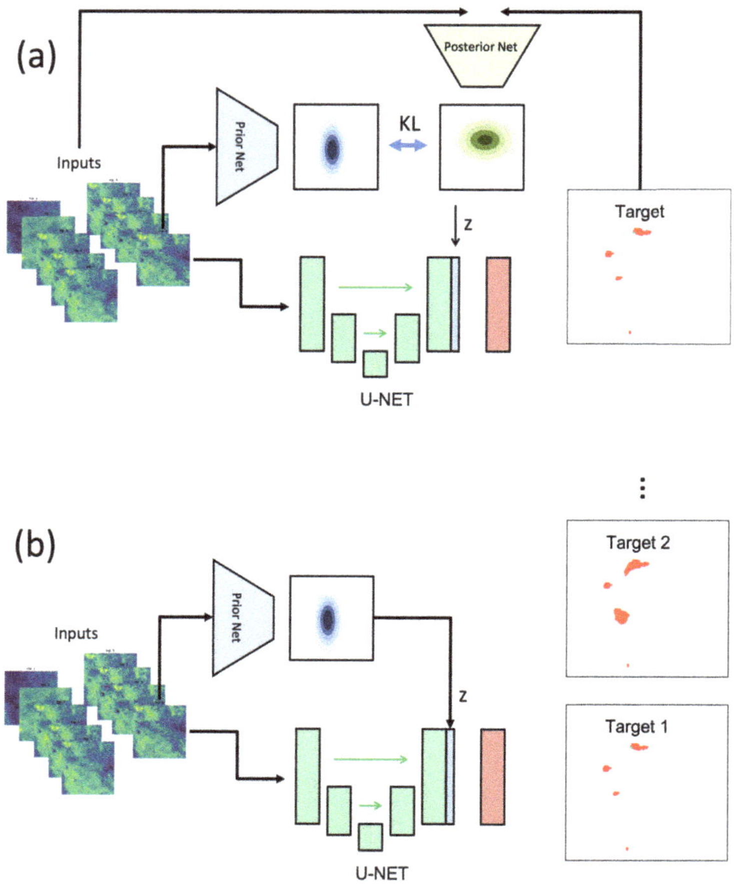

Figure 1. Graphical illustration of the proposed Probabilistic U-Net framework. The inputs are NDVI, NDVI difference with long-term NDVI, and MODIS MCD43A4 channels for Land/Cloud/Aerosols. (**a**) presents the training scheme where the prior network encodes inputs and the posterior network encodes the inputs and target data together into multivariate Gaussian distributions. The samples from the unified multivariate Gaussian distribution are concatenated with U-Net outputs to produce stochastic events of target data. (**b**) demonstrates the inference scheme where samples are drawn from the prior network.

The output of the U-NET (the green block) and the drawn sample from latent space (the blue block labeled as z) are concatenated and passed to the red block $\mathcal{F}$, which generates the corresponding segmentation $S_i = \mathcal{F}(f_{\text{U-NET}}(X,\theta), z_i; \psi)$, where S_i represents the segmentation corresponding to the latent space sample z_i, and θ and ψ are the model parameters of the U-NET model and the $\mathcal{F}$, respectively. The model is trained using two objectives, namely (1) generating accurate wildfire segmentation from the input data and (2) generalizing well to unseen or rare scenarios. The model is enforced to meet the first objective by minimizing the supervised cross-entropy loss between the generated segmentation, $S(X,z)$, and the ground truth, Y. The model generalizes its understanding by minimizing the Kullback–Leibler divergence between the prior, $P(z \mid X)$, and posterior, $Q(z \mid Y, X)$, distributions of the variables in the latent feature space. Thus, the total loss function is a combination of the two losses as follows:

$$\begin{aligned} \mathcal{L}(Y,X) =& \mathbb{E}_{z\sim Q(.|Y,X)}\big[-\log P(Y \mid S(X,z))\big]+ \\ & \beta \mathrm{KL}\big[Q(x \mid Y,X)||P(z \mid X)\big] \end{aligned} \tag{3}$$

The parameter β serves as a hyper-parameter that governs the extent to which the KL-divergence term, also known as the regularization term, influences the model's output. The model (illustrated in detail in Figure 2) is trained end-to-end Hyper-parameter optimization was performed using logarithmic scaling from 10^{-6} to 10^{-3} and the optimum value for β is 0.0001.

2.3. Baseline Methods

To compare the performance of our proposed approach and evaluate its capabilities in generating consistent and contextual information, we developed two baseline methods with the similar stochastic nature. These baseline models are similar to the proposed model in capturing distributions over multi-modal segmentation. The introduced models are designed to accommodate the similarity in network architecture and to investigate the nature of stochasticity. By developing these baselines, we were able to analyze the effect of each stochasticity approach on (1) learning the underlying distribution of the wildfires, and (2) the performance of the network architectures under similar conditions. The baselines will shed light on the efficacy of the different stochasticity similar varieties of U-Net.

2.3.1. U-Net with Dropout

Dropout in a U-Net architecture can perform as a special case of the delta rule in which we introduce noise in the transmission of information [38] by randomly masking weights of the network. Dropout is presented as an especial case of delta rule called stochastic delta rule [39] in which each weight in the model is assigned as a random variable from a Gaussian distribution with the mean $\mu_{w_{ij}}$ and standard deviation of $\sigma_{w_{ij}}$ [38]. Dropout, as an special case of stochastic delta rule, introduces a form of regularization that aids in escaping poor local minima. By randomly deactivating a subset of neurons during each training iteration, dropout prevents the network from relying too heavily on specific neurons or features. This selective deactivation encourages the remaining neurons to compensate and learn more robust representations, leading to a broader exploration of the weight space and increasing the odds of finding the optimum solution [38]. Additionally, keeping the dropout in the inference process will introduce stochasticity by generating results from a randomly selected sub-network and will result in an approximation of posterior distribution [40]. Dropout obtains these advantages by removing hidden neurons according to a Bernoulli distribution with a probability parameter p. The dropout probability of the baseline model is set to be $p = 0.3$ meaning that at each pass of the network, only 70% of the neurons will be activated via a random selection.

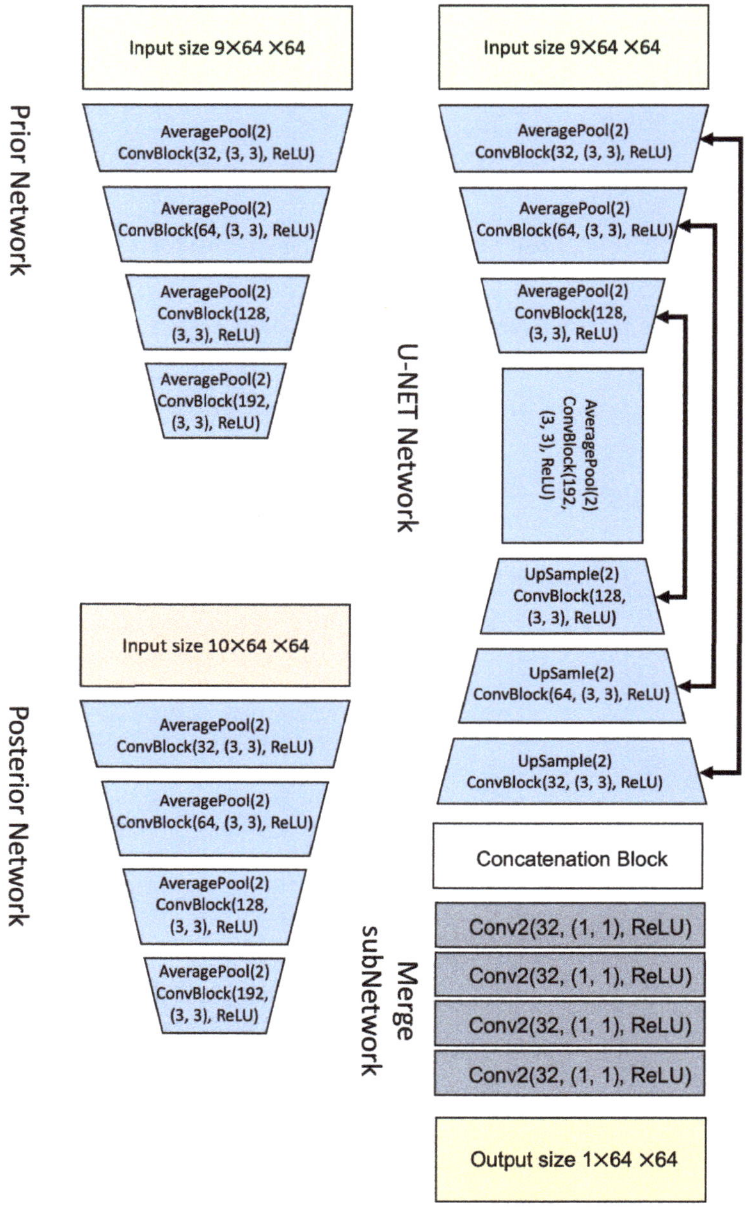

Figure 2. Network Architecture of Probabilistic U-Net consisting of three main networks: U-Net (plus a merge sub-network), prior network, and posterior network. The Convolution Blocks in the blue hidden layers consist of three sub-blocks of 2D convolutional layers with the same size features and kernel size and ReLU activation. The darker blue layers in U-Net represent the merger sub-network where the samples from latent distribution are concatenated with U-Net output and flow through the merger network to generate wildfire masks.

2.3.2. U-Net with Stochastic Activations

The concept of stochastic non-linear activations was first proposed by [41] to improve models by resolving the degenerative behavior of deterministic activation functions. Another study by Shridhar et al., 2020 [42] introduced a probabilistic activation definition which makes the model behavior stochastic. The activation function, regardless of its type, will gain stochasticity by introducing Gaussian noise to its value [42]. In this architecture, instead of using a deterministic activation (e.g., ReLU), a Gaussian noise trick will apply perturbation to the forward and backward processes Figure 3. The parameters of the Gaussian perturbation can stay fixed or trained as a trainable parameter via backpropagation. Obtained from several experiments, the optimum sigma is found to be 5.

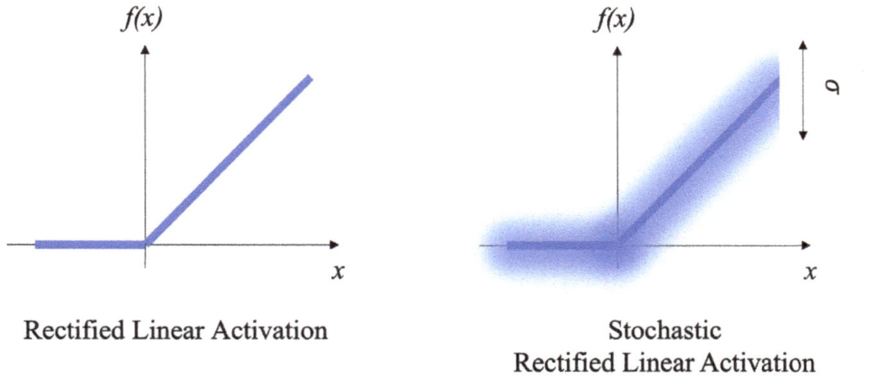

Figure 3. Rectified Linear Activation (ReLU) and stochastic ReLU. σ represents the standard deviation of the Gaussian noise.

2.4. Statistical Metrics

We have used multiple statistical metrics to evaluate the segmentation quality and assign lower and higher bounds for multiple draws of the same events. In particular, we used

$$Precision = \frac{TP}{TP + FP} \tag{4}$$

$$Recall = \frac{TP}{TP + FN} \tag{5}$$

$$F1\text{-}score = \frac{2\,TP}{2\,TP + FP + FN} \tag{6}$$

$$Jaccard\ Index = \frac{TP}{TP + FP + FN} \tag{7}$$

where TP, TN, FP, and FN are the true positive, true negative, false positive, and false negative, respectively. Precision, Recall, and F1-score are popular metrics that provide valuable Jaccard Index, also known as Intersect of Union (IoU), which is a good measure to calculate the overlap of predicted and target wildfire segmentation. Due to the stochasticity of each sample, we represented the statistics with lower and upper bounds of performance for each metric.

3. Experiments

In this section, we present the segmentation performance of the proposed method along with the two baselines, but first, we introduce the dataset used in this study and then explain the statistical metrics used in comparisons.

3.1. Dataset

In this study, we focus on the discrepancies in the fire products of MODIS constellation and VIIRS instruments onboard the joint NASA/NOAA Suomi National Polar-Orbiting

Partnership (Suomi NPP) and NOAA-20 satellites [43]. We aim to frame this problem to (1) offer an alternative fire product to resolve the MODIS' patchy and inconsistent segmentation, and (2) develop a distribution-over-event-based model to obtain epistemic uncertainty quantification and run what-if scenarios on input variables. For this purpose, we have collected MODIS MCD43A4, a daily product with 250 m spatial resolution, and collocated VIIRS fire product, with a daily 375 m spatial resolution, as target data. We used the Land/Cloud/Aerosol boundaries and properties channels with bandwidths of 620-670, 841–876, 459–479, 545–565, 1230–1250, 1628–1652, 2105–2155 nanometer (Table 1). We added the Normalized Difference Vegetation Index (NDVI) as a reliable proxy for estimating the fuel loads available for fires [44,45] using the following equation:

$$NDVI = \frac{NIR - RED}{NIR + RED} \quad (8)$$

The $NDVI$ ranges from -1 (not vegetation) to 1 (healthy vegetation) and is obtained from near-infrared (841–876 nm) and red (620–670 nm) bands. Multiple alternatives to NDVI have tried to address some of its issues using additional band [46,47]; however, due to the less noise sensitivity of NDVI and wide application of NDVI in the literature [48–51], NDVI has been considered to be the reference index for fuel-load analysis. Despite the usefulness of NDVI, it cannot be useful directly for fire detection due to the location dependency of NDVI values. For instance, NDVI values can be lower in arid zones compared to subtropical regions, but still, wildfires happen in subtropical regions due to abnormally low vegetation moisture. To tackle this issue, relative NDVI is calculated by subtracting the NDVI of each day from the mean NDVI of the same location for the whole period of study. This will give us a sense of abnormal vegetation conditions potent for wildfires.

Table 1. List of data used as inputs in the model.

Input	Bandwidth
Land/Cloud/Aerosols Boundary	620–670 841–876
Land/Cloud/Aerosols Properties	459–79 545–565 1230–1250 1628–1652 2105–2155
NDVI	N/A
NDVI Derivation	N/A

The target fire dataset is obtained from thermal anomalies/active fire products with two fire-associated properties; brightness temperatures (in Kelvin), and fire radiative power (in Megawatts) among others. The dataset is provided in individual point locations with a spatial resolution of 375 m which is converted into gridded maps using the nearest neighbor method.

The training, validation, and testing sets consist of patches of data described above over wildfire events detected across the Continental United States. We shifted patches randomly to generate augmented patches and prevent the artifact of always having wildfire pixels in the center of the patch. The data were collected and patched for 2018 and filtered to only keep events with more than 20 pixels of wildfire inside. We used a 60-20-20 percentage for training, validation, and testing to obtain the hyper-parameter values. Then we retrained the models using training and validation sets and then evaluated the unbiased estimate of the performance using the testing set.

To generate multiple inference segmentations from the same input data, we feed the inputs to the U-NET model to obtain relevant spatial features. Simultaneously the inputs are fed into the prior network and obtain latent space samples (z in Figure 1b). Combining the U-NET features with each sample z will provide a unique variation of corresponding segmentation. Multiple samples drawn from prior networks will provide multiple segmentations for that specific event.

3.2. Results

Throughout this section, we focus on evaluating the proposed Probabilistic U-Net model and compare the performances to the two baseline models: Dropout U-Net and Stochastic ReLU U-Net. We will first present a visual comparison benchmark for wildfire detection and quantify the visual uncertainty, and then present a more comprehensive performance using the metrics discussed in Section 3.3. Figure 4 consists of two independent wildfire incidents that describe two different wildfire dynamics. Each incident ((a) and (b)) demonstrates the visual consistency of the Probabilistic U-Net and the two baselines by drawing five random samples for a specific event. In Figure 4a, first five columns from left for the first, second, and third rows present the samples from Dropout U-Net, Stochastic ReLU U-Net, and Probabilistic U-Net models, respectively. Overall, all samples from all models are consistent with the target segmentation (last row, far left column). It is noticed that Dropout U-Net has less spatial coherency compared to the other two. Stochastic ReLU U-Net detects consistent wildfire in the circular area but misses the bottom left region of fire. The Probabilistic U-Net on the other hand shows a diverse range of detections capturing both patches of circular and bottom left fires. Comparing the detected wildfires by all models with NDVI indicates that all models understand the dynamics of vegetation and wildfire, where fire spreads in the surrounding of low vegetation (burned area). The NDVI deviation from the historical mean is very similar to the current NDVI in terms of burned area shape and size meaning the NDVI has not significantly changed, but the region is still experiencing wildfire activities. The far right column illustrates the spatial stochasticity of each model from 1000 independent samples. The Dropout U-Net model demonstrates low confidence in the bottom left region and left semi-circle of the circular region. Stochastic ReLU U-Net is confident in its detections and does not anticipate any fire in the bottom left region. The Probabilistic U-Net model produces a reasonable uncertainty map, covering most of the observed region with high confidence; however, the model is uncertain about the wildfire shape, specifically in the bottom left region.

Figure 4b demonstrates second independent incident where, similar to Figure 4a, the first, second and third rows belong to Dropout U-Net, Stochastic U-Net and Probabilistic U-Net, respectively. Performing similar to Figure 4a, all the segmentations are consistently close to target mask (bottom row, left column). Dropout U-Net presents higher variability compared to the other two models and result in higher uncertainty, especially in the wildfire border areas. Stochastic U-Net detects a more consistent segmentation pattern with less variability. It is noteworthy that Stochastic U-Net segmentations are undercomplete and do not fully cover the target segmentation area. Probabilistic U-Net demonstrates coherent patterns as target data, with uncertainty in the boundary of burning regions. The incident is slightly different in dynamics compared to Figure 4 due to NDVI behavior. In this incident, the NDVI deviation from historical mean is different, meaning the area is lossing vegetation health quality due to the wildfire.

Figure 5 indicates the statistical performance of the three models over 1000 runs. We present the model performances for the testing set (including 1500 non-overlapping wildfire events) in box plots to incorporate the uncertainty level for each model. The precision statistics show similar sentiment to the visual samples, where Dropout U-Net under-detects the fire pixels (causing lower True Positives) with high variability, Stochastic ReLU U-Net detects a significant area of wildfire with high confidence and Probabilistic U-Net that has moderate detection capability with a similar range of Dropout U-Net variability. However, the results shift in the recall, showing a higher range for Probabilistic U-Net compared to the baselines. This stems from the lower FN values of this model, compared to the other two baselines. As a result of this, the F1-score which is a harmonic mean of precision and recall shows higher performance for the proposed Probabilistic U-Net compared to the baselines. Lastly, the Jaccard index or IOU indicates higher agreement between the target segmentations and the Probabilistic U-Net segmentation variants. Stochastic ReLU U-Net is the second-best model with low variability, and the lowest IOU belongs to the Dropout U-Net model.

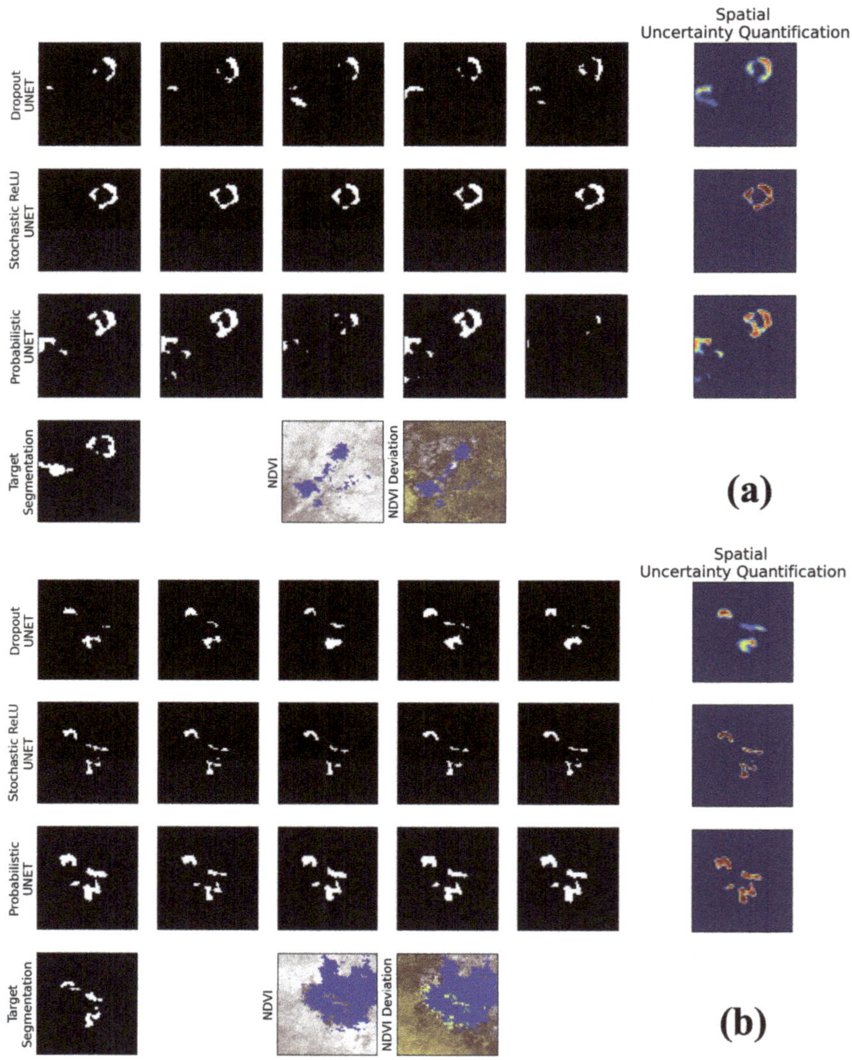

Figure 4. The figure demonstrates two independent wildfire incidents (subplot (**a**,**b**)) consisting of 5 drawn samples (first 5 columns) from the proposed Prob. U-Net (first row) and the baseline models (second and third rows) along with spatial uncertainty quantification for the same event using 1000 runs (last column). The last row, shows the target segmentation, corresponding NDVI and NDVI deviation from historical, from left to right, respectively.

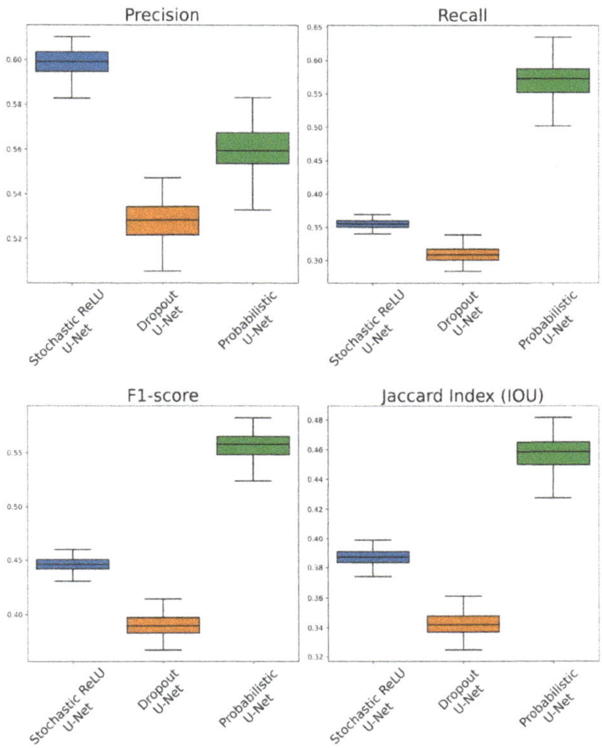

Figure 5. Statistical comparison between Probabilistic U-Net, Dropout U-Net, and Stochastic ReLU U-Net over 1000 runs.

3.3. Discussion

Furthermore, we investigated the semantic variation of the models by changing the dynamic of NDVI. In this experiment, we aimed to better understand the grasp of each model in understanding the NDVI dynamics. The experiments continued to investigate the physical comprehension of each model by changing NDVI and observing the changes in wildfire segmentations. We follow the following reasonings: (1) an increase in NDVI will not trigger wildfire (at least not as severe as before), (2) a spotty decrease in NDVI allows the wildfire to spread toward lower NDVI (unhealthy vegetation) area, (3) a significant decrease in NDVI in a region will not provide enough fuel for the fire to spread. We tested these hypotheses on the sample data we had in Figure 6. In the first three rows, we have the model detections from original NDVI values which are similar to the samples shown in Figure 4. The second three rows demonstrate the model responses to an increase of NDVI within and surrounding low NDVI area (burned region). Based on the results, we see that Dropout and Probabilistic U-Nets will not detect a burning segment and Stochastic ReLU U-Net will detect smaller segmentations. The third three rows investigate the idea of sparsely lowering the vegetation in regions close to burning scars. The results show that Dropout U-Net and Stochastic ReLU U-Nets will not capture the ignitions toward new places, especially in the bottom left region. However, Probabilistic U-Net is understanding spread reasoning and detecting segments in the bottom left area. Lastly, we show significant NDVI reduction for a large area in the last three rows. The NDVI decrease mainly impacts the bottom left region and spotty locations in the circular segment region. All models are correctly ruling out the possibility of wildfire in the bottom left region. Dropout U-Net has difficulty understanding the circular shape affected by spotty NDVI decreases. Stochastic ReLU U-Net is persistently detecting the circular segment, but Probabilistic

U-Net has slightly adjusted the circular segment according to the spotty changes. It is noteworthy that the model hyperparameters (σ, β, and dropout rate for Stochastic ReLU, Probabilistic, and Dropout U-Nets) are selected based on best precision and recall. It seems that Stochastic ReLU U-Net performs better under lower variability and deteriorates under higher σ values.

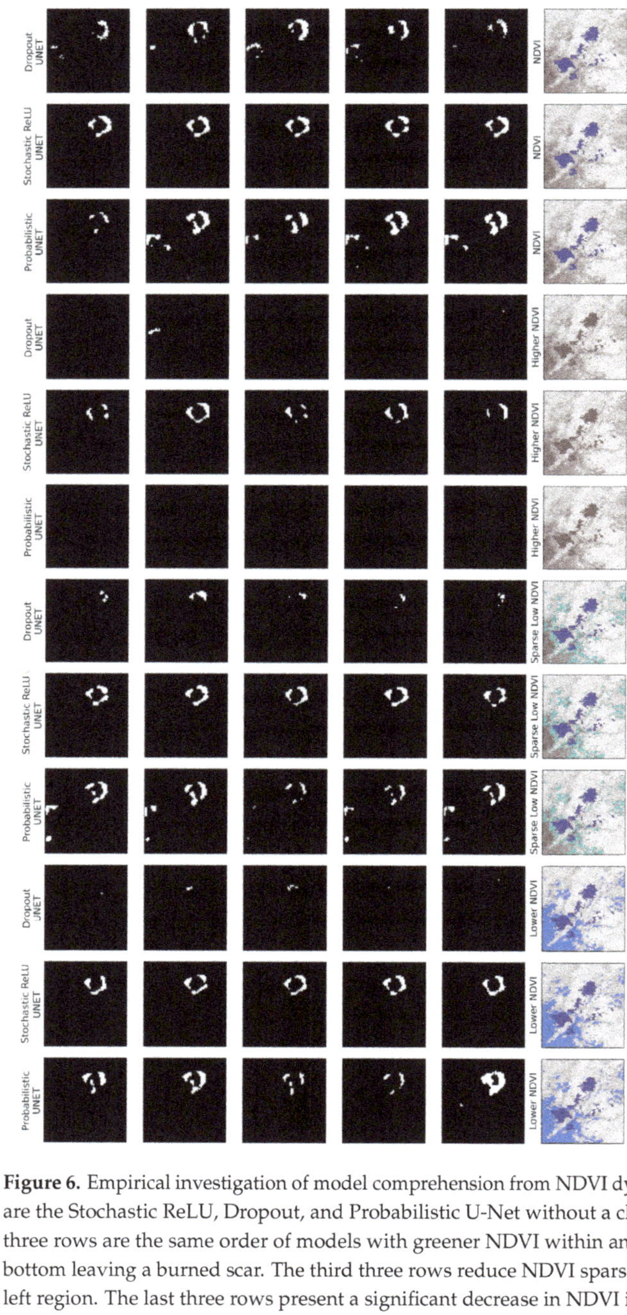

Figure 6. Empirical investigation of model comprehension from NDVI dynamics. The first three rows are the Stochastic ReLU, Dropout, and Probabilistic U-Net without a change in NDVI. The second three rows are the same order of models with greener NDVI within and in the surrounding of the bottom leaving a burned scar. The third three rows reduce NDVI sparsely, especially in the bottom left region. The last three rows present a significant decrease in NDVI in the vicinity of the bottom left region and spotty locations close to the circular scar.

4. Conclusions

In this study, we proposed a stochastic machine-learning approach that learns a latent distribution of wildfire events in a supervised manner and addresses the uncertainty quantification and inter-dataset discrepancies. We investigated the proposed method by segmenting active wildfires using the seven bands from MODIS and two derivatives (NDVI and historical deviation of NDVI) as inputs. The proposed model was compared with two stochastic baseline machine-learning models called Dropout U-NET, a U-NET with dropouts in training and test phases, and Stochastic ReLU U-NET, a U-NET with Stochastic ReLU activations. It was discovered through the conducted experiments that Probabilistic U-Net is more accurate and flexible compared to the other two models. The Stochastic ReLU U-Net seems to perform more accurately with lower variability, and Dropout U-Net is less accurate but demonstrates a wider range of variability. Additionally, we performed a scenario-based experiment to analyze the impact of physical changes on the response of the models. The probabilistic model showed a more comprehensive understanding of the physical relationship between NDVI and wildfire. However, the other two baseline models demonstrated partial alignment with the scenarios.

Author Contributions: Conceptualization, A.A.A., M.M. and P.A.L.; methodology, M.M. and A.A.A.; modeling and implementation, A.A.A.; formal analysis, A.A.A.; investigation, A.A.A.; data curation, A.A.A.; writing—original draft preparation, A.A.A. and M.M.; writing—review and editing, A.A.A., M.M., P.A.L., E.R. and S.G.; visualization, A.A.A.; supervision, P.A.L., E.R. and S.G. All authors have read and agreed to the published version of the manuscript.

Funding: This research is funded and made possible by NASA ROSES AIST-QRS-21. A.A.A., M.M. and P.A.L. are thankful for the support from NASA Academic Mission Services, Contract No. NNA16BD14C.

Conflicts of Interest: The authors declare no conflict of interest.

References

1. Agee, J.K.; Finney, M.; Gouvenain, R.D. Forest fire history of desolation peak, Washington. *Can. J. For. Res.* **1990**, *20*, 350–356. [CrossRef]
2. Alizadeh, M.R.; Abatzoglou, J.T.; Luce, C.H.; Adamowski, J.F.; Farid, A.; Sadegh, M. Warming enabled upslope advance in western US forest fires. *Proc. Natl. Acad. Sci. USA* **2021**, *118*, e2009717118. [CrossRef]
3. Barros, A. M.; Ager, A. A.; Day, M. A.; Krawchuk, M. A.; Spies, T. A. Wildfires managed for restoration enhance ecological resilience *Ecosphere* **2018**, *9*, e02161. [CrossRef]
4. Calkin, D.E.; Thompson, M.P.; Finney, M.A. Negative consequences of positive feedbacks in US wildfire management. *For. Ecosyst.* **2015**, *2*, 1–10. [CrossRef]
5. Chas-Amil, M.L.; Prestemon, J.P.; McClean, C.J.; Touza, J. Human-ignited wildfire patterns and responses to policy shifts. *Appl. Geogr.* **2015**, *56*, 164–176. [CrossRef]
6. Dennison, P.E.; Brewer, S.C.; Arnold, J.D.; Moritz, M.A. Large wildfire trends in the western United States, 1984–2011. *Geophys. Res. Lett.* **2014**, *41*, 2928–2933. [CrossRef]
7. Hoover, K.; Hanson, L.A. *Wildfire Statistics*; Technical Report; Congressional Research Service: Washington, DC, USA, 2021.
8. Giglio, L.; Boschetti, L.; Roy, D.P.; Humber, M.L.; Justice, C.O. The Collection 6 MODIS burned area mapping algorithm and product. *Remote Sens. Environ.* **2018**, *217*, 72–85. [CrossRef]
9. Jain, P.; Coogan, S.C.; Subramanian, S.G.; Crowley, M.; Taylor, S.; Flannigan, M.D. A review of machine learning applications in wildfire science and management. *Environ. Rev.* **2020**, *28*, 478–505. [CrossRef]
10. Carmo, M.; Moreira, F.; Casimiro, P.; Vaz, P. Land use and topography influences on wildfire occurrence in northern Portugal. *Landsc. Urban Plan.* **2011**, *100*, 169–176. [CrossRef]
11. Narayanaraj, G.; Wimberly, M.C. Influences of forest roads on the spatial patterns of human-and lightning-caused wildfire ignitions. *Appl. Geogr.* **2012**, *32*, 878–888. [CrossRef]
12. Ghali, R.; Akhloufi, M.A.; Jmal, M.; Souidene Mseddi, W.; Attia, R. Wildfire segmentation using deep vision transformers. *Remote Sens.* **2021**, *13*, 3527. [CrossRef]
13. Green, M.E. *Some Results on a Set of Data Driven Stochastic Wildfire Models*; The University of Vermont and State Agricultural College: Burlington, VT, USA, 2020.
14. Khryashchev, V.; Larionov, R. Wildfire segmentation on satellite images using deep learning. In Proceedings of the 2020 Moscow Workshop on Electronic and Networking Technologies (MWENT), Moscow, Russia, 11–13 March 2020; pp. 1–5.
15. Rashkovetsky, D.; Mauracher, F.; Langer, M.; Schmitt, M. Wildfire detection from multisensor satellite imagery using deep semantic segmentation. *IEEE J. Sel. Top. Appl. Earth Obs. Remote Sens.* **2021**, *14*, 7001–7016. [CrossRef]

16. Poole, D.; Raftery, A.E. Inference for deterministic simulation models: The Bayesian melding approach. *J. Am. Stat. Assoc.* **200(** *95*, 1244–1255. [CrossRef]
17. Cencerrado, A.; Cortés, A.; Margalef, T. Response time assessment in forest fire spread simulation: An integrated methodolog for efficient exploitation of available prediction time. *Environ. Model. Softw.* **2014**, *54*, 153–164. [CrossRef]
18. Toan, N.T.; Cong, P.T.; Hung, N.Q.V.; Jo, J. A deep learning approach for early wildfire detection from hyperspectral satellit images. In Proceedings of the 2019 7th International Conference on Robot Intelligence Technology and Applications (RiTA Daejeon, Republic of Korea, 1–3 November 2019; pp. 38–45.
19. Sayad, Y.O.; Mousannif, H.; Al Moatassime, H. Predictive modeling of wildfires: A new dataset and machine learning approacl *Fire Saf. J.* **2019**, *104*, 130–146. [CrossRef]
20. Couce, E.; Knorr, W. Statistical parameter estimation for a cellular automata wildfire model based on satellite observations. *WI Trans. Ecol. Environ.* **2010**, *137*, 47–55.
21. Quill, R.; Sharples, J.J.; Wagenbrenner, N.S.; Sidhu, L.A.; Forthofer, J.M. Modeling wind direction distributions using a diagnosti model in the context of probabilistic fire spread prediction. *Front. Mech. Eng.* **2019**, *5*, 5. [CrossRef]
22. Palmer, T.; Shutts, G.; Hagedorn, R.; Doblas-Reyes, F.; Jung, T.; Leutbecher, M. Representing model uncertainty in weather anc climate prediction. *Annu. Rev. Earth Planet. Sci.* **2005**, *33*, 163–193. [CrossRef]
23. Artés, T.; Cencerrado, A.; Cortés, A.; Margalef, T. Time aware genetic algorithm for forest fire propagation prediction: exploitin multi-core platforms. *Concurr. Comput. Pract. Exp.* **2017**, *29*, e3837. [CrossRef]
24. Denham, M.; Laneri, K. Using efficient parallelization in graphic processing units to parameterize stochastic fire propagatior models. *J. Comput. Sci.* **2018**, *25*, 76–88. [CrossRef]
25. Ramirez, J.; Monedero, S.; Silva, CA; Cardil, A. Stochastic decision trigger modelling to assess the probability of wildland fir impact *Sci. Total Environ.* **2019**, *694*, 133505. [CrossRef]
26. Valero, M. M.; Jofre, L.; Torres, R. Multifidelity prediction in wildfire spread simulation: Modeling, uncertainty quantificatior and sensitivity analysis *Environ. Model. Softw.* **2021**, *141*, 105050 . [CrossRef]
27. Li, F.; Zhang, X.; Kondragunta, S.; Csiszar, I. Comparison of fire radiative power estimates from VIIRS and MODIS observation *J. Geophys. Res. Atmos.* **2018**, *123*, 4545–4563. [CrossRef]
28. Allison, R.S.; Johnston, J.M.; Craig, G.; Jennings, S. Airborne optical and thermal remote sensing for wildfire detection anc monitoring. *Sensors* **2016**, *16*, 1310. [CrossRef]
29. Schultz, C.J.; Nauslar, N.J.; Wachter, J.B.; Hain, C.R.; Bell, J.R. Spatial, temporal and electrical characteristics of lightning ir reported lightning-initiated wildfire events. *Fire* **2019**, *2*, 18. [CrossRef]
30. Bengio, Y.; Courville, A.; Vincent, P. Representation learning: A review and new perspectives. *IEEE Trans. Pattern Anal. Mach Intell.* **2013**, *35*, 1798–1828. [CrossRef]
31. Akbari Asanjan, A.; Das, K.; Li, A.; Chirayath, V.; Torres-Perez, J.; Sorooshian, S. Learning instrument invariant characteristic for generating high-resolution global coral reef maps. In Proceedings of the 26th ACM SIGKDD International Conference or Knowledge Discovery & Data Mining, Virtual Event, 6–10 July 2020; pp. 2617–2624.
32. Ronneberger, O.; Fischer, P.; Brox, T. U-net: Convolutional networks for biomedical image segmentation. In Medica Image Computing and Computer-Assisted Intervention—MICCAI 2015: 18th International Conference, Munich, German 5–9 October 2015; Springer International Publishing: Berlin/Heidelberg, Germany, 2015; pp. 234–241.
33. Shu, R.; Brofos, J.; Zhang, F.; Bui, H.H.; Ghavamzadeh, M.; Kochenderfer, M. Stochastic video prediction with conditional densit estimation. In Proceedings of the ECCV Workshop on Action and Anticipation for Visual Learning, Amsterdam, The Netherland 8–10 October 2016; Volume 2, p. 2.
34. Kohl, S.; Romera-Parades, B.; Meyer, C.; De Fauw, J.; Ledsam, J.; Maier-Hein, K.; Eslami, S.; Rezende, D.; Ronneberger, O. A Probabilistic U-Net for Segmentation of Ambiguous Images. *NeurIPS* **2018**, *31*, 6965–6975.
35. Kingma, D.; Welling M. Auto-encoding variational bayes. *arXiv* **2013**, arXiv:1312.6114.
36. Kingma, D.; Rezende, D.; Mohamed, S.; Welling, M. Semi-supervised learning with deep generative models. *Adv. Neural In Process. Syst. (NeurIPS)* **2014**, *27*.
37. Sohn, K.; Lee, H.; Yan, X. Learning structured output representation using deep conditional generative models. *Adv. Neural In Process. Syst. (NeurIPS)* **2015**, *28*, 3483–3491.
38. Frazier-Logue, N.; Hanson, S.J. Dropout is a special case of the stochastic delta rule: Faster and more accurate deep learning *arXiv* **2018**, arXiv:1808.03578.
39. Hanson, S.J. A stochastic version of the delta rule. *Phys. D Nonlinear Phenom.* **1990**, *42*, 265–272. [CrossRef]
40. Kendall, A.; Badrinarayanan, V.; Cipolla, R. Bayesian segnet: Model uncertainty in deep convolutional encoder-decode architectures for scene understanding. *arXiv* **2015**, arXiv:1511.02680.
41. Gulcehre, C.; Moczulski, M.; Denil, M.; Bengio, Y. Noisy activation functions. In Proceedings of the International Conference or Machine Learning, New York, NY, USA, 20–22 June 2016; pp. 3059–3068.
42. Shridhar, K.; Lee, J.; Hayashi, H.; Mehta, P.; Iwana, B.K.; Kang, S.; Uchida, S.; Ahmed, S.; Dengel, A. Probact: A probabilistic activation function for deep neural networks. *arXiv* **2019**, arXiv:1905.10761.
43. Schroeder, W.; Oliva, P.; Giglio, L.; Csiszar, I.A. The New VIIRS 375 m active fire detection data product: Algorithm descriptior and initial assessment. *Remote Sens. Environ.* **2014**, *143*, 85–96. [CrossRef]

44. Liu, Y.; Stanturf, J.; Goodrick, S. Wildfire potential evaluation during a drought event with a regional climate model and NDVI. *Ecol. Inform.* **2010**, *5*, 418–428. [CrossRef]
45. Dasgupta, S.; Qu, J.J.; Hao, X.; Bhoi, S. Evaluating remotely sensed live fuel moisture estimations for fire behavior predictions in Georgia, USA. *Remote Sens. Environ.* **2007**, *108*, 138–150. [CrossRef]
46. Huete, A.R. A soil-adjusted vegetation index (SAVI). *Remote Sens. Environ.* **1988**, *25*, 295–309. [CrossRef]
47. Matsushita, B.; Yang, W.; Chen, J.; Onda, Y.; Qiu, G. Sensitivity of the enhanced vegetation index (EVI) and normalized difference vegetation index (NDVI) to topographic effects: A case study in high-density cypress forest. *Sensors* **2007**, *7*, 2636–2651. [CrossRef]
48. Pereira-Pires, J.E.; Aubard, V.; Ribeiro, R.A.; Fonseca, J.M.; Silva, J.M.; Mora, A. Fuel Break Vegetation Monitoring with Sentinel-2 NDVI Robust to Phenology and Environmental Conditions. In Proceedings of the 2021 IEEE International Geoscience and Remote Sensing Symposium IGARSS, Brussels, Belgium, 11–16 July 2021; pp. 6264–6267.
49. Mazeh, F.; El Sahili, J.; Zaraket, H. Low-Cost NDVI Platform for Land Operation: Passive and Active. *IEEE Sens. Lett.* **2021**, *5*, 1–4. [CrossRef]
50. Quan, X.; Xie, Q.; He, B.; Luo, K.; Liu, X. Corrigendum to: Integrating remotely sensed fuel variables into wildfire danger assessment for China. *Int. J. Wildland Fire* **2021**, *30*, 822–822. [CrossRef]
51. Holsinger, L.; Parks, S.A.; Miller, C. Weather, fuels, and topography impede wildland fire spread in western US landscapes. *For. Ecol. Manag.* **2016**, *380*, 59–69. [CrossRef]

remote sensing

MDPI

Article

A Pattern Classification Distribution Method for Geostatistical Modeling Evaluation and Uncertainty Quantification

Chen Zuo [1], Zhuo Li [1], Zhe Dai [1,*], Xuan Wang [2] and Yue Wang [3]

[1] Department of Big Data Management and Applications, Chang'an University, Xi'an 710064, China; chenzuo@chd.edu.cn (C.Z.); zhuoli803@chd.edu.cn (Z.L.)
[2] School of Computer and Control Engineering, Yantai University, Yantai 264005, China; xuanwang91@ytu.edu.cn
[3] Xi'an Key Laboratory of Digital Construction and Management for Transportation Infrastructure, Xi'an 710064, China; ywang@chd.edu.cn
* Correspondence: zhedai@chd.edu.cn

Abstract: Geological models are essential components in various applications. To generate reliable realizations, the geostatistical method focuses on reproducing spatial structures from training images (TIs). Moreover, uncertainty plays an important role in Earth systems. It is beneficial for creating an ensemble of stochastic realizations with high diversity. In this work, we applied a pattern classification distribution (PCD) method to quantitatively evaluate geostatistical modeling. First, we proposed a correlation-driven template method to capture geological patterns. According to the spatial dependency of the TI, region growing and elbow-point detection were launched to create an adaptive template. Second, a combination of clustering and classification was suggested to characterize geological realizations. Aiming at simplifying parameter specification, the program employed hierarchical clustering and decision tree to categorize geological structures. Third, we designed a stacking framework to develop the multi-grid analysis. The contribution of each grid was calculated based on the morphological characteristics of TI. Our program was extensively examined by a channel model, a 2D nonstationary flume system, 2D subglacial bed topographic models in Antarctica, and 3D sandstone models. We activated various geostatistical programs to produce realizations. The experimental results indicated that PCD is capable of addressing multiple geological categories, continuous variables, and high-dimensional structures.

Keywords: geostatistical modeling; multiple-point statistics; uncertainty quantification; subglacial topographic model; hydrological model

Citation: Zuo, C.; Li, Z.; Dai, Z.; Wang, X.; Wang, Y. A Pattern Classification Distribution Method for Geostatistical Modeling Evaluation and Uncertainty Quantification. *Remote Sens.* **2023**, *15*, 2708. https://doi.org/10.3390/rs15112708

Academic Editor: Gwanggil Jeon

Received: 20 April 2023
Revised: 15 May 2023
Accepted: 20 May 2023
Published: 23 May 2023

1. Introduction

Geological models play an important role in a wide range of real-world applications. Recent developments in the Earth surface dynamics have highlighted the importance of high-quality hydrological models [1]. A set of stochastic realizations comprises fundamental materials to express the spatiotemporal evolution of delta and flume systems [2,3]. Moreover, there has been increasing interest in the high-resolution subglacial topography models [4,5]. The roughness of bedrock has a substantial influence on subglacial flow behaviors in Arctic and Antarctica [6,7]. With the development of computing platforms, multiple-point statistics (MPS) has gained considerable attention [8,9]. With the aim of creating realistic models, MPS concentrates on the relationship between one target point and neighboring points. Viewing the training image (TI) as a prior material, the spatial patterns are constantly extracted and reproduced in the simulation grid (SG). With the objective of improving the simulation quality as well as computational efficiency, a range of image-processing and machine-learning techniques have been introduced into the MPS framework [10–12]. For example, spatial correlation is used to create an adaptive template and conserve patterns [13,14]. In order to save running time, clustering is a feasible way to organize patterns in TI and find representatives [7,15–17]. The medoid of each group has a high rank in the downstream procedure. In addition, multi-grid analysis is employed to capture spatial structures across different resolutions [18,19]. During the MPS simulation, the long-range connectivity is regenerated before the fine-grain characteristics. In addition, the development of the generative adversarial network (GAN) technique has received considerable attention in the geostatistics community [20,21]. Based on a large amount of TIs, two neural networks are simultaneously trained through an adversarial competition. A generator network attempts to produce an image associated with similar characteristics to TIs. By contrast, the discriminator is responsible for distinguishing real and simulated models. The expanding applications of GAN include geological facies [22,23], probability inversion [24], and porous media [25].

One major challenge for MPS, GAN, and other TI-based modeling programs is to quantitatively evaluate simulation quality. Therefore, numerous descriptors have been presented. As a classical two-point statistics metric, variogram focuses on calculating the expected squared difference between two points divided by a certain distance [26]. By contrast, the two-point correlation function and the lineal-path function are broadly used to characterize microstructures [27]. The former concentrates on the probability that two randomly chosen points have the same material phase. The latter approach is defined as the possibility that a straight line is entirely in a certain facies. Moreover, the connectivity functions are devised to compute the probability that two points in SG belong to the same connected component [28]. While two-point approaches are used in a variety of applications, a shortcoming is that geometrically and morphologically complicated structures cannot be finely represented with these methods.

The limitations of two-point statistics motivate researchers to develop high-order methods. From the geostatistical point of view, there are two variabilities within simulated realizations: pattern reproduction and spatial uncertainty [29]. On one hand, the core task of the geostatistical simulation method is to reproduce spatial patterns in SG. This is favorable for exhibiting consistencies between TI and the generated models. On the other hand, spatial uncertainty plays an essential role in understanding Earth systems. The use of a group of stochastic realizations is helpful to represent uncertainty and randomness. Therefore, competitive methods not only create similar realizations to TI but also enrich the diversity within generated models. Furthermore, the observation variable is important prior knowledge in conditional simulations. For example, the borehole interpretation is directly sampled from the subsurface system [17]. Produced by gound penetrating radar, the geophysical data describe the trend of the geological structure under investigation [30]. It is necessary to respect conditioning data during geological modeling. These conflicting objectives create a challenge for simulation programs.

Based on the variabilities mentioned above, multiple-point histogram (MPH) is reported to assess the quality of the unconditional simulation [31]. First, the program extracts spatial patterns from a geological model. Second, a probability distribution is created according to the frequency of each pattern. Third, MPH views the difference between two pattern distributions as a measure of the distance between two geological models. In particular, Jensen–Shannon (JS) divergence is applied to distinguish two distributions. The pattern reproduction is expressed by the average distance between the TI and the simulated realizations. By contrast, the mean distance between the generated models implies the spatial uncertainty. However, one primary drawback of MPH is the ability to describe complicated structures. Within the MPH framework, the template is the key to capturing geological patterns. Since MPH records every possible pattern configuration, the dimension of the pattern distribution grows rapidly. There is a tradeoff between the running speed and the evaluation accuracy. On one hand, an extending template is useful for identifying complex structures. On the other hand, high-dimensional distributions have a negative effect on the calculational efficiency. It is time-consuming to apply large templates in MPH.

With the purpose of improving evaluation performance, the analysis of distance (ANODI) was designed by Tan et al. [32]. The technical developments were as follows:

(1) Patterns in the TI are organized by a clustering method. The medoid of each group becomes the representative instance. (2) The program classifies patterns in the SG on the basis of their distances with representative patterns. A cluster-based pattern histogram is created according to the number of members in each group. (3) Similar to MPH, JS divergence is employed to quantify the difference between two histograms. The program performs multi-dimensional scaling (MDS) to visualize the affinity between geological models. (4) A multi-grid strategy is utilized to analyze geological structures across difference scales. The program individually captures long-range structures as well as fine-grain patterns. A weighted aggregation is conducted to combine JS divergences from several resolutions.

In recent years, ANODI has been used to examine the simulation quality in various applications. However, two primary concerns are the parameter specification and the evaluation accuracy. There are three noticeable technical limitations. First, the size of the template and the number of clusters are two user-defined parameters. Inappropriate configurations bring uncorrelated and redundant knowledge into the pattern analysis. Second, the program organizes patterns in the SG by calculating their distances with the medoid patterns in the TI. Given a TI with complicated structures, a large number of prototypes are found during the clustering step. It is time-demanding to compare the patterns and every representative. The time consumption constrains the dimension of the template and pattern groups. Third, the weight of each resolution is fixed and constant in the multi-grid analysis. The intrinsic characteristics of the geological structure are not taken into account.

In this paper, we provide a valuable alternative to quantitatively evaluate geostatistical modeling and quantify uncertainty. With the objective of improving the evaluation accuracy and simplifying the parameter specification, a pattern classification distribution (PCD) program is proposed to compare geostatistical realizations. First, our program applies an irregular template of adaptive size to extract geological patterns. According to the spatial correlation in the TI, the template points are sequentially gathered by a region-growing program. The computer controls the number of conditioning points based on the elbow point of the entropy function. Second, a clustering-and-classification program is designed to characterize the geological models. Aiming to customize the parameter setting, we apply hierarchical clustering to group the training patterns. Our program applies a decision tree to classify geological patterns and creates a pattern classification distribution from a geological realization. The similarity between two geological models is defined by the JS divergence between two distributions. Third, we devise a stacking framework to develop the multi-grid analysis. To improve the aggregation accuracy, the importance of each grid is calculated according to the intrinsic characteristics of the TI. A large weight is assigned to the coarse grid when there is an intensive long-range dependency in the TI.

We conducted four practical applications with the intention of comprehensively examining the proposed method. In the first test, a benchmark channel model was utilized. We ran a range of MPS programs to generate hydrological models. MPH, ANODI, and our PCD are applied to rank the realization sets. Compared with the existing methods, the key advantage of our PCD is the automatic parameter specification according to the simulation scenario. The geological models are reasonably distinguished by the proposed method. Further applications include non-stationary flume realizations, subglacial digital elevation models in Antarctica, and three-dimensional sandstone models. PCD exhibits a versatile ability to solve multiple geological categories, continuous variables, morphologically complex structures, and high-dimensional structures.

The rest of this paper is organized as follows. Section 2 establishes the context of the geostatistical evaluation methods and provides detailed procedures within MPH and ANODI. Our proposed PCD is explained in Section 3. Section 4 presents four real-world applications. The experimental results and findings are discussed in Section 5. Finally, conclusions are drawn in Section 6.

2. Background of the Geostatistical Evaluation Methods

2.1. Multiple-Point Histogram

Prior to explaining the proposed program, we provide a brief overview of the MPH and ANODI methods. There are three basic steps within MPH [31]. (1) Based on a geological model, spatial patterns are extracted by a predefined template. (2) The program records the frequency of each pattern. The pattern histogram becomes a tool to describe TI and geological realizations. (3) Jensen–Shannon divergence is utilized to measure the similarity between two distributions.

The MPH program is illustrated in Figure 1. To simplify the explanation, a template with five points is applied. As Figure 1a shows, the program visits point a and creates a pattern $p(a) = (Z(a+u_1), Z(a+u_2), Z(a+u_3), Z(a+u_4), Z(a+u_5)) = (1,0,1,0,1)$. Here, $Z(a)$ denotes the geological state of the point a. The program continuously visits every available point in TI. In this case, 36 patterns were found.

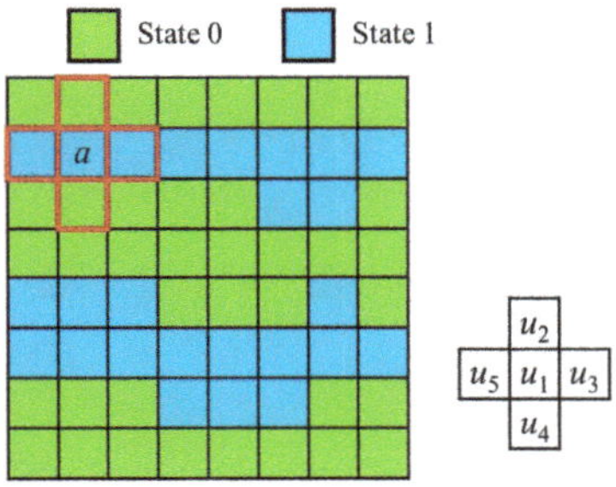

(a) (b)

Index	Pattern	Frequency
1	(0,0,0,0,0)	0.056
2	(0,0,0,0,1)	0.000
3	(0,0,0,1,0)	0.083
......	......	......
31	(1,1,1,1,0)	0.000
32	(1,1,1,1,1)	0.000

(c)

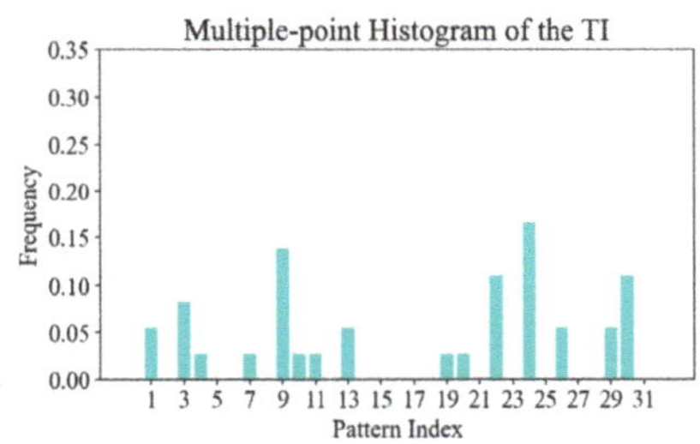

(d)

Figure 1. Multiple-point histogram based on a conceptual image. (**a**) A training image with size of 8 × 8. There are two geological states within the TI. The first pattern centered on the point a is highlighted in red; (**b**) a template with five conditioning points; (**c**) pattern frequency table; (**d**) the resulting multiple-point histogram.

Next, MPH focuses on analyzing patterns and creating a descriptor. Let f_1 denote the frequency of pattern p_1. As shown in Figure 1c,d, a pattern frequency table and a histogram are produced by recording the occurrence of each pattern. The number of possible patterns is $2^5 = 32$ becuase there are five conditioning points in the template and two geological categories in TI. Accordingly, the dimension of the pattern histogram is specified as 32.

After counting the frequency of each pattern, MPH regards the distance between two distributions as a measure of the dissimilarity between two geological models. In particular, JS divergence is an appropriate way to compare two histograms [33]. Suppose that there are two distributions $P_1 = \left\{f_1^{(1)}, f_2^{(1)}, \ldots, f_{32}^{(1)}\right\}$ and $P_2 = \left\{f_1^{(2)}, f_2^{(2)}, \ldots, f_{32}^{(2)}\right\}$. The JS divergence is defined as follows:

$$dis_{JS}(P_1, P_2) = \frac{1}{2}\sum_{i=1}^{32} f_i^{(1)} \log\left(\frac{f_i^{(1)}}{f_i^{(2)}}\right) + \frac{1}{2}\sum_{j=1}^{32} f_i^{(2)} \log\left(\frac{f_j^{(2)}}{f_i^{(1)}}\right) \quad (1)$$

As mentioned above, there are two variabilities within the geological models: pattern reproduction and spatial uncertainty. On one hand, the simulation program repeatedly extracts structures from the TI and reproduces proper patterns in the SG. The consistency between the TI and the generated realizations plays a key role in the assessment of geological modeling quality. On the other hand, uncertainty is a fundamental factor in exploring the surface and subsurface system. In general, the stochastic simulation method creates an ensemble of realizations. The distance between two geological realizations should be sufficiently large to represent uncertainty and randomness. Within the MPH framework, two variabilities are individually measured by the mean distance. Suppose that there is one training image TI and several geological realizations RE. The pattern reproduction ability of the geostatistical modeling method is quantified as follows:

$$dis_{RE}^{within} = \frac{1}{L}\sum_{l=1}^{L} dis_{JS}\left(P_{TI}, P_{RE}^{(l)}\right) \tag{2}$$

where L is the number of geological models and $P_{RE}^{(l)}$ is the pattern histogram computed from the l-th realization.

By contrast, the average distance between two geological models becomes a measure of uncertainty. The computation detail is shown below:

$$dis_{RE}^{between} = \frac{1}{L(L-1)}\sum_{l=1}^{L}\sum_{l'=1}^{L} dis_{JS}\left(P_{RE}^{(l)}, P_{RE}^{(l')}\right) \tag{3}$$

Next, MPH applies the preceding distances to compare two modeling methods. Suppose that there are two realization sets, A and B. The output ratios are defined as follows:

$$r_{A,B}^{between} = \frac{dis_{A}^{between}}{dis_{B}^{between}} \tag{4}$$

$$r_{A,B}^{within} = \frac{dis_{A}^{within}}{dis_{B}^{within}} \tag{5}$$

$$r_{A,B}^{overall} = \frac{r_{A,B}^{between}}{r_{A,B}^{with}} \tag{6}$$

The first ratio $r_{A,B}^{between}$ focuses on the extent of uncertainty. A high value of $r_{A,B}^{between}$ indicates that set A has greater uncertainty and diversity than set B. In comparison, the pattern reproduction ability is quantified by $r_{A,B}^{within}$. A small value of $r_{A,B}^{within}$ reveals that set A has a close affinity with the TI. The last ratio $r_{A,B}^{overall}$ summarizes previous two aspects. The $r_{A,B}^{overall} > 1.0$ implies that set A has better quality than set B.

One key limitation of MPH is the tradeoff between the evaluation accuracy and the running speed. It is worth noting that MPH identifies the relationship between multiple points. For a geometrically complex structure, an extending template has a substantial effect on the characterization quality. However, a large template dramatically increases the number of possible patterns. For example, supposing that a template of 5×5 is used to analyze a TI with two geological categories, the dimension of multiple-point histogram becomes $2^{25} = 33,554,432$. It is time-consuming and memory-intensive to handle high-dimensional histograms. Furthermore, the pattern histogram becomes a sparse vector when a large template is applied. Numerous zero values not only affect the effectiveness of JS divergence but also lead to considerable computation costs. Therefore, MPH generally applies small templates. Morphologically complicated structures cannot be represented effectively.

2.2. Analysis of Distance

Aiming to overcome the limitations of MPH, Tan et al. proposed the analysis of distance [32]. Their core contribution was the application of k-means clustering to group training patterns. Figure 2 provides a conceptual example. Similar to MPH, the first step in ANODI is to extract spatial patterns from the TI. Using a template with five points, 36 patterns are captured. Next, the program performs k-means clustering to inspect the underlying proximity within these patterns. The instances with strong similarity are allocated into one group. Since there is a categorical variable in the TI, Hamming distance is applied to distinguish between patterns. Moreover, the computer performs multidimensional scaling (MDS) to facilitate the visualization. In MDS feature space, one node represents a pattern. Two similar patterns have a small distance in the feature space. As shown in Figure 2e, four clusters are detected by the k-means program. Next, ANODI concentrates on finding the medoid of each group. The medoid is defined as the pattern closest to the geometrical center of a cluster. In addition, the program records the size of each cluster. As shown in Figure 2g, a cluster-based histogram of patterns (CHP) is created. The key benefit of ANODI is the dimension of the pattern distribution. It is convenient to control the dimension of the pattern histogram by specifying the number of pattern clusters. The high-dimension issue in MPH is therefore significantly alleviated.

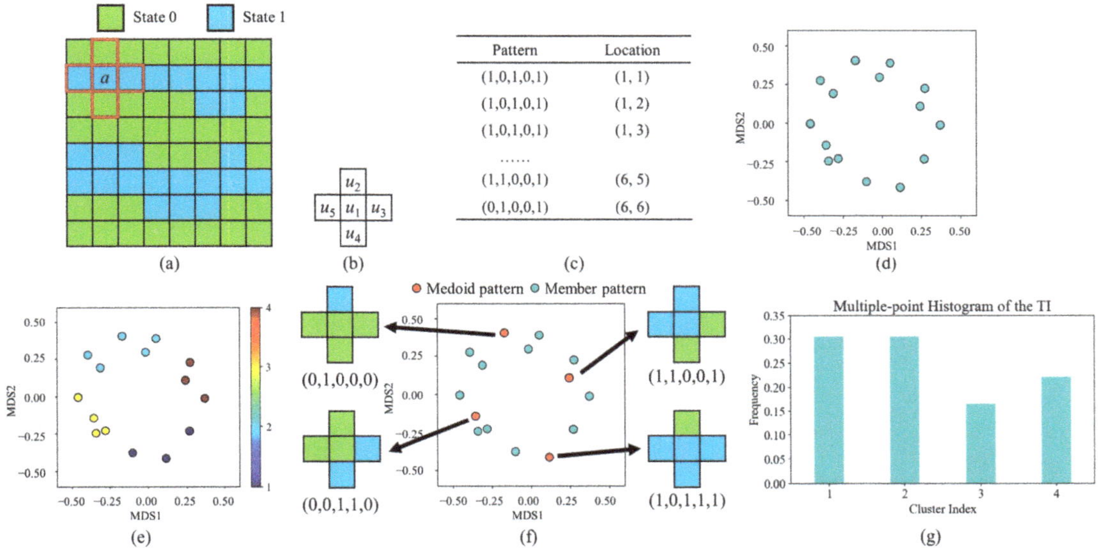

Figure 2. Analysis of distance based on a conceptual image. (**a**) Training image; (**b**) template; (**c**) pattern dataset. (**d**) patterns in the MDS feature space. (**e**) 4 pattern groups created by k-means clustering; (**f**) representative patterns; (**g**) cluster-based histogram created by TI.

Based on the medoid patterns, ANODI attempts to extract the morphological characteristics and rank the geological models. The characterization procedure is composed of three steps. (1) The patterns are extracted from the geological realizations. (2) The program classifies the pattern examples according to their distances from the representatives. (3) The number of member instances in each pattern category is output as a pattern histogram. Next, JS divergence is carried out to compare two distributions. Within the ANODI framework, a close similarity between two distributions suggests that there is a strong agreement between two geological realizations. Moreover, MDS is introduced to visualize the relationship between realizations. Based on a distance matrix, MDS projects data points into the low-dimensional feature space. The geological models which show matching structures are close in the MDS space. The technical details of MDS are elaborated in [1].

In addition, long-range correlations and connectivity are common in geostatistical simulations. It is difficult to extract long-scale structures in accordance with small templates. Therefore, the multi-grid strategy is incorporated into the ANODI framework. As shown in Figure 3, the program creates a pyramid of multi-resolution views. Inspired by MPS, the coarse grid is recursively generated by subsampling the fine grid. In this conceptual case, we implement a down-sampling procedure of stride 2. Starting from the bottom-left corner, the pixels in the fine grid are sequentially checked. The computer removes every even-numbered row and column to produce a small grid. For complex scenarios, a Gaussian pyramid and facies-frequency-based methods are favorable to preserve important geological structures. Let G denote the number of grids. Therefore, one training image TI can be expanded into a set of multi-resolution grids $\{TI_1, TI_2, \ldots, TI_G\}$. In the similar manner, a geological realization RE is extended into $\{RE_1, RE_2, \ldots, RE_G\}$. Based on a specified grid g, the computer performs pattern extraction, k-means clustering, and pattern classification to create a cluster-based histogram $P^{(l)}_{RE,g}$ from the realization RE_g. In a multi-grid context, the average distances in Equations (2) and (3) are modified as follows:

$$dis^{within}_{RE,G} = \frac{1}{L}\sum_{g=1}^{G} w_g \sum_{l=1}^{L} dis_{JS}\left(P_{TI,g}, P^{(l)}_{RE,g}\right) \quad (7)$$

$$dis^{between}_{RE,G} = \frac{1}{L(L-1)}\sum_{g=1}^{G} w_g \sum_{l=1}^{L}\sum_{l'=1}^{L} dis_{JS}\left(P^{(l)}_{RE}, P^{(l')}_{RE}\right) \quad (8)$$

where w_g is the weight of the g-th grid. In ANODI, a fixed weight $w_g = 1/2^g$ is applied. The high-resolution images and the fine grid are assigned high contributions. There are two assumptions behind this design. First, that there is less information and variability in low-resolution grids. Second, that short-scale patterns are more important than large-scale structures.

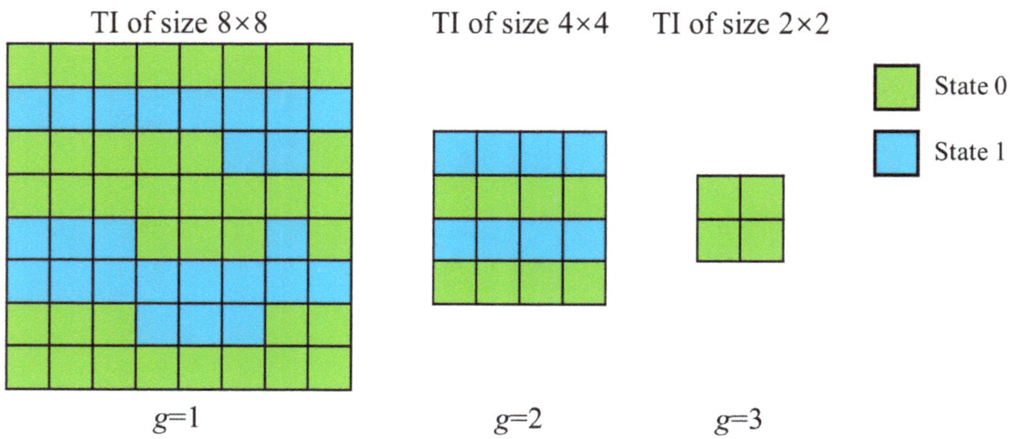

Figure 3. Multi-resolution TIs in the multi-grid analysis.

Although a range of practical applications are performed, there are three key technical limitations in ANODI. (1) The parameter specification is a key step to ensure the evaluation accuracy. Prior to the pattern-extraction step, the user has to specify the shape and size of template. Moreover, it is necessary to set the number of groups in the k-means clustering. An unsuitable setting has a negative influence on the computation quality. (2) It is time consuming to generate the cluster-based histogram from the geological realization. In order to classify patterns, ANODI has to calculate the distance with every representative pattern in the TI. (3) The multi-grid analysis suffers from the fixed weight. The long-range structure and connectivity do not receive sufficient attentions.

3. The Key Principles of Pattern-Classification Distribution

3.1. The Correlation-Driven Template-Design Program

The first step in our method is to design a reasonable template to extract geological patterns. Compared with the two-point statistics method, one key benefit of MPH and ANODI is the application of a template to explore the relationship between multiple points. However, these two methods apply a fixed and user-defined template. In order to improve the modeling quality, there are many adaptive template-design methods in the MPS community. For example, Honarkhah and Caers (2010) presented a template-selection method in their distance of pattern (DISPAT) [16]. The entropy is a measure of the information needed to encode a pattern. The program finds the optimal template size using elbow-point detection. The key drawback is that the DISPAT template is always square. The program cannot change the shape of the template according to the TI of interest. It is difficult to address anisotropic structures. By comparison, correlation-driven direct sampling (CDS) is an applicable way to quantify the contribution of each template point [14]. At first, the spatial dependency of the TI is analyzed by the correlation coefficient. With the purpose of removing the effects of noise and geologic cyclicity, CDS employs a Gaussian function to approximate the correlogram. Next, the weight of each template point is calculated by the Gaussian function. Nevertheless, CDS cannot automatically control the number of conditioning points. The template size is a user-defined parameter.

In this work, we combine the strengths of DISPAT and CDS together. An irregular template of adaptive size is devised to capture patterns. Based on the inherent characteristics of the TI, the computer automatically determines not only the shape but also the size of the template. Figure 4 provides an example to discuss our correlation-driven template design program. There are four basic steps. (1) Motivated by CDS, we compute the correlation coefficient between the template center and each neighboring point. In this conceptual case, a template with a size of 3×3 is employed to extract patterns. In Figure 4b, the deep purple indicates a strong dependency. Since there are two channels in the TI, an intensive correlation is presented in the horizontal direction. (2) A region-growing program is activated to sequentially collect template points. Viewing the template center as the seed point, we iteratively incorporate the neighboring point with the maximum correlation into the template. The template points that exhibit strong relationships with the center are given the highest priority. As shown in Figure 4c, the program creates several candidate templates with irregular shapes. (3) Our program assesses the template by means of the entropy. Based on a specified template, a group of patterns are extracted from the TI. Inspired by DISPAT, the entropy of the pattern distribution becomes a measure of the information captured by the template. We repeatedly perform steps 2 and 3 until every template configuration is examined. In this case, an entropy function of the pattern histogram is displayed in Figure 4d. (4) An elbow-point-detection technique is utilized to find the optimal parameter. In this work, we apply the profile log-likelihood approach [34]. Let $E = \{e_1, e_2, \ldots, e_N\}$ denote the entropy set. N is the number of template points. For every instance of entropy, we define two groups: $\{e_1, e_2, \ldots, e_n\}$ and $\{e_{n+1}, e_{n+2}, \ldots, e_N\}$. Next, the profile log-likelihood function $l(n)$ is defined as:

$$l(n) = -n \log\left(\frac{1}{\sqrt{2\pi\sigma^2}}\right) \sum_{i=1}^{n} \frac{(e_i - \mu_1)^2}{2\sigma^2} + (n - N) \log\left(\frac{1}{\sqrt{2\pi\sigma^2}}\right) \sum_{i=n+1}^{N} \frac{(e_i - \mu_2)^2}{2\sigma^2} \quad (9)$$

$$\sigma^2 = \frac{(n-1)\sigma_1^2 + (N-n-1)\sigma_2^2}{N-2} \quad (10)$$

where μ_1 and μ_2 are the means of the two groups, respectively. In contrast, σ_1 and σ_2 are sample variances. The common scale variance is denoted by σ. Consequently, the position with the maximum values of $l(n)$ is the optimal choice. In this case, the program specifies the template size as 4.

3.2. Geological Model Characterization Using Hierarchical Clustering and Decision Tree

Based on the template described above, we extracted a range of patterns from the TI. A key technical problem is the organization of these training instances. In this work, our program applies the agglomerative hierarchical clustering method [35]. The advantages of this method include: (1) There is no strong assumption on the distribution of the clusters. As a bottom-up approach, the program constantly merges similar groups. The local connectivity plays an essential role in the clustering step. Thus, the use of hierarchical clustering makes it possible to tackle data groups with varying densities and affinities. (2) It is easy to address categorical variables as well as continuous variables. Our program employs Hamming distance to deal with categorical data. By contrast, continuous variables are handled by the normalized Euclidean distance. (3) Rather than the number of groups, the distance threshold becomes the user-defined parameter in the hierarchical clustering. The program recursively performs the merging step until there is no group whose distances from others are shorter than the predefined tolerance. For a TI with diverse structures, the hierarchical clustering method automatically adopts numerous groups to organize pattern instances. In comparison, few clusters are produced when there are repetitive and redundant structures in the TI.

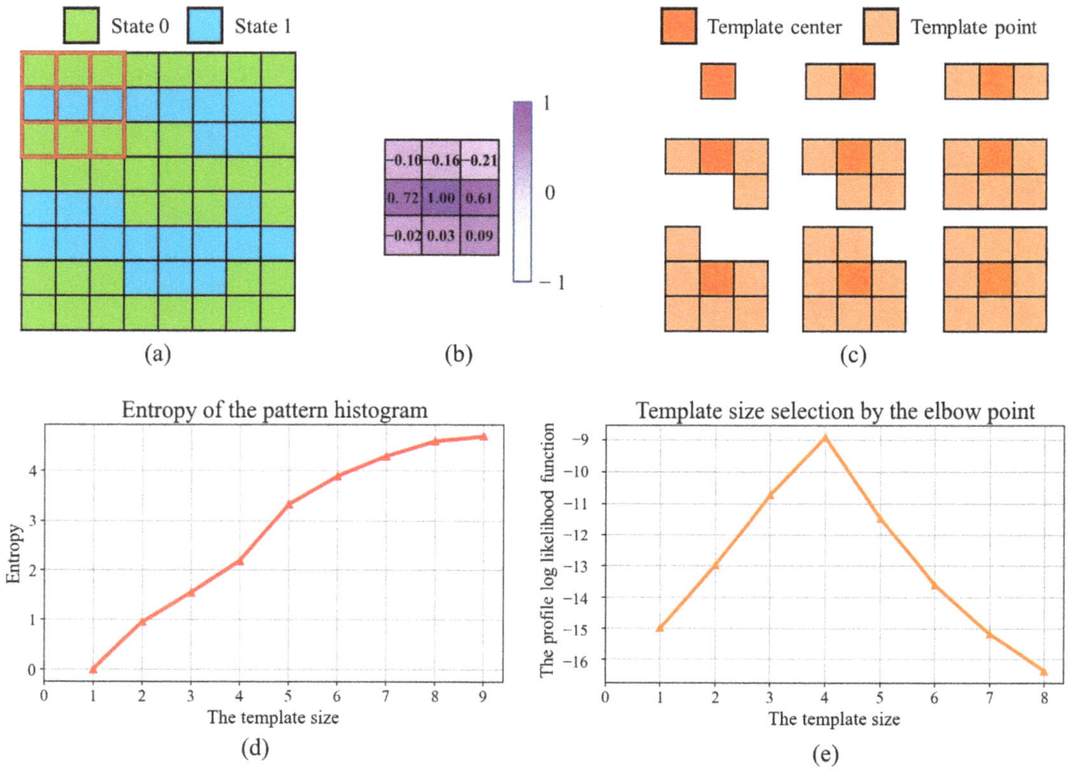

Figure 4. An irregular template of adaptive size. (**a**) Training image. The first pattern captured by a template of 3×3 is shown in red; (**b**) the correlation coefficient between the template center and each neighboring point; (**c**) candidate templates of different sizes; (**d**) entropy curve of the pattern histogram; (**e**) the optimal template size specified by the elbow-point detection.

Figure 5 provides a conceptual example to explain how to extract the morphological characteristics from the TI. Prior to the clustering step, the computer extracts every pattern with a template. Next, hierarchical clustering is performed to analyze the patterns. As the initial condition, each instance is regarded as an individual group. Subsequently, the computer combines the two groups with the shortest distances. It is worth noting that the distance between two pattern groups is a key point in hierarchical clustering. To prevent the influence of outliers, we select the average linkage. In other words, the mean distance between all members in the two groups is the similarity metric. The program successively performs the merging function until a stopping condition is met. In this simplified case, we specify the distance threshold as 0.34. Namely, our program does not combine groups whose distance is larger than 0.34. Thus, five pattern groups are detected in this case. Based on the clustering results, the program records the number of members in each pattern group. As displayed in Figure 5f, a pattern distribution is generated as an indicator of the TI.

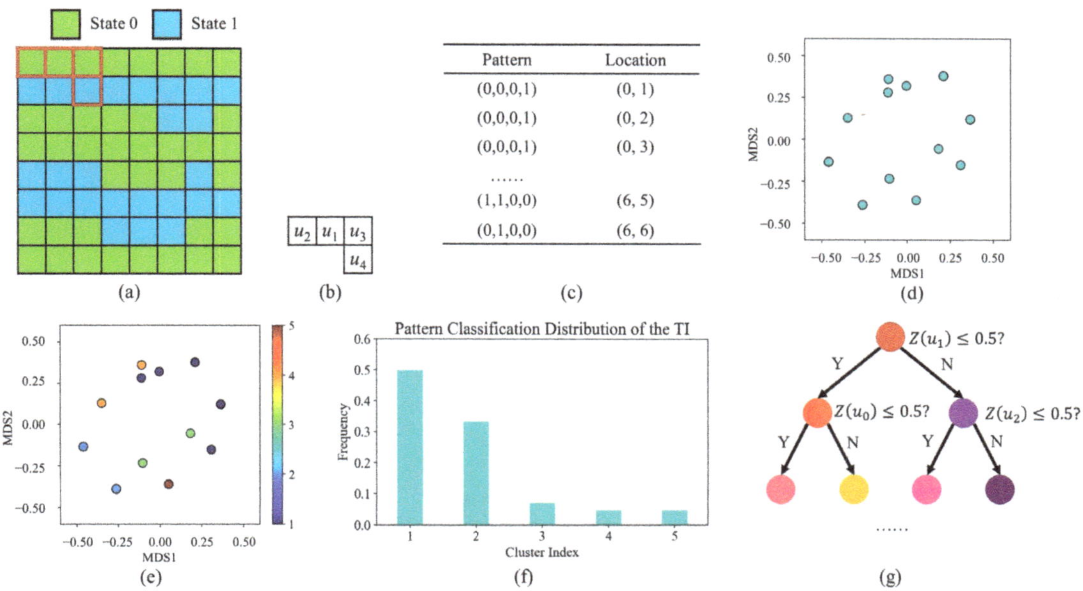

Figure 5. The hierarchical clustering of training patterns. (**a**) Training image and the first spatial pattern; (**b**) the irregular template designed by the correlation-driven method; (**c**) pattern dataset; (**d**) patterns in MDS feature space; (**e**) hierarchical clustering result in MDS feature space; (**f**) pattern-classification distribution of the TI; (**g**) decision-tree classifier trained by the clustering result.

After the clustering procedure, each pattern in the TI has a label. Next, our program focuses on characterizing the geological model in accordance with the clustering results. Similar to ANODI, a classification program is launched to create a pattern distribution from the realizations. With the objective of improving the performance, there are two important developments. First, the proposed method employs a decision tree [35] as the classifier. Compared with the nearest-neighbor approach in ANODI, the benefit of the decision tree lies in the efficiency to complete classification tasks. As an eager learning technique, the decision tree attempts to extract valuable information from the training data. Based on a flowchart-like structure, the prediction is achieved by continuously examining an attribute of the query instance. The time complexity is heavily dependent on the number of features. In comparison, the nearest-neighbor classifier is a lazy learner. The interpretation of the training data is generally delayed until the computer inputs a query. Given the large amount of patterns, the distance computation with every representative pattern leads to considerable time consumption.

The second improvement in our classification method is that every pattern in the TI is considered during the classifier training step. There is no pattern-selection step in the proposed program. It is noticeable that the number of training examples does not have a substantial influence on the speed of the decision tree classifier. In comparison, ANODI identifies the pattern that is closest to the group centroid as a representative. To save running time, the rest of the observations are removed by the classification program. The geological model characterization program suffers from information loss.

A simplified example with which to discuss the geological model characterization is shown in Figure 6. It can be seen that the first realization has a similar structure to the TI in Figure 5a. Two channels flow from the left to the right side. In contrast, the RE_2 exhibits different behavior. There are no connected components between the three blue areas. Our program contains four procedures to distinguish two realizations. (1) The computer extracts geological patterns with a template. As shown in Figure 6b, a dataset is created to store all the patterns. (2) Each pattern is classified with the decision tree. In this case, we allocate the geological patterns into five groups. (3) The program records the size of each pattern group. As shown in Figure 6c,f, a pattern classification distribution is independently generated. (4) Based on Equation (1), Jensen–Shannon divergence is applied to measure the similarity. A large value for distance indicates that there is a huge difference between two geological models. In this case, the divergence between the *TI* and the first realization is $dis_{JS}(TI, RE_1) = 0.11$. By comparison, $dis_{JS}(TI, RE_2) = 0.24$ is yielded when the computer focuses on the *TI* and the second realization. These computation results are consistent with the morphological characteristics of the three models.

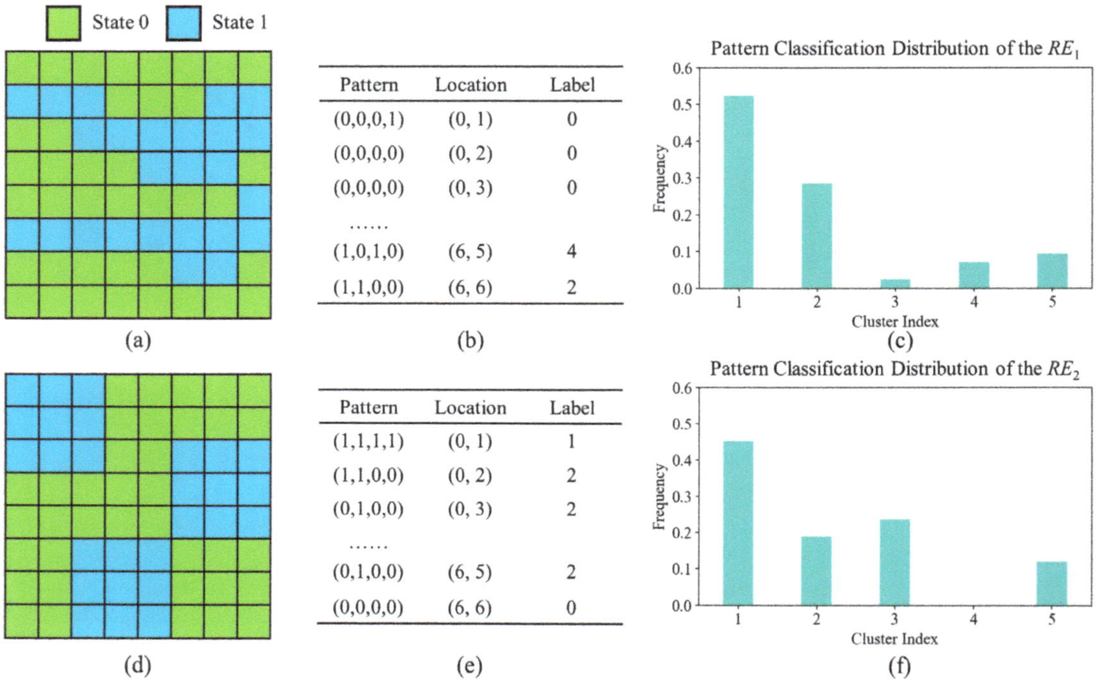

Figure 6. Geological model characterization with the decision-tree classifier. (**a**) The first realization RE_1 with two channels; (**b**) dataset pattern constructed by RE_1; (**c**) pattern classification distribution of RE_1; (**d**) the second realization RE_2 with three isolated patches; (**e**) dataset pattern constructed by RE_2; (**f**) pattern classification distribution of RE_2.

Based on the JS divergence, our PCD method is able to evaluate the geostatistical variability and rank modeling methods. As with MPH and ANODI, the average distance between the TI and the realizations becomes a descriptor of the pattern reproduction ability. On the other hand, the spatial uncertainty is quantified by the average distance between geological realizations. In addition, we insisted on applying the ratio to compare the two simulation methods. According to Equations (4)–(6), three ratios were output as the comparative results.

3.3. *Automatic Resolution Importance Assignment with a Stacking Strategy*

On the basis of the template design and model characterization procedures described above, our program has the ability to measure geometrical similarity. One technical limitation is that the template used above can only observe short-radius structures. As noted in Section 2.2, ANODI applies a multi-resolution pyramid to analyze small-scale as well as long-range structures. According to Equations (7) and (8), the morphological similarity is combined across different resolutions. However, the geostatistical evaluation program suffers from the fixed weights in the aggregation formula. The underlying assumption behind the current weight assignment is that the long-scale structure is less important than the fine-grain pattern in the geostatistical modeling. However, this assumption is not always true. For example, the long-range structure is a contributing factor in water resource management. With the aim of creating high-quality realizations, MPS is dedicated to reproducing the long-distance connectivity in flume systems and subsurface aquifer systems [10,17]. In petroleum engineering, the permeability of reservoir rocks heavily relies on the pore space [36]. Large-scale connected components play an important role in geostatistical modeling.

With the objective of developing the multi-grid strategy and improving the evaluation accuracy, a stacking framework was proposed to automatically assign the importance of each grid. The motivation for this proposal was the ensemble-learning method in the field of machine learning [35]. As a meta-learning framework, the stacking method takes advantage of two or more base machine-learning programs. There are two basic steps. First, a collection of base machine-learning programs are trained based on the same dataset. In general, it is helpful to use a diverse range of learning techniques. Second, the computer applies a meta-estimator to assess the effectiveness of each base program. The meta-estimator focuses on exploring the relationship between the predictions made by the base learner and the ground-truth labels.

Figure 7 provides an example to illustrate the stacking framework in our PCD. In this case, we specify $G = 2$ due to the limited size of the TI. There are two kinds of classifier in the proposed method. On one hand, the base classifier focuses on identifying the relationship between the template center and neighboring points. Multiple-point statistics information is effectively captured. For example, the short-range structure in the TI is analyzed by the first base classifier. By comparison, base classifier 2 attempts to capture long-range patterns with an extending template. On the other hand, the computer applies a meta-classifier to evaluate the strength of each base classifier. For a TI with long-term connectivity, there is a strong correlation between the center point and the conditioning points gathered by a large template. Base classifier 2 plays an influential role in the point prediction. A bigger contribution is assigned to the large-scale grid.

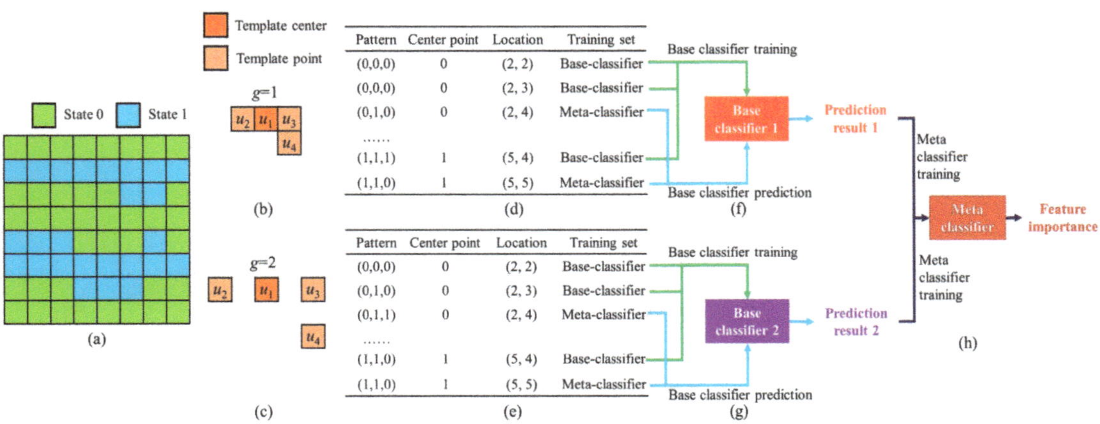

Figure 7. A stacking framework to compute the importance of each grid. (**a**) Training image; (**b**) a compact template with four points; (**c**) a sparse template with an extending radius; (**d**) patterns captured by the compact template; (**e**) patterns captured by the extending template; (**f**) the first base classifier. The green lines represent the instances used to train the classifier. By comparison, the patterns that are used to make point predictions are highlighted in blue; (**g**) the second base classifier; (**h**) the meta-classifier, which outputs the importance of each grid.

Furthermore, our program applies two key modifications to the traditional stacking framework. First, the base classifier is trained with different data. Based on the template with varying receptive fields, the computer captures spatial patterns across different resolutions. The multi-grid features are individually input to each base classifier. Second, the base classifiers share the same classification technique. In our program, the decision tree is applied. By comparison, the stacking program previously used in the machine -learning community encourages the utilization of heterogeneous classification techniques.

The detailed steps in the proposed method are as follows. (1) Viewing the irregular template as the prototype, our program creates a set of expanding templates to extract patterns from different grids. (2) A dataset is generated to collect training patterns. A noticeable phenomenon is that our program individually stores the center point and the template points. The conditioning points gathered by the template are the input features in the classification task. By comparison, the center point is viewed as the target variable. (3) The computer divides each pattern dataset into two subsets. Our program randomly assigns 70% instances into the first set. Accordingly, 11 patterns are selected in this case. The five remaining patterns comprise the second subset. (4) Two base classifiers are trained by the first pattern subset. The proposed method applies the decision tree to build the bridge between the template center and the neighboring points. After the training step, the patterns in the second subset are fed into these two base classifiers. Thus, five predictions are separately produced. (5) Our program trains the meta-classifier. In this framework, random forest is employed as the meta-classifier. The meta-classifier focuses on determining the relationship between the base-classifier outputs and the center points. (6) The meta-classifier outputs the feature importance as the contribution of each base classifier. In this case, the importance values of the two resolutions are 0.51 and 0.49, respectively. These computational results imply that the role of long-range structures is close to that of the small-range patterns. In other words, reproducing large-scale connectivity is an important aspect in this modeling scenario. However, the current MPH and ANODI cannot automatically control the weight assignment according to the simulation scenario. The significance of long-range structures is underestimated in this conceptual case.

4. Applications

4.1. A 2D Benchmark Channel Model with Anisotropic Structures

As the first application, the benchmark channel model was employed to examine the proposed PCD. Based on the TI shown in Figure 8a, we independently launched a group of MPS programs to generate 200 realizations. At first, single normal equation simulation (SNESIM) [18] and filter-based simulation (FILTERSIM) [15] were individually performed by Stanford Geostatistics Modeling Software (SGeMS) [37]. We applied a template with a size of 9×9 and a multigrid strategy of $G = 3$ in SNESIM. By comparison, the default setting was utilized by FILTERSIM. The sizes of the searching template and the pasting template were specified as 11×11 and 7×7, respectively. Next, we implemented three database-based MPS programs. Improved parallelization (IMPALA) [38], column-oriented simulation (CSSIM) [39], and nearest-neighbor simulation (NNSIM) [10] were performed to create the channel models. The parameters in IMPALA and CSSIM were the same as those in SNESIM. In NNSIM, our program specifies the cosine distance threshold and the number of teachers as 0.1 and 5, respectively. Finally, direct sampling (DS) [40] and tree-based direct sampling (TDS) [17] were carried out in this stochastic simulation scenario. According to the experiments conducted by Meerschman et al. [41], three predefined parameters of DS are $N^{DS} = 30$, $t^{DS} = 0.05$ and $f^{DS} = 0.5$. In addition, TDS is activated by a clustering tree with a height of 9. The first realization created by these methods is shown in Figure 8.

Based on the geological models discussed above, our PCD was performed to quantitatively assess the MPS simulation quality. As the first step, our program generated a template according to the intrinsic characteristics of TI. The computation results are shown in Figure 9. It is clear that there was strong anisotropy in the channel image. The spatial correlation and connectivity in the horizontal direction were more intensive than in the vertical direction. In order to conserve the channel structures, the proposed template design method sequentially collects points with strong correlations. The pattern entropy function and the elbow point detection technique were employed to determine the template size. As displayed in Figure 9d, 26 points were involved in our irregular template.

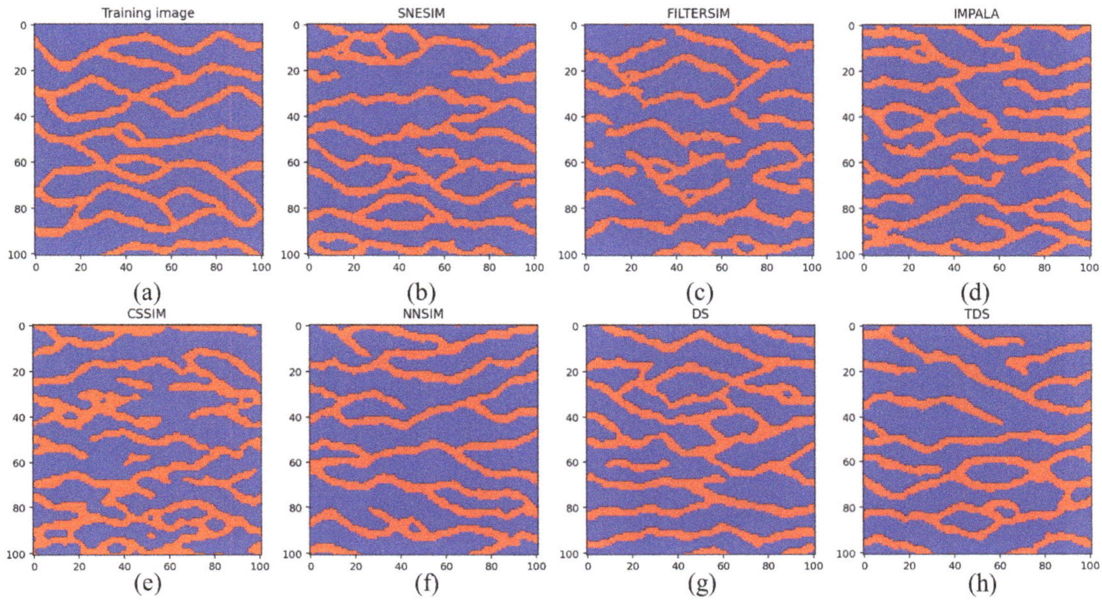

Figure 8. Channel realizations created by MPS programs. (**a**) Training image; (**b**) SNESIM model; (**c**) FILTERSIM model; (**d**) IMPALA model; (**e**) CSSIM model; (**f**) NNSIM model; (**g**) DS model; and (**h**) TDS model.

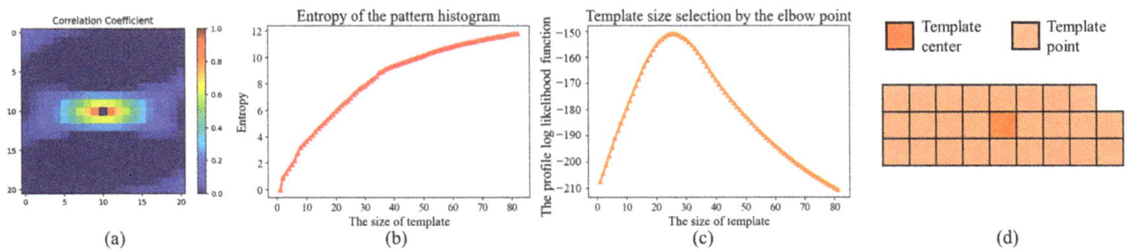

Figure 9. The correlation-driven template design in the channel simulation. (**a**) The correlation coefficient of each template point; (**b**) entropy curve of the pattern histogram; (**c**) template size selection by the elbow point detection. (**d**) the optimal template with 26 points.

Next, the stacking framework was launched to assign the resolution importance. In this case, we specified the number of $G = 4$. As the base classifiers, four decision trees focused on predicting the state of the template center according to the neighboring points across different resolutions. Moreover, the random forest technique was used as the meta-classifier to validate each base classifier. Consequently, the resulting weight vector was $[0.56, 0.24, 0.17, 0.03]$.

Next, the hierarchical clustering method was used to organize the training patterns. For instance, our program extracted 9207 patterns from the finest grid $g = 1$. Based on the hierarchical clustering program associated with a distance threshold of 0.1, 339 pattern groups were found. Subsequently, our program performed a decision tree to characterize the MPS realizations. The frequency of each category became a descriptor of the morphological characteristics. Finally, the JS divergence and the weighted aggregation technique were carried out to distinguish the two geological models. Based on Equations (7) and (8), we quantified the pattern reproduction and spatial uncertainty by the average distance. The relative behavior of each MPS program was compared by Equations (4)–(6). In this case, we applied the SNESIM realizations as the method B. The computation results are highlighted in Figure 10a. Furthermore, we activated multi-dimensional scaling (MDS) to visualize the calculation results. In the feature space, each node represents a geological model. The two close points imply that there was intensive compatibility between two MPS realizations. Figure 11a displays the MDS visualization results. In order to avoid visual confusion, we partitioned the point cloud into three parts. First, the SNESIM, IMPALA, and CSSIM realizations are emphasized. Second, the blue points display the dispersals of NNSIM, DS, and TDS models. Third, the FILTERSIM realizations are presented in yellow.

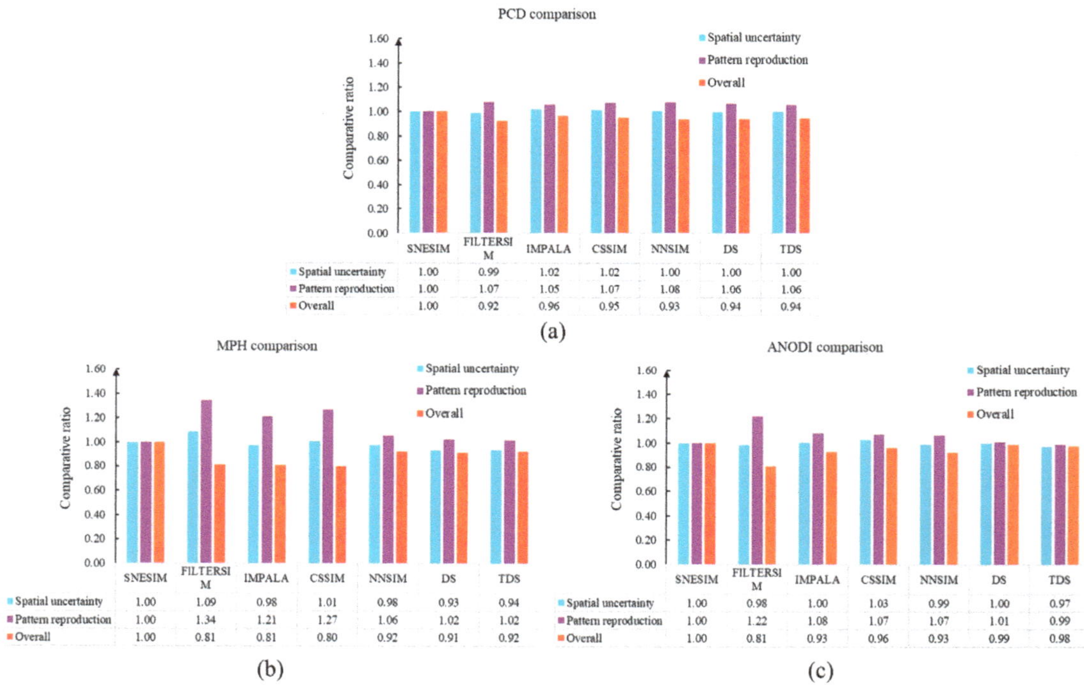

Figure 10. Geostatistical-quality-evaluation results from (**a**) PCD; (**b**) MPH; and (**c**) ANODI.

With the aim of performing an extensive comparison, we implemented MPH and ANODI in this case. In MPH, a template of size 3×3 was applied to extract the patterns. Therefore, there were 512 possible values in the multiple-point histogram. Furthermore, we employed a template of size 7×7 and a multi-grid strategy with $G = 4$ in ANODI. The number of pattern clusters was specified as 40. The computation results of MPH and ANODI are shown in Figures 10 and 11.

Two key observations were made based on the PCD results. (1) According to the comparative ratios, SNESIM exhibited a competitive performance. In Figure 10a, all the overall ratios are lower than 1.0. The main reason is that the postprocessing step in SNESIM plays a positive role in improving simulation quality. Mismatching structures are upgraded by the re-simulation step. (2) Based on Figure 11a, the preceding MPS programs can be partitioned into three groups. First, there was a strong similarity between the SNESIM, IMPALA, and CSSIM realizations. The orange, green, and turquoise points are located at the top. Next, the NNSIM, DS, and TDS models had relatively small distances. It is clear that the blue points constitute a group in the bottom-left. Finally, the FILTERSIM realizations in yellow were in disagreement with the other methods. A similar phenomenon can be observed in Figure 11b. However, ANODI did not highlight a significant difference between the MPS realizations. The main reason for this finding lies in the pattern-matching mechanism in the MPS framework. SNESIM, IMPALA, and CSSIM employ the pruning strategy to find a compatible instance. When a completely matching pattern does not exist, the program discards the conditioning point with the maximum distance. By comparison, distance computation is an essential component in NNSIM, DS, and TDS. These three programs apply Hamming distance to distinguish between patterns. Furthermore, the core idea in FILTERSIM is to utilize a set of filters to organize training patterns. The program classifies 2D patterns based on the convolution scores with six predefined kernels.

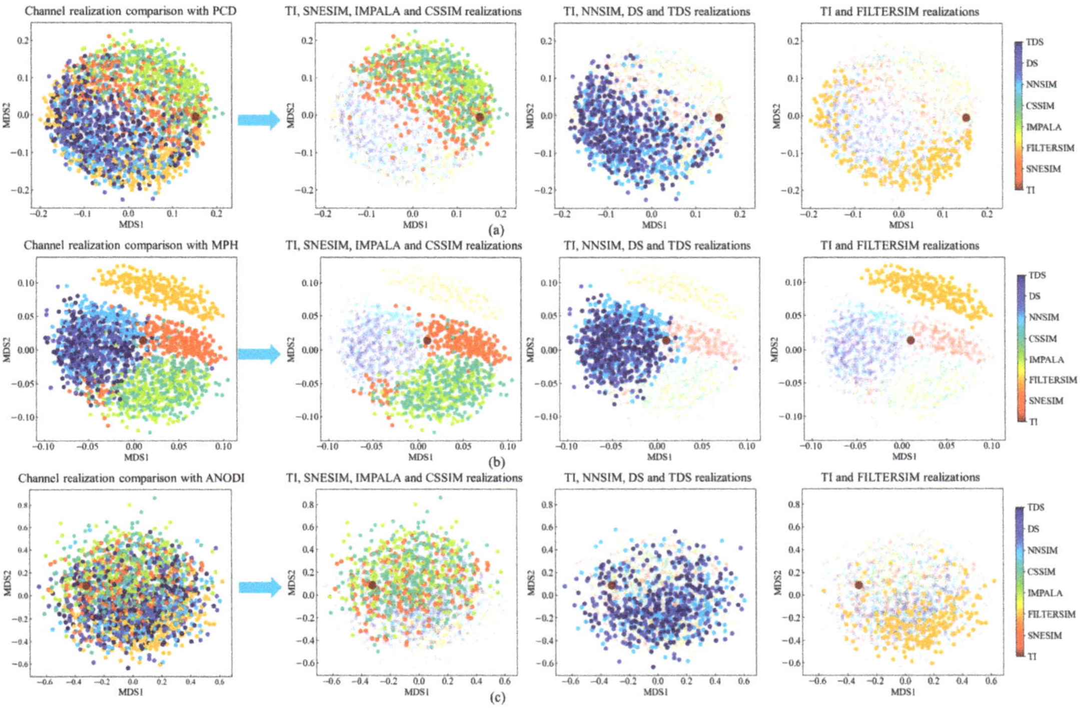

Figure 11. Uncertainty-quantification results based on (**a**) PCD; (**b**) MPH; and (**c**) ANODI.

With the purpose of improving practicability, we investigated the parameter sensitivity of PCD, MPH, and ANODI. SNESIM and DS models were adopted as method *A* and method *B*, respectively. On one hand, we studied the influence of the distance threshold in the PCD. The comparison results are shown in Figure 12a. As the only user-defined parameter, the distance threshold in the hierarchical clustering had a small effect on the evaluation result. Three ratios did not demonstrate intensive variation. On the other hand, parameter setting is a key module within MPH and ANODI. There were strong fluctuations in the comparative ratio curves. The changing behavior creates difficulties in the quantitative evaluation of the MPS modeling quality.

4.2. A 2D Non-Stationary Flume System with Morphologically Complex Structures

Autogenic variability is a fundamental aspect of numerous Earth surface systems. Schedit et al. [1] and Hoffimann et al. [3] simulated a delta evolution in laboratory experiments. Based on the overhead snapshots, a group of flume realizations were created to express spatiotemporal uncertainty in a channelized transport system. In this section, we focus on examining the performance of PCD in the context of multiple geological categories.

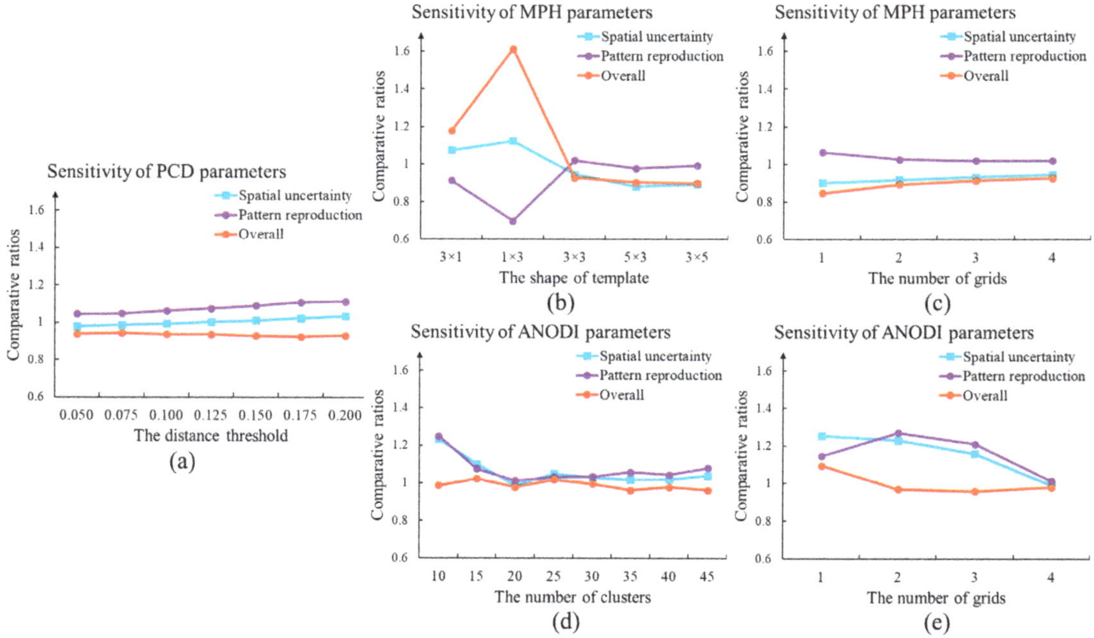

Figure 12. Parameter sensitivity of three evaluation methods. (**a**) The influence of distance threshold within PCD; (**b**) the influence of template within MPH; (**c**) the influence of the grid number within MPH; (**d**) the influence of the cluster number within ANODI; (**e**) the influence of the grid number within ANODI.

Figure 13a exhibits three TIs. The sediment is expressed by the blue area, while the intensity of flow is reflected by other colors. In the first TI, there are only two geological categories. By contrast, three and four states are presented in the second and third TIs, respectively. We independently implemented NNSIM and TDS to generate the non-stationary flume model. 10 flume realizations were individually created by two programs. With the aim of addressing non-stationarity, the computer introduces the auxiliary variable into the MPS framework. Based on the proximity to the original point, the simulation grid is split into four subareas. Therefore, NNSIM and TDS create an independent pattern dataset for each area. With the purpose of improving the simulation quality, NNSIM applies a template of size 9×9 and a multi-grid strategy of grid 4. We specified the parameters in TDS as $N^{DS} = 30$, $t^{DS} = 0.05$ and $f^{DS} = 0.5$. Moreover, the height of the clustering tree was configured as 9. Figure 13c,d provide the first realization of the methods described above. Detailed explanations of the TI and MPS simulation are provided in [1,10].

The proposed PCD was applied to rank the MPS programs. TDS and NNSIM were specified as methods A and B, respectively. In order to address geometrically complex structures, we specified the distance threshold in the hierarchical clustering as 0.15. Within the multi-grid framework, the number of grids was set as 4. The comparative ratios and MDS visualization are shown in Figure 14. The overall ratios in the three scenarios were close to 1.00. In addition, the point clouds had comparable dispersal. These findings suggest that NNSIM and TDS had similar accuracy in this simulation task.

We carefully investigated the parameter setting issue. One key advantage of the proposed method is that PCD applies the correlation-driven template, a combination of hierarchical clustering and decision tree, and a stacking framework. A set of parameters are automatically computed according to the morphological characteristics of the TI. As shown in Figure 15, the distance threshold did not play an influential role in the evaluation results. The adaptive parameter configuration within PCD is beneficial to quantitatively assess the progress of each MPS program. By contrast, ANODI is heavily dependent on the parameter specification. In order to extract complex patterns, ANODI applies a template with a size of 7×7. The effects of the number of clusters and the number of grids were checked. As displayed in Figure 16, there were remarkable differences between the ANODI outputs. It is challenging to objectively compare the strengths of NNSIM and TDS.

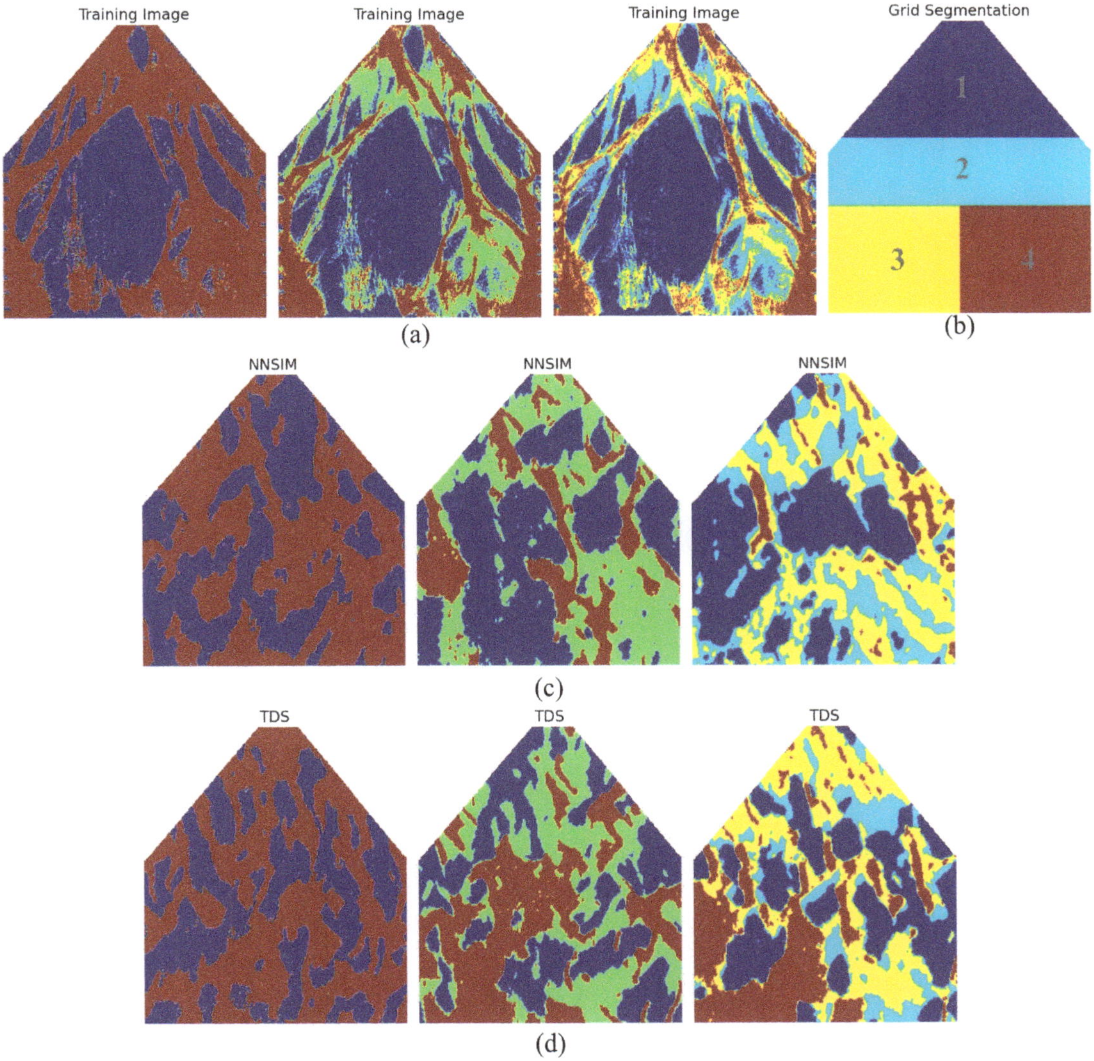

Figure 13. MPS realizations in the flume modeling. (**a**) Three training images with multiple geological categories; (**b**) auxiliary variable in MPS simulation. The numbers indicate the indices of four subareas; (**c**) NNSIM realizations in three modeling scenarios; (**d**) TDS realizations in three modeling scenarios.

Moreover, we investigated the sensitivity of the control parameters in relation to the NNSIM simulation quality. On one hand, the template is an essential component in the collection of conditioning points. An expanding template has a positive effect on the reproduction of complicated structures. NNSIM applies templates with sizes of $3 \times 3, 5 \times 5, 7 \times 7, 9 \times 9, 11 \times 11$, and 13×13. Furthermore, $G = 3$ is utilized to extract patterns across different scales. On the other hand, the multi-grid strategy provides a valuable tool to simulate patterns across different resolutions. The numbers of grids were configurated as 1, 2, 3, and 4, respectively. A template with a size of 9 × 9 was employed. For each parameter specification, 20 flume models were independently generated by NNSIM.

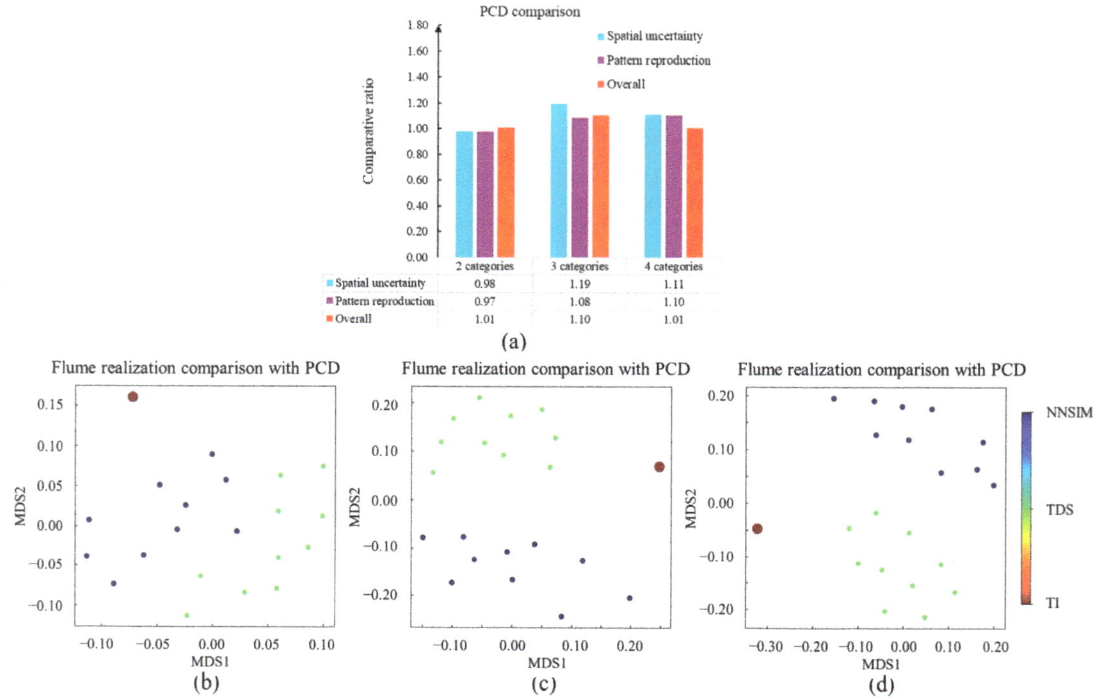

Figure 14. PCD calculation results. (**a**) Comparative ratios between NNSIM and TDS; (**b**) uncertainty quantification in the two-facies flume simulation; (**c**) uncertainty quantification in the three-facies flume simulation; (**d**) uncertainty quantification in the four-facies flume simulation.

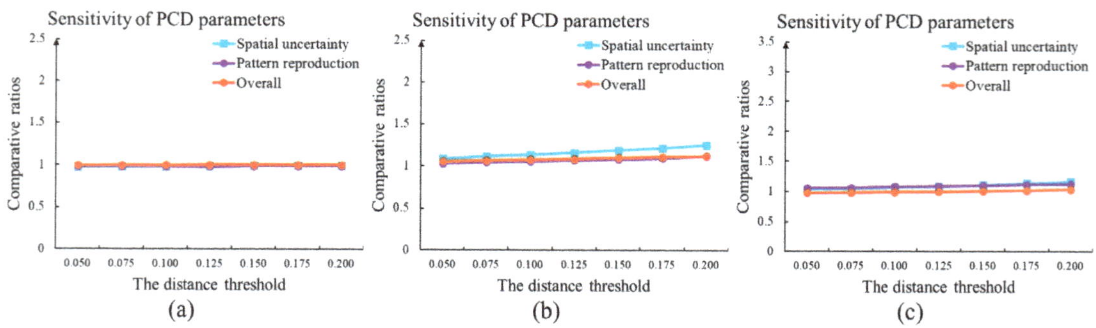

Figure 15. Parameter sensitivity of PCD in (**a**) two-facies flume simulation; (**b**) three-facies flume simulation; (**c**) four-facies flume simulation.

We applied PCD to find reliable realizations and quantify the uncertainty. The model set produced by the first parameter setting was defined as method *B*. The evaluation results are shown in Figure 17. The findings were as follows. (1) MPS simulation benefits from an extending template. In Figure 17a–c, the purple columns have decreased with the development of the template size. This reveals that MPS methods are able to effectively reproduce patterns in the simulation domain. Furthermore, the expanding template contributed to the increase in the overall ratios, which are expressed in orange. (2) The multi-grid strategy is a key module to improve MPS quality. According to Figure 17d–f, a high grid value not only enriches the spatial uncertainty but also reduces the differences between the TI and the realizations.

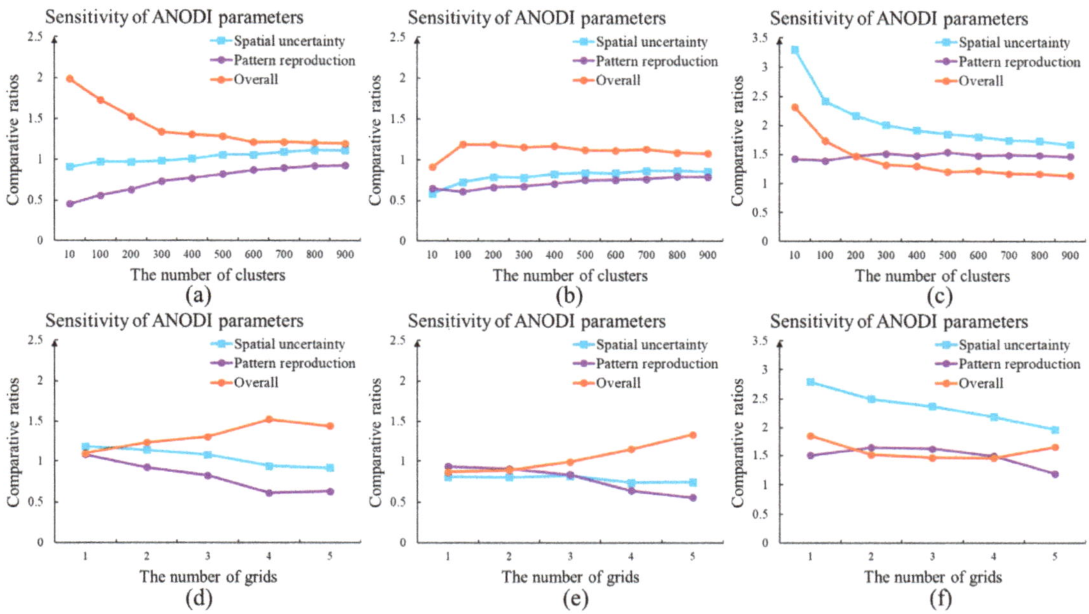

Figure 16. Parameter sensitivity of ANODI in the three modeling scenarios. (**a**) The influence of cluster number on two-facies simulation; (**b**) the influence of cluster number on three-facies simulation; (**c**) the influence of cluster number on four-facies simulation; (**d**) the influence of grid number on two-facies simulation; (**e**) the influence of grid number on three-facies simulation; (**f**) the influence of grid number on four-facies simulation.

In order to complete an extensive comparison, we inspected more parameter combinations within NNSIM. The influences of the template size and the multi-grids were simultaneously investigated. In this case, our PCD selected the realizations created by a template 9×9 and two grids as method B. The computational results are shown in Figure 18. First, the red color in the spatial uncertainty map indicates that the model sets demonstrated a high level of diversity. Second, the close proximity to the TI is emphasized by the white and blue at the second column of Figure 18. Third, the competitive programs are highlighted by the red in the overall ratio matrix. Apparently, the use of a large template and a multi-grid approach are helpful to create realistic realizations. The program provides its best performance when a template of size 9×9 and three grids are employed. In addition, it should be noted that there are blue squares in the bottom-right area at the last column in Figure 18. This implies that NNSIM does not provide reliable models when a template of size 13×13 and $G = 4$ is utilized. The main reason for this is that the large template and grids contain numerous conditioning points in MPS simulations. The uncorrelated points not only provide redundant information but also create computational burden. Therefore, the parameter selection is a key component within the MPS framework. An unsuitable extension in template size and grids prevents the MPS program from outputting favorable models.

4.3. A 2D Subglacial-Bedrock-Elevation Model with Continuous Variable

Knowledge of the topography beneath Antarctica and Greenland ice sheets is essential for a wide range of glaciological investigations. With the aim of better predicting subglacial flow behavior, it is necessary to create a collection of high-resolution topographic models and express spatial uncertainty. In particular, the characteristics of the subglacial topography of the Thwaites Glacier in the Amundsen Sea Embayment have received considerable attention [42]. The accelerating ice loss in Thwaites Glacier has a substantial influence on the stability of the West Antarctic Ice Sheet [43]. Based on the non-stationary multiple-point geostatistics method, Yin et al. generated a set of realistic topographic models [7]. On one hand, the stochastic modeling method is guided by 166 high-quality topographic training images, which are extensively sampled from the deglaciated regions in Arctic and Antarctica. On the other hand, ice penetrating radar data become the hard data in this simulation task.

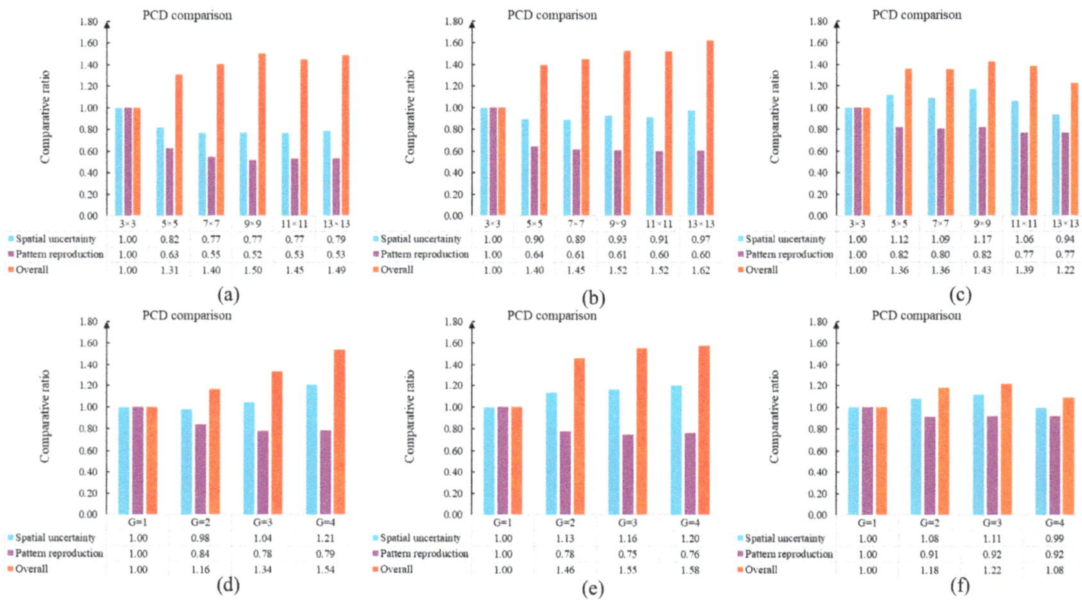

Figure 17. PCD evaluation to check the parameter sensitivity within NNSIM. (**a**) PCD comparison between different templates in two-facies simulation; (**b**) PCD comparison between different templates in three-facies simulation; (**c**) PCD comparison between different templates in four-facies simulation; (**d**) PCD comparison between different grids in two-facies simulation; (**e**) PCD comparison between different grids in three-facies simulation; (**f**) PCD comparison between different grids in four-facies simulation.

Aiming at generating diverse models, three geostatistical simulation methods were applied in this case. (1) We used the Kriging method to generate a subglacial topographic model according to the radar data. As a deterministic method, Kriging produces only one realization. (2) Sequential Gaussian Simulation (SGSIM) was carried out to create 10 stochastic realizations. The program estimates the variogram on the basis of radar lines. (3) We implemented non-stationary multiple-point geostatistics to create digital elevation models. There were two major steps in this process. First, a training-image-transition method was used to find the optimal prior model for each local subarea. Second, the computer activated direct sampling (DS) to complete the gap-filling task. In this case, MPS generated 10 subglacial topographic models. The first realization of three geostatistical modeling programs is shown in Figure 19. The authors of [7] discuss the technical details of previous methods.

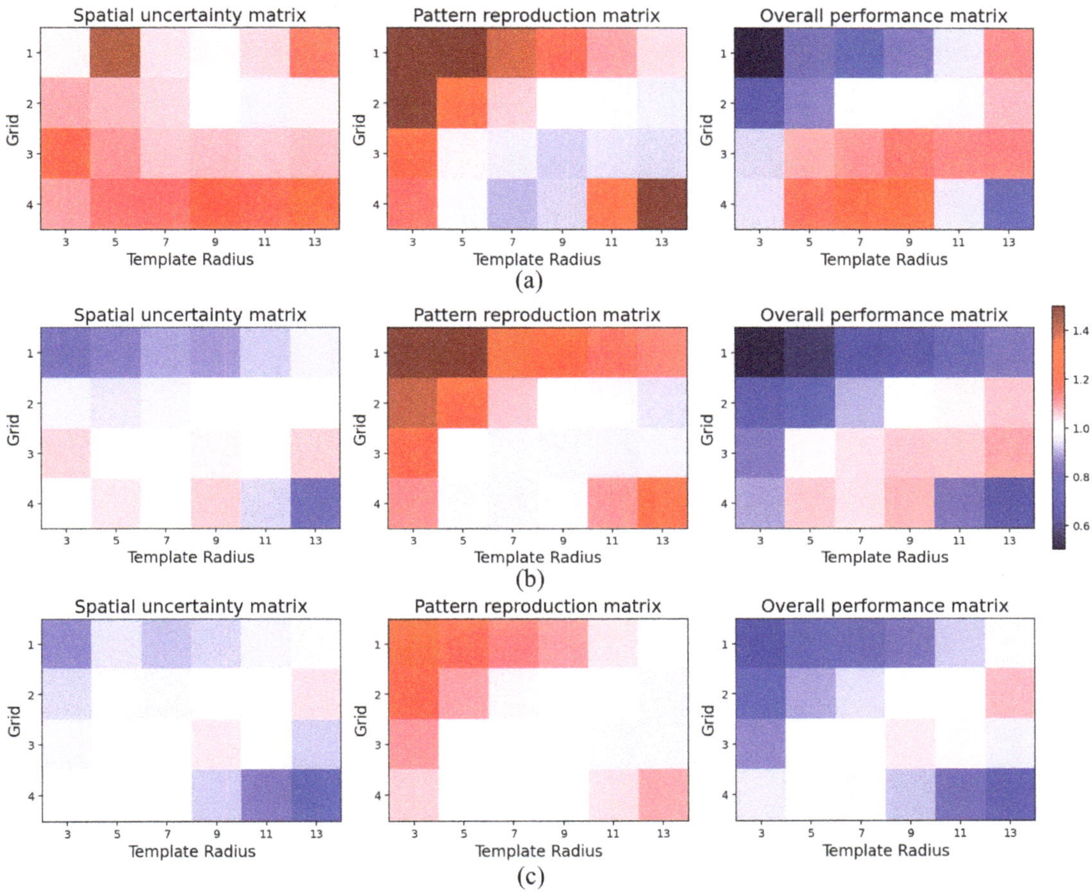

Figure 18. PCD evaluation of various NNSIM parameter combinations. (**a**) Two-facies simulation; (**b**) three-facies simulation; (**c**) four-facies simulation.

In this section, the proposed PCD is examined by a large-scale stochastic simulation with continuous variable. A noticeable phenomenon is that there was no training image in this evaluation task. Kriging and SGSIM were used to explore the spatial dependency in accordance with the radar lines. In contrast, the non-stationary MPS featured a min–max normalization on 166 TIs. The standardization of the bedrock elevation helped MPS concentrate on reproducing the morphological structures. Accordingly, it was not reasonable to directly compare the 166 TIs and geostatistical realizations in this case.

With the aim of mitigating the absence of TI, our PCD was used to analyze the spatial patterns in the first MPS realization. As the first step, the correlation-driven template design method was launched. To calculate the entropy curve, we applied the multi-level thresholding program to tackle the continuous variable. The bedrock elevation was uniformly partitioned into several bins. Therefore, the number of geological categories was an important parameter. Figure 20b provides the segmentation results. As shown in Figure 20c, our program computed the template size according to the intrinsic characteristics of the categorical models. There were two findings. (1) The template size ranged from 18 to 22 when the number of facies was lower than 6. (2) With the development of the geological categories, the template size progressively decreased. The main reason is that there was an exponential relationship between the number of geological states and the amount of possible pattern configurations. A high value of geological categories leads to a sparse pattern histogram and a slow increase in the entropy function. Therefore, we selected the template that was determined by the MPS realization with three categories. A template with 18 conditioning points was applied to extract the spatial patterns in the downstream steps.

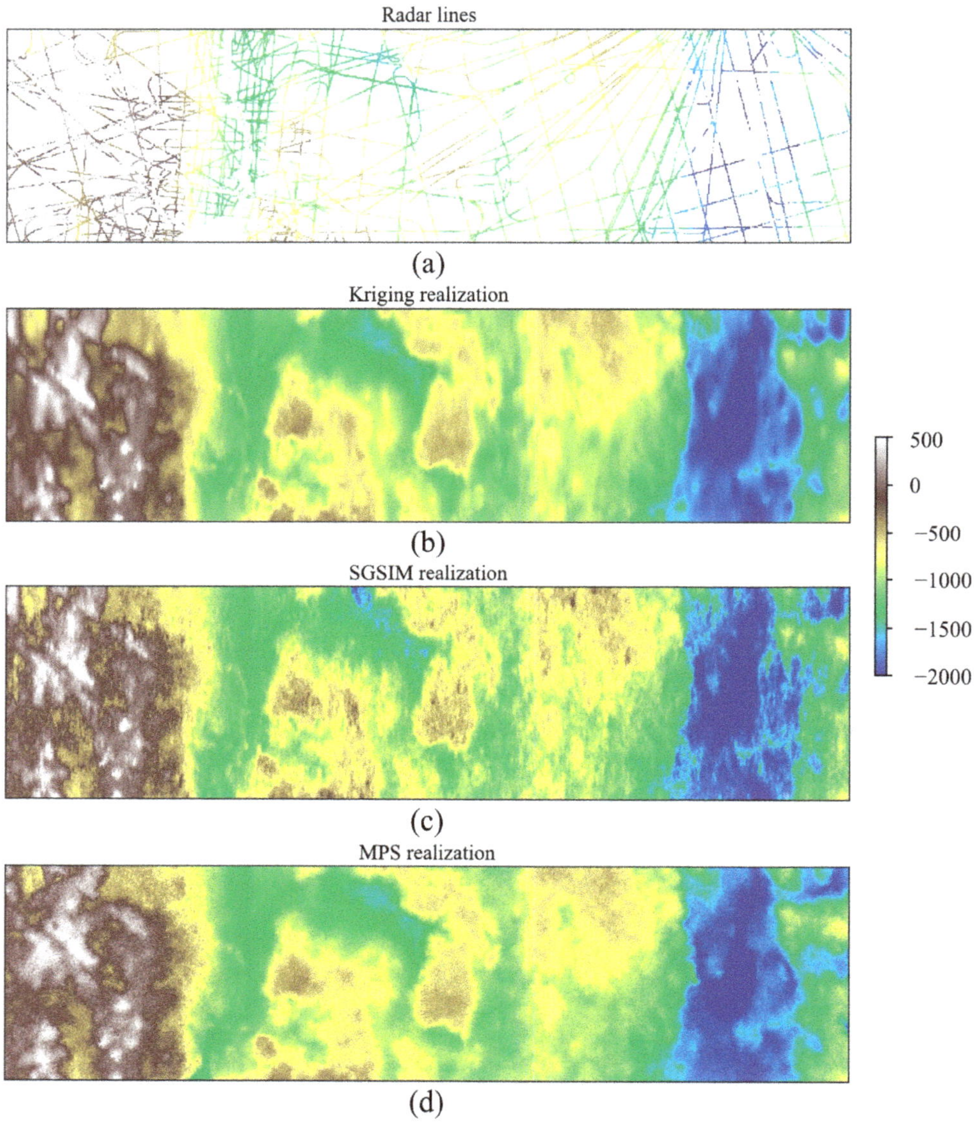

Figure 19. Antarctica topographic models created by the geostatistical methods. (**a**) Radar lines are the hard data in the modeling task; (**b**) Kriging realization; (**c**) the first SGSIM realization; (**d**) the first MPS realization.

Next, the stacking strategy was activated to quantify the importance of long-range and small-scale structures. With the template mentioned above, our program conducted a multi-grid analysis based on the first realization of the three geostatistical modeling methods. The grid importance is shown in Figure 21. To emphasize the key findings, we do not depict the contribution of the finest grid. It is apparent that there is a significant difference between the three realizations. On one hand, long-distance structures are not well conserved by Kriging and SGSIM. The conditioning points collected by the large templates had a weak correlation with the template center. On the other hand, there was a strong relationship between the template center and the surrounding points in MPS realization. Compared with two-point statistics, MPS exhibits better performance in terms of long-range pattern reproduction.

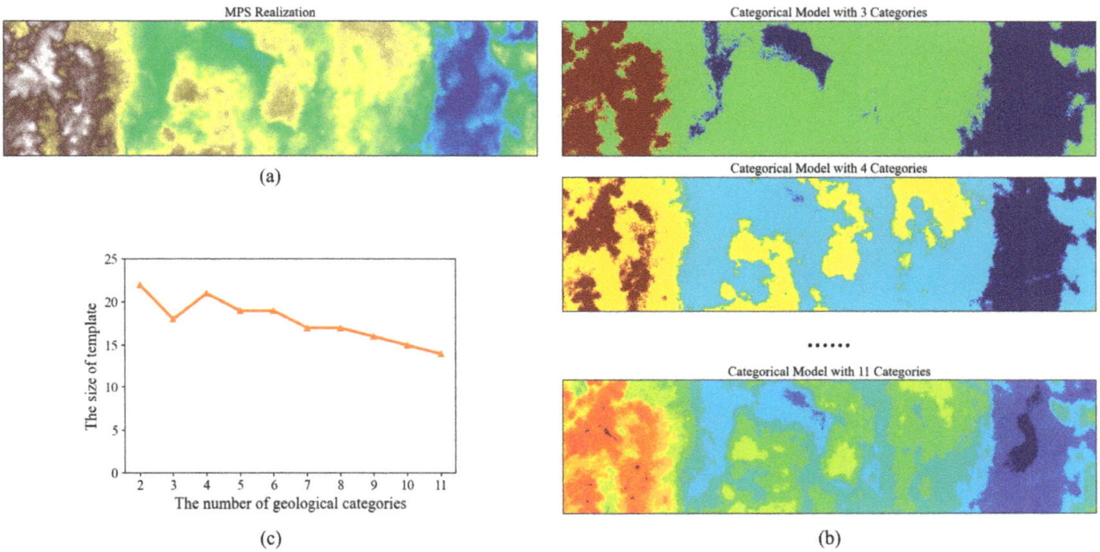

Figure 20. Correlation-driven template design method in the subglacial modeling case. (**a**) MPS realization; (**b**) categorical models with different numbers of geological states; (**c**) the template sizes computed by multiple categorical models.

In accordance with the adaptive template and grid importance, we performed hierarchical clustering and the decision tree classifier to characterize the subglacial models. The JS divergence and multi-dimensional scaling was carried out to quantify the morphological similarity. Figure 22a provides the evaluation results. The topographic realizations generated by MPS are highlighted in red. By contrast, blue and green are used to represent the Kriging and SGSIM models, respectively. In the feature space, two distant points indicate that there was a large mismatch between the two topographic models. Therefore, the three geostatistical methods have different behaviors in terms of their morphological characteristics. On one hand, Kriging and SGSIM are two-point-statistics modeling methods. The linear assumption is an important concept in Kriging. SGSIM applies a multi-Gaussian random function to describe spatial structures. Unknown points in SG are estimated according to a weighted combination of the surrounding points. In order to mimic the prior material, both Kriging and SGSIM utilize a variogram to allocate the weight of each conditioning point. It is challenging to reproduce geometrically complex structures in this way. On the other hand, MPS applies a template to extract spatial patterns. The relationship between the template center and the neighboring points is a core component in the simulation of geological models. Therefore, one key advantage of MPS is its ability to generate realistic realizations.

Furthermore, our PCD was used to examine the effectiveness of the DS parameter. According to the experiment conducted by Meerschman et al. [41], the DS performance largely relies on the neighboring points as well as distance tolerance. Increases in the number of neighbors have positive effects on the simulation of complicated patterns. In comparison, pattern reproduction quality can be improved by reducing the distance tolerance. Thus, we created two realization sets. First, the influence of the distance toleration was investigated. The program fixed the neighbors $N^{DS} = 30$ and $f^{DS} = 0.1$. The tolerance t^{DS} was configurated as 0.025, 0.050, 0.075, and 0.100, respectively. Second, the program concentrated on the function of the neighboring points. The neighbor N^{DS} varied from 10 to 40, while the t^{DS} and f^{DS} were specified as 0.05 and 0.1, respectively. Figure 23 displays the first realization produced by these parameter combinations.

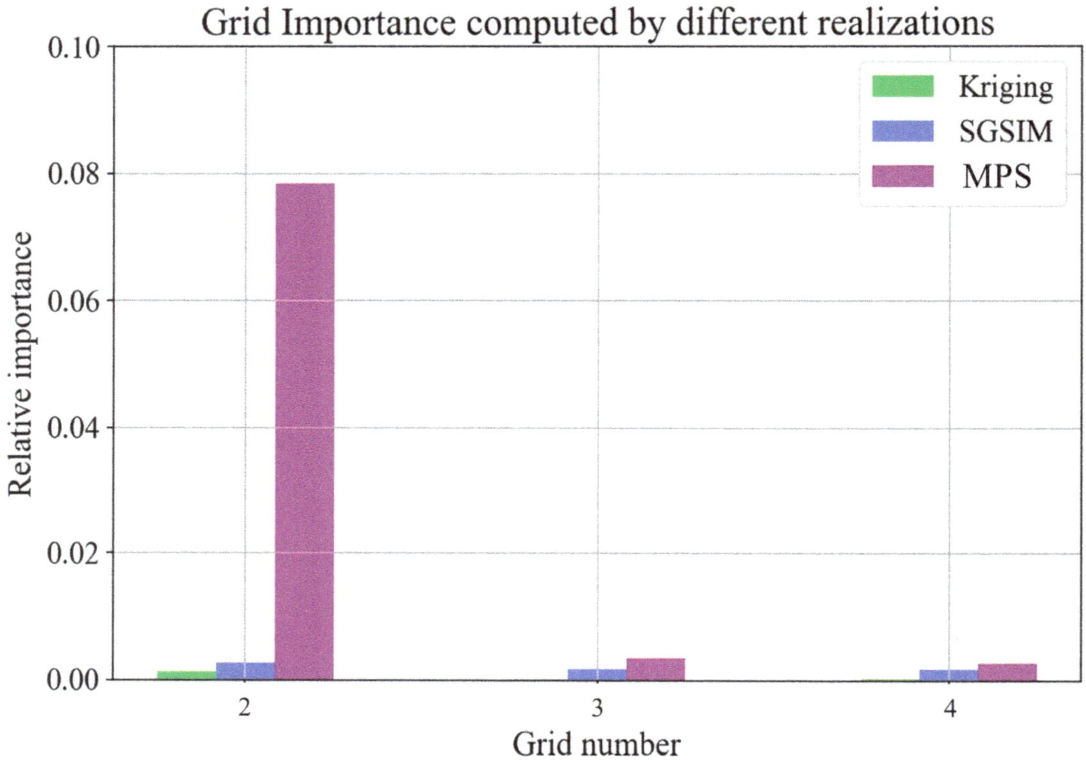

Figure 21. The grid importance computed by three geostatistical models.

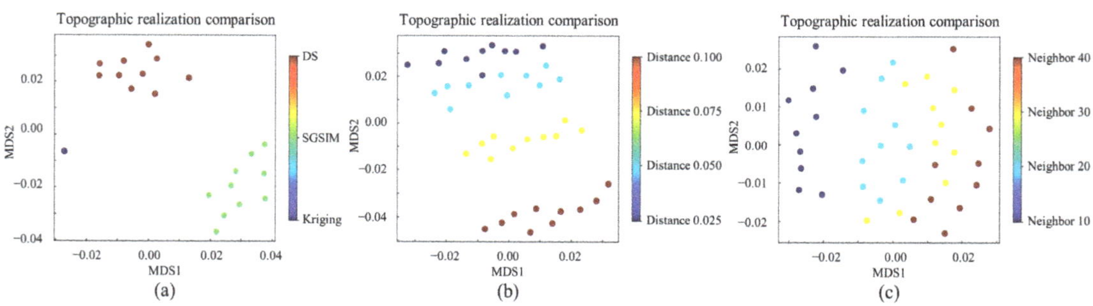

Figure 22. Uncertainty quantification based on 2D topographic realizations. (**a**) Model comparison between Kriging, SGSIM, and MPS; (**b**) model comparison between MPS simulations performed with different distance tolerances; (**c**) model comparison between MPS simulations performed with different neighbors.

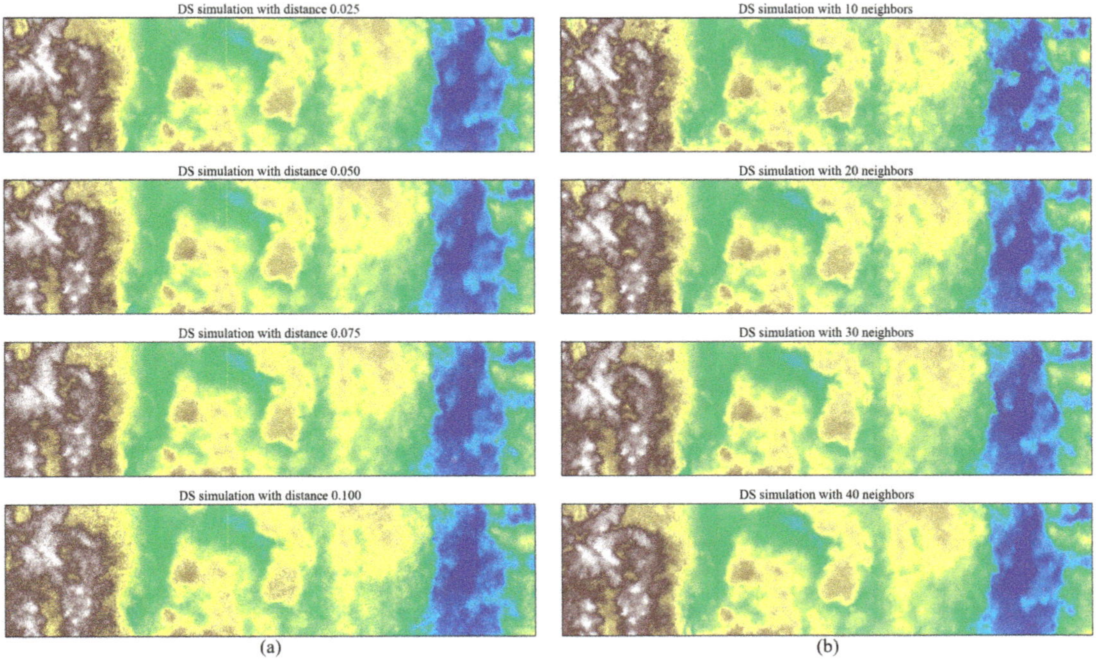

Figure 23. DS realizations with various parameter combinations. (**a**) Subglacial topographic models generated with different distance tolerances; (**b**) subglacial topographic models generated with different neighbors.

The proposed PCD was applied to measure the similarities between the DS models. The patterns in the first realization created by $N^{DS} = 30$, $t^{DS} = 0.05$, and $f^{DS} = 0.1$ were applied to train the decision-tree classifier. PCD calculation results are shown in Figure 22b,c. Two notable phenomena were observed. (1) The distance tolerance has a substantial effect on the modeling quality. According to Figure 22b, the red and yellow clouds had a large mismatch with other realizations. By comparison, the two blue groups were relatively close. This indicates that the topographic realizations created by $t^{DS} = 0.050$ had similar morphological characteristics to the models produced by $t^{DS} = 0.025$. However, a low value of the distance tolerance creates computational burden. Given that $t^{DS} = 0.025$, 4.07 h is necessary to create one realization. In contrast, the computer requires 0.86 h to generate a model when t^{DS} is set as 0.050. Considering the time performance, 0.050 is an appropriate choice in this simulation scenario. (2) the neighboring point is a contributing factor to the simulation quality. In Figure 22c, the deep cloud is distant from the others. A small number of neighboring points in DS is not sufficient to reproduce morphologically complex structures.

4.4. A 3D Sandstone Model from a 2D Slice

In petroleum engineering, three-dimensional sandstone models are important materials with which to study the geometrical and physical properties of rocks [44,45]. In this section, we evaluate the PCD performance in a high-dimension sandstone system. As shown in Figure 24a, a 2D sandstone slice of 128 × 128 was the TI motivating a geostatistical simulation. This image was generated by computed tomography (CT) with a resolution of 10 μm. There were two geological categories. The pore and grain are expressed by the white and black areas, respectively.

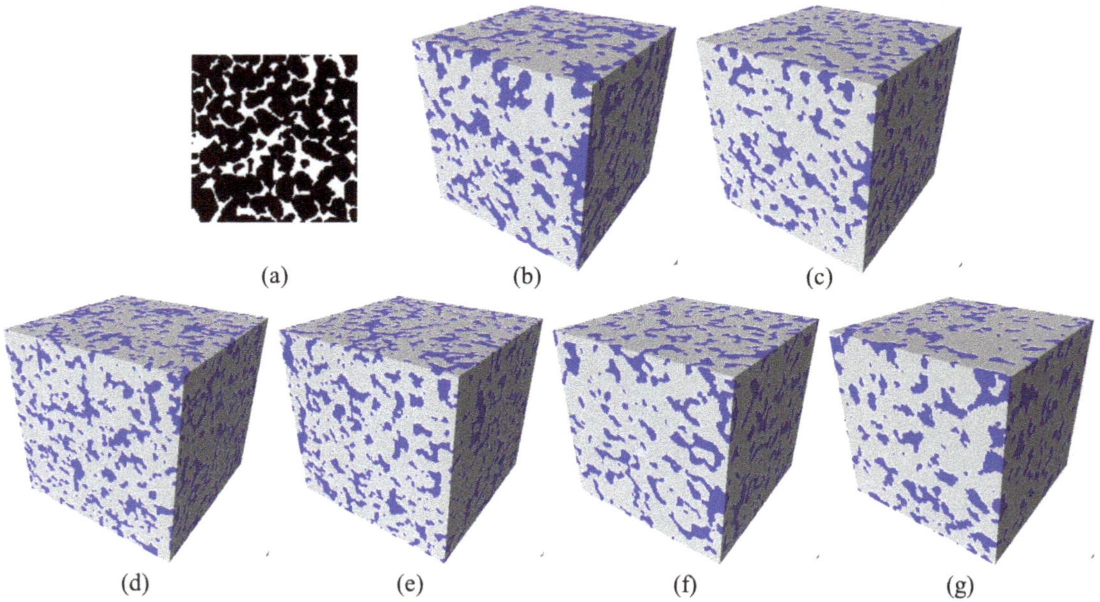

Figure 24. The 2D sandstone slice and 3D models. (**a**) Training image; (**b**) IMPALA realization; (**c**) NNSIM realization; (**d**) DS realization; (**e**) CDS realization; (**f**) PDFSIM realization; (**g**) SGSIM realization.

Aiming at outputting realistic models, we implemented a series of modeling programs to create 3D models on the basis of the 2D slice. First, IMPALA [38] and NNSIM [10] were activated. Because of the absence of 3D TI, these two programs were straightforwardly introduced into the probability aggregation framework. Instead of 3D patterns, the probability aggregation focuses on calculating a conditional probability based on several 2D patterns. The detailed workflow is described in [46]. We specified a template with a size of 5×5 and a multi-grid strategy with $G = 3$. Second, we implemented direct sampling (DS) [40,47] and correlation-driven direct sampling (CDS) [14]. With the intention of reproducing complex structures, CDS employed the weighted Hamming distance to define the compatible patterns. The other parameters were specified as $N^{DS} = 30$, $t^{DS} = 0.0$ and $f^{DS} = 1.0$. Third, a pattern-density-function simulation (PDFSIM) [48,49] was carried out. The core idea was to adopt the pattern density function to characterize the geological models and guide the modeling procedure. Moreover, the program designed a cascaded polymorphic method, which directly pasted a matching patch between cascaded grids within the multi-grid framework. Fourth, we performed a sequential Gaussian simulation (SGSIM) with SGeMS [37]. As a two-point statistics method, the variogram was a key tool in the description of the microstructure in the TI. Based on the parameters stated above, 10 sandstone realizations were individually created by the previous methods. The first realization is shown in Figure 24. The pore space is highlighted in blue while the grain is represented by the gray area.

The proposed PCD was launched to validate the simulation quality. To organize the spatial patterns, we specified the distance threshold as 0.1 in the hierarchical clustering. Moreover, a multi-grid strategy with $G = 4$ was adopted to capture the patterns across different scales. PCD focused on comparing IMPALA with the other modeling programs. The calculation results are shown in Figure 25.

There were three phenomena in the PCD results. (1) The CDS and PDFSIM realizations had small distances to the TI. In order to better reproduce the pore space, CDS assigned the weights of the conditioning points according to the visual features of the TI. The high correlation points had a strong influence on the MPS simulation. By comparison, PDFSIM is an iterative program. The sandstone model is continuously upgraded until the difference from the TI is lower than the predefined threshold. (2) The high-dimension modeling strategy played an important role in the pattern reproduction as well as the spatial uncertainty. On one hand, the NNSIM and IMPALA exhibited comparable simulation qualities, as shown in Figure 25b. Although different datasets were employed, both two programs were incorporated into the probability aggregation framework. The computer employed Bordely formula to calculate the global probability according to three 2D conditional probabilities. By contrast, DS and CDS applied the majority vote to predict an unknown point in the 3D domain. There were no conditional probability computations within the DS and CDS frameworks. (3) There was a noticeable mismatch between the TI and SGSIM models. Compared with the MPS framework, the SGSIM suffers from the limited ability of the variogram. In this sandstone application, it was difficult for the two-point statistics to express the pore microstructure.

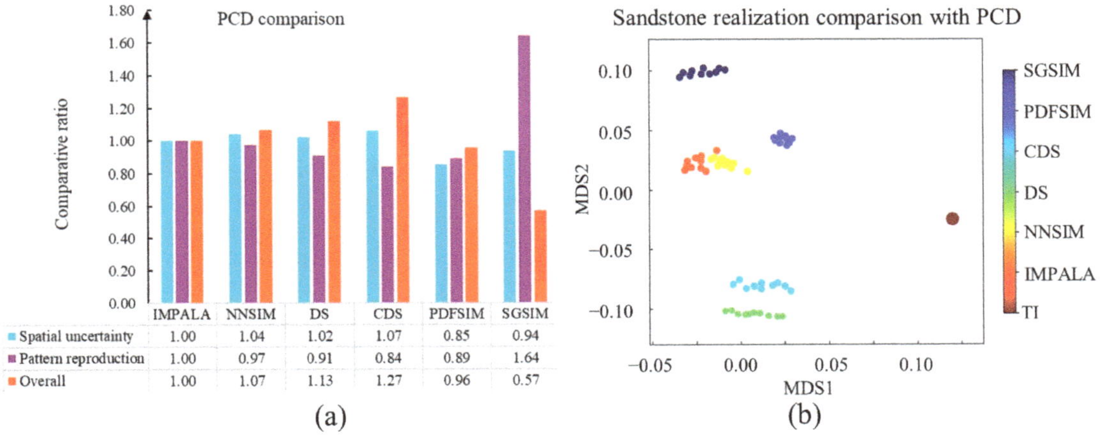

Figure 25. PCD evaluation of the sandstone models. (**a**) Three comparative ratios between IMPALA and other programs. (**b**) Uncertainty quantification based on sandstone models.

5. Discussion

As mentioned in the Introduction, pattern reproduction and spatial uncertainty are two important variabilities within geostatistical modeling. MPH focuses on recording the frequency of each pattern configuration. Pattern distributions are descriptors of geological models. By contrast, ANODI activates a clustering step to control the dimension of the pattern distribution. A multi-grid strategy is carried out to analyze spatial patterns across different scales. In this paper, we proposed a PCD method to evaluate the modeling quality and perform the uncertainty quantification. As reported in Section 4, four practical applications were conducted to examine the previous three methods. A key finding was that the performances of MPH and ANODI were heavily dependent on the parameter configuration. On one hand, the accuracy of MPH relies on the template setting and the number of grids. As shown in Figure 12b, the comparative ratio between SNESIM and DS drastically changed with the variations in the templates. With the development of the grids, the three ratios between the two realization sets approached 1.00. On the other hand, the numbers of clusters and grids were contributing factors in ANODI. According to Figures 12 and 16, the program could not output coherent results in the channel and flume simulation. The fluctuation in the comparative ratios created difficulties in the quantitative analysis of the modeling accuracy.

One key advantage of the proposed PCD is the adaptive parameter specification. In accordance with the morphological characteristics of the TI, our method automatically specifies the template configuration, the number of pattern clusters, and the importance of each resolution. The distance threshold in the hierarchical clustering step is the only user-defined parameter. As displayed in Figure 12a, the comparison between SNESIM and DS was not significantly influenced by the value of the distance threshold. A similar phenomenon is presented in Figure 15. The comparative ratios indicated that NNSIM and TDS had comparable accuracies.

In addition, the strengths of two-point statistics and multiple-point statistics were discussed. One core concept in Kriging and SGSIM is the utilization of the variogram to explain the expected difference between two points. The simulation procedure estimates an unknown point through a weighted sum of the surrounding pixels. In contrast, MPS takes advantage of a template to explore the relationship between the template center and the neighboring points. Therefore, MPS provides a powerful way to reproduce geometrically complicated structures. As shown in Figures 19 and 21, the long-range structures were well-conserved in MPS realizations. This finding supports the investigations conducted by Yin et al. [7] and Zuo et al. [17]. According to their research, MPS has the ability to produce a group of realistic topographic models with suitable diversity. The benefits of multiple-point information are further examined in Sections 4.2 and 4.3. An extending template has a positive effect on improving the consistency between the TI and the simulated realizations. As reported in Section 4.4, we performed a 3D sandstone reconstruction. The pore microstructure was properly recreated in the MPS models. By contrast, the deep blue points corresponding to the SGSIM realizations were at large distances from the TI in Figure 25.

Next, we focused on the sensitive factors within the MPS simulation. An important finding was that the pattern similarity metric had a substantial influence on the modeling quality. In Figure 11a, the MPS programs can be divided into three groups. First, SNESIM, IMPALA, and CSSIM shared similar dispersals. The main reason is that they employ the pruning strategy to find desired pattern instances. The program removes the farthest points in the conditioning pattern if the matching pattern is not presented in the TI. Therefore, the points that are close to the template center play an important role in the pattern-searching step. Second, the Hamming distance is used by NNSIM, DS, and TDS to distinguish patterns. The template points have the same importance. The similarity between NNSIM and TDS was further validated by the flume models. Third, FILTERSIM applies six convolutional kernels to characterize 2D patterns. A compatible image patch is pasted into the simulation domain. However, one key drawback of these MPS programs is that the intrinsic characteristics of the TI are not considered in the pattern similarity metric. The contribution of each template point is fixed and constant. Accordingly, it would be interesting to assess the effect of each conditioning point during MPS simulations. Adaptive weight assignment is a promising way to further improve geostatistical modeling programs.

6. Conclusions

In this work, a pattern classification distribution method was proposed to assess geostatistical modeling and quantify spatial uncertainty. With the objective of improving the evaluation accuracy, a set of machine-learning techniques were employed to overcome the technical limitations in the previous multiple-point histogram and the analysis of distance methods. First, a correlation-driven template design approach was suggested to extract spatial patterns. With a region-growing program, the computer sequentially collects conditioning points according to their correlations with the template center. The number of template points is automatically determined by the elbow point of the entropy function. An irregular template of adaptive size has a positive effect on preserving the structure in the TI. Second, the proposed PCD utilizes the clustering and classification programs to characterize the geological realizations. In order to simplify the parameter setting, hierarchical clustering is launched to organize patterns in the TI. On the basis of the clustering results, a decision tree is trained to classify each pattern in geostatistical models. The program outputs a pattern distribution according to the number of member instances in each pattern category. The Jensen–Shannon divergence within the pattern distribution becomes a measure of the similarity between two realizations. Third, a stacking framework was applied to develop the multi-grid analysis. The base classifier focuses on exploring the relationship between the template center and the neighboring points in different resolutions. By comparison, a meta-classifier was employed to evaluate the effectiveness of each base-classifier. The importance of each resolution was adaptively assigned according to the morphological characteristics of the TI.

We examined the proposed PCD by using benchmark channel models, non-stationary flume models, subglacial topographic realizations in Antarctica, and three-dimensional sandstone models. With the intention of facilitating an extensive comparison, various multiple-point statistics methods were implemented to generate geological models. The computational results indicated that our method is capable of addressing multiple geological categories, continuous variables, and high-dimensional structure. Compared with MPH and ANODI methods, the proposed PCD benefits from the automatic parameter-specification step. The underlying relationship between geostatistical realizations is efficiently recognized by our method. As the only predefined parameter, the distance threshold in the hierarchical clustering does not have a significant effect on the computational results. The findings indicate that our PCD provides a feasible way to find reliable geostatistical models and quantify spatial uncertainty.

Author Contributions: Conceptualization, C.Z. and Z.D.; methodology, C.Z. and Z.L.; software, C.Z. and X.W.; validation, X.W. and Y.W.; writing—original draft preparation, C.Z. and Z.L.; writing—review and editing, Z.D., X.W. and Y.W. All authors have read and agreed to the published version of the manuscript.

Funding: This research was funded by Natural Science Foundation of Shaanxi Province, grant number 2022JQ-227; the Postdoctoral Science Foundation of China, grant number 2022M710482; and the Department of Transportation Science and Technology Project of Zhejiang Province, grant number 2023016.

Data Availability Statement: All data generated or analyzed during this study are included in this published article.

Conflicts of Interest: The authors declare no conflict of interest.

References

1. Scheidt, C.; Fernandes, A.M.; Paola, C.; Caers, J. Quantifying natural delta variability using a multiple-point geostatistics prior uncertainty model. *J. Geophys. Res. Earth Surf.* **2016**, *121*, 1800–1818. [CrossRef]
2. Hoffimann, J.; Scheidt, C.; Barfod, A.; Caers, J. Stochastic simulation by image quilting of process-based geological models. *Comput. Geosci.* **2017**, *106*, 18–32. [CrossRef]
3. Hoffimann, J.; Bufe, A.; Caers, J. Morphodynamic analysis and statistical synthesis of geomorphic data: Application to a flume experiment. *J. Geophys. Res. Earth Surf.* **2019**, *124*, 2561–2578. [CrossRef]
4. Zhang, T.; Liu, D. Reconstructing digital terrain models from ArcticDEM and worldview-2 imagery in Livengood, Alaska. *Remote Sens.* **2023**, *15*, 2061. [CrossRef]
5. Leong, W.J.; Horgan, H.J. DeepBedMap: A deep neural network for resolving the bed topography of Antarctica. *Cryosphere* **2020**, *14*, 3687–3705. [CrossRef]
6. MacKie, E.J.; Schroeder, D.M.; Zuo, C.; Yin, Z.; Caers, J. Stochastic modeling of subglacial topography exposes uncertainty in water routing at Jakobshavn Glacier. *J. Glaciol.* **2021**, *67*, 75–83. [CrossRef]
7. Yin, Z.; Zuo, C.; MacKie, E.J.; Caers, J. Mapping high-resolution basal topography of West Antarctica from radar data using non-stationary multiple-point geostatistics (MPS-bedmappingV1). *Geosci. Model Dev.* **2022**, *15*, 1477–1497. [CrossRef]
8. Hadjipetrou, S.; Mariethoz, G.; Kyriakidis, P. Gap-filling sentinel-1 offshore wind speed image time series using multiple-point geostatistical simulation and reanalysis data. *Remote Sens.* **2023**, *15*, 409. [CrossRef]
9. Mariethoz, G.; Caers, J. *Multiple-Point Geostatistics: Stochastic Modeling with Training Image*; Wiley: New York, NY, USA, 2014.
10. Zuo, C.; Pan, Z.; Yin, Z.; Guo, C. A nearest neighbor multiple-point statistics method for fast geological modeling. *Comput. Geosci.* **2022**, *167*, 105208. [CrossRef]
11. Liu, G.; Fang, H.; Chen, Q.; Cui, Z.; Zeng, M. A feature-enhanced MPS approach to reconstruct 3D deposit models using 2D geological cross sections: A case study in the Luodang Cu deposit, Southwestern China. *Nat. Resour. Res.* **2022**, *31*, 3101–3120. [CrossRef]
12. Gravey, M.; Mariethoz, G. QuickSampling v1.0: A robust and simplified pixel-based multiple-point simulation approach. *Geosci. Model Dev.* **2020**, *13*, 2611–2630. [CrossRef]
13. Bai, H.; Ge, Y.; Mariethoz, G. Utilizing spatial association analysis to determine the number of multiple grids for multiple-point statistics. *Spat. Stat.* **2016**, *17*, 83–104. [CrossRef]

14. Zuo, C.; Pan, Z.; Gao, Z.; Gao, J. Correlation-driven direct sampling method for geostatistical simulation and training image evaluation. *Phys. Rev. E* **2019**, *99*, 053310. [CrossRef]
15. Zhang, T.; Switzer, P.; Journel, A. Filter-based classification of training image patterns for spatial simulation. *Math. Geol.* **2006**, *38*, 63–80. [CrossRef]
16. Honarkhah, M.; Caers, J. Stochastic simulation of patterns using distance-based pattern modeling. *Math. Geosci.* **2010**, *42*, 487–517. [CrossRef]
17. Zuo, C.; Yin, Z.; Pan, Z.; MacKie, E.J.; Caers, J. A tree-based direct sampling method for surface and subsurface hydrological modeling. *Water Resour. Res.* **2020**, *56*, e2019WR026130. [CrossRef]
18. Strebelle, S.; Cavelius, C. Solving speed and memory issues in multiple-point statistics simulation program SNESIM. *Math. Geosci.* **2014**, *46*, 171–186. [CrossRef]
19. Straubhaar, J.; Malinverni, D. Addressing conditioning data in multiple-point statistics simulation algorithms based on a multiple grid approach. *Math. Geosci.* **2014**, *46*, 187–204. [CrossRef]
20. Song, S.; Mukerji, T.; Hou, J. Bridging the gap between geophysics and geology with generative adversarial networks (GANs). *IEEE Trans. Geosci. Remote. Sens.* **2022**, *60*, 1–11.
21. Li, X.; Li, B.; Liu, F.; Li, T.; Nie, X. Advances in the application of deep learning methods to digital rock technology. *Adv. Geo-Energy Res.* **2022**, *8*, 5–18. [CrossRef]
22. Song, S.; Mukerji, T.; Hou, J. GANSim: Conditional facies simulation using an improved progressive growing of generative adversarial networks (GANs). *Math. Geosci.* **2021**, *53*, 1413–1444. [CrossRef]
23. Zhang, T.F.; Tilke, P.; Dupont, E.; Zhu, L.C.; Liang, L.; Bailey, W. Generating geologically realistic 3D reservoir facies models using deep learning of sedimentary architecture with generative adversarial networks. *Pet. Sci.* **2019**, *16*, 541–549. [CrossRef]
24. Laloy, E.; Hérault, R.; Jacques, D.; Linde, N. Training-image based geostatistical inversion using a spatial generative adversarial neural network. *Water Resour. Res.* **2018**, *54*, 381–406. [CrossRef]
25. Chen, Q.; Cui, Z.; Liu, G.; Yang, Z.; Ma, X. Deep convolutional generative adversarial networks for modeling complex hydrological structures in Monte-Carlo simulation. *J. Hydrol.* **2022**, *610*, 127970. [CrossRef]
26. Gringarten, E.; Deutsch, C.V. Teacher's aide variogram interpretation and modeling. *Math. Geol.* **2001**, *33*, 507–534. [CrossRef]
27. Sahimi, M.; Tahmasebi, P. Reconstruction, optimization, and design of heterogeneous materials and media: Basic principles, computational algorithms, and applications. *Phys. Rep.* **2021**, *939*, 1–82. [CrossRef]
28. Renard, P.; Allard, D. Connectivity metrics for subsurface flow and transport. *Adv. Water Resour.* **2013**, *51*, 168–196. [CrossRef]
29. Scheidt, C.; Li, L.; Caers, J. *Quantifying Uncertainty in Subsurface Systems*; Wiley: New York, NY, USA, 2018.
30. Song, S.; Mukerji, T.; Hou, J.; Zhang, D.; Lyu, X. GANSim-3D for conditional geomodelling: Theory and field application. *Water Resour. Res.* **2022**, *58*, e2021WR031865. [CrossRef]
31. Boisvert, J.B.; Pyrcz, M.J.; Deutsch, C.V. Multiple-point statistics for training image selection. *Nat. Resour. Res.* **2007**, *16*, 313–321. [CrossRef]
32. Tan, X.J.; Tahmasebi, P.; Caers, J. Comparing training-image based algorithms using an analysis of distance. *Math. Geosci.* **2014**, *46*, 149–169. [CrossRef]
33. Endres, D.M.; Schindelin, J.E. A new metric for probability distributions. *IEEE Trans. Inf. Theory.* **2003**, *49*, 1858–1860. [CrossRef]
34. Zhu, M.; Ghodsi, A. Automatic dimensionality selection from the scree plot via the use of profile likelihood. *Comput. Stat. Data Anal.* **2006**, *51*, 918–930. [CrossRef]
35. Geron, A. *Hands-On Machine Learning with Scikit-Learn, Keras & Tensorflow*, 2nd ed.; OReilly: Sebastopol, CA, USA, 2019.
36. Wu, Y.; Tahmasebi, P.; Lin, C.; Dong, C. A comprehensive investigation of the effects of organic-matter pores on shale properties: A multicomponent and multiscale modeling. *J. Nat. Gas Eng.* **2020**, *81*, 103425. [CrossRef]
37. Remy, N.; Boucher, A.; Wu, J. *Applied Geostatistics with SGeMS: A User's Guide*; Cambridge University Press: Cambridge, UK, 2009.
38. Straubhaar, J.; Renard, P.; Mariethoz, G.; Froidevaux, R.; Besson, O. An improved parallel multiple-point algorithm using a list approach. *Math. Geosci.* **2011**, *43*, 305–328. [CrossRef]
39. Guo, C.; Zhang, H.; Zuo, C. A column searching-based multiple-point statistics for efficient image generation. In Proceedings of the 2022 4th International Conference on Communications, Information System and Computer Engineering (CISCE), Shenzhen, China, 17 August 2022.
40. Mariethoz, G.; Renard, P.; Straubhaar, J. The direct sampling method to perform multiple-point geostatistical simulations. *Water Resour. Res.* **2010**, *46*, W11536. [CrossRef]
41. Meerschman, E.; Pirot, G.; Mariethoz, G.; Straubhaar, J.; Meirvenne, M.V.; Renard, P. A practical guide to performing multiple-point statistical simulations with the direct sampling algorithm. *Comput. Geosci.* **2013**, *52*, 307–324. [CrossRef]
42. Goff, J.A.; Powell, E.M.; Young, D.A.; Blankenship, D.D. Conditional simulation of thwaites glacier (antarctica) bed topography for flow models: Incorporating inhomogeneous statistics and channelized morphology. *J. Glaciol.* **2014**, *60*, 635–646. [CrossRef]
43. Rignot, E.; Mouginot, J.; Scheuchl, B.; Broeke, M.; Wessem, M.J.; Morlighem, M. Four decades of Antarctic Ice sheet mass balance from 1979–2017. *Proc. Natl. Acad. Sci. USA* **2019**, *116*, 1095–1103. [CrossRef]
44. Wu, Y.; Tahmasebi, P.; Yu, H.; Lin, C.; Wu, H.; Dong, C. Pore-scale 3D dynamic modeling and characterization of shale samples: Considering the effects of thermal maturation. *J. Geophys. Res. Solid Earth* **2020**, *125*, e2019JB01830. [CrossRef]
45. Li, B.; Nie, X.; Cai, J.; Zhou, X.; Wang, C.; Han, D. U-Net model for multi-component digital rock modeling of shales based on CT and QEMSCAN images. *J. Pet. Sci. Eng.* **2020**, *216*, 110734. [CrossRef]

46. Comunian, A.; Renard, P.; Straubhaar, J. 3D multiple-point statistics simulation using 2D training images. *Comput. Geosci.* **2012**, *40*, 49–65. [CrossRef]
47. Gueting, N.; Caers, J.; Comunian, A.; Vanderborght, J.; Englert, A. Reconstruction of three-dimensional aquifer heterogeneity from two-dimensional geophysical data. *Math. Geosci.* **2018**, *50*, 53–75. [CrossRef]
48. Zhang, D.; Gao, M.; Liu, F.; Qin, X.; Yin, X.; Fang, W.; Luo, Y. Reconstruction of anisotropic 3D medium using multiple 2D images. *J. Pet. Sci. Eng.* **2022**, *219*, 111048. [CrossRef]
49. Liu, F.; Gao, M.; Li, X.; Lin, H.; Deng, K.; Xu, Y.; Jiang, J. Reconstruction of 3D porous medium using a type of cascaded polymorphic method. *Microporous Mesoporous Mater.* **2021**, *326*, 111356. [CrossRef]

remote sensing

MDPI

Article

Spectral-Swin Transformer with Spatial Feature Extraction Enhancement for Hyperspectral Image Classification

Yinbin Peng [1], Jiansi Ren [1,2,*], Jiamei Wang [1] and Meilin Shi [1]

[1] School of Computer Science, China University of Geosciences, Wuhan 430078, China
[2] Hubei Key Laboratory of Intelligent Geo-Information Processing, China University of Geosciences, Wuhan 430078, China
* Correspondence: renjsv@cug.edu.cn

Abstract: Hyperspectral image classification (HSI) has rich applications in several fields. In the past few years, convolutional neural network (CNN)-based models have demonstrated great performance in HSI classification. However, CNNs are inadequate in capturing long-range dependencies, while it is possible to think of the spectral dimension of HSI as long sequence information. More and more researchers are focusing their attention on transformer which is good at processing sequential data. In this paper, a spectral shifted window self-attention based transformer (SSWT) backbone network is proposed. It is able to improve the extraction of local features compared to the classical transformer. In addition, spatial feature extraction module (SFE) and spatial position encoding (SPE) are designed to enhance the spatial feature extraction of the transformer. The spatial feature extraction module is proposed to address the deficiency of transformer in the capture of spatial features. The loss of spatial structure of HSI data after inputting transformer is supplemented by proposed spatial position encoding. On three public datasets, we ran extensive experiments and contrasted the proposed model with a number of powerful deep learning models. The outcomes demonstrate that our suggested approach is efficient and that the proposed model performs better than other advanced models.

Keywords: transformer; shifted window; spatial feature extraction (SFE); spatial position encoding (SPE); hyperspectral image (HSI) classification

Citation: Peng, Y.; Ren, J.; Wang, J.; Shi, M. Spectral-Swin Transformer with Spatial Feature Extraction Enhancement for Hyperspectral Image Classification. *Remote Sens.* **2023**, *15*, 2696. https://doi.org/10.3390/rs15102696

Academic Editor: Gwanggil Jeon

Received: 15 April 2023
Revised: 14 May 2023
Accepted: 20 May 2023
Published: 22 May 2023

1. Introduction

Because of the rapid advancement of hyperspectral sensors, the resolution and accuracy of hyperspectral images (HSI) have also increased greatly. HSI contains a wealth of spectral information, collecting hundreds of bands of electron spectrum at each pixel. Its rich information allows for excellent performance in classifying HSI, and thus its application has great potential in several fields such as precision agriculture [1] and Jabir et al. [2] used machine learning algorithm for weed detection, medical imaging [3], object detection [4], urban planning [5], environment monitoring [6], mineral exploration [7], dimensionality reduction [8] and military detection [9].

Numerous conventional machine learning methods have been used to the classification of HSI in the past decade or so, such as K-nearest neighbors (KNN) [10], support vector machines (SVM) [11–14], random forests [15,16]. Navarro et al. [17] used neural network for hyperspectral image segmentation. However, as the size and complexity of the training set increases, the fitting ability of traditional methods can show weakness for the task, and the performance often encounters bottlenecks. Song et al. [18] proposed a HSI classification method based on the sparse representation of KNN, but it cannot effectively apply the spatial information in HSI. Guo et al. [19] used a fused SVM of spectral and spatial features for HSI classification, but it is still difficult to extract important features from high-dimensional HSI data. Deep learning have developed rapidly in recent years, and their powerful fitting ability can extract features from multivariate data. Inspired by this, the designed deep learning models have proposed in HSI classification tasks, such as recurrent neural network (RNN) [20–22], convolutional neural network (CNN) [23–28], graph convolutional network (GCN) [29,30], capsule network (CapsNet) [31,32], long short term memory (LSTM) networks [33–35]. Although these deep learning models show good performance in several different domains, they have certain shortcomings in HSI classification tasks.

For CNNs, which are good at natural image tasks, Its benefit is that the image's spatial information can be extracted during the convolution operation. HSI-CNN [36] stacks multi-dimensional data from HSI into two-dimensional data and then extracts features efficiently. 2D-CNN [37] can capture spatial features in HSI data to improve classification accuracy. However, HSI has rich information in the spectral dimension, and if it is not exploited, the performance of the model is bound to be difficult to break through. Although the advent of 3D-CNN [38–41] enables the extraction of both spatial and spectral features, the convolution operation is localized, so the extracted features lack the mining and representation of the global Information.

Recently, transformer has evolved rapidly and shown good performance when performing tasks like natural language processing. Based on its self-attention mechanism, it is very good at processing long sequential information and extracting global relations. Vision transformer (ViT) [42] makes it perform well in several vision domains by dividing images into patches and then inputting them into the model. Swin-transformer [43] enhances the capability of local feature extraction by dividing the image into windows and performing multi-head self-attention (MSA) separately within the windows, and then enabling the exchange of information between the windows by shifting the windows. It improves the accuracy in natural image processing tasks and effectively reduces the computational effort in the processing of high-resolution images. Due to transformer's outstanding capabilities for natural image processing, more and more studies are applying it to the classification of HSI [44–50]. However, if ViT is applied directly to the HSI classification, there will be some problems that will limit the performance improvement, specifically as follows.

(1) The transformer performs well at handling sequence data(spectral dimension information), but lacks the use of spatial dimension information.
(2) The multi-head self-attention (MSA) of transformer is adept at resolving the global dependencies of spectral information, but it is usually difficult to capture the relationships for local information.
(3) Existing transformer models usually map the image to linear data to be able to input into the transformer model. Such an operation would destroy the spatial structure of HSI.

HSI can be regarded as a sequence in the spectral dimension, and the transform is effective at handling sequence information, so the transformer model is suitable for HSI classification. The research in this paper is based on tranformer and considers the above mentioned shortcomings to design a new model, called spectral-swin transformer (SSWT) with spatial feature extraction enhancement, and apply it in HSI classification. Inspired by swin-transformer and the characteristics of HSI data which contain a great deal of information in the spectral dimension, we design a method of dividing and shifting windows in the spectral dimension. MSA is performed within each window separately, aiming to improve the disadvantage of transformer to extract local features. We also design two modules to enhance model's spatial feature extraction. In summary, the following are the contributions of this paper.

(1) Based on the characteristics of HSI data, a spectral dimensional shifted window multi-head self-attention is designed. It enhances the model's capacity to capture local information and can achieve multi-scale effect by changing the size of the window.
(2) A spatial feature extraction module based on spatial attention mechanism is designed to improve the model's ability to characterize spatial features.
(3) A spatial position encoding is designed before each transformer encoder to deal with the lack of spatial structure of the data after mapping to linear.

(4) Three publicly accessible HSI datasets are used to test the proposed model, which is compared with advanced deep learning models. The proposed model is extremely competitive.

The rest of this paper is organized as follows: Section 2 discusses the related work on HSI classification using deep learning, which includes transformer. Section 3 describes the proposed model and the design method for each component. Section 4 presents the three HSI datasets, as well as the experimental setup, results, corresponding analysis. Section 5 concludes with a summary and outlook of the full paper.

2. Related Work

2.1. Deep-Learning-Based Methods for HSI Classification

Deep learning has developed quickly, more and more researchers are using deep learning methods(e.g., RNNs, CNNs, GCNs, CapsNet, LSTM) to the classification tasks of HSI [20,22,23,29–31,33,34]. Mei et al. [51] constructed a network based on bidirectional long short-term memory (Bi-LSTM) for HSI classification. Zhu et al. [52] proposed an end-to-end residual spectral–spatial attention network (RSSAN) for HSI classification, which consists of spectral and spatial attention modules for spectral band and spatial information adaptive selection. Song et al. [53] created a deep feature fusion network (DFFN) to solve the negative effects of excessively increasing network depth.

Due to CNN's excellent capability of taking the local spatial context information and it's outstanding capabilities in natural picture processing, many CNN-based HSI classification models have emerged. For example, Hang et al. [54] proposed two CNN sub-networks based on the attention mechanism for extracting the spectral and spatial features of HSI, respectively. Chakraborty et al. [55] designed a wavelet CNN that uses layers of wavelet transforms to display spectral features. Gong et al. [56] proposed a hybrid model that combines 2D-CNN and 3D-CNN in order to include more in-depth spatial and spectral features while using fewer learning samples. Hamida et al. [57] introduced a new 3-D DL method that permits the processing of both spectral and spatial information simultaneously.

However, each of these deep learning approaches has some respective drawbacks that can limit the model performance when processing HSI classification tasks. For CNN, it is good at handling two-dimensional spatial features, but since the data of HSI is stereoscopic and contains a large amount of information in the spectral dimension. It's possible that CNN will have trouble extracting the spectral features. Moreover, although CNNs have achieved good results by relying on their local feature focus, the inability to deal with global dependencies limits their performance when processing spectral information in the form of long sequences. These shortcomings will be addressed in the transformer.

2.2. Vision Transformers for Image Classification

With the increasing use of transformers in computer vision, researchers have begun to consider images in terms of sequential data, such as ViT [42] and Swin-transformer [43] etc. Fang et al. [58] proposed MSG-Transformer, which presents a specialized token in each region as a messenger (MSG). Information can be transmitted flexibly among areas and computational cost is decreased by manipulating these MSG tokens. Guo et al. [59] proposed CMT, which combines the advantages of CNN and ViT, a new hybrid transformer-based network that captures long-range dependencies using transformers and extracts local information using CNN. Chen et al. [60] designed MobileNet and transformer in parallel, connected in the middle by a two-way bridge. This structure benefits from MobileNet for local processing and Transformer for global communication.

An increasing number of researchers are applying transformer to HSI classification tasks. Hong et al. [44] proposed a model called SpectralFormer (SF) for HSI classification, which divides neighboring bands into the same token for learning features and connects encoder blocks across layers, but the spatial information in HSI was not considered. Sun et al. [45] proposed the Spectral-Spatial Feature Tokenization Transformer (SSFTT) to capture high-level semantic information and spectral-spatial features, resulting in a large performance improvement. Ayas et al. [61] designs a spectal-swin module in front of the swin transformer, which extracts spatial and spectral features and fuses them with Conv 2-D operation and Conv 3-D operation, respectively. Mei et al. [47] proposed the Group-Aware Hierarchical Transformer (GAHT) to restrict the MSA to a local spatial-spectral range by using a new group pixel embedding module, which enables the model to have improved capability of local feature extraction. Yang et al. [46] proposed a hyperspectral image transformer (HiT) classification network that captures subtle spectral differences and conveys local spatial context information by embedding convolutional operations in the transformer structure, however it is not effective in capturing local spectral features. Transformer is increasingly used in the field of HSI classification and we believe it has great potential for the future.

3. Methodology

In this section, we will introduce the proposed spectral-swin transformer (SSWT) with spatial feature extraction enhancement, which will be described in four aspects: the overall architecture, spatial feature extraction module(SFE), spatial position encoding(SPE), and spectral swin-transformer module.

3.1. Overall Architecture

In this paper, we design a new transformer-based method SSWT for the HSI classification. SSWT consists of two major Components for solving the challenges in HSI classification, namely, spatial feature extraction module(SFE) and spectral swin(S-Swin) transformer module. An overview of the proposed SSWT for the HSI classification is shown in Figure 1. The input to the model is a patch of HSI. the data is first input to SFE to perform initial spatial feature extraction, the module consists of convolution layers and spatial attention. In Section 3.2, it is explained in further detail. The data is then flattened and entered into the s-swin transformer module. A spatial position encoding is added in front of each s-swin transformer layer to add spatial structure to the data. This part will be described in Section 3.3. The s-swin transformer module uses the spectral-swin self attention, which will be introduced in Section 3.4. The final classification results are obtained by linear layers.

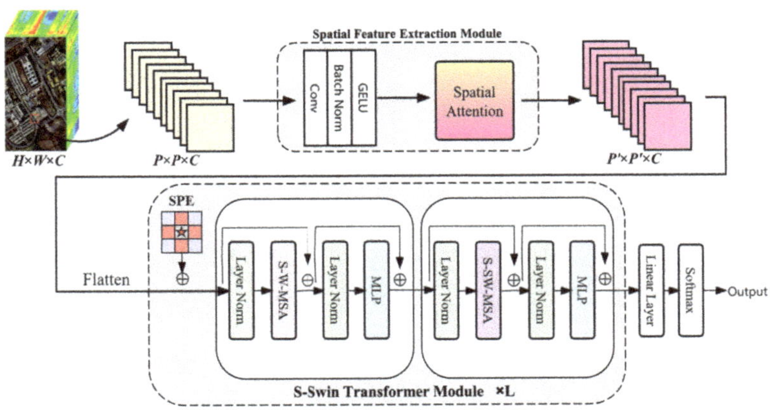

Figure 1. Overall structure of the proposed SSWT model for HSI classification.

3.2. Spatial Feature Extraction Module

Due to transformer's lack of ability in handling spatial information and local features, we designed a spatial feature extraction (SFE) module to compensate. It consists of two parts, the first one consists of convolutional layers to preliminary extraction of spatial features and batch normalization to prevent overfitting. The second part is a spatial attention mechanism, which aims to enable the model to learn the important spatial locations in the data. The structure of SFE is shown in Figure 1.

For the input HSI patch cube $I \in \mathbb{R}^{H \times W \times C}$, where $H \times W$ is the spatial size and C is the number of spectral bands. Each pixel space in I consists of C spectral dimensions and forms a one-hot category vector $S = [s_1, s_2, s_3, \cdots, s_n] \in \mathbb{R}^{1 \times 1 \times n}$, where n is the number of ground object classes.

Firstly, the spatial features of HSI are initially extracted by CNN layers, and the formula is shown as follows:

$$X = GELU\Big(BN\Big(Conv(I)\Big)\Big) \tag{1}$$

where $Conv(\cdot)$ represents the convolution layer. $BN(\cdot)$ represents batch normalization. $GELU(\cdot)$ denotes the activation function. The formula for the convolution layer is shown below:

$$Conv(I) = \mathop{||}_{j=0}^{J} (I * W_j^{r1 \times r2} + b_j) \tag{2}$$

where I is the input, J is the number of convolution kernels, $W_j^{r1 \times r2}$ is the jth convolution kernel with the size of $r1 \times r2$, and b_j is the jth bias. $||$ denotes concatenation, and $*$ is convolution operation.

Then, the model may learn important places in the data thanks to a spatial attention mechanism (SA). The structure of SA is shown in Figure 2. For an intermediate feature map $X \in \mathbb{R}^{H' \times W' \times C}(H' \times W'$ is the spatial size of $X)$, the process of SA is shown in the following formula:

$$S_M = MaxPooling(X) \tag{3}$$

$$S_A = AvgPooling(X) \tag{4}$$

$$X_{SA} = \sigma\Big(Conv\Big(Concat(S_M, S_A)\Big)\Big) \otimes X \tag{5}$$

MaxPooling and AvgPooling are global maximum pooling and global average pooling along the channel direction. Concat denotes concatenation in the channel direction. σ is activation function. $\otimes$ denotes the elementwise multiplication.

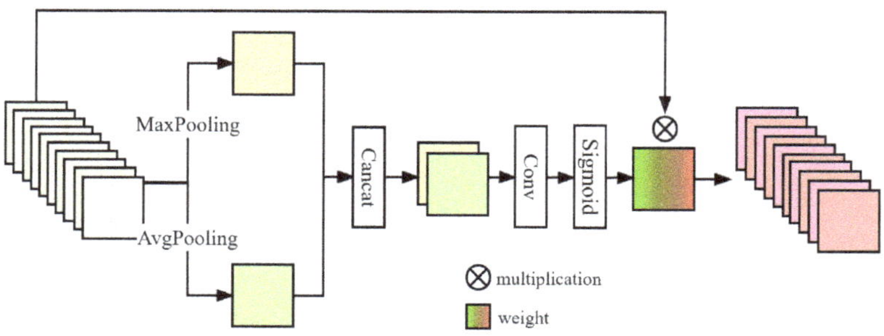

Figure 2. The structure of the spatial attention in SFE.

3.3. Spatial Position Encoding

The HSI of the input transformer is mapped to linear data, which can damage the spatial structure of HSI. To describe the relative spatial positions between pixels and to maintain the rotational invariance of samples, a spatial position encoding (SPE) is added before each transformer module.

The input to HSI classification is a patch of a region, but only the label of the center pixel is the target of classification. The surrounding pixels can provide spatial information for the classification of center pixel, and their importance tends to decrease with the distance to the center. SPE is to learn such a center-important position encoding. The pixel positions of a patch is defined as follows.

$$pos(x_i, y_i) = |x_i - x_c| + |y_i - y_c| + 1 \tag{6}$$

where (x_c, y_c) denotes the coordinate of central position of the sample, that is the pixel to be classified. (x_i, y_i) denotes the coordinates of other pixels in the sample. The visualization of SPE when the spatial size of the sample is 7×7 can be seen in Figure 3. The pixel in the central position is unique and most important, and the other pixels are given different position encoding depending on the distance from the center.

To flexibly represent the spatial structure in HSI, the learnable position encoding are embedded in the data:

$$Y = X + spe(P) \tag{7}$$

where X is the HSI data, and P represents the position matrix (like Figure 3) constructed according to Equation (6). $spe(\cdot)$ is a learnable array that takes the position matrix as a subscript to get the final spatial position encoding. Finally, the position encoding is added to the HSI data.

7	6	5	4	5	6	7
6	5	4	3	4	5	6
5	4	3	2	3	4	5
4	3	2	1	2	3	4
5	4	3	2	3	4	5
6	5	4	3	4	5	6
7	6	5	4	5	6	7

Figure 3. SPE in a sample with the spatial size is 7 × 7.

3.4. Spectral Swin-Transformer Module

The structure of the spectral swin-transformer (S-SwinT) module is shown in Figure 1. Transformer is good at processing long dependencies and lacks the ability to extract local features. Inspired by swin-transformer [43], window-based multi-head self-attention (MSA) is used in our model. Because the input of HSI is a patch which is usually small in spatial size, it cannot divide the window in space as Swin-T does. Considering the rich data of HSI in the spectral dimension, a window of spectral shift was designed for MSA, called spectral window multi-head self-attention (S-W-MSA) and spectral shifted window multi-head self-attention (S-SW-MSA). MSA within windows can effectively improve local feature capturing, and window shifting allows information to be exchanged in the neighboring windows. MSA can be expressed by the following formula:

$$Z = Attn(Q, K, V) = softmax\left(\frac{QK^T}{\sqrt{d_K}}\right)V \tag{8}$$

$$\psi = Concat(Z_1, Z_2, \cdots, Z_h)W \tag{9}$$

Q, K, V are matrices mapped from the input matrices called queries, keys and values. d_K is the dimension of K. The attention scores are calculated from Q and K. h is the head number of MSA, W denotes the output mapping matrix., and ψ represents the output of MSA.

As shown in Figure 4, the size of input is assumed to be $H \times W \times C$, where $H \times W$ is the space size and C is the number of spectral bands. Given that all windows' size is set to

$C/4$, the window is divided uniformly for the spectral dimension. The size of each window after division is $[C/4, C/4, C/4, C/4]$. Then MSA is performed in each window. Next the window is moved half a window in the spectral direction, The size of each window at this point is $[C/8, C/4, C/4, C/4, C/8]$. MSA is again performed in each window. Wherefore the process of S-W-MSA with m windows is:

$$Y^{(m)} = \left[\psi(y^{(1)}) \oplus \psi(y^{(2)}) \oplus, \cdots, \oplus \psi(y^{(m)})\right] \tag{10}$$

where $\oplus$ means concat, $y^{(i)}$ is the data of the i-th window.

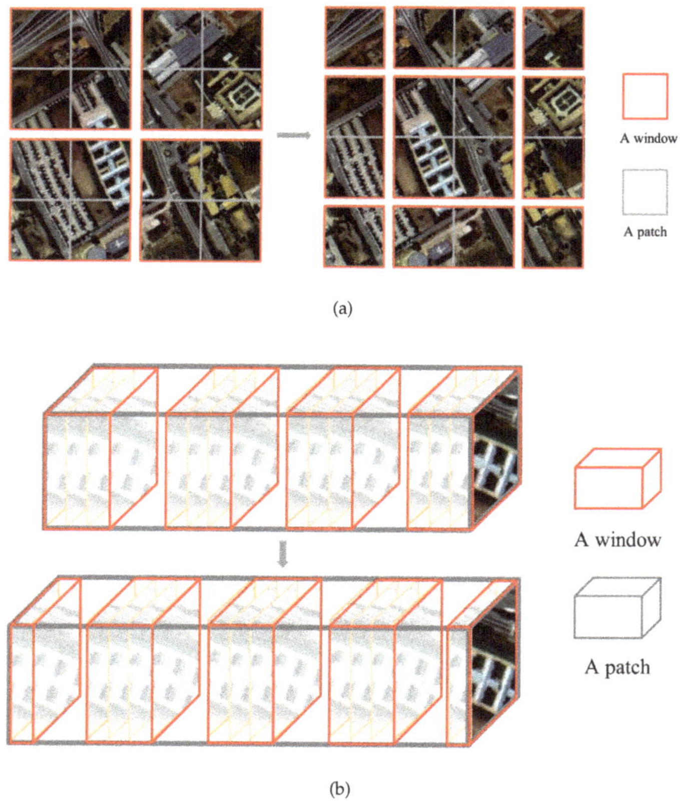

Figure 4. The structure of (**a**) S(W)-MSA of SwinT and (**b**) S-(S)W-MSA of SSWT (ours).

Compared to SwinT, the other components of the S-SwinT module remain the same except for the design of the window, such as MLP, layer normalization (LN) and residual connections. Figure 1 describes two nearby S-SwinT modules in each stage, which can be represented by the following formula.

$$\hat{Y}^l = \text{S-W-MSA}(\text{LN}(Y^{l-1})) + Y^{l-1} \tag{11}$$

$$Y^l = \text{MLP}(\text{LN}(\hat{Y}^l)) + \hat{Y}^l \tag{12}$$

$$\hat{Y}^{l+1} = \text{S-SW-MSA}(\text{LN}(Y^l)) + Y^l \tag{13}$$

$$Y^{l+1} = \text{MLP}(\text{LN}(\hat{Y}^{l+1})) + \hat{Y}^{l+1} \tag{14}$$

where S-W-MSA and S-SW-MSA denote the spectral window based and spectral shifted window based MSA, $\hat{Y}^l$ and Y^l are the outputs of S-(S)W-MSA and MLP in block l.

4. Experiment

In this section, we conducted extensive experiments on three benchmark datasets to demonstrate the effectiveness of the proposed method, including Pavia University (PU), Salinas (SA) and Houston2013 (HU).

4.1. Dataset

The three datasets that utilised in the experiments are detailed here.

(1) Pavia University:The Reflective Optics System Imaging Spectrometer (ROSIS) sensor acquired the PU dataset in 2001. It comprises 115 spectral bands with wavelengths ranging from 380 to 860 nm. Following the removal of the noise bands, there are now 103 open bands for investigation. The image measures 610 pixels in height and 340 pixels in width. The collection includes 42,776 labelled samples of 9 different land cover types.

(2) Salinas: The Airborne Visible/Infrared Imaging Spectrometer (AVIRIS) sensor acquired the SA dataset in 1998. The 224 bands in the original image have wavelengths between 400 and 2500 nm. 204 bands are used for evaluating after the water absorption bands have been removed. The data has 512 and 217 pixels of height and width, respectively. There are 16 object classes represented in the dataset's 54,129 marked samples.

(3) Houston2013: The Hyperspectral Image Analysis Group and the NSF-funded Airborne Laser Mapping Center (NCALM) at the University of Houston in the US provided the Houston 2013 dataset. The 2013 IEEE GRSS Data Fusion Competition used the dataset initially for scientific research. It has 144 spectral bands with wavelengths between 0.38 and 1.05 m. This dataset contains 15 classes and measures 349 × 1905 pixels with a 2.5 m spatial resolution.

We divided the label samples in different ways for each dataset. Tables 1–3 provide specifics on the number of each class for the three dataset training, validation, and testing sets. False-color map and ground-truth map of three datasets are shown in Figures 5–7.

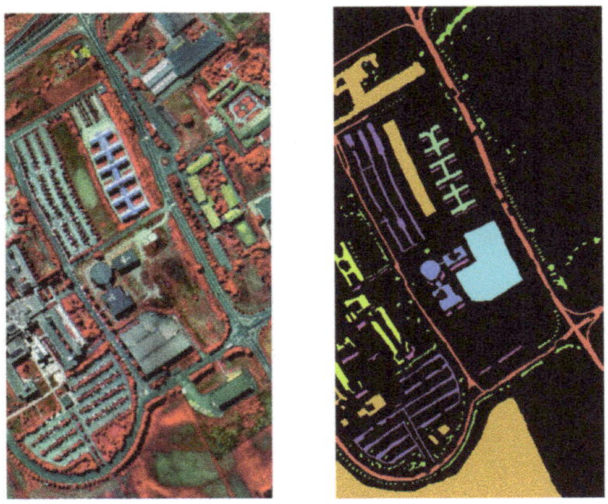

(a) (b)

Figure 5. Visualization of PU Datasets. (**a**) False-color map. (**b**) Ground-truth map.

(a) (b)

Figure 6. Visualization of SA Datasets. (**a**) False-color map. (**b**) Ground-truth map.

(a)

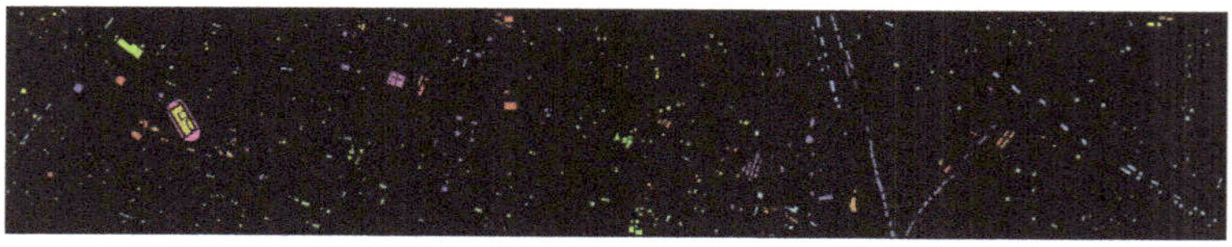

(b)

Figure 7. Visualization of HU Datasets. (**a**) False-color map. (**b**) Ground-truth map.

Table 1. Number of training, validation and testing samples for the PU dataset.

No.	Name	Train.	Val.	Test.
1	Asphalt	83	83	6465
2	Meadows	233	233	18,183
3	Gravel	26	26	2047
4	Trees	38	38	2987
5	Painted metal sheets	17	17	1311
6	Bare Soil	63	63	4903
7	Bitumen	17	17	1297
8	Self-Blocking Bricks	46	46	3590
9	Shadows	12	12	923
-	Total	535	535	41,706

Table 2. Number of training, validation and testing samples for the SA dataset.

No.	Name	Train.	Val.	Test.
1	Brocoli_green_weeds_1	25	25	1959
2	Brocoli_green_weeds_2	47	47	3633
3	Fallow	25	25	1927
4	Fallow_rough_plow	17	17	1358
5	Fallow_smooth	33	33	2611
6	Stubble	49	49	3860
7	Celery	45	45	3490
8	Grapes_untrained	141	141	10,989
9	Soil_vinyard_develop	78	78	6048
10	Corn_senesced_green_weeds	41	41	3196
11	Lettuce_romaine_4wk	13	13	1041
12	Lettuce_romaine_5wk	24	24	1879
13	Lettuce_romaine_6wk	11	11	893
14	Lettuce_romaine_7wk	13	13	1043
15	Vinyard_untrained	91	91	7086
16	Vinyard_vertical_trellis	23	23	1762
-	Total	676	676	52,775

Table 3. Number of training, validation and testing samples for the HU dataset.

No.	Name	Train.	Val.	Test.
1	Healthy grass	31	31	1188
2	Stressed grass	31	31	1191
3	Synthetic grass	17	17	662
4	Trees	31	31	1182
5	Soil	31	31	1180
6	Water	8	8	309
7	Residential	32	32	1205
8	Commercial	31	31	1182
9	Road	31	31	1189
10	Highway	31	31	1166
11	Railway	31	31	1173
12	Parking Lot 1	31	31	1171
13	Parking Lot 2	12	12	446
14	Tennis Court	11	11	407
15	Running Track	17	17	627
-	Total	376	376	14,278

4.2. Experimental Setting

(1) Evaluation Indicators: To quantitatively analyse the efficacy of the suggested method and other methods for comparison, four quantitative evaluation indexes are introduced: overall accuracy (OA), average accuracy (AA), kappa coefficient (κ), and the classification accuracy of each class. A better classification effect is indicated by a higher value for each indicator.

(2) Configuration: All verification experiments for the proposed technique were performed in the PyTorch environment using a desktop computer with an Intel(R) Core(TM) i7-10750H CPU, 16GB of RAM, and an NVIDIA Geforce GTX 1660Ti 6-GB GPU. The learning rate was initially set to 1×10^{-3} and the Adam optimizer was selected as the initial optimizer. The size of each training batch was set to 64. Each dataset received 500 training epochs.

4.3. Parameter Analysis

4.3.1. Influence of Patch Size

Patch size is the spatial size of the input patches, which determines the spatial information that the model can utilize when classifying HSIs. Therefore, The model's performance is influenced by the patch size. A too large patch size will increase the computational burden of the model. In this section we compare a set of patch sizes $\{3, 5, 7, 9, 11, 13\}$ to explore the effect of patch size on the model. The experimental results about patch size on the three datasets are shown in Figure 8. A similar trend was observed in all three datasets: OA first increased and then stabilized with increasing patch size. Specifically, the highest value of OA is achieved when the patch size is 9 in the PU and HU datasets, and the highest value of OA is achieved when the patch size is 11 in the SA dataset.

The size of patch is positively correlated with the spatial information contained in the patch. Increasing the patch means that the model can learn more spatial information, which will be beneficial to improve OA. And when the patch increases to a certain size, the distance between the pixels in the newly region and the center pixel is too far, and the spatial information that can be provided is of little value. So the improvement of OA is not much, and the OA will tend to be stable at this time.

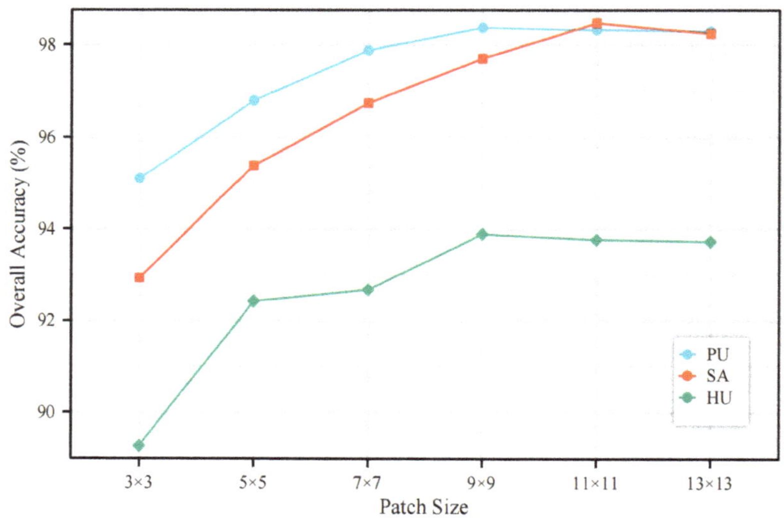

Figure 8. Overall accuracy(%) with different patch sizes on the three datasets. The window numbers in transformer layers is set to [1, 2, 2, 4].

4.3.2. Influence of Window Number

In proposed S-SW-MSA, the number of windows is a parameter that can be set depending on the characteristics of the dataset. Moreover, the number of windows can be different for each transformer layer in order to extract multiple scales of features. We set up six sets of experiments, the model contains four transformer layers in the first four sets and five transformer layers in the last two sets. the numbers in [] indicate the number of windows of S-SW-MSA in each transformer layer. The experimental results on the three datasets are shown in Table 4. According to the experimental results, the best OA for each dataset was obtained for different window number settings, and the best OA was obtained for the PU, SA and HU datasets in the 4th, 2nd and 6th group settings, respectively. We also found that increasing the number of transformer layers does not necessarily increase the performance of the model. For example, the best OA is achieved when the number of transformer layers is 4 for the PU and SA datasets and 5 for the HU dataset. Because the features of each dataset are different, the parameter settings will change accordingly.

Table 4. Overall accuracies (%) of proposed model with different number of windows in transformer layers on SA, PU and HU datasets. The patch size is set to 9.

Windows Size	PU	SA	HU
[1, 1, 2, 2]	97.05	97.56	93.24
[1, 2, 2, 4]	97.86	**97.80**	93.35
[2, 2, 4, 4]	98.33	96.93	93.31
[2, 2, 4, 8]	**98.37**	97.70	93.58
[1, 1, 2, 4, 8]	98.20	96.25	93.38
[2, 2, 4, 4, 8]	98.25	96.31	**93.69**

4.4. Ablation Experiments

To sufficiently demonstrate that proposed method is effective, we conducted ablation experiments on the Pavia University dataset. With ViT as the baseline, the components of the model are added separately: S-Swin, SPE and SFE. In total, there are 5 combinations. The experimental results are shown in the Table 5. The classification overall accuracy of ViT without any improvement was 84.43%. SPE, SFE and S-Swin are proposed improvements for the ViT backbone network, which can respectively increase classification overall accuracy of 1.69%, 7.21% and 7.87% after adding into the model. The classification overall accuracy of applying the two improvements to the model together can reach 93.78%, which is higher than baseline by 9.35%. It is considered to be a great result for the improved pure transformer, but it's a little lower than our final result. After the SFE was added to the model, the classification overall accuracy improved by 4.59%, eventually reaching 98.37.

Table 5. Ablation experiments in PU.

Method	Module (%)			Metric (%)		
	S-Swin	SPE	SFE	OA(%)	AA(%)	$\kappa \times 100$ (%)
ViT(Baseline)	✗	✗	✗	84.43	78.06	78.95
ViT	✗	✓	✗	86.12	80.18	81.31
ViT	✗	✗	✓	91.64	90.43	88.97
SSWT(Ours)	✓	✗	✗	92.30	89.58	89.75
SSWT(Ours)	✓	✓	✗	93.78	91.17	91.74
SSWT(Ours)	✓	✓	✓	98.37	97.25	97.84

4.5. Classification Results

The proposed model's outcomes are compared with those of the advanced deep learning models: a LSTM based network (Bi-LSTM) [51], a 3-D CNN-based deep learning network (3D-CNN) [57], a deep feature fusion network (DFFN) [53], a RSSAN [52], and some transformer based model include a Vit, Swin-transformer (SwinT) [43], a SpectralFormer (SF) [44], a Hit [46] and a SSFTT [45].

Tables 6–8 show the OA, AA, κ and the accuracy of each category for each model's classification on the three public datasets. Each result is the average of repeating the experiment five times. The best results are shown in bold. As the results show, proposed SSWT performs the best. On the PU dataset, SSWT is 1.02% higher than SSFTT, 3.85% higher than HiT, 9.01% higher than SwinT and 1.51% higher than RSSAN in terms of OA. Moreover, SSWT outperforms other models in terms of AA and κ. SSWT achieved the highest classification accuracy in 7 out of 9 categories. On the SA dataset, the advantage of SSWT is more prominent. SSWT is 3.22% higher than SSFTT, 3.99% higher than HiT, 7.10% higher than SwinT, 2.64% higher than RSSAN, and 3.01% higher than DFFN in terms of OA. The same advantage was achieved for SSWT in AA and κ. SSWT achieved the highest classification accuracy in 11 out of 16 categories. Similar results can be observed in HU dataset, where SSWT achieved significant advantages in all three metrics of OA, AA and κ. SSWT achieved the highest classification accuracy in 6 out of 15 categories.

Table 6. Classification results of the PU dataset.

Class	Bi-LSTM	3D-CNN	RSSAN	DFFN	Vit	SwinT	SF	Hit	SSFTT	SSWT
1	91.67 ± 0.83	95.16 ± 1.56	97.12 ± 0.57	96.66 ± 0.81	87.96 ± 1.80	93.05 ± 5.32	89.41 ± 2.23	93.72 ± 1.44	97.31 ± 1.12	**98.06 ± 0.24**
2	96.96 ± 1.60	98.31 ± 0.96	99.46 ± 0.11	99.05 ± 0.51	96.56 ± 3.00	96.98 ± 1.43	97.22 ± 0.76	98.66 ± 0.48	99.37 ± 0.26	**99.91 ± 0.08**
3	70.65 ± 9.73	36.91 ± 6.18	85.74 ± 5.05	70.37 ± 12.56	53.18 ± 19.35	29.49 ± 23.08	77.28 ± 3.19	80.42 ± 7.56	87.25 ± 5.43	**94.59 ± 2.40**
4	92.88 ± 2.78	95.52 ± 1.58	96.92 ± 1.32	94.22 ± 3.16	89.76 ± 2.25	92.09 ± 1.41	90.80 ± 1.92	94.74 ± 1.84	97.59 ± 1.15	**97.70 ± 1.05**
5	99.10 ± 0.60	99.83 ± 0.34	99.86 ± 0.17	99.97 ± 0.06	**100.00 ± 0.00**	99.16 ± 0.59	**100.00 ± 0.00**	99.95 ± 0.04	99.95 ± 0.06	99.85 ± 0.27
6	67.03 ± 14.76	49.91 ± 12.17	97.00 ± 1.09	95.07 ± 3.03	51.97 ± 7.05	88.47 ± 5.03	82.13 ± 6.02	95.54 ± 2.05	97.00 ± 1.60	**98.37 ± 1.63**
7	82.67 ± 3.31	46.74 ± 14.05	84.15 ± 5.66	74.68 ± 7.86	47.59 ± 8.36	45.18 ± 31.47	52.80 ± 6.23	75.17 ± 8.05	91.43 ± 3.70	**91.95 ± 5.61**
8	83.17 ± 3.25	89.73 ± 3.00	92.49 ± 1.51	87.38 ± 4.34	78.79 ± 8.88	92.76 ± 1.65	81.81 ± 4.44	85.83 ± 4.19	93.81 ± 1.51	**95.35 ± 5.61**
9	98.94 ± 0.51	98.66 ± 0.62	98.37 ± 0.96	99.57 ± 0.22	96.71 ± 1.00	76.85 ± 12.09	96.32 ± 1.38	97.16 ± 1.29	**99.72 ± 0.20**	99.48 ± 0.88
OA(%)	89.52 ± 1.91	86.63 ± 1.43	96.86 ± 0.36	94.74 ± 1.40	84.43 ± 1.56	89.36 ± 3.14	90.16 ± 0.89	94.52 ± 1.03	97.35 ± 0.45	**98.37 ± 0.24**
AA(%)	87.01 ± 1.97	78.97 ± 2.05	94.57 ± 0.84	90.77 ± 2.46	78.06 ± 2.56	79.34 ± 7.46	85.31 ± 1.20	91.24 ± 1.94	95.94 ± 0.73	**97.25 ± 0.64**
$\kappa \times 100$	85.94 ± 2.63	81.79 ± 2.04	95.84 ± 0.48	93.00 ± 1.87	78.95 ± 2.03	85.82 ± 4.23	86.87 ± 1.21	92.74 ± 1.37	96.49 ± 0.60	**97.84 ± 0.32**

Table 7. Classification results of the SA dataset.

Class	Bi-LSTM	3D-CNN	RSSAN	DFFN	Vit	SwinT	SF	Hit	SSFTT	SSWT
1	79.24 ± 39.63	97.09 ± 1.46	99.58 ± 0.48	97.12 ± 0.89	90.19 ± 2.51	72.30 ± 1.87	95.05 ± 1.49	98.69 ± 2.05	99.44 ± 0.90	**99.79 ± 0.43**
2	98.94 ± 0.55	**99.90 ± 0.08**	99.36 ± 0.87	99.58 ± 0.14	98.05 ± 1.17	97.24 ± 1.92	99.32 ± 0.18	99.32 ± 0.35	99.80 ± 0.34	99.80 ± 0.17
3	85.20 ± 12.23	88.23 ± 4.35	97.01 ± 1.63	95.01 ± 3.54	87.52 ± 1.83	89.31 ± 2.96	92.89 ± 1.29	95.51 ± 2.29	98.41 ± 1.04	**98.48 ± 1.54**
4	97.79 ± 1.21	98.22 ± 1.10	98.56 ± 0.70	96.67 ± 1.39	94.11 ± 1.43	96.12 ± 1.50	94.05 ± 2.02	98.82 ± 0.51	**99.59 ± 0.56**	98.53 ± 1.23
5	96.40 ± 1.22	93.41 ± 2.41	96.06 ± 1.37	96.87 ± 1.04	82.59 ± 2.93	97.68 ± 0.76	93.24 ± 1.83	96.03 ± 2.17	98.28 ± 0.77	**98.74 ± 0.80**
6	99.46 ± 0.37	99.79 ± 0.32	99.36 ± 1.00	99.84 ± 0.30	99.44 ± 0.64	98.89 ± 1.29	99.68 ± 0.36	**99.99 ± 0.02**	99.98 ± 0.02	99.96 ± 0.06
7	98.84 ± 0.36	99.47 ± 0.23	99.28 ± 0.40	99.62 ± 0.28	98.05 ± 0.71	97.79 ± 0.92	98.81 ± 0.47	98.88 ± 0.62	99.44 ± 0.46	**99.72 ± 0.42**
8	83.66 ± 3.85	82.53 ± 2.36	90.93 ± 2.87	89.16 ± 1.74	82.79 ± 1.93	87.64 ± 1.38	85.03 ± 2.46	88.55 ± 1.73	90.08 ± 4.06	**95.87 ± 1.47**
9	97.84 ± 1.34	98.51 ± 1.11	99.66 ± 0.26	98.88 ± 0.80	96.38 ± 0.57	99.16 ± 0.63	98.05 ± 0.64	99.62 ± 0.37	99.53 ± 0.24	**99.92 ± 0.06**
10	81.10 ± 8.62	89.40 ± 2.50	95.58 ± 2.48	95.39 ± 1.01	75.44 ± 3.81	89.52 ± 3.74	91.23 ± 2.28	93.74 ± 2.38	95.73 ± 2.58	**97.07 ± 1.88**
11	83.59 ± 6.83	73.95 ± 4.65	93.37 ± 5.75	92.56 ± 5.81	70.47 ± 15.29	83.99 ± 14.49	89.86 ± 4.74	91.16 ± 6.19	94.66 ± 4.66	**95.64 ± 4.52**
12	98.84 ± 0.61	99.21 ± 0.56	99.36 ± 0.79	**99.97 ± 0.03**	98.67 ± 1.31	95.76 ± 0.75	98.45 ± 1.46	99.30 ± 0.64	99.80 ± 0.28	99.78 ± 0.45
13	94.78 ± 2.72	99.66 ± 0.07	98.92 ± 0.99	**99.98 ± 0.04**	96.28 ± 2.05	94.92 ± 6.31	98.61 ± 0.92	98.99 ± 1.12	99.06 ± 1.66	99.87 ± 0.18
14	90.20 ± 2.51	97.24 ± 1.05	96.63 ± 0.57	98.52 ± 0.76	96.51 ± 1.38	94.47 ± 1.04	95.03 ± 2.32	97.16 ± 0.77	95.61 ± 2.88	**99.23 ± 0.55**
15	78.87 ± 9.66	73.91 ± 2.47	86.60 ± 3.27	87.97 ± 2.81	72.03 ± 5.50	86.75 ± 6.26	79.87 ± 3.00	81.79 ± 3.34	81.36 ± 6.09	**94.10 ± 2.05**
16	90.27 ± 9.62	92.36 ± 1.46	96.67 ± 1.27	95.16 ± 2.32	91.57 ± 0.75	92.77 ± 3.30	95.35 ± 0.99	96.79 ± 1.67	97.20 ± 1.02	**98.40 ± 1.08**
OA(%)	89.66 ± 3.03	90.22 ± 0.70	95.16 ± 0.35	94.79 ± 0.80	87.58 ± 0.37	90.70 ± 2.38	91.81 ± 0.73	93.81 ± 0.56	94.58 ± 0.41	**97.80 ± 0.25**
AA(%)	90.94 ± 3.31	92.68 ± 0.71	96.68 ± 0.49	96.39 ± 0.57	89.38 ± 0.51	90.02 ± 3.99	94.03 ± 0.48	95.90 ± 0.24	96.75 ± 0.26	**98.43 ± 0.35**
$\kappa \times 100$	88.49 ± 3.39	89.11 ± 0.77	94.61 ± 0.39	94.20 ± 0.89	86.17 ± 0.41	89.63 ± 2.67	90.89 ± 0.81	93.10 ± 0.62	93.97 ± 0.46	**97.55 ± 0.28**

Table 8. Classification results of the HU dataset.

Class	Bi-LSTM	3D-CNN	RSSAN	DFFN	Vit	SwinT	SF	Hit	SSFTT	SSWT
1	84.09 ± 4.77	89.90 ± 6.62	95.05 ± 2.77	94.71 ± 5.79	90.72 ± 6.21	94.56 ± 2.55	95.05 ± 5.10	93.37 ± 4.54	93.96 ± 4.32	**95.13 ± 4.45**
2	90.60 ± 7.71	81.28 ± 6.08	98.05 ± 1.19	97.75 ± 1.06	83.93 ± 9.70	93.93 ± 5.83	93.53 ± 3.77	97.78 ± 0.87	98.71 ± 1.11	**98.77 ± 1.18**
3	75.14 ± 17.70	91.81 ± 4.04	98.67 ± 0.81	99.49 ± 0.74	88.01 ± 8.50	96.68 ± 1.98	97.19 ± 2.01	98.64 ± 0.91	**99.52 ± 0.89**	99.46 ± 0.67
4	90.83 ± 3.70	91.91 ± 0.35	94.06 ± 1.91	91.34 ± 0.74	85.63 ± 3.35	94.42 ± 2.77	89.54 ± 1.79	95.35 ± 1.99	**96.65 ± 2.55**	95.75 ± 1.55
5	92.86 ± 2.93	95.97 ± 1.83	98.29 ± 0.77	98.44 ± 0.74	95.86 ± 1.75	97.99 ± 0.74	96.97 ± 0.92	98.69 ± 0.98	99.54 ± 0.49	**99.93 ± 0.08**
6	52.43 ± 31.32	72.69 ± 2.15	80.58 ± 6.70	86.15 ± 6.72	6.93 ± 7.34	71.20 ± 14.20	63.88 ± 5.20	81.49 ± 2.85	90.42 ± 6.32	**92.62 ± 5.67**
7	72.93 ± 9.32	84.15 ± 2.50	87.09 ± 3.56	84.60 ± 3.98	64.32 ± 11.11	71.84 ± 14.62	74.67 ± 4.06	81.16 ± 5.29	86.22 ± 5.43	**88.70 ± 4.61**
8	55.74 ± 5.24	55.87 ± 6.14	78.88 ± 3.64	79.10 ± 3.82	66.84 ± 6.80	73.69 ± 9.90	76.31 ± 2.76	78.85 ± 2.03	82.79 ± 2.81	**85.08 ± 3.38**
9	73.05 ± 5.75	81.90 ± 2.13	81.77 ± 4.72	84.24 ± 4.75	66.24 ± 5.56	73.28 ± 2.75	72.94 ± 6.60	83.62 ± 5.81	**89.96 ± 4.24**	87.47 ± 3.31
10	39.43 ± 20.49	48.10 ± 12.51	89.76 ± 0.52	90.22 ± 5.12	63.29 ± 5.92	78.56 ± 2.66	81.13 ± 5.79	86.14 ± 5.11	93.60 ± 1.29	**96.05 ± 3.71**
11	66.55 ± 10.85	60.66 ± 2.63	82.85 ± 4.35	82.46 ± 3.64	58.67 ± 3.08	76.21 ± 0.37	68.80 ± 6.54	79.52 ± 4.94	86.36 ± 2.82	**87.55 ± 5.08**
12	67.21 ± 9.90	58.29 ± 10.86	92.13 ± 2.73	93.10 ± 2.00	61.69 ± 6.32	87.50 ± 3.52	85.02 ± 4.18	90.96 ± 3.22	88.95 ± 5.90	**97.83 ± 1.12**
13	19.96 ± 14.65	59.10 ± 10.82	71.21 ± 8.17	**92.47 ± 1.57**	40.09 ± 16.86	71.60 ± 2.70	50.85 ± 9.67	79.28 ± 2.86	92.33 ± 2.81	90.76 ± 3.15
14	89.93 ± 8.82	93.12 ± 3.76	92.38 ± 3.91	94.74 ± 2.65	77.49 ± 4.47	89.03 ± 7.12	78.28 ± 3.02	93.96 ± 2.88	**96.46 ± 2.24**	94.55 ± 3.68
15	90.91 ± 8.78	**99.39 ± 0.77**	95.82 ± 2.62	98.88 ± 0.88	91.48 ± 3.12	96.65 ± 1.75	95.15 ± 2.84	98.47 ± 1.03	98.66 ± 1.13	98.63 ± 1.56
OA(%)	72.60 ± 3.03	76.73 ± 1.69	89.76 ± 0.39	90.62 ± 0.79	72.80 ± 1.54	84.78 ± 2.39	82.97 ± 0.99	89.16 ± 1.03	92.47 ± 0.97	**93.69 ± 1.07**
AA(%)	70.78 ± 4.49	77.61 ± 1.80	89.11 ± 0.61	91.18 ± 0.86	69.41 ± 0.73	84.48 MSA 1.67	81.29 ± 1.16	89.15 ± 0.88	92.94 ± 1.01	**93.89 ± 1.07**
$\kappa \times 100$	70.34 ± 3.29	74.84 ± 1.83	88.93 ± 0.42	89.86 ± 0.85	70.58 ± 1.64	83.55 MSA 2.58	81.58 ± 1.07	88.28 ± 1.11	91.86 ± 1.05	**93.18 ± 1.16**

We visualized the prediction results of each model on the samples to compare the performance of the models, and the visualization results of each model on the three datasets are shown in Figures 9–11 Proposed SSWT has less noise in all three datasets compared to other models, and the classification result of SSWT are closest to the ground truth. In the

PU dataset, the blue area in the middle is misclassified by many models, and the SSWT result in the fewest errors. In the SA dataset, the pink area and the green area on the top left show a number of errors in the classification results of other models, and the SSWT classification results are the smoothest. A similar situation is observed in the HU dataset. The superiority of proposed model is further demonstrated.

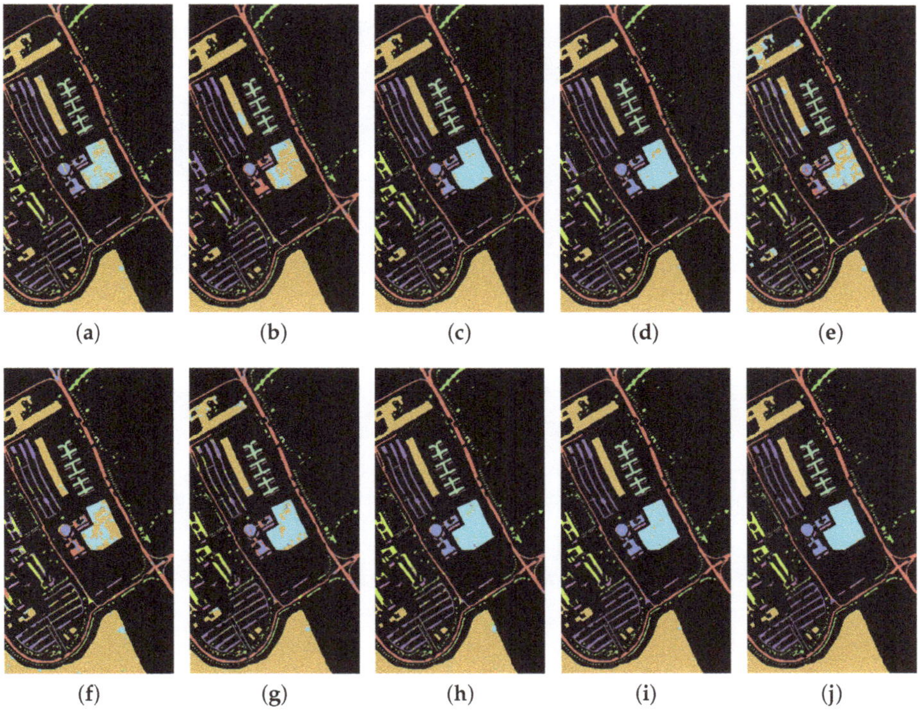

Figure 9. Classification maps of different methods in PU dataset. (**a**) Bi-LSTM. (**b**) 3D-CNN. (**c**) RSSAN. (**d**) DFFN. (**e**) Vit. (**f**) SwinT. (**g**) SF. (**h**) Hit. (**i**) SSFTT. (**j**) Proposed SSWT.

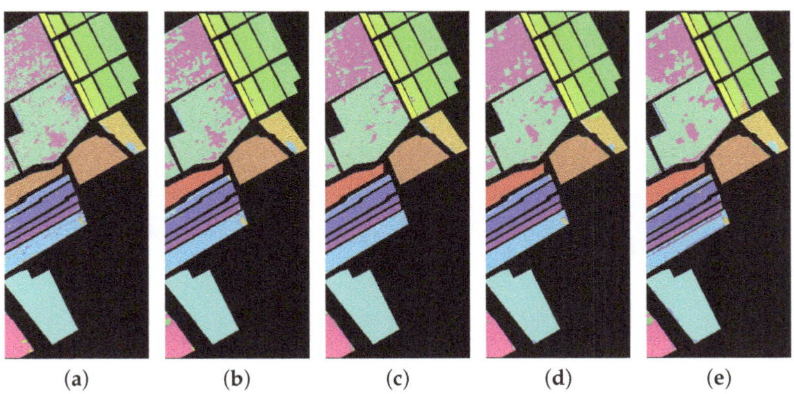

Figure 10. *Cont.*

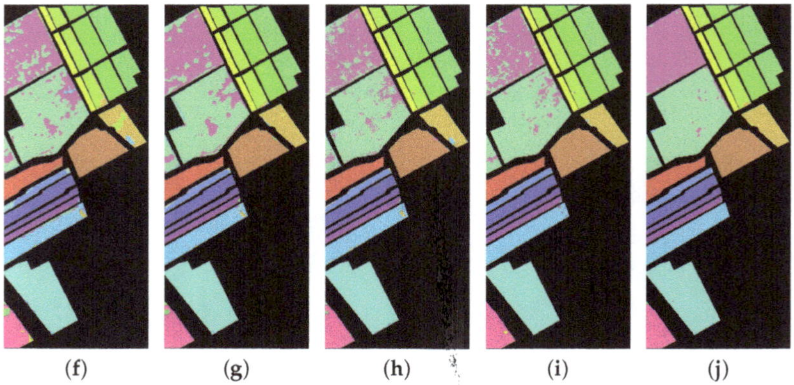

Figure 10. Classification maps of different methods in SA dataset. (**a**) Bi-LSTM. (**b**) 3D-CNN (**c**) RSSAN. (**d**) DFFN. (**e**) Vit. (**f**) SwinT. (**g**) SF. (**h**) Hit. (**i**) SSFTT. (**j**) Proposed SSWT.

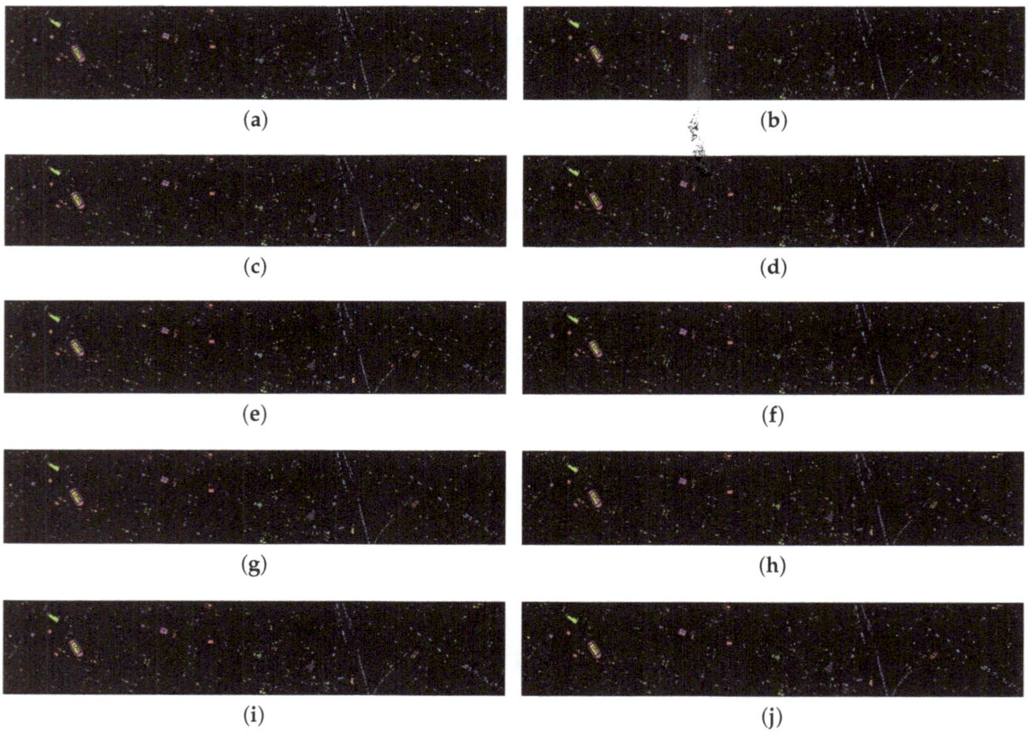

Figure 11. Classification maps of different methods in HU dataset. (**a**) Bi-LSTM. (**b**) 3D-CNN (**c**) RSSAN. (**d**) DFFN. (**e**) Vit. (**f**) SwinT. (**g**) SF. (**h**) Hit. (**i**) SSFTT. (**j**) Proposed SSWT.

4.6. Robustness Evaluation

In order to evaluate the robustness of the proposed model, we conducted experiments with the proposed model and other models under different numbers of training samples Figure 12 shows the experimental results on three datasets, we selected 0.5%, 1%, 2%, 4%, and 8% of the samples in turn as training data for the PU and SA dataset , while 2%, 4%, 6%, 8% and 10% for the HU dataset. It can be observed that the proposed SSWT is performing best in every situation, especially in the case of few training samples. The robustness of proposed SSWT and its superiority in the case of small samples can be demonstrated. Taking the PU dataset as an example, most of the models achieve high accuracy at 8% of the training percent, with SSWT having a small advantage. And as the training percent decreases, SSWT has higher accuracy compared to other models. Similar results were found on the SA and HU datasets, where SSWT showed excellent performance for all training percents.

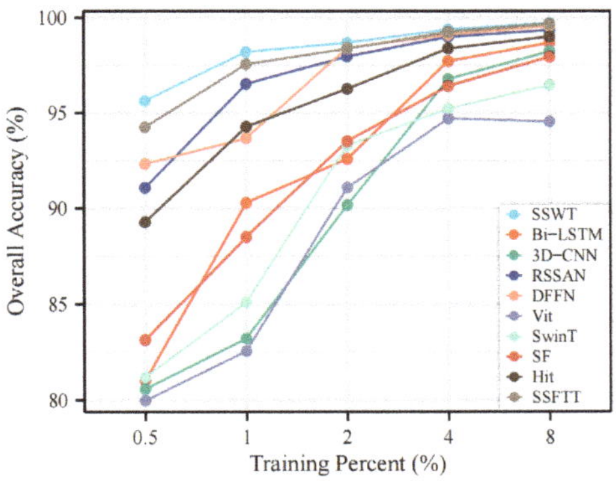

(**a**)

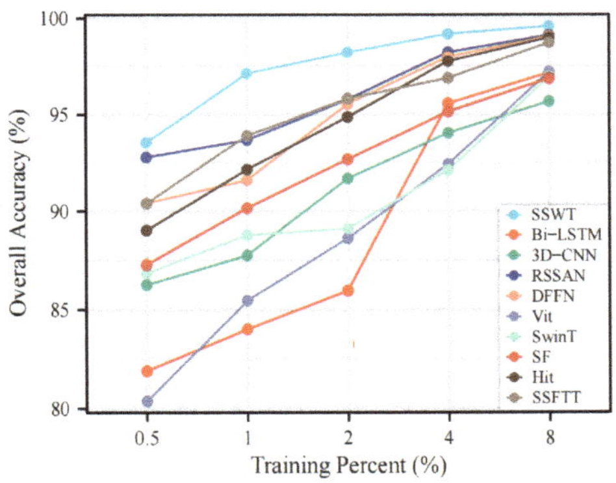

(**b**)

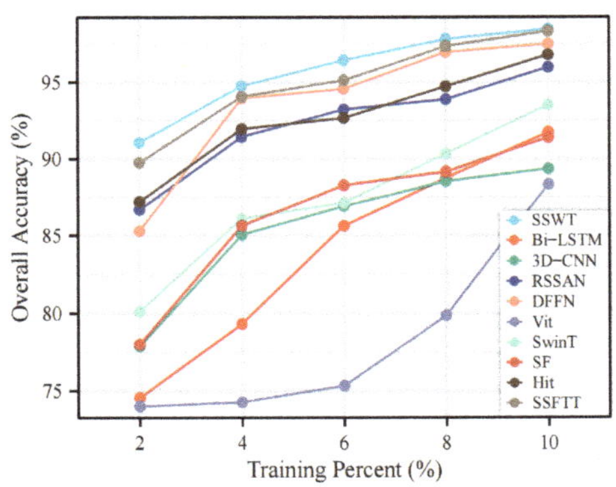

(**c**)

Figure 12. Classification results in different training percent of samples on the three datasets. (**a**) PU. (**b**) SA. (**c**) HU.

5. Conclusions

In this paper, we summarize the shortcomings of the existing ViT for HSI classification tasks. For the lack of ability to capture local contextual features, we use the self-attentive mechanism of shifted windows. The corresponding design is made for the characteristics of HSI, i.e., the spectral shifted window self-attention, which effectively improves the local feature extraction capability. For the insensitivity of ViT to spatial features and structure, we designed the spatial feature extraction module and spatial position encoding to compensate. The superiority of the proposed model has been verified by experimental results across three public HSI datasets.

In future work, we will improve the calculation of S-SW-MSA to reduce its time complexity. In addition, we will continue our research based on the transformer and try to achieve higher performance with a model of pure transformer structure.

Author Contributions: All the authors made significant contributions to the work. Y.P., J.R. and J.W. designed the research, analyzed the results, and accomplished the validation work. M.S. provided advice for the revision of the paper. All authors have read and agreed to the published version of the manuscript.

Funding: This research received no external funding.

Conflicts of Interest: The authors declare no conflict of interest.

References

1. Gevaert, C.M.; Suomalainen, J.; Tang, J.; Kooistra, L. Generation of spectral–temporal response surfaces by combining multispectral satellite and hyperspectral UAV imagery for precision agriculture applications. *IEEE J. Sel. Top. Appl. Earth Observ. Remote Sens.* **2015**, *8*, 3140–3146. [CrossRef]
2. Jabir, B.; Falih, N.; Rahmani, K. Accuracy and Efficiency Comparison of Object Detection Open-Source Models. *Int. J. Online Biomed. Eng.* **2021**, *17*, 165–184. [CrossRef] [CrossRef]
3. Lu, G.; Fei, B. Medical hyperspectral imaging: A review. *J. Biomed. Opt.* **2014**, *19*, 010901. [CrossRef]
4. Lone, Z.A.; Pais, A.R. Object detection in hyperspectral images. *Digit. Signal Process.* **2022**, *131*, 103752. [CrossRef] [CrossRef]
5. Weber, C.; Aguejdad, R.; Briottet, X.; Avala, J.; Fabre, S.; Demuynck, J.; Zenou, E.; Deville, Y.; Karoui, M.S.; Benhalouche, F.Z.; et al. Hyperspectral imagery for environmental urban planning. In Proceedings of the IGARSS 2018–2018 IEEE International Geoscience and Remote Sensing Symposium, Valencia, Spain, 22–27 July 2018; IEEE: New York, NY, USA, 2018; pp. 1628–1631.
6. Li, N.; Lü, J.S.; Altermann, W. Hyperspectral remote sensing in monitoring the vegetation heavy metal pollution. *Spectrosc. Spectr. Anal.* **2010**, *30*, 2508–2511. [CrossRef]
7. Saralıoğlu, E.; Görmüş, E.T.; Güngör, O. Mineral exploration with hyperspectral image fusion. In Proceedings of the 2016 24th Signal Processing and Communication Application Conference (SIU), Zonguldak, Turkey, 16–19 May 2016; IEEE: New York, NY, USA, 2016; pp. 1281–1284.
8. Ren, J.; Wang, R.; Liu, G.; Feng, R.; Wang, Y.; Wu, W. Partitioned relief-F method for dimensionality reduction of hyperspectral images. *Remote Sens.* **2020**, *12*, 1104. [CrossRef] [CrossRef]
9. Ke, C. Military object detection using multiple information extracted from hyperspectral imagery. In Proceedings of the 2017 International Conference on Progress in Informatics and Computing (PIC), Nanjing, China, 15–17 December 2017; IEEE: New York, NY, USA, 2017; pp. 124–128.
10. Cariou, C.; Chehdi, K. A new k-nearest neighbor density-based clustering method and its application to hyperspectral images. In Proceedings of the 2016 IEEE International Geoscience and Remote Sensing Symposium (IGARSS), Beijing, China, 10–15 July 2016; IEEE: New York, NY, USA, 2016; pp. 6161–6164.
11. Ren, J.; Wang, R.; Liu, G.; Wang, Y.; Wu, W. An SVM-based nested sliding window approach for spectral–spatial classification of hyperspectral images. *Remote Sens.* **2020**, *13*, 114. [CrossRef] [CrossRef]
12. Yaman, O.; Yetis, H.; Karakose, M. Band Reducing Based SVM Classification Method in Hyperspectral Image Processing. In Proceedings of the 2020 Zooming Innovation in Consumer Technologies Conference (ZINC), Novi Sad, Serbia, 26–27 May 2020; IEEE: New York, NY, USA, 2020; pp. 21–25.
13. Chen, Y.; Zhao, X.; Lin, Z. Optimizing subspace SVM ensemble for hyperspectral imagery classification. *IEEE J. Sel. Top. Appl. Earth Observ. Remote Sens.* **2014**, *7*, 1295–1305. [CrossRef] [CrossRef]
14. Shao, Z.; Zhang, L.; Zhou, X.; Ding, L. A novel hierarchical semisupervised SVM for classification of hyperspectral images. *IEEE Geosci. Remote Sens. Lett.* **2014**, *11*, 1609–1613. [CrossRef] [CrossRef]
15. Zhang, Y.; Cao, G.; Li, X.; Wang, B. Cascaded random forest for hyperspectral image classification. *IEEE J. Sel. Top. Appl. Earth Observ. Remote Sens.* **2018**, *11*, 1082–1094. [CrossRef] [CrossRef]

16. Belgiu, M.; Drăguţ, L. Random forest in remote sensing: A review of applications and future directions. *ISPRS J. Photogramm. Remote Sens.* **2016**, *114*, 24–31. [CrossRef] [CrossRef]
17. Navarro, A.; Nicastro, N.; Costa, C.; Pentangelo, A.; Cardarelli, M.; Ortenzi, L.; Pallottino, F.; Cardi, T.; Pane, C. Sorting biotic and abiotic stresses on wild rocket by leaf-image hyperspectral data mining with an artificial intelligence model. *Plant Methods* **2022**, *18*, 45. [CrossRef] [CrossRef] [PubMed]
18. Song, W.; Li, S.; Kang, X.; Huang, K. Hyperspectral image classification based on KNN sparse representation. In Proceedings of the 2016 IEEE International Geoscience and Remote Sensing Symposium (IGARSS), Beijing, China, 10–15 July 2016; IEEE: New York, NY, USA, 2016; pp. 2411–2414. [CrossRef]
19. Guo, Y.; Yin, X.; Zhao, X.; Yang, D.; Bai, Y. Hyperspectral image classification with SVM and guided filter. *EURASIP J. Wirel. Commun. Netw.* **2019**, *2019*, 56. [CrossRef] [CrossRef]
20. Wu, H.; Prasad, S. Convolutional recurrent neural networks for hyperspectral data classification. *Remote Sens.* **2017**, *9*, 298. [CrossRef] [CrossRef]
21. Luo, H. Shorten spatial-spectral RNN with parallel-GRU for hyperspectral image classification. *arXiv* **2018**, arXiv:1810.12563. [CrossRef]
22. Mou, L.; Ghamisi, P.; Zhu, X.X. Deep recurrent neural networks for hyperspectral image classification. *IEEE Trans. Geosci. Remote Sens.* **2017**, *55*, 3639–3655. [CrossRef] [CrossRef]
23. Lee, H.; Kwon, H. Contextual deep CNN based hyperspectral classification. In Proceedings of the 2016 IEEE International Geoscience and Remote Sensing Symposium (IGARSS), Beijing, China, 10–15 July 2016; IEEE: New York, NY, USA, 2016; pp. 3322–3325.
24. Chen, Y.; Zhu, L.; Ghamisi, P.; Jia, X.; Li, G.; Tang, L. Hyperspectral images classification with Gabor filtering and convolutional neural network. *IEEE Geosci. Remote Sens. Lett.* **2017**, *14*, 2355–2359. [CrossRef] [CrossRef]
25. Zhao, X.; Tao, R.; Li, W.; Li, H.C.; Du, Q.; Liao, W.; Philips, W. Joint classification of hyperspectral and LiDAR data using hierarchical random walk and deep CNN architecture. *IEEE Trans. Geosci. Remote Sens.* **2020**, *58*, 7355–7370. [CrossRef] [CrossRef]
26. Chen, Y.; Jiang, H.; Li, C.; Jia, X.; Ghamisi, P. Deep feature extraction and classification of hyperspectral images based on convolutional neural networks. *IEEE Trans. Geosci. Remote Sens.* **2016**, *54*, 6232–6251. [CrossRef] [CrossRef]
27. He, M.; Li, B.; Chen, H. Multi-scale 3D deep convolutional neural network for hyperspectral image classification. In Proceedings of the 2017 IEEE International Conference on Image Processing (ICIP), Beijing, China, 17–20 September 2017; IEEE: New York, NY, USA, 2017; pp. 3904–3908.
28. Makantasis, K.; Karantzalos, K.; Doulamis, A.; Doulamis, N. Deep supervised learning for hyperspectral data classification through convolutional neural networks. In Proceedings of the 2015 IEEE International Geoscience and Remote Sensing Symposium (IGARSS), Milan, Italy, 26–31 July 2015; IEEE: New York, NY, USA, 2015; pp. 4959–4962.
29. Wan, S.; Gong, C.; Zhong, P.; Du, B.; Zhang, L.; Yang, J. Multiscale dynamic graph convolutional network for hyperspectral image classification. *IEEE Trans. Geosci. Remote Sens.* **2019**, *58*, 3162–3177. [CrossRef] [CrossRef]
30. Mou, L.; Lu, X.; Li, X.; Zhu, X.X. Nonlocal graph convolutional networks for hyperspectral image classification. *IEEE Trans. Geosci. Remote Sens.* **2020**, *58*, 8246–8257. [CrossRef] [CrossRef]
31. Paoletti, M.E.; Haut, J.M.; Fernandez-Beltran, R.; Plaza, J.; Plaza, A.; Li, J.; Pla, F. Capsule networks for hyperspectral image classification. *IEEE Trans. Geosci. Remote Sens.* **2018**, *57*, 2145–2160. [CrossRef] [CrossRef]
32. Yin, J.; Li, S.; Zhu, H.; Luo, X. Hyperspectral image classification using CapsNet with well-initialized shallow layers. *IEEE Geosci. Remote Sens. Lett.* **2019**, *16*, 1095–1099. [CrossRef] [CrossRef]
33. Zhou, F.; Hang, R.; Liu, Q.; Yuan, X. Hyperspectral image classification using spectral-spatial LSTMs. *Neurocomputing* **2019**, *328*, 39–47. [CrossRef] [CrossRef]
34. Gao, J.; Gao, X.; Wu, N.; Yang, H. Bi-directional LSTM with multi-scale dense attention mechanism for hyperspectral image classification. *Multimed. Tools Appl.* **2022**, *81*, 24003–24020. [CrossRef]
35. Xu, Y.; Du, B.; Zhang, L.; Zhang, F. A band grouping based LSTM algorithm for hyperspectral image classification. In *Computer Vision: Second CCF Chinese Conference, CCCV 2017, Tianjin, China, 11–14 October 2017, Proceedings, Part II*; Springer: Berlin/Heidelberg, Germany, 2017; pp. 421–432.
36. Luo, Y.; Zou, J.; Yao, C.; Zhao, X.; Li, T.; Bai, G. HSI-CNN: A novel convolution neural network for hyperspectral image. In Proceedings of the 2018 International Conference on Audio, Language and Image Processing (ICALIP), Shanghai, China, 16–17 July 2018; IEEE: New York, NY, USA, 2018; pp. 464–469. [CrossRef]
37. Haut, J.M.; Paoletti, M.E.; Plaza, J.; Plaza, A.; Li, J. Hyperspectral image classification using random occlusion data augmentation. *IEEE Geosci. Remote Sens. Lett.* **2019**, *16*, 1751–1755. [CrossRef] [CrossRef]
38. Sun, K.; Wang, A.; Sun, X.; Zhang, T. Hyperspectral image classification method based on M-3DCNN-Attention. *J. Appl. Remote Sens.* **2022**, *16*, 026507. [CrossRef]
39. Xu, H.; Yao, W.; Cheng, L.; Li, B. Multiple spectral resolution 3D convolutional neural network for hyperspectral image classification. *Remote Sens.* **2021**, *13*, 1248. [CrossRef] [CrossRef]
40. Li, W.; Chen, H.; Liu, Q.; Liu, H.; Wang, Y.; Gui, G. Attention mechanism and depthwise separable convolution aided 3DCNN for hyperspectral remote sensing image classification. *Remote Sens.* **2022**, *14*, 2215. [CrossRef]
41. Sellami, A.; Abbes, A.B.; Barra, V.; Farah, I.R. Fused 3-D spectral-spatial deep neural networks and spectral clustering for hyperspectral image classification. *Pattern Recognit. Lett.* **2020**, *138*, 594–600. [CrossRef]

42. Dosovitskiy, A.; Beyer, L.; Kolesnikov, A.; Weissenborn, D.; Zhai, X.; Unterthiner, T.; Dehghani, M.; Minderer, M.; Heigold, G. Gelly, S.; others. An image is worth 16x16 words: Transformers for image recognition at scale. *arXiv* **2020**, arXiv:2010.11929 [CrossRef]
43. Liu, Z.; Lin, Y.; Cao, Y.; Hu, H.; Wei, Y.; Zhang, Z.; Lin, S.; Guo, B. Swin transformer: Hierarchical vision transformer using shifted windows. *arXiv* **2013**, arXiv:2103.14030. [CrossRef]
44. Hong, D.; Han, Z.; Yao, J.; Gao, L.; Zhang, B.; Plaza, A.; Chanussot, J. SpectralFormer: Rethinking hyperspectral image classification with transformers. *IEEE Trans. Geosci. Remote Sens.* **2021**, *60*, 1–15. [CrossRef] [CrossRef]
45. Sun, L.; Zhao, G.; Zheng, Y.; Wu, Z. Spectral-spatial feature tokenization transformer for hyperspectral image classification. *IEEE Trans. Geosci. Remote Sens.* **2022**, *60*, 5522214. [CrossRef] [CrossRef]
46. Yang, X.; Cao, W.; Lu, Y.; Zhou, Y. Hyperspectral image transformer classification networks. *IEEE Trans. Geosci. Remote Sens.* **2022**, *60*, 1–15. [CrossRef]
47. Mei, S.; Song, C.; Ma, M.; Xu, F. Hyperspectral image classification using group-aware hierarchical transformer. *IEEE Trans. Geosci. Remote Sens.* **2022**, *60*, 5539014. [CrossRef]
48. Xue, Z.; Xu, Q.; Zhang, M. Local transformer with spatial partition restore for hyperspectral image classification. *IEEE J. Sel. Top. Appl. Earth Observ. Remote Sens.* **2022**, *15*, 4307–4325. [CrossRef]
49. Hu, X.; Yang, W.; Wen, H.; Liu, Y.; Peng, Y. A lightweight 1-D convolution augmented transformer with metric learning for hyperspectral image classification. *Sensors* **2021**, *21*, 1751. [CrossRef] [CrossRef] [PubMed]
50. Qing, Y.; Liu, W.; Feng, L.; Gao, W. Improved transformer net for hyperspectral image classification. *Remote Sens.* **2021**, *13*, 2216 [CrossRef] [CrossRef]
51. Mei, S.; Li, X.; Liu, X.; Cai, H.; Du, Q. Hyperspectral image classification using attention-based bidirectional long short-term memory network. *IEEE Trans. Geosci. Remote Sens.* **2021**, *60*, 1–12. [CrossRef]
52. Zhu, M.; Jiao, L.; Liu, F.; Yang, S.; Wang, J. Residual spectral-spatial attention network for hyperspectral image classification. *IEEE Trans. Geosci. Remote Sens.* **2020**, *59*, 449–462. [CrossRef] [CrossRef]
53. Song, W.; Li, S.; Fang, L.; Lu, T. Hyperspectral image classification with deep feature fusion network. *IEEE Trans. Geosci. Remote Sens.* **2018**, *56*, 3173–3184. [CrossRef] [CrossRef]
54. Hang, R.; Li, Z.; Liu, Q.; Ghamisi, P.; Bhattacharyya, S.S. Hyperspectral image classification with attention-aided CNNs. *IEEE Trans. Geosci. Remote Sens.* **2020**, *59*, 2281–2293. [CrossRef] [CrossRef]
55. Chakraborty, T.; Trehan, U. Spectralnet: Exploring spatial-spectral waveletcnn for hyperspectral image classification. *arXiv* **2021**, arXiv:2104.00341. [CrossRef]
56. Gong, H.; Li, Q.; Li, C.; Dai, H.; He, Z.; Wang, W.; Li, H.; Han, F.; Tuniyazi, A.; Mu, T. Multiscale information fusion for hyperspectral image classification based on hybrid 2D-3D CNN. *Remote Sens.* **2021**, *13*, 2268. [CrossRef] [CrossRef]
57. Hamida, A.B.; Benoit, A.; Lambert, P.; Amar, C.B. 3-D deep learning approach for remote sensing image classification. *IEEE Trans. Geosci. Remote Sens.* **2018**, *56*, 4420–4434. [CrossRef]
58. Fang, J.; Xie, L.; Wang, X.; Zhang, X.; Liu, W.; Tian, Q. MSG-transformer: Exchanging local spatial information by manipulating messenger tokens. *arXiv* **2022**, arXiv:2105.15168. [CrossRef]
59. Guo, J.; Han, K.; Wu, H.; Tang, Y.; Chen, X.; Wang, Y.; Xu, C. Cmt: Convolutional neural networks meet vision transformers. *arXiv* **2022**, arXiv:2103.14030. [CrossRef]
60. Chen, Y.; Dai, X.; Chen, D.; Liu, M.; Dong, X.; Yuan, L.; Liu, Z. Mobile-former: Bridging mobilenet and transformer. *arXiv* **2022**, arXiv:2108.05895. [CrossRef]
61. Ayas, S.; Tunc-Gormus, E. SpectralSWIN: A spectral-swin transformer network for hyperspectral image classification. *Int. J. Remote Sens.* **2022**, *43*, 4025–4044. [CrossRef] [CrossRef]

remote sensing

MDPI

Article

Moving Point Target Detection Based on Temporal Transient Disturbance Learning in Low SNR

Weihua Gao [1,2], Wenlong Niu [1,*], Pengcheng Wang [1,2], Yanzhao Li [1,2], Chunxu Ren [1], Xiaodong Peng [1] and Zhen Yang [1]

1 National Space Science Center, Chinese Academy of Sciences, Beijing 100190, China
2 School of Computer Science and Technology, University of Chinese Academy of Sciences, Beijing 100049, China
* Correspondence: niuwenlong@nssc.ac.cn

Abstract: Moving target detection in optical remote sensing is important for satellite surveillance and space target monitoring. Here, a new moving point target detection framework under a low signal-to-noise ratio (SNR) that uses an end-to-end network (1D-ResNet) to learn the distribution features of transient disturbances in the temporal profile (TP) formed by a target passing through a pixel is proposed. First, we converted the detection of the point target in the image into the detection of transient disturbance in the TP and established mathematical models of different TP types. Then, according to the established mathematical models of TP, we generated the simulation TP dataset to train the 1D-ResNet. In 1D-ResNet, the structure of CBR-1D (Conv1D, BatchNormalization, ReLU) was designed to extract the features of transient disturbance. As the transient disturbance is very weak, we used several skip connections to prevent the loss of features in the deep layers. After the backbone, two LBR (Linear, BatchNormalization, ReLU) modules were used for further feature extraction to classify TP and identify the locations of transient disturbances. A multitask weighted loss function to ensure training convergence was proposed. Sufficient experiments showed that this method effectively detects moving point targets with a low SNR and has the highest detection rate and the lowest false alarm rate compared to other benchmark methods. Our method also has the best detection efficiency.

Keywords: moving point target; low SNR; transient disturbance; temporal profile; skip connection

Citation: Gao, W.; Niu, W.; Wang, P.; Li, Y.; Ren, C.; Peng, X.; Yang, Z. Moving Point Target Detection Based on Temporal Transient Disturbance Learning in Low SNR. *Remote Sens.* **2023**, *15*, 2523. https://doi.org/10.3390/rs15102523

Academic Editor: Gwanggil Jeon

Received: 25 February 2023
Revised: 24 April 2023
Accepted: 7 May 2023
Published: 11 May 2023

1. Introduction

The detection of moving targets has important applications in security monitoring, military reconnaissance, and satellite detection [1–3]. In some scenarios, such as early warning against space debris [4] and small faint bodies in near-Earth space or against naval ships and fighters, optical remote sensing detection has the characteristics of long distance and large field of view [5]. In this condition, the fast-moving target is more like a point in the image. The point target does not have shape, size, texture, or other spatial information and may even be submerged in background and clutter, resulting in a very low space-time signal-to-noise ratio (SNR) of the target and making it difficult to detect. Therefore, the problem of moving target detection in optical remote sensing images at a long distance and under a large field of view can be transformed into the problem of moving point target detection under a low SNR, which is important for effective detection.

There are currently three detection methods based on the temporal and spatial features of moving point targets: spatial-based detection, temporal-based detection, and spatiotemporal-based detection.

1.1. Spatial-Based Detection Methods

Spatial-based detection mainly realizes detection by enhancing small targets and suppressing the background or by converting the detection problem into an optimization problem of separating sparse and low-rank matrices. For example, the top-hat algorithm first calculates an image to estimate the background and then subtracts the background from the original image to obtain small targets [6]. The max-mean filter and max-median filter suppress clutter by filtering in four directions and then subtracting the background to obtain candidate targets [7]. Local contrast measure (LCM) and its improved algorithms, such as MPCM, HWLCM, MLCM-LEF, and WVCLCM, use local contrast information to enhance the point target and suppress the background [8–12]. In contrast to the above-mentioned methods, IPI-based methods use the background non-local self-correlation property to transform the small target detection problem into an optimization problem of the recovery of low-rank and sparse matrices and use principal component pursuit to solve the problem [13–16]. Xia et al. considered both the global sparsity and local contrast of small targets and proposed a modified graph Laplacian model (MGLM) with local contrast and consistency constraints [17]. Because a point target with a low SNR lacks effective spatial information, the above methods cannot separate the target from the background.

In recent years, with the development of deep learning, point target detection algorithms based on convolutional neural networks have emerged endlessly, including ALCNet, GLFM, ISTDU, ISTNet, MLCL, and APANet [18–23]. The principles of these CNN-based methods are predominantly similar to those of traditional methods. Multilayer neural networks are used to enhance the point targets, suppress the background, and box the target position. Although the CNN-based method has improved the feature extraction ability of the target, it still cannot achieve excellent detection for low-SNR point targets lacking spatial information. In addition, the track of the target cannot be obtained by detecting a single image. In early warning systems, it is still necessary to detect image sequences. Because CNN-based detection methods take a long time to detect image sequences, they are inefficient.

1.2. Temporal-Based Detection Methods

Temporal-based detection refers to the detection of image sequences using the target's movement information in temporal terms, such as optical flow [24], temporal difference [25], dynamic background modeling (DBM) [26,27], and tracking before detection (TBD) [28]. Optical flow uses the correlation between adjacent frames in the image sequence and the changes of pixels over time to find the corresponding relationship between moving targets in the frames in order to calculate the motion information of moving targets. This method assumes that the brightness of the target is constant, that the motion between adjacent frames is derivable, and that the motion of adjacent pixels is similar. There are numerous constraints and few scenes that satisfy this assumption. In addition, the optical flow method is time-consuming and struggles to meet real-time requirements. The temporal difference method makes use of the gradual change of the background in the image sequence to directly identify differences in the adjacent frames. If there are moving targets in the sequence, this will lead to a large difference in the intensity of the adjacent frames. However, the temporal difference is sensitive to background noise and has a poor detection effect for point targets with a low SNR. The DBM models the background in the image sequence and determines whether the pixel belongs to the foreground or background according to the established model to segment the moving target. The detection performance of this method depends on the modeling accuracy. It is difficult to distinguish the moving point target from the background under a low SNR, and the target is easily misjudged as background. Thus, this method's robustness is poor. TBD is a commonly used algorithm for detecting the traces of small moving targets. This algorithm accumulates multiple frames, searches for every possible trace of targets, and finally decides on the searched trace. Therefore, it does not need to detect every single image, but it directly outputs the target's motion trace. However, this method requires excessive time to search. Moreover, if the target is weak, the target cannot be found effectively.

1.3. Spatiotemporal-Based Detection Methods

Researchers have proposed spatiotemporal-based detection methods that combine spatial and temporal information. For example, Zhang et al. proposed a three-dimensional filtering detection method, which takes a segment of an image sequence as the input and uses multiple matching filters to suppress the background in order to ultimately obtain point targets [29,30]. Deng et al. proposed a filtering method based on spatiotemporal local contrast, which calculates spatial and temporal local contrast, respectively, and then performs filtering mapping on spatiotemporal local contrast to obtain detection results [31]. Lin et al. used the Pearson correlation coefficient to suppress the background in the time-domain window and then used the target detection algorithm based on the regional gray level to suppress the residual background and finally obtained a target motion track [32]. Zhu et al. filtered the frame first, then detected the frame's gradient to obtain the candidate targets, and finally supplemented local contrast information in temporal terms for spatiotemporal joint judgment. Yan et al. used the top-hat algorithm to separate small targets from the background, a grid-based density peak search algorithm and gray area growth algorithm to identify false alarm points, and an improved KCF algorithm to achieve target tracking for continuous frames [33]. These algorithms use the spatiotemporal information of point targets to improve the detection effect, but their assumptions on small targets are too strong and require considerable prior information.

1.4. TP-Based Detection Methods

These temporal-based or spatiotemporal-based methods only use a few frames and do not fully use the temporal information of the target, and so they do not exhibit good detection performance. Under the observation condition of staring imaging, the intensity change of a single pixel in the image sequence over time can be regarded as a profile. If a target passes a pixel, it will produce a transient disturbance in the temporal profile (TP) of that pixel. If the transient disturbance can be detected, the target will be detected. Thus, the point target detection in an image can be converted into the detection of transient disturbances in the TP. Methods based on TP have been proposed. Liu et al. estimated the background signal from the original TP and then subtracted it to obtain the target signal [34,35]. Subsequently, Liu et al. performed the nonlinear adaptive filtering of TP to extract the target signal [36]. Recently, Liu et al. used FFT and KL to calculate the similarity between the TP and waveform to detect the target signal [37]. Niu et al. proposed detection methods based on statistical distribution distance involving high-frame-rate detection [38–40]. These methods are effective for TPs with a high SNR, but for TPs with a low SNR, the target signal cannot be separated from the background signal, and the time when the target appears in the TP cannot be identified.

The transient disturbance of the target formed by the pixels can be regarded as a pattern that can be recognized by CNN-1D. Therefore, to overcome the problems of the previous methods and achieve effective moving point target detection under a low SNR, we proposed a detection framework based on transient disturbance distribution feature learning. The framework takes the image sequence as the input and directly outputs the track of the point target.

The main contributions of our work are as follows:

1. We converted the point target detection in the image into the detection of transient disturbance in the TP formed by a pixel and propose a low-SNR point target detection framework based on transient disturbance distribution feature learning.
2. In the detection framework, we designed a 1D-ResNet for transient disturbance feature learning. The 1D-ResNet can learn the distribution features of the transient disturbance and realize the classification of the TP and the location of the transient disturbance. In 1D-ResNet, skip connections are used to prevent the loss of the target signal feature. To prevent gradient disappearance and gradient explosion, the structure of the CBR-1D was designed to extract the features of the weak transient disturbance. The specially designed weighted multitask loss function ensures

training convergence. In addition, we verified the effect of network depth on the detection performance of 1D-ResNet and trained two networks: 1D-ResNet-8 and 1D-ResNet-16. The two networks deal with detection speed priority and detection rate priority, respectively.

3. We formulated the TP formed by pixels and generated a simulation dataset according to the TP formula. By combining the simulation data and real-world data, a training and verification dataset satisfying the research of moving point target detection with a low SNR is generated. Compared to other spatial-based and temporal-based methods the proposed method exhibits the best performance in terms of its detection rate, false alarm rate, and computing efficiency. The biggest advantage of our method is that it exhibits excellent detection performance under extremely low SNRs.

The remainder of this paper is organized as follows. Section 2 analyzes the components of the TP and establishes mathematical models for each part. The mathematical expressions for the target TP, background TP, and clutter TP are presented in Section 2. Section 3 details the moving point target detection framework, including the network architectures, model training, and the entire detection process. Section 4 presents the experimental scheme and results. We designed experiments based on four aspects and compared our method with other benchmark methods on test sequences. Section 5 discusses our method in detail and compares it with other methods, followed by network ablation experiments and visualization studies. Section 6 presents the conclusions of this study.

2. Temporal Profile Analysis

2.1. The Components of the Temporal Profile

Under the condition of staring imaging, each pixel in the image will form a *TP*, which tracks the change in pixel intensity value over time. Each *TP* is different. What is most important is the transient disturbance formed by the target passing through the pixel. Therefore, all *TP*s can be divided into two categories: background *TP* and target *TP* [34]. The *TP* of any pixel under ideal clutter-free conditions can be described as follows:

$$\overline{TP_{i,j}(k)} = \begin{cases} t_{i,j}(k), \; k_1 < k < k_2 \\ b_{i,j}(k), \; others \end{cases} \quad (1)$$

where $t_{i,j}$ and $b_{i,j}$ represent the distribution of the target *TP* and background *TP*, respectively; *i* and *j* represent the row and column index of the pixel in the image, respectively; *k* represents time; and k_1 and k_2 are the times when the target enters and leaves the pixel, respectively. The *TP* formation process is illustrated in Figure 1.

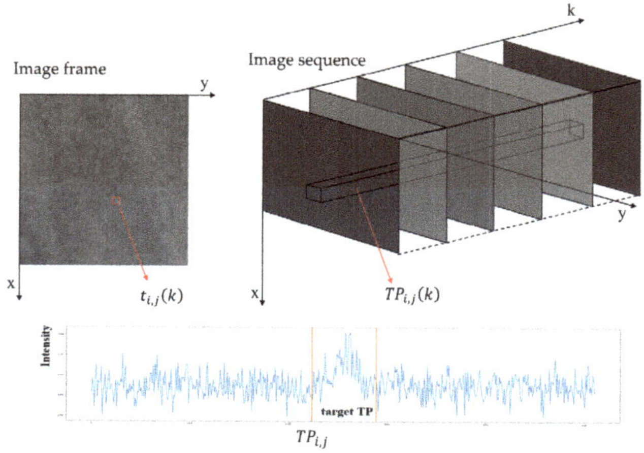

Figure 1. The formation process of *TP* in the image sequence; *x* and *y* are the horizontal and vertical coordinates of the image, respectively, and *k* is the frame number.

Under ideal clutter-free conditions, because the view of the detector is fixed, the background pixel intensity is constant for a short time, and the background *TP* can be considered a short-time stationary signal. However, in real image processing, the imaging results are affected by noise from different sources, including shot noise, thermal noise, photon noise, etc. In [35], additive white Gaussian noise (AWGN) was used to model these different noises. Thus, the actual *TP* can be expressed as follows:

$$TP_{i,j}(k) = AN + \overline{TP_{i,j}(k)} \tag{2}$$

where *AN* represents the AWGN.

2.2. The Target Temporal Profile

The TP of a target passing through a pixel can be regarded as a transient disturbance, and the following formula is used to describe the target *TP*:

$$t(k) = \begin{cases} s(k), & k_1 < k < k_2 \\ 0, & others \end{cases} \tag{3}$$

where $s(k)$ represents the transient disturbance caused by the appearance of the target.

The ideal imaging model of the optical system is pinhole imaging, and the light diffracts when mapping the object through the pinhole, forming a series of light–dark alternating diffraction rings. Therefore, a point in the real world will be a circle with a certain radius after imaging. Academia describes this phenomenon with a point spread function, and Pentland uses a two-dimensional Gaussian distribution to model it [41], which is defined as follows:

$$g(x,y) = Ae^{-a[(x-x_0)^2+(y-y_0)^2]} \tag{4}$$

where A represents the intensity of the target in the imaging, a represents the optical parameters of the detector, and (x_0, y_0) represents the center position of the target.

When the point target passes through a pixel, the intensity of the pixel first increases and then decreases, and a bell-shaped transient disturbance then appears on the *TP* of the pixel. The bell-shaped *TP* can be described by the following formula:

$$s(k) = Ae^{-a[v(k-k_0)]^2}, \quad k_1 < k < k_2 \tag{5}$$

where v is the moving speed of the target and $k_0 = k_1 + (k_2 - k_1)/2$ represents the time when the target center passes through the pixel.

The formation process and specific shape of target *TP* are shown in Figure 2.

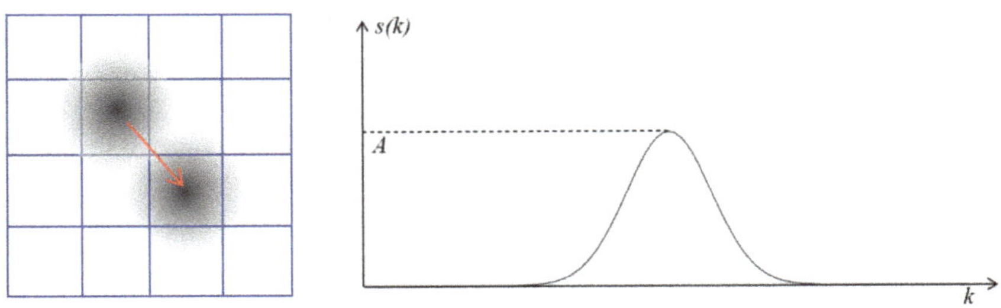

Figure 2. The formation process and the specific shape of target *TP*; *k* is the frame number and *s*(*k*) is the intensity value of the transient disturbance; *A* is the maximum intensity of the point target.

Because the size of the target is smaller than the imaging spatial resolution, the target cannot completely cover the background, and the intensity of the target in imaging will be affected by the background. Therefore, the formula of *TP* can be expressed as background distribution plus target distribution, as shown below:

$$TP_{i,j}(k) = n_{i,j}(k) + t_{i,j}(k) + b_{i,j}(k) \quad (6)$$

where $n_{i,j}(k)$ is the distribution of *AN*.

3. Detection Method

3.1. The Framework of Temporal Transient Disturbance Learning

We used CNN-1D to detect transient disturbances formed by the target. Because the transient disturbance is extremely weak, the feature extraction of the transient disturbance is difficult and the extracted features are easily lost in the network. The skip connection can directly transfer the shallow feature to deeper layers so that the network can fully learn the distribution feature of the transient disturbance and achieve high-accuracy detection. The detection framework of our method is shown in Figure 3, which includes two modules: training and detection. In the training part, we first generated the simulated *TP* by adding a bell-shaped signal to the background signal. Noise was then added to the *TP* to simulate a real situation. Next, a training dataset containing 160,000 *TP*s was generated under the experimental parameters. Subsequently, the proposed networks, 1D-ResNet-8 and 1D-ResNet-16, were trained under the same super-parameter settings. In the detection part, the trained model was used to detect the transient disturbance in *TP*s formed by pixels to detect the moving track of the point target.

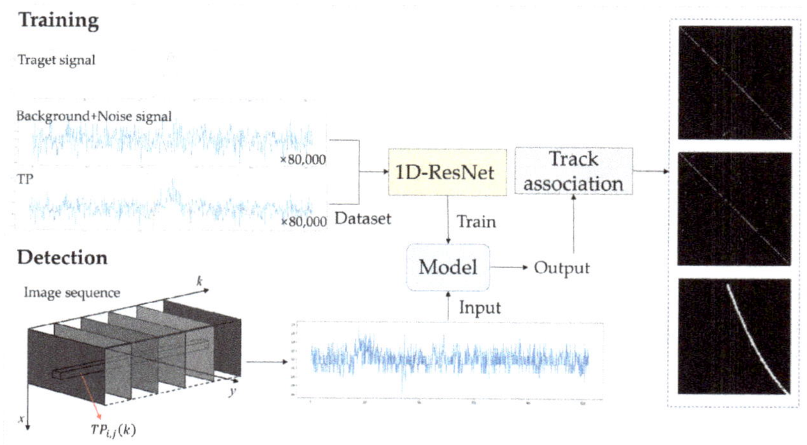

Figure 3. The detection framework of our method.

3.2. Architectures of 1D-ResNet

There are two tasks for detecting transient disturbances in a *TP*. One involves classifying the target *TP* containing a bell-shaped signal and the background *TP*. The other involves obtaining information on transient disturbances, such as the time of occurrence and the duration of the bell-shaped signal. Therefore, for these two detection tasks, inspired by classical ResNet and Darknet, we use one-dimensional ResNet as the backbone feature extraction network and CBR-1D as the basic feature extraction unit [42,43] to propose the 1D-ResNet. To verify the impact of the network layers on the detection performance, 1D-ResNet-8 and 1D-ResNet-16 were designed. The architectures of these 1D-ResNet are shown in Figure 4.

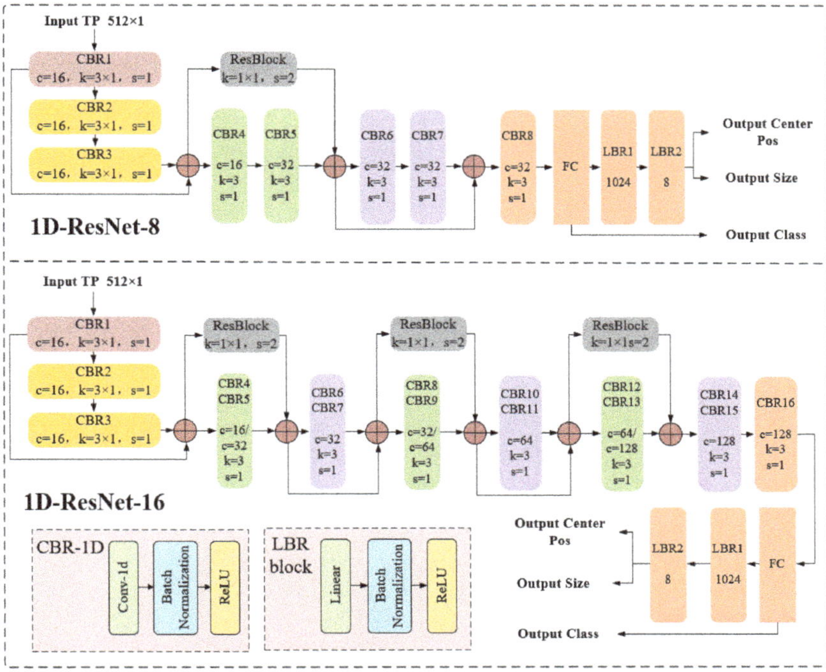

Figure 4. The architectures of 1D-ResNet.

Both networks are composed of input, backbone, neck, and output. A *TP* with a size of 512 × 1 was the input for the network. Backbone was used to extract the features of the *TP*. The neck connects the backbone and the output and to provide higher-dimensional features for the output. Finally, three outputs are obtained. If a bell-shaped signal exists, the outputs are the class of *TP*, the center position, and the size of the bell-shaped signal. Otherwise, we obtain three zero outputs.

In the training network stage, the *TP* class is easy to identify, as it is the first output. Meanwhile, identifying the center position and size is difficult. Therefore, two LBR blocks are set behind the convolution layer as the neck to further extract the features. Each LBR block includes a linear layer, batch normalization (BN), and ReLU.

Several skip connections were used to transmit the feature from the shallow layer to the deeper layer in order to avoid the loss of the transient disturbance feature. The CBR-1D includes a one-dimensional convolution layer (Conv1D), BN, and ReLU. Conv1D was used to extract local features in the *TP* and then normalize the extracted features. Finally, ReLU was used to activate the features. This can inhibit the change in the data distribution, accelerate the convergence speed, and avoid the problems of gradient disappearance and gradient explosion.

3.3. *Training the 1D-ResNet*

3.3.1. Generate the Dataset

Point targets with a SNR below 3 dB will have no obvious spatial features; therefore, the SNR research range was established as −3 dB to 3 dB. Because the actual *TP* under a specific SNR is difficult to obtain and label, the features of the target and ground *TP* were combined to generate a dataset through simulation. To enable the network to fully learn the features of *TP* within the research SNR range, *TP*s were generated between −4 dB and 4 dB.

During *TP* simulation, a bell-shaped target signal is generated according to Formula (5) and the location where the target signal appears is set randomly. To verify the effect of the target signal size on the detection performance, the target signal size range was set to 10~110 and signals of different sizes were generated in equal proportions. A constant was randomly set as the background signal. The two signals were superimposed to obtain the simulated *TP*. Finally, AWGN was added to the *TP* simulation. To ensure that the model exhibits good performance on *TP*s with different SNRs and different target signal sizes, we set the number of TPs to be equal for each SNR and size range.

After generating the dataset, we divided it into training and validation sets at a ratio of 8:2, respectively. The composition of the *TP*s in the dataset is shown in Figure 5. The left figure shows the distribution of *TP*s under different SNRs and the right figure shows the distribution of *TP*s with different sizes under the same SNR.

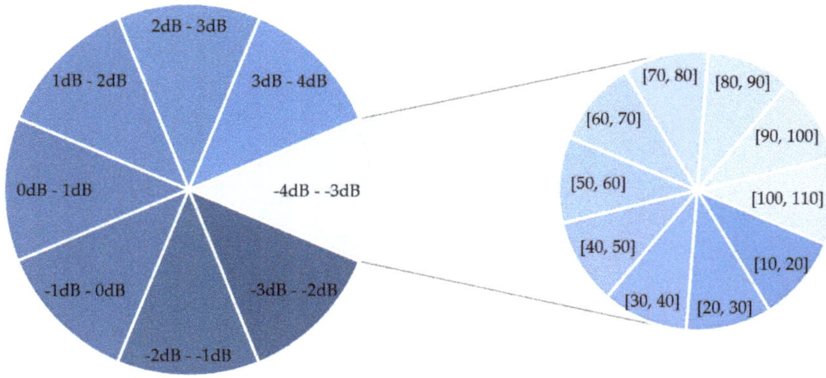

Figure 5. The composition of *TP*s in the dataset.

3.3.2. Loss Function

As the network trained in this study is a multitask learning network, the loss function is composed of three parts: classification loss, center position loss, and size loss. The classification loss uses binary cross-entropy loss, and the center position loss and size loss use the mean square error loss. Because there are significant differences in the order of magnitude of these three parts of the loss function, it is necessary to manually set their weights to prevent imbalance loss, and the final weighted loss function is shown in Formula (7).

$$Loss = w_1 * Loss_C + w_2 * Loss_P + w_3 * Loss_s \tag{7}$$

where $Loss_C$ represents the classification loss, $Loss_P$ represents the center position loss, and $Loss_s$ represents size loss. The formulas for these three parts are as follows:

$$\&Loss_C = -\frac{1}{N}\sum_{i=1}^{N}[C_i \log(\widehat{C}_i) + (1 - C_i)\log(1 - \widehat{C}_i)] \tag{8}$$

$$Loss_P = -\frac{1}{2N}\sum_{i=1}^{N}\left(P_i - \widehat{P}_i\right)^2 \tag{9}$$

$$Loss_W = -\frac{1}{2N}\sum_{i=1}^{N}\left(W_i - \widehat{W_i}\right)^2 \tag{10}$$

where C_i represents the category label, $\widehat{C}_i$ represents the predicted category, P_i represents the center point position label, $\widehat{P}_i$ represents the predicted center position, W_i represents the size label, and $\widehat{W_i}$ represents the predicted size label. N represents the number of TPs in a batch.

3.3.3. Training the Networks

PyTorch was used to build the network architectures and the training environment. The training equipment used was a workstation with an NVIDIA GeForce GTX 1080 ti GPU and 32 GB of memory.

During training, the random seed was set to 3407, the Adam optimizer was used, the parameter penalty coefficient was set to 1×10^{-5}, the learning rate was initially set to 1×10^{-4}, and the batch size was set to 2000. During training, rough training was first conducted for 10 epochs, then the learning rate was reduced 10-fold and fine-tuning was performed. If the loss of the validation set did not decrease within 10 epochs, the training ended. The loss optimization of network training is shown in Figure 6.

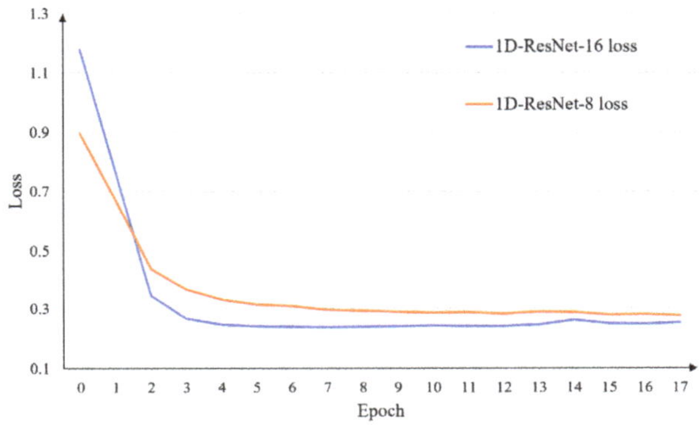

Figure 6. The loss optimization of network training. Epoch represents the number of iterations in the two networks' training.

Figure 6 shows that the two networks converged after five epochs of training, and the training effect of 1D-ResNet-16 was slightly better than that of 1D-ResNet-8.

3.4. The Moving Point Target Trajectory Detection Process

The detection process of our proposed framework is as follows:

1. Input an image sequence and obtain its *TP* for each pixel.
2. Pre-process the *TP*s, standardize the *TP*s, and divide the *TP* segments according to the network input size.
3. Load the trained model, input the *TP*s into 1D-ResNet in batches, and obtain the outputs.
4. Determine whether the *TP*s exist in the transient disturbance caused by the target according to the specified threshold value. If a *TP* exists, its pixel is considered to be in the foreground; otherwise, it is considered to be in the background.
5. Unify all foreground pixels and output the motion track of the target.

4. Experiments and Analysis

To evaluate the feasibility and performance of the proposed method, extensive experiments were conducted, including a *TP* simulation experiment, image-sequence simulation experiment, real-world experiment, and comparison experiment.

- The *TP* simulation experiment directly detects the simulated *TP* and evaluates the classification and positioning performance of the method under ideal conditions using the accuracy of the receiver operating characteristic (ROC) and intersection over union (IOU).
- To further fit the real scene and test the performance of the detection framework, we established image-sequence simulation experiments. A simulated moving point target was added to the real background image sequence. Simulation sequences were used as the input data of the detection framework.

- We shot the movement process of the point target outdoors and conducted a real-world experiment based on these data.
- To verify the performance of the proposed method, we compared it with that of other benchmark methods.

4.1. Details of Image Sequences in Experiments

In the experiments, we used seven image sequences, three of which were simulated. The other four were real-world data taken outdoors. The details of the image sequences used in the experiments are listed in Table 1.

Table 1. The details of image sequences.

Sequences	Resolution	Scenes	Speed (Pixels/Frame)	Frames	SNR (dB)
Sequence 1	128 × 128	Asphalt Road	0.0125	10,240	1.22
Sequence 2	128 × 128	Pure Sky	0.0125	10,240	0.6
Sequence 3	128 × 128	Complex Scene	0.0625	2048	1.09
Sequence 4	100 × 100	Asphalt Road	0.0122	8192	1–5
Sequence 5	100 × 100	Asphalt Road	0.0244	4096	1–5
Sequence 6	100 × 100	Asphalt Road	0.0488	2048	1–5
Sequence 7	100 × 100	Asphalt Road	0.0977	1024	1–5

In the image-sequence simulation experiment, we used asphalt roads, pure sky, and a complex scene to simulate space-based and ground-based detection. The backgrounds of sequence 1 and sequence 2 are simple, while the background of sequence 3 is more complex and has scenes such as sky, mountains, buildings, etc., in the background. In sequences 1–3, we added a point moving target that was 1–3 pixels in size to these background image sequences. This point target moves from the upper-left corner to the lower-right corner of the image sequence.

To verify the performance of the proposed method in a real image sequence, we used a high-speed camera to capture outdoor image sequences. We tracked the movement of a glass ball from a height of approximately 50 m at 20,000 fps. The diameter of the glass ball was 1.5 cm, and the SNR was approximately 1–5 dB. To facilitate the experimental analysis, we obtained 8192 frames from the original sequence and established a window of 100 × 100 pixels for the target to pass through.

After obtaining the original sequence 4, to verify the impact of the target's stay time on the detection effect on a single pixel, we down-sampled sequence 4 to obtain sequences 5–7.

4.2. TP Simulation Experiment

The experiments in this section were conducted in two ways to verify the detection effect of our method on *TP*s with different SNRs and target signals of different sizes. ROC and IOU accuracy rates were used to evaluate the classification and positioning capabilities of the method, respectively.

The ROC curve is a graphical representation of the performance of a binary classification model as the discrimination threshold is varied. The x-axis represents the false positive rate (FPR), which is the ratio of false positives (incorrectly classified negative samples) to the total number of negative samples. The y-axis represents the true positive rate (TPR), which is the ratio of true positives (correctly classified positive samples) to the total number of positive samples. In this paper, positive samples refer to *TP*s containing the target signal, while negative samples refer to TPs without the target signal. Each point on the ROC curve reflects the sensitivity of the classifier to different discrimination thresholds. The larger the area under the curve (AUC) covered under the ROC curve, the better the detection performance of the method.

The center position and size of the transient disturbance form the bounding box. If the IOU is greater than 0.5, the positioning is considered correct. The calculation method of the

IOU of the predicted and true bounding boxes is shown in Equation (11). The higher the accuracy of the IOU, the better the positioning performance of the method.

$$IOU = \frac{\min(E_T, E_P) - \max(S_T, S_P)}{\max(E_T, E_P) - \min(S_T, S_P)} \quad (11)$$

where E_T and E_P represent the right boundary of the true bounding box and predicted bounding box, respectively, and S_T and S_P represent the left boundary of the true bounding box and predicted bounding box, respectively.

4.2.1. The Detection Performance under Difference SNR

To verify the influence of SNR on detection performance, simulation *TPs* under different SNRs were generated. Under each SNR, the size of the target signal is set between 10 and 110 in equal proportion. The ROC curves drawn using the two networks under different SNRs are shown in Figure 7, and the AUC and accuracy of the IOU are shown in Table 2.

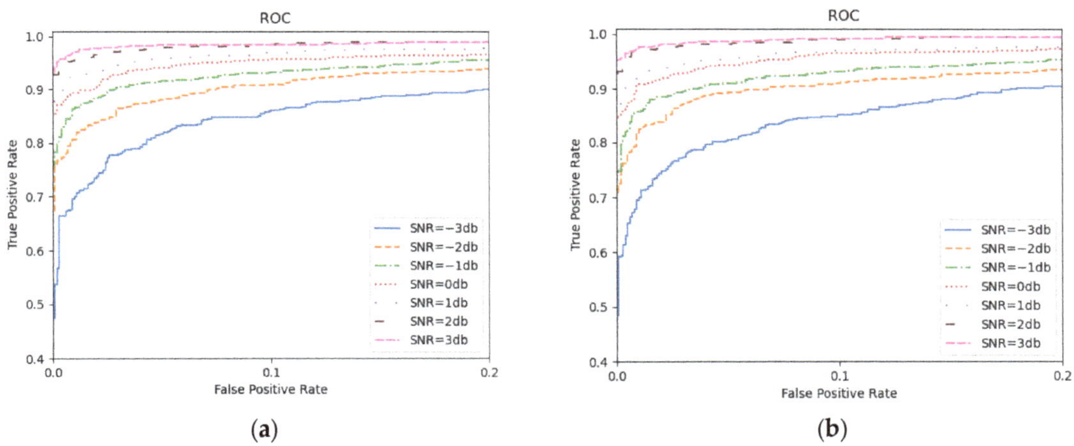

Figure 7. The detailed ROC of two networks under different SNRs. (**a**) ROC of 1D-ResNet-8. (**b**) ROC of 1D-ResNet-16.

Table 2. AUC and accuracy of IOU under different SNRs.

SNR	AUC		Accuracy of IOU	
	1D-ResNet-8	**1D-ResNet-16**	**1D-ResNet-8**	**1D-ResNet-16**
−3 dB	0.9393	0.9389	0.5350	0.6460
−2 dB	0.9591	0.9593	0.6130	0.7000
−1 dB	0.9724	0.9719	0.6840	0.7320
0 dB	0.9807	0.9832	0.7350	0.7770
1 dB	0.9867	0.9859	0.7340	0.7670
2 dB	0.9941	0.9957	0.7730	0.7810
3 dB	0.9955	0.9963	0.7880	0.8020

As is shown in Figure 7 and Table 2, the classification performance of the two models reached a good level, and all ROCs covered over 90% of the area. With a decrease in the SNR, the classification performance worsens. The accuracy of the IOU also decreases with a decrease in the SNR. This is because transient disturbances under low SNR are very weak and can easily be submerged in the background. During the detection process, the target signal is prone to clutter interference, resulting in classification and positioning errors.

4.2.2. The Detection Performance under Different Target Signal Sizes

To verify the influence of target signal size on detection performance, in the experiment simulated TPs with different sizes were generated, in which the SNR was set at an equal ratio of −3 dB to 3 dB under each size. The ROC drawn by the two networks under different target signal sizes are shown in Figure 8, and the AUC and accuracy of IOU are shown in Table 3.

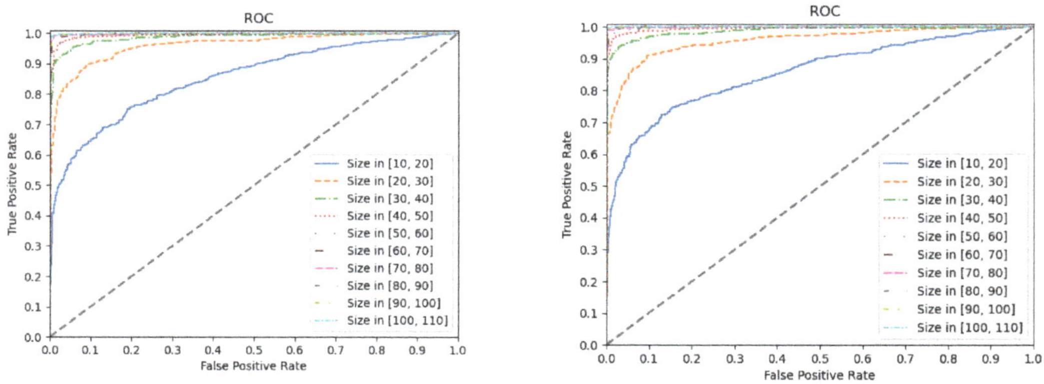

(**a**) (**b**)

Figure 8. The ROC of two networks under different target signal sizes. (**a**) ROC of 1D-ResNet-8 (**b**) ROC of 1D-ResNet-16.

Table 3. AUC and accuracy of IOU under different target signal sizes.

Size	AUC		Accuracy of IOU	
	1D-ResNet-8	1D-ResNet-16	1D-ResNet-8	1D-ResNet-16
[10, 20]	0.8541	0.8574	0.0614	0.1214
[20, 30]	0.9614	0.9575	0.2486	0.3143
[30, 40]	0.9873	0.9882	0.4371	0.4914
[40, 50]	0.9937	0.9951	0.6586	0.6643
[50, 60]	0.9966	0.9979	0.7543	0.7671
[60, 70]	0.9986	0.9993	0.8314	0.8771
[70, 80]	0.9988	0.9991	0.9129	0.9343
[80, 90]	0.9980	0.9993	0.9529	0.9529
[90, 100]	0.9991	0.9997	0.9614	0.9771
[100, 110]	0.9997	0.9999	0.9914	0.9871

As is shown in Figure 8 and Table 3, with an increase in the target signal size, the classification and positioning capabilities of the two models show a significant improvement trend. For classification tasks, when the target signal size was less than 20, the classification performance was very poor, whereas when the target signal size increased to 40, the AUC of both models reached over 99%.

For positioning tasks, the IOU accuracy exhibited a more obvious trend with an increase in the target signal size. When the size was increased to 70, the accuracy increased to over 90%.

From the experimental results, we can see that the size of the target signal is a crucial factor for our methods. The longer the moving target stays on a single pixel, the more sufficient are the motion features and the better the performance of the proposed method Therefore, the detection performance can be improved by increasing the frame rate of the detector.

4.2.3. *TP* Simulation Experiment Analysis

The SNR and size of the target signal are important factors that affect detection performance. The higher the SNR and the larger the proportion of the target signal, the better the model detection performance. The proportion of the target signal has a greater impact on the detection effect than the SNR. The SNR cannot be significantly improved; however, the proportion of the target signal can be further improved by increasing the frame rate of the detection equipment.

Among the two networks, although 1D-ResNet-16 has an additional eight layers of CBR-1D and 228,160 parameters compared to 1D-ResNet-8, the improvement of the model's detection performance is very small, the classification performance gap is small, and the IOU accuracy rate is less than two percentage points higher than that of 1D-ResNet-8. This proves that for weak transients, deeper network layers do not lead to greater performance improvement; however, deeper networks lead to greater computing consumption, which is contradictory to real-time detection performance.

4.3. *Image-Sequence Simulation Experiment*

We used our method to detect sequences 1–3. The detection results for the two networks are presented in Figure 9 and Table 4.

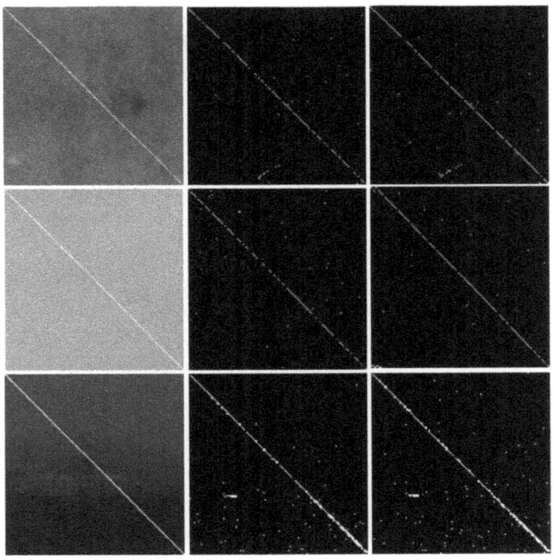

Figure 9. The detection results of the two networks in simulated image sequences. (**a**) The ground truth of the image sequences. (**b**) The detection results of 1D-ResNet-8. (**c**) The detection results of 1D-ResNet-16.

Table 4. The detection results of the simulated image sequences.

Sequences	Detection Rate		False Alarm Rate		Efficiency (ms/Frame)	
	1D-ResNet-8	1D-ResNet-16	1D-ResNet-8	1D-ResNet-16	1D-ResNet-8	1D-ResNet-16
Sequence 1	72.67%	88.28%	0.15%	0.17%	1.77	2.97
Sequence 2	69.53%	87.50%	0.16%	0.11%	1.83	2.96
Sequence 3	78.12%	79.69%	0.58%	0.51%	3.01	4.19

From Figure 9 and Table 4, we can observe that both networks show good detection performance for all three sequences. Although there were some false alarm points, the moving track (main diagonal) of the target was clear. Additionally, these false alarm points can be removed through post-processing.

Compared to 1D-ResNet-8, the detection rate of 1D-ResNet-16 is higher, but the time consumption of 1D-ResNet-8 is lower. In an actual detection task, we should use 1D-ResNet-16 if the detection rate is more important. However, if the detection speed is more important, 1D-ResNet-8 should be used.

4.4. Real-World Experiment

We used our method to detect real-world sequences. The detection results are shown in Figure 10 and Table 5.

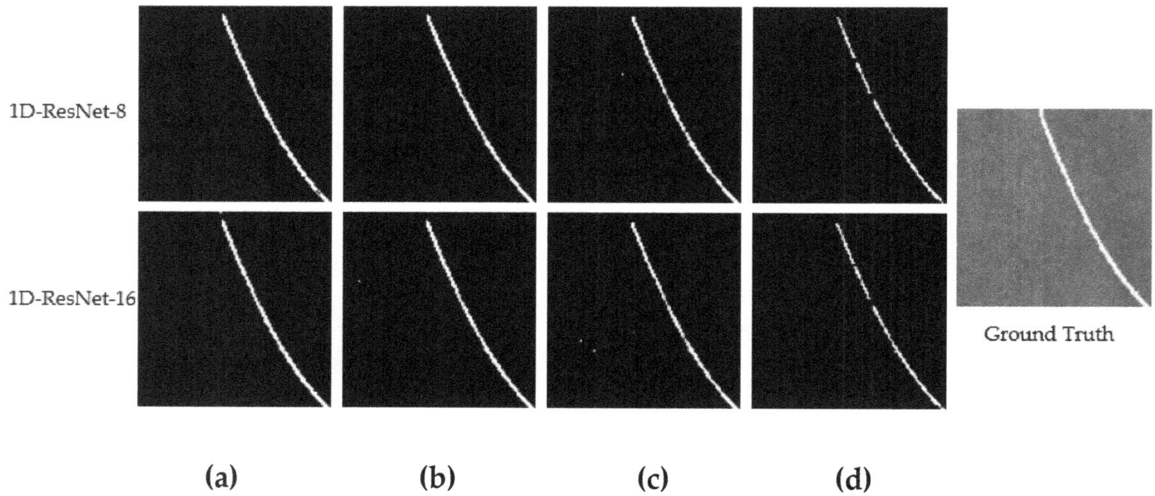

Figure 10. The detection results of the two networks on real-data sequences. (**a**) The detection results of the two networks on sequence 4. (**b**) The detection results of the two networks on sequence 5. (**c**) The detection results of the two networks on sequence 6. (**d**) The detection results of the two networks on sequence 7.

Table 5. The detection results on real data sequences.

Sequences	Detection Rate		False Alarm Rate		Efficiency (ms/Frame)	
	1D-ResNet-8	1D-ResNet-16	1D-ResNet-8	1D-ResNet-16	1D-ResNet-8	1D-ResNet-16
Sequence 4	96.00%	97.00%	0.00%	0.00%	1.63	2.27
Sequence 5	96.00%	96.00%	0.00%	0.01%	1.79	2.58
Sequence 6	95.00%	95.00%	0.01%	0.02%	2.55	3.15
Sequence 7	90.00%	91.00%	0.00%	0.00%	3.62	4.49

The results show that both networks have relatively good detection performance on the sequences, both of which completely detect the moving track of the glass ball. With an increase in the de-sampling fold, the stay frames of the target in a single pixel become shorter and the detection performance worsens. In this experiment, 1D-ResNet-16 had no significant advantage over 1D-ResNet-8 in terms of detection performance. Therefore, 1D-ResNet-8 can meet the detection requirements when the SNR is high.

4.5. Contrast Experiments with the Benchmark Methods

To verify the performance of the proposed method, we compared it with some benchmark methods, including MaxMean [7], IPI [13], LCM [8], Kernel [38], ICLSP [35], NAF [36], and TRLCM [44]. MaxMean, IPI, and LCM are spatial-based methods, whereas Kernel, ICLSP, NAF, and TRLCM are temporal-based methods.

Sequences 1–4 were used for comparison. The results are presented in Figure 11 and Tables 6 and 7.

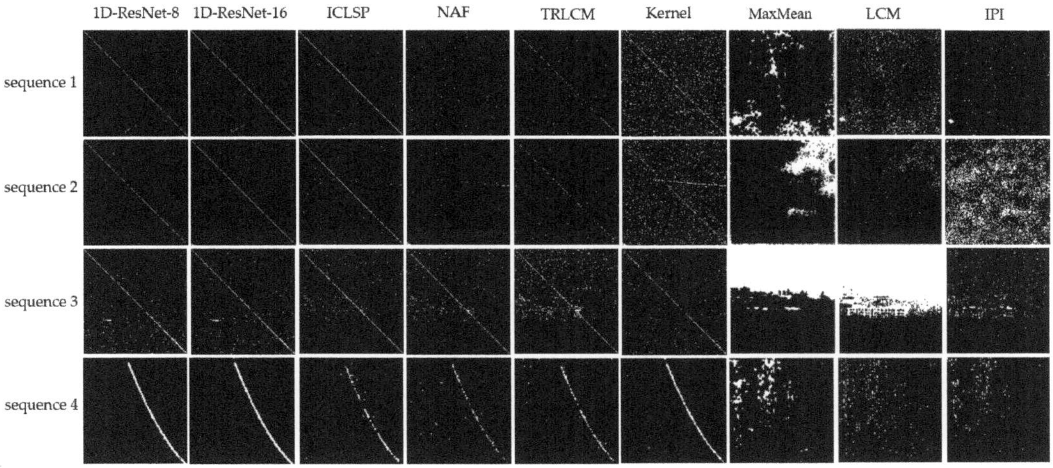

Figure 11. The detection results of the proposed methods and benchmark methods.

Table 6. The detection rates of the proposed methods and the benchmark methods.

Method	Detection Rate			
	Sequence 1	Sequence 2	Sequence 3	Sequence 4
1D-ResNet-16	**88.28%**	**87.50%**	**79.69%**	**97.00%**
1D-ResNet-8	72.67%	69.53%	78.13%	96.00%
ICLSP	88.28%	82.81%	78.13%	68.00%
NAF	1.56%	0.78%	73.43%	52.00%
TRLCM	35.16%	29.69%	71.88%	68.00%
Kernel	67.97%	60.16%	66.41%	95.00%
MaxMean	10.16%	1.56%	46.09%	8.00%
LCM	6.25%	0.78%	28.12%	9.00%
IPI	1.56%	23.43%	4.69%	3.00%

Table 7. The false alarm rates of the proposed methods and the benchmark methods.

Method	False Alarm Rate			
	Sequence 1	Sequence 2	Sequence 3	Sequence 4
1D-ResNet-16	0.15%	0.16%	**0.51%**	**0.00%**
1D-ResNet-8	0.17%	**0.11%**	0.58%	**0.00%**
ICLSP	**0.14%**	0.19%	0.54%	0.13%
NAF	0.29%	0.23%	0.61%	0.15%
TRLCM	0.20%	0.18%	2.22%	0.14%
Kernel	5.38%	4.71%	0.61%	0.02%
MaxMean	8.87%	13.81%	43.21%	4.80%
LCM	5.73%	1.29%	33.03%	2.13%
IPI	0.70%	26.00%	1.51%	1.54%

Figure 11 shows that the temporal-based methods can detect low-SNR point targets in an image sequence, whereas the spatial-based methods cannot detect the target track.

Among the temporal-based methods, our method has the best performance, followed by ICLSP. Although ICLSP exhibits similar performance to our method on simulation sequences, its detection effect is far inferior to that of our method on real-world low-SNR sequences. The Kernel method can better detect a real sequence with a high SNR, but there are many false alarm points for the simulation sequence with a low SNR. This shows that our method not only has excellent detection ability for moving point targets with a low SNR but also has good robustness for real point targets.

Table 8 shows the computational efficiency of all methods. These methods are implemented on a computer with an AMD Ryzen 7 1700 CPU and a Nvidia GeForce GTX 1080 ti GPU. From Table 8, it can be seen that our method has the fastest detection speed. The detection speed of 1D-ResNet-8 is faster than that of 1D-ResNet-16, as 1D-ResNet-8 has fewer parameters. In the future, we will improve the network by proposing lightweight networks to further improve the detection speed.

Table 8. The computing efficiency of all methods.

Method	Environment	Computing Efficiency (ms/Frame)			
		Sequence 1	Sequence 2	Sequence 3	Sequence 4
1D-ResNet-16	python3.9+cuda11.7	**2.97**	**2.96**	**4.19**	**2.27**
1D-ResNet-8	python3.9+cuda11.7	**1.77**	**1.83**	**3.01**	**1.63**
ICLSP	python3.9	30.84	32.35	29.82	18.58
NAF	python3.9	1194.15	1203.66	1060.80	761.13
TRLCM	python3.9	580.75	528.75	478.37	355.44
Kernel	python3.9	1586.91	1287.25	3589.67	1481.46
MaxMean	matlab2018	246.33	244.61	226.59	147.11
LCM	matlab2018	429.53	428.21	418.99	258.24
IPI	matlab2018	612.23	528.25	760.69	237.29

5. Discussion

In this section, we discuss our method in detail and compare it with other methods to illustrate its advantages and disadvantages. After that, we discuss the results of our ablation experiments to verify the effects of various parts of 1D-ResNet. Finally, we discuss the results of our visualization research on the network to verify whether it learned the features of transient disturbances.

5.1. Analysis of All Methods

In this section, we analyze the characteristics, advantages, and disadvantages of all methods, as shown in Table 9.

Table 9. The characteristics, advantages, and disadvantages of all methods.

Method	Characteristics	Advantages	Disadvantages
1D-ResNet-16	Batch detection of transient disturbances in *TP* using 1D-ResNet-16 on GPU	Best detection ability and fast detection speed for low-SNR point targets; Few hyperparameters	The detection speed is slower than that of 1D-ResNet-8
1D-ResNet-8	Batch detection of transient disturbances in *TP* using 1D-ResNet-8 on GPU	Good detection ability and fastest detection speed for low-SNR point targets; Few hyperparameters	The detection performance is slightly worse than that of 1D-ResNet-16
ICLSP	Calculate the deviation distribution between *TP* and CLSP on the CPU to detect the target *TP*	Good detection ability for low-SNR point targets	Poor detection performance for real data; Slow detection speed; More hyperparameters
NAF	Using a nonlinear filter to extract the target *TP* on CPU	Moving point target with a higher SNR can be detected	Very slow detection speed; Unable to detect low-SNR targets; More hyperparameters
TRLCM	Using temporal local contrast information to detect target *TP* on CPU	Can detect low-SNR point targets	Very slow detection speed; Poor detection performance for low-SNR targets; More hyperparameters

Table 9. *Cont.*

Method	Characteristics	Advantages	Disadvantages
Kernel	Calculate the statistical distribution distance between the target and background to detect target *TP* on CPU	Can detect low-SNR point targets; Few hyperparameters	Very slow detection speed; Poor detection performance for low-SNR targets
MaxMean	Detect each image based on local maximum mean on CPU	Simple detection theory; Few hyperparameters	Very slow detection speed; Completely unable to detect low-SNR targets
LCM	Detect each image based on local contrast on CPU	Simple detection theory; Few hyperparameters	Very slow detection speed; Completely unable to detect low-SNR targets
IPI	Based on the non-local autocorrelation characteristics of the background, transform the small target detection into an optimization problem of recovering low-rank and sparse matrices and use stable principal component pursuit to solve this problem on CPU	Simple detection theory; Few hyperparameters	Very slow detection speed; Completely unable to detect low-SNR targets

The two networks we propose have the best detection performance and fastest detection speed for low-SNR moving point targets. Of the two networks, 1D-ResNet-16 has the best detection performance, while 1D-ResNet-8 has the fastest detection speed.

Other *TP*-based detection methods (ICLSP, NAF, TRLCM, and Kernel) can also detect the motion trajectory of targets, but their detection rate and false alarm rate are not as good as those of our methods, and these methods require more time for detection.

Other spatial-based methods (MaxMean, LCM, and IPI) are completely unable to detect point targets under a low SNR.

5.2. Ablation Experiments

In this section, we conducted ablation experiments to verify the superiority of the 1D-ResNet and CBR-1D. Due to the similarity of the two network structures (1D-ResNet-16 only has eight more layers than 1D-ResNet-8), this section is based on 1D-ResNet-8 only.

5.2.1. Network Structure Study

We first removed the skip connections from the network and then replaced the basic structural unit CBR with Conv-1D (Conv1D and ReLU) and CBL (Conv1D, BN, and LeakyReLU). We then removed the LBR module from the network. The ablation experiment we designed is shown in Table 10.

Table 10. The networks of ablation experiments.

Network	Skip Connection	Basic Structural Unit	LBR
CNN-1D	✗	Conv-1D	✗
ResNet-1D	✓	Conv-1D	✗
ResNet-1D LBR	✓	Conv-1D	✓
ResNet-1D CBL+LBR	✓	CBL-1D	✓
Ours	✓	CBR-1D	✓

We did not weigh the loss function to see the performance of these networks. The loss optimization of all networks is shown in Figure 12.

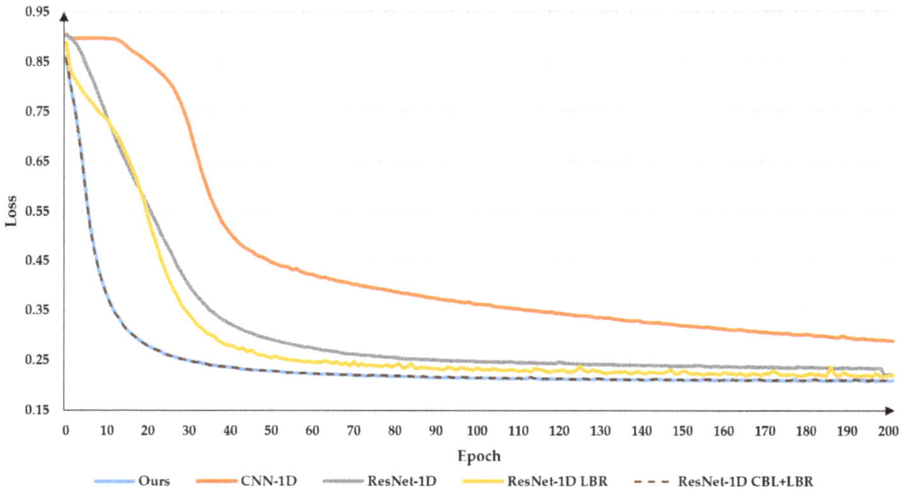

Figure 12. The loss optimization of all networks in training.

From Figure 12, we can see that the performance of the CNN-1D network is the worst, but after adding skip connections, the performance of ResNet-1D is significantly improved. This indicates that skip connections are very helpful for optimizing network loss. After adding the LBR module, the loss of ResNet-1D LBR further decreased. The addition of BN to the basic structural unit accelerates the convergence speed of the network. However, we can also see that replacing the activation function (ReLU or LeakyReLU) does not affect the network optimization.

Next, we use these networks to test the TP and verify its detection performance. The experimental data are *TP* with SNR = 0 dB and target signal size = 80. The experimental results are shown in Figure 13 and Table 11.

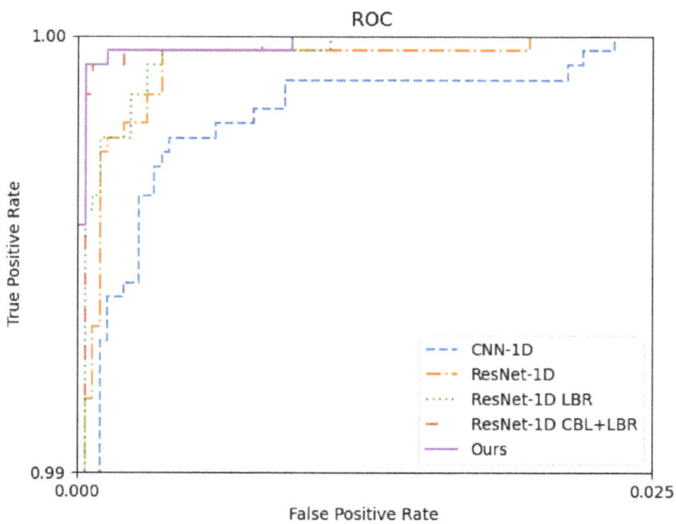

Figure 13. The detailed ROC of different networks.

Table 11. The AUC and accuracy of IOU of different networks.

Network	AUC	Accuracy of IOU
CNN-1D	$0.9999 + 0.3956 \times 10^{-4}$	64.58%
ResNet-1D	$0.9999 + 0.7801 \times 10^{-4}$	70.64%
ResNet-1D LBR	$0.9999 + 0.8678 \times 10^{-4}$	87.14%
ResNet-1D CBL+LBR	$0.9999 + 0.9389 \times 10^{-4}$	82.51%
Ours	$0.9999 + 0.9522 \times 10^{-4}$	86.97%

From the experimental results, it can be seen that all networks have good classification ability, but our network has the highest AUC. The positioning ability of networks without skip connections and LBR modules is poor. After adding skip connections, the network positioning ability is improved but not by very much. The addition of the LBR module greatly improves the positioning performance of the network. This indicates that skip connections can transfer the transient disturbance features extracted from shallow layers to deeper layers, preventing feature loss. Additionally, the LBR module can extract higher dimensional features, which helps to better locate transient disturbances. ResNet-1D LBR with no BN in its basic structural unit has the best positioning performance, but it is only 0.17% higher than that of our network. Adding BN will not affect the performance of the network in theory, but it can accelerate the convergence speed of the network.

5.2.2. Network Visualization

In this section, we conduct visualization research on the network to verify whether it has learned the distribution features of transient disturbances. Grad-CAM [45] (Gradient-weighted Class Activation Mapping) was used to visualize the network in order to verify whether the network has learned the features of the *TP*. The intensity of the target signal was set to 3, the size was set to 60, and its SNR was controlled at 3 dB. The chosen visualization layers were CBR5, CBR9, CBR13, and CBR16. The visualization results are shown in Figure 14, where the blue line is the original *TP* and the orange line is the heatmap calculated using Grad-CAM. The larger the value of the heatmap, the more interested the network is.

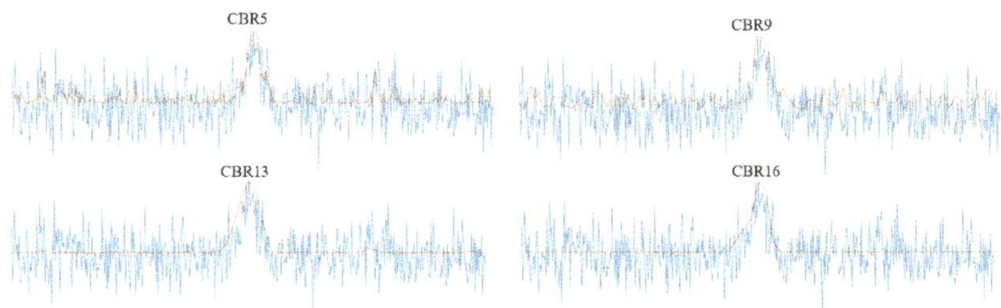

Figure 14. The heatmap of 1D-ResNet-16.

Figure 14 shows that the heatmap has the highest value at the target signal, proving that the network has fully learned the distribution features of the transient disturbance; however, the heatmap of the shallow layers also contains a lot of clutter. With an increase in the network depth, the clutter gradually decreases and the network learns more features of the transient disturbance. Therefore, the Grad-CAM visualization of the network shows that the network proposed in this study has interpretability.

6. Conclusions

To resolve the problem of moving point target detection at a low SNR, we converted the problem of point target detection into the problem of transient disturbance detection in the *TP* formed by each pixel. For the transient disturbance detection problem, we propose a detection framework to learn the distribution features of the transient disturbances. In this framework, we first formulated different types of *TP* and generated a training dataset. Then, two networks, 1D-ResNet-8 and 1D-ResNet-16, were designed, which can adapt to the situation of detection speed priority and detection rate priority. Of the two networks, 1D-ResNet-16 has better detection performance than 1D-ResNet-8, but it requires more time. For detection tasks with high real-time requirements, 1D-ResNet-8 is a better choice. Adequate experiments showed that our *TP* model is correct and that our method is effective. Compared to other benchmark methods, the proposed method has obvious advantages when it comes to improving the detection rate and reducing the false alarm rate at a low SNR. Our method also has the fastest detection speed. In addition, we conducted ablation experiments to verify the superiority of our network and the CBR-1D structure, and the experimental results showed that all the modules of our proposed network were necessary. Network visualization research proved that our network learned the features of transient disturbances well.

Moreover, we studied the factors that affect detection performance and found that the size of the target signal had a greater impact on the detection results than the SNR of the *TP*. The detection performance of our method can be improved by increasing the sampling frame rate of the camera.

The method proposed in this study has the potential to be deployed in space-based or ground-based intelligent detection equipment. In the future, we will continue to study the problem of moving point target detection to propose a more efficient and stable detection method in order to make further contributions to this research field.

Author Contributions: Conceptualization, W.G. and P.W.; methodology, W.G.; software, W.G. and P.W.; validation, W.G., P.W. and Y.L.; formal analysis, W.G.; investigation, W.G., Y.L. and C.R. resources, W.N.; data curation, W.G. and P.W.; writing—original draft preparation, W.G. and P.W. writing—review and editing, W.G. and P.W.; visualization, W.G. and Y.L.; supervision, W.N., X.P. and Z.Y.; project administration, W.N., X.P. and Z.Y.; funding acquisition, W.N. and X.P. All authors have read and agreed to the published version of the manuscript.

Funding: This work was partly supported by the Youth Innovation Promotion Association, Grant NO. E1213A02, and the Key Research Program of Frontier Sciences, CAS, Grant NO. 22E0223301.

Data Availability Statement: Not applicable.

Conflicts of Interest: The authors declare no conflict of interest.

References

1. Zhou, Q.; Yao, X.; Wang, C.; Hu, J.; Liu, P.; Lin, J. Adaptive Moving Ground-Target Detection Method Based on Seismic Signal. *IEEE Geosci. Remote Sens. Lett.* **2022**, *19*, 2503705. [CrossRef]
2. Du, J.; Lu, H.; Zhang, L.; Hu, M.; Deng, Y.; Shen, X.; Li, D.; Zhang, Y. DP-MHT-TBD: A Dynamic Programming and Multiple Hypothesis Testing-Based Infrared Dim Point Target Detection Algorithm. *Remote Sens.* **2022**, *14*, 5072. [CrossRef]
3. Eysa, R.; Hamdulla, A. Issues on infrared dim small target detection and tracking. In Proceedings of the 2019 International Conference on Smart Grid and Electrical Automation (ICSGEA), Xiangtan, China, 10–11 August 2019; pp. 452–456.
4. Bernhard, P.; Deschamps, M.; Zaccour, G. Large Satellite Constellations and Space Debris: Exploratory Analysis of Strategic Management of the Space Commons. *Eur. J. Oper. Res.* **2023**, *304*, 1140–1157. [CrossRef]
5. Chen, L.; Chen, X.; Rao, P.; Guo, L.; Huang, M. Space-Based Infrared Aerial Target Detection Method via Interframe Registration and Spatial Local Contrast. *Opt. Lasers Eng.* **2022**, *158*, 107131. [CrossRef]
6. Zhou, J.; Lv, H.; Zhou, F. Infrared small target enhancement by using sequential top-hat filters. In Proceedings of the International Symposium on Optoelectronic Technology and Application 2014: Image Processing and Pattern Recognition, Beijing, China, 13–15 May 2014; Sharma, G., Zhou, F., Eds.; Spie-Int Soc Optical Engineering: Bellingham, WA, USA, 2014; Volume 9301, p. 93011L.
7. Deshpande, S.D.; Er, M.H.; Ronda, V.; Chan, P. Max-Mean and Max-Median Filters for Detection of Small-Targets. *Proc. SPIE Int Soc. Opt. Eng.* **1999**, *3809*, 74–83. [CrossRef]
8. Chen, C.L.P.; Li, H.; Wei, Y.; Xia, T.; Tang, Y.Y. A Local Contrast Method for Small Infrared Target Detection. *IEEE Trans. Geosci Remote Sens.* **2014**, *52*, 574–581. [CrossRef]
9. Wei, Y.; You, X.; Li, H. Multiscale Patch-Based Contrast Measure for Small Infrared Target Detection. *Pattern Recognit.* **2016**, *58*, 216–226. [CrossRef]

10. Du, P.; Hamdulla, A. Infrared Small Target Detection Using Homogeneity-Weighted Local Contrast Measure. *IEEE Geosci. Remote Sens. Lett.* **2020**, *17*, 514–518. [CrossRef]
11. Xia, C.; Li, X.; Zhao, L.; Shu, R. Infrared Small Target Detection Based on Multiscale Local Contrast Measure Using Local Energy Factor. *IEEE Geosci. Remote Sens. Lett.* **2020**, *17*, 157–161. [CrossRef]
12. He, Y.; Li, M.; Wei, Z.; Cai, Y. Infrared small target detection based on weighted variation coefficient local contrast measure. In Proceedings of the Pattern Recognition and Computer Vision, Pt. III, Beijing, China, 29 October–1 November 2021; Ma, H., Wang, L., Zhang, C., Wu, F., Tan, T., Wang, Y., Lai, J., Zhao, Y., Eds.; Springer International Publishing Ag: Cham, Switzerland, 2021; Volume 13021, pp. 117–127.
13. Gao, C.; Meng, D.; Yang, Y.; Wang, Y.; Zhou, X.; Hauptmann, A.G. Infrared Patch-Image Model for Small Target Detection in a Single Image. *IEEE Trans. Image Process.* **2013**, *22*, 4996–5009. [CrossRef]
14. Dai, Y.; Wu, Y.; Song, Y.; Guo, J. Non-Negative Infrared Patch-Image Model: Robust Target-Background Separation via Partial Sum Minimization of Singular Values. *Infrared Phys. Technol.* **2017**, *81*, 182–194. [CrossRef]
15. Guo, J.; Wu, Y.; Dai, Y. Small Target Detection Based on Reweighted Infrared Patch-Image Model. *IET Image Process.* **2018**, *12*, 70–79. [CrossRef]
16. Rawat, S.S.; Verma, S.K.; Kumar, Y. Reweighted Infrared Patch Image Model for Small Target Detection Based on Non-ConvexScript Capital Lp-Norm Minimisation and TV Regularisation. *IET Image Process.* **2020**, *14*, 1937–1947. [CrossRef]
17. Xia, C.; Li, X.; Zhao, L.; Yu, S. Modified Graph Laplacian Model with Local Contrast and Consistency Constraint for Small Target Detection. *IEEE J. Sel. Top. Appl. Earth Obs. Remote Sens.* **2020**, *13*, 5807–5822. [CrossRef]
18. Dai, Y.; Wu, Y.; Zhou, F.; Barnard, K. Attentional Local Contrast Networks for Infrared Small Target Detection. *IEEE Trans. Geosci. Remote Sens.* **2021**, *59*, 9813–9824. [CrossRef]
19. Ma, T.; Yang, Z.; Wang, J.; Sun, S.; Ren, X.; Ahmad, U. Infrared Small Target Detection Network with Generate Label and Feature Mapping. *IEEE Geosci. Remote Sens. Lett.* **2022**, *19*, 6505405. [CrossRef]
20. Hou, Q.; Zhang, L.; Tan, F.; Xi, Y.; Zheng, H.; Li, N. ISTDU-Net: Infrared Small-Target Detection U-Net. *IEEE Geosci. Remote Sens. Lett.* **2022**, *19*, 7000805. [CrossRef]
21. Ju, M.; Luo, J.; Liu, G.; Luo, H. ISTDet: An Efficient End-to-End Neural Network for Infrared Small Target Detection. *Infrared Phys. Technol.* **2021**, *114*, 103659. [CrossRef]
22. Yu, C.; Liu, Y.; Wu, S.; Hu, Z.; Xia, X.; Lan, D.; Liu, X. Infrared Small Target Detection Based on Multiscale Local Contrast Learning Networks. *Infrared Phys. Technol.* **2022**, *123*, 104107. [CrossRef]
23. Lv, G.; Dong, L.; Liang, J.; Xu, W. Novel Asymmetric Pyramid Aggregation Network for Infrared Dim and Small Target Detection. *Remote Sens.* **2022**, *14*, 5643. [CrossRef]
24. Hossen, M.K.; Tuli, S.H. A surveillance system based on motion detection and motion estimation using optical flow. In Proceedings of the 2016 5th International Conference on Informatics, Electronics and Vision (ICIEV), Beijing, China, 29 October–1 November 2016; pp. 646–651.
25. Singla, N. Motion Detection Based on Frame Difference Method. *Int. J. Inf. Comput. Technol.* **2014**, *4*, 1559–1565.
26. Sun, T.; Qi, Y.; Geng, G. Moving Object Detection Algorithm Based on Frame Difference and Background Subtraction. *J. Jilin University. Eng. Technol. Ed.* **2016**, *46*, 1325–1329. [CrossRef]
27. Yi, K.M.; Yun, K.; Kim, S.W.; Chang, H.J.; Choi, J.Y. Detection of moving objects with non-stationary cameras in 5.8 ms: Bringing motion detection to your mobile device. In Proceedings of the 2013 IEEE Conference on Computer Vision and Pattern Recognition Workshops, Portland, OR, USA, 23–28 June 2023; IEEE: Piscataway, NJ, USA; pp. 27–34.
28. Aprile, A.; Grossi, E.; Lops, M.; Venturino, L. Track-before-Detect for Sea Clutter Rejection: Tests with Real Data. *IEEE Trans. Aerosp. Electron. Syst.* **2016**, *52*, 1035–1045. [CrossRef]
29. Li, M.; Zhang, T.X.; Yang, W.D.; Sun, X.C. Moving Weak Point Target Detection and Estimation with Three-Dimensional Double Directional Filter in IR Cluttered Background. *Opt. Eng.* **2005**, *44*, 107007. [CrossRef]
30. Zhang, T.; Li, M.; Zuo, Z.; Yang, W.; Sun, X. Moving Dim Point Target Detection with Three-Dimensional Wide-to-Exact Search Directional Filtering. *Pattern Recognit. Lett.* **2007**, *28*, 246–253. [CrossRef]
31. Deng, L.; Zhu, H.; Tao, C.; Wei, Y. Infrared Moving Point Target Detection Based on Spatial-Temporal Local Contrast Filter. *Infrared Phys. Technol.* **2016**, *76*, 168–173. [CrossRef]
32. Ping-yue, L.; Lin, C.; Sun, S. Dim Small Moving Target Detection and Tracking Method Based on Spatial-Temporal Joint Processing Model. *Infrared Phys. Technol.* **2019**, *102*, 102973. [CrossRef]
33. Zhu, S.; Yang, D.; Jia, P.; Li, J.; Chai, X. Design and Implementation of Space-Time Combined Infrared Small Target Detection Algorithm. *Laser Infrared* **2021**, *51*, 388–392.
34. Liu, D.; Zhang, J.; Dong, W. Temporal Profile Based Small Moving Target Detection Algorithm in Infrared Image Sequences. *Int. J. Infrared. Milli Waves* **2007**, *28*, 373–381. [CrossRef]
35. Liu, D.; Li, Z. Temporal Noise Suppression for Small Target Detection in Infrared Image Sequences. *Optik* **2015**, *126*, 4789–4795. [CrossRef]
36. Liu, D.; Li, Z.; Wang, X.; Zhang, J. Moving Target Detection by Nonlinear Adaptive Filtering on Temporal Profiles in Infrared Image Sequences. *Infrared Phys. Technol.* **2015**, *73*, 41–48. [CrossRef]
37. Liu, X.; Li, L.; Liu, L.; Su, X.; Chen, F. Moving Dim and Small Target Detection in Multiframe Infrared Sequence with Low SCR Based on Temporal Profile Similarity. *IEEE Geosci. Remote Sens. Lett.* **2022**, *19*, 7507005. [CrossRef]

38. Wu, Y.; Yang, Z.; Niu, W.; Zheng, W. A Weak Moving Point Target Detection Method Based on High Frame Rate Image Sequences. In Proceedings of the IGARSS 2018—2018 IEEE International Geoscience and Remote Sensing Symposium, Valencia, Spain, 22–27 July 2018.
39. Niu, W.; Zheng, W.; Yang, Z.; Wu, Y.; Vagvolgyi, B.; Liu, B. Moving Point Target Detection Based on Higher Order Statistics in Very Low SNR. *IEEE Geosci. Remote Sens. Lett.* **2018**, *15*, 217–221. [CrossRef]
40. Niu, W.; Fan, M.; Han, X.; Deng, H.; Guo, Y.; Zheng, W.; Yang, Z.; Peng, X. Moving point target detection based on temporal analysis of pixels in very low SNR. In Proceedings of the Seventh Symposium on Novel Photoelectronic Detection Technology and Applications, Kunming, China, 5–7 November 2021; Su, J., Chu, J., Jiang, H., Yu, Q., Eds.; SPIE Internationl Society of Optical Engineering: Bellingham, WA, USA, 2021; Volume 11763, p. 11763A7.
41. Pentland, A.P. A New Sense for Depth of Field. *IEEE Trans. Pattern Anal. Mach. Intell.* **1987**, *PAMI-9*, 523–531. [CrossRef]
42. He, K.; Zhang, X.; Ren, S.; Sun, J. Deep residual learning for image recognition. In Proceedings of the 2016 IEEE Conference on Computer Vision and Pattern Recognition, CVPR 2016, Las Vegas, NV, USA, 27–30 June 2016; IEEE Computer Society: Washington, DC, USA, 2016; pp. 770–778.
43. Redmon, J.; Farhadi, A. YOLOv3: An Incremental Improvement 2018. *arXiv* **2018**, arXiv:1804.02767.
44. Han, J.; Zhang, X.; Jiang, Y.; Dong, X.; Li, Z.; Li, N. Small moving target detection in infrared sequences by using the multi-scale temporal relative local contrast. In Proceedings of the Advances in Guidance, Navigation and Control, Tianjin, China, 23–25 October 2020; Yan, L., Duan, H., Yu, X., Eds.; Springer: Singapore, 2022; pp. 4433–4445.
45. Selvaraju, R.R.; Cogswell, M.; Das, A.; Vedantam, R.; Parikh, D.; Batra, D. Grad-CAM: Visual explanations from deep networks via gradient-based localization. In Proceedings of the 2017 IEEE International Conference on Computer Vision (ICCV), Venice, Italy, 22–29 October 2017; pp. 618–626.

remote sensing

MDPI

Article

SSANet: An Adaptive Spectral–Spatial Attention Autoencoder Network for Hyperspectral Unmixing

Jie Wang [1], Jindong Xu [1], Qianpeng Chong [1], Zhaowei Liu [1], Weiqing Yan [1], Haihua Xing [2], Qianguo Xing [3] and Mengying Ni [1,*]

[1] School of Computer and Control Engineering, Yantai University, Yantai 264005, China
[2] School of Information Science and Technology, Hainan Normal University, Haikou 571158, China
[3] Yantai Institute of Coastal Zone Research, Chinese Academy of Sciences, Yantai 264003, China
* Correspondence: nimengying@ytu.edu.cn

Abstract: Convolutional neural-network-based autoencoders, which can integrate the spatial correlation between pixels well, have been broadly used for hyperspectral unmixing and obtained excellent performance. Nevertheless, these methods are hindered in their performance by the fact that they treat all spectral bands and spatial information equally in the unmixing procedure. In this article, we propose an adaptive spectral–spatial attention autoencoder network, called SSANet, to solve the mixing pixel problem of the hyperspectral image. First, we design an adaptive spectral–spatial attention module, which refines spectral–spatial features by sequentially superimposing the spectral attention module and spatial attention module. The spectral attention module is built to select useful spectral bands, and the spatial attention module is designed to filter spatial information. Second, SSANet exploits the geometric properties of endmembers in the hyperspectral image while considering abundance sparsity. We significantly improve the endmember and abundance results by introducing minimum volume and sparsity regularization terms into the loss function. We evaluate the proposed SSANet on one synthetic dataset and four real hyperspectral scenes, i.e., Samson, Jasper Ridge, Houston, and Urban. The results indicate that the proposed SSANet achieved competitive unmixing results compared with several conventional and advanced unmixing approaches with respect to the root mean square error and spectral angle distance.

Keywords: hyperspectral unmixing; spectral–spatial attention mechanism; deep learning; autoencoder

Citation: Wang, J.; Xu, J.; Chong, Q.; Liu, Z.; Yan, W.; Xing, H.; Xing, Q.; Ni, M. SSANet: An Adaptive Spectral–Spatial Attention Autoencoder Network for Hyperspectral Unmixing. *Remote Sens.* **2023**, *15*, 2070. https://doi.org/10.3390/rs15082070

Academic Editor: Gwanggil Jeon

Received: 21 February 2023
Revised: 9 April 2023
Accepted: 12 April 2023
Published: 14 April 2023

1. Introduction

Hyperspectral image (HSI) analysis has attracted a large amount of attention in the domain of remote sensing because of the rich content information contained in HSI [1,2]. Despite this, because of the inadequate spatial resolution of satellite sensors, atmospheric mixed effects, and complex ground targets, a pixel in an HSI typically includes multiple spectral features. Such pixels are known as "mixed pixels". The presence of a large quantity of mixed pixels causes serious issues for further research on HSI [3–5]. Hyperspectral unmixing (HU) aims to separate the mixed pixels into a set of pure spectral signatures (endmembers) and relative mixing coefficients (abundances) [6–8].

Recently, with its impressive learning ability and data fitting capability, deep learning (DL) has undergone rapid development in the HU domain [9,10]. The autoencoder (AE), which is a typical representation of unsupervised DL, has been extensively applied to HU tasks. The AE framework is mainly divided into two parts: the encoder, which aims to automatically learn the low-dimensional embeddings (i.e., abundances) of input pixels, and the decoder, which aims to reconstruct input pixels with the associated basis (i.e., endmembers) [11,12]. Moreover, to achieve satisfying unmixing performance, numerous refinements have been made to the existing AE-based unmixing framework. For example, Qu and Qi [13] developed a sparse denoising AE unmixing network that introduces denoising constraints and sparsity constraints to the encoder and decoder, respectively. Zhao et al. [14] presented an AE network that uses two constraints to optimize the spectral unmixing task. Min et al. [12] designed a joint metric AE framework, which uses the Wasserstein distance and feature matching as constraints in the objective function. Jin et al. [15] designed a two-stream AE architecture, which introduces a stream to solve the problem of lacking effective guidance for the endmembers. A deep matrix factorization model was developed in [16], which constructs a multilayer nonlinear structure and employs a self-supervised constraint. Ozkan et al. [17] proposed a two-staged AE architecture that combines spectral angle distance (SAD) with multiple regularizers as the final objective. Su et al. adopted stacked AEs to handle outliers and noise, and employed a variational AE to pose the proper constraint on abundances. An end-to-end unmixing framework was proposed in [18,19], which combines the benefits of learning-based and model-based approaches. However, these methods, which receive one mixed pixel at a time during training, only use the spectral information in an HSI, thereby ignoring the spatial correlation between neighboring pixels.

Importantly, an HSI contains both rich spectral feature information and a degree of spatial information [6]. Incorporating spatial correlation in the unmixing process has been confirmed to significantly improve unmixing performance [20,21]. Therefore, many researchers have introduced convolutional neural networks (CNN) into the traditional AE structure to compensate for the absence of spatial features. For instance, Hong et al. [22] proposed a self-supervised spatial–spectral unmixing method, which incorporates an extra sub-network to guide the endmember information to obtain good unmixing results. Gao et al. [23] developed a cycle-consistency unmixing architecture and designed a self-perception loss to refine the detailed information. Rasti et al. [24] proposed a minimum simplex CNN unmixing approach that incorporates the spatial contextual structure and exploits the geometric properties of endmembers. Ayed et al. [25] presented an approach that uses extended morphological profiles, which combines the spatial correlation between pixels. In [26], a Bayesian fully convolutional framework was developed, which considers the noise, endmembers, and spatial information. Most recently, a perceptual loss-constrained adversarial AE was designed in [27], which takes into account factors such as reconstruction errors and spatial information. Hadi et al. [28] presented a hybrid 3-D and 2-D architecture to leverage the spectral and spatial features. A dual branch AE framework was constructed in [29] to incorporate spatial–contextual information.

Although the above CNN-based AE achieves satisfactory unmixing results, how to adaptively adjust the weights of spectral and spatial features that influence the unmixing performance is a new challenge. Humans can distribute their finite resources to the parts that are most significant, informative, or salient. Inspired by visual attention mechanisms, we propose a spectral–spatial attention AE network for HU and introduce a spectral–spatial attention module (SSAM) to strengthen useful information and suppress information that is unnecessary. Additionally, the absence of both abundance sparsity and endmember geometric information are also responsible for limiting unmixing performance. Thus, we combine a minimum volume constraint and sparsity constraint in the loss function. Specifically, the primary contributions of our proposed SSANet are as follows:

1. We design an unsupervised unmixing network, which is based on a combination of a learnable SSAM and convolutional AE. The SSAM plays two roles. First, the spectral attention module (SEAM) adaptively learns the weights of spectral bands in input data to enhance the representation of spectral information. Second, the spatial attention module (SAAM) adaptively yields the attention weight assigned to each adjacent pixel to derive useful spatial information.
2. We combine the prior knowledge that two regularizers (minimum volume regularization and sparsity regularization) are applied to endmembers and abundances, respectively. Additionally, to acquire high-quality endmember spectra, we design a new minimum volume constraint.

3. We apply the proposed unmixing network to one synthetic dataset and four real hyperspectral scenes—i.e., Samson, Jasper Ridge, Houston, and Urban—and compare it with several classical and advanced approaches. Furthermore, we investigate the performance gain of SSANet with ablation experiments, involving the objective functions and network modules.

The remainder of this paper is structured as follows: In Section 2, we describe the theoretical knowledge of the AE-based unmixing approach simply. In Section 3, we explain the SSANet method in detail. In Section 4, we evaluate SSANet using synthetic and real datasets. In Section 5, we summarize the study.

2. AE-Based Unmixing Model

In the linear mixing model (LMM) [30], the observed spectral reflectance can be given by

$$Y = EA + N \tag{1}$$

where $Y = \{y_i | i = 1, 2, ..., P\} \in \mathbb{R}^{B\times P}$ denotes the observed HSI with B bands and P pixels, and y_i denotes the ith pixel. $N \in \mathbb{R}^{B\times P}$ denotes an additive noise matrix. $E = \{e_k | k = 1, 2, ..., R\} \in \mathbb{R}^{B\times R}$ denotes the endmember matrix with R endmember signatures and needs to satisfy the nonnegative constraint. $A = \{a_i | i = 1, 2, ..., P\} \in \mathbb{R}^{R\times P}$ is the corresponding abundance matrix, where a_i denotes the abundance percentage of the ith pixel, and should be subjected to the abundance nonnegative constraint (ANC) and abundance sum-to-one constraint (ASC)—that is,

$$\begin{cases} a_i \geq 0 \\ \sum_{k=1}^{R} a_{ki} = 1 \end{cases} \tag{2}$$

The fundamental workflow of classic AE unmixing is shown in Figure 1 and is mainly divided into two parts.

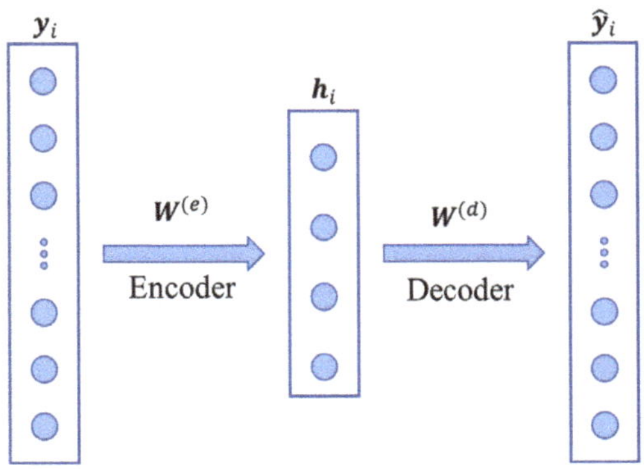

Figure 1. Workflow of the conventional AE unmixing network.

(1) An encoder $En(\cdot)$ transforms the input data $\{y_i\}_{i=1}^{P} \in \mathbb{R}^B$ into a hidden representation h_i, which can be described as

$$h_i = En(y_i) = f(W^{(e)T} y_i + b^{(e)}) \tag{3}$$

where $W^{(e)}$ and $b^{(e)}$ denote the weight and bias of the eth encoder layer, respectively. $f(\cdot)$ denotes the nonlinear activation function.

(2) A decoder $De(\cdot)$ reconstructs the data $\{\hat{y}_i\}_{i=1}^{P} \in \mathbb{R}^B$ using h_i, which is formalized as

$$\hat{y}_i = De(h_i) = W^{(d)T} h_i \tag{4}$$

where $W^{(d)}$ is a matrix that denotes the weights of the hidden and output layers.

Because of the characteristic of Equation (4), the output of the $En(\cdot)$ result is considered as the predicted abundance vector, that is, $\hat{a}_i \leftarrow h_i$, and the estimated endmember is represented by the weights of $De(\cdot)$, that is, $\hat{E} \leftarrow W^{(d)}$. In this framework, the reconstruction loss of the training process is mathematically formulated as

$$Loss_{AE} = \frac{1}{P}\sum_{i=1}^{P}\|\hat{y}_i - y_i\|^2 \quad (5)$$

3. Spectral–Spatial Attention Unmixing Network

To leverage the spectral and spatial information in HSI, we first divide the HSI Y into a set of 3-D neighboring patches $M = \{m_i | i = 1, 2, ..., P\} \in \mathbb{R}^{s \times s \times B}$, where s is the width of patches. In SSANet, each patch m_i in M is fed into the proposed network. In each patch m_i, the central pixel y_i is the target pixel to be unmixed. The framework of SSANet is shown in Figure 2. Its structure consists of three core components: the SSAM, encoder, and decoder. The SSAM, which aims to provide meaningful spectral–spatial priors, helps to solidify feature extraction at later stages. The encoder is designed to extract features and reduce dimensionality. The role of the decoder is to reconstruct the learned features according to the LMM. We provide details on the aforementioned components in Section 3.1, Section 3.2, and Section 3.3, respectively.

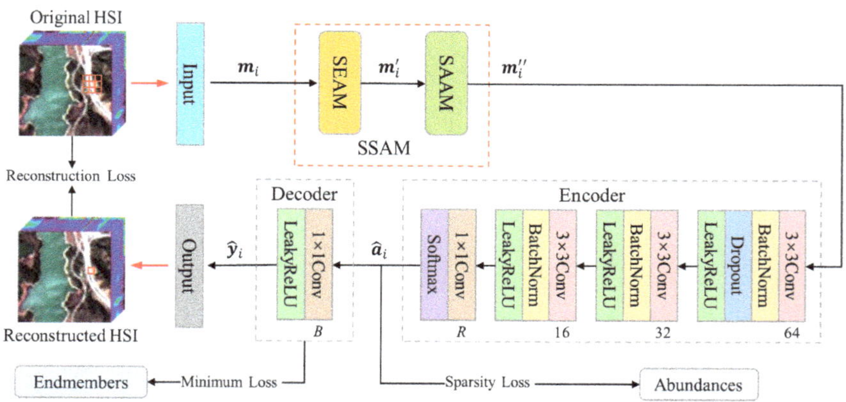

Figure 2. Network architecture of the proposed SSANet.

3.1. Spectral–Spatial Attention Module

The SSAM contains two core modules—that is, the SEAM and SAAM—which are arranged sequentially to perform the selection of spectral bands and spatial features in the HSI, respectively. We describe the SEAM and SAAM in the following.

3.1.1. Spectral Attention Module

The SEAM [31] is introduced into the SSANet, aiming to adaptively learn the weights of spectral bands in the HSI in an end-to-end manner. It generates a spectral weight vector that reflects the significance of different spectral bands. The spectral bands modulated by this vector can significantly improve unmixing performance. The framework of the SEAM is shown in Figure 3.

Given the input $m_i \in \mathbb{R}^{s \times s \times B}$, first, global max pooling (GMP) and global average pooling (GAP) are used to acquire spectral feature vectors $\alpha_i \in \mathbb{R}^{1 \times 1 \times B}$ and $\beta_i \in \mathbb{R}^{1 \times 1 \times B}$, respectively. Next, the corresponding weight vectors $\gamma_i \in \mathbb{R}^{1 \times 1 \times B}$ and $\delta_i \in \mathbb{R}^{1 \times 1 \times B}$ can be derived using a multilayer perceptron (MLP) that can extract the weight information of each band. γ_i and δ_i are then summed, and the sigmoid function is applied to obtain the spectral weight coefficients $\boldsymbol{v}_i \in \mathbb{R}^{1\times1\times B}$. The spectral attention formulation can be defined as

$$\boldsymbol{v}_i = \sigma(MLP(GMP(\boldsymbol{m}_i)) + MLP(GAP(\boldsymbol{m}_i))) \tag{6}$$

where $\sigma(\cdot)$ denotes the sigmoid function. Finally, the output of SEAM $\boldsymbol{m}_i'$ is calculated by the following equation:

$$\boldsymbol{m}_i' = \boldsymbol{v}_i \bigotimes \boldsymbol{m}_i \tag{7}$$

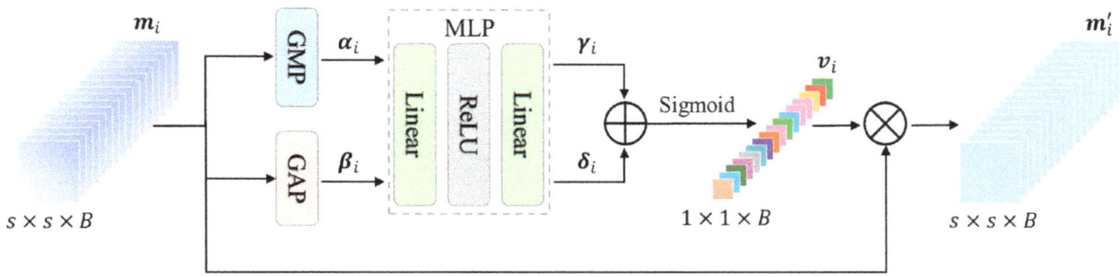

Figure 3. Detailed workflow of the SEAM.

3.1.2. Spatial Attention Module

In this part, we design the SAAM to evaluate the adjacent dependence between pixels. Similar to the SEAM, the SAAM also learns in an end-to-end manner and adaptively selects spatial features from the pixels in the neighborhood. The module generates a spatial weight matrix that expresses the importance of adjacent pixels. The recalibration of spatial features using this matrix leads to an obvious improvement in the unmixing accuracy. The framework of the SAAM is shown in Figure 4.

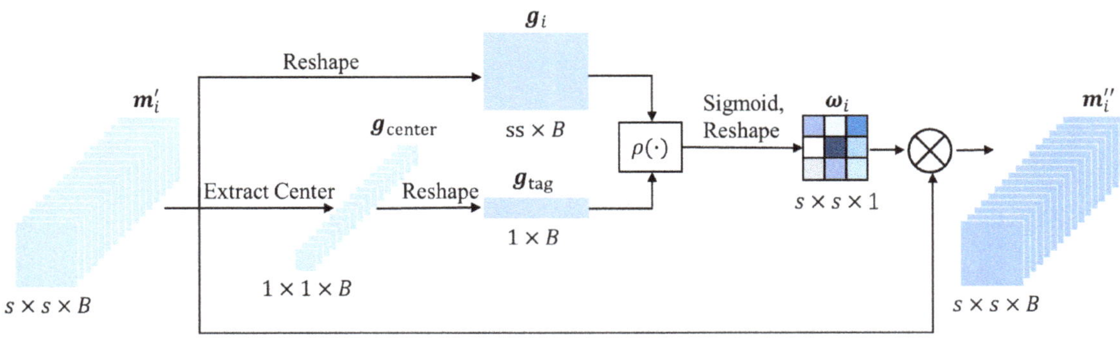

Figure 4. Detailed workflow of the SAAM.

Specifically, given the input $\boldsymbol{m}_i' \in \mathbb{R}^{s\times s\times B}$, in order to facilitate the calculation of the similarity between neighboring pixels and the central pixel, the input $\boldsymbol{m}_i'$ is reshaped into $g_i \in \mathbb{R}^{ss\times B}(ss = s\times s)$. The center pixel $g_{center} \in \mathbb{R}^{1\times1\times B}$ is extracted from the center of $\boldsymbol{m}_i'$; then, g_{center} is reshaped into $g_{tag} \in \mathbb{R}^{1\times B}$. Next, both g_i and g_{tag} are fed into the scoring function $\rho(\cdot)$ to compute the spatial similarity scores between them. The $\rho(\cdot)$ is produced as follows:

$$\rho(h_i) = \varphi\left(\sum_{i=1}^{ss} h_i \mathbf{W}\right) \tag{8}$$

$$h_i = \exp\left(-\frac{1}{B}\left\|g_i - g_{tag}\right\|_2^2\right) \tag{9}$$

where h_i is used to compute the correlation between g_i and g_{tag}. $\rho(\cdot)$ is implemented by a full connection layer, parameterized by a weight matrix $W \in \mathbb{R}^{ss \times ss}$. The spatial similarity scores are derived by multiplying all the h_i with W and the results are activated by a rectified linear unit (ReLU) function $\varphi(\cdot)$. Subsequently, a sigmoid function is adopted to compute the spatial weight matrix $\omega_i \in \mathbb{R}^{s \times s \times 1}$. Finally, we perform elementwise multiplication of ω_i with m_i' to implement the recalibration of spatial information:

$$m_i'' = \omega_i \bigotimes m_i' \quad (10)$$

where m_i'' represents the output of SAAM.

3.2. Encoder

As shown in Figure 2, the encoder consists of four convolutional layers, and the number of convolution kernels diminishes with the depth of the layer, which can be formulated as

$$En\left(m_i''\right) = softmax(W^4 \circledast LR(BN(W^3 \circledast LR(BN(W^2 \circledast LR(DO(BN(W^1 \circledast m_i'' + b^1))) + b^2)) + b^3)) + b^4) \quad (11)$$

where W^e and b^e denote the weights and biases, respectively, at the eth level of the encoder for $e = 1,2,3,4$. $\circledast$ denotes the convolution operation. $BN(\cdot)$ represents batch normalization, which is used to enhance the performance and stability of the network, and speed up the learning of the network. $LR(\cdot)$ denotes the leaky ReLU (LReLU) function, which aims to promote nonlinearity. $DO(\cdot)$ represents the dropout function, which is currently the key technique for preventing network overfitting. The purpose of the softmax function is to satisfy two physical constraints on abundance: ANC and ASC.

3.3. Decoder

The decoder contains a 1×1 convolutional layer and uses LReLU as the activation function. It is formulated as

$$De\left(En\left(m_i''\right)\right) = LR\left(W \circledast En\left(m_i''\right) + b\right) \quad (12)$$

where W and b denote the weights and biases of the decoder, respectively. It should be noted that, in our experiments, to help the training of the decoder, we used the endmembers extracted using the vertex component analysis (VCA) [32] approach to initialize the weights W.

3.4. Objective Functions

The overall loss function of SSANet consists of the following three terms.

Numerous AE-based works have adopted the SAD with the scale invariance as the reconstruction loss [33,34]. Therefore, we apply the SAD measurement as the reconstruction loss of SSANet, which is denoted as follows:

$$Loss_{AE} = \frac{1}{P}\sum_{i=1}^{P} arccos(\frac{\hat{y}_i^T y_i}{\|\hat{y}_i\|_2 \|y_i\|_2}) \quad (13)$$

The softmax function does not yield sparse abundance maps. Qian et al. [35] demonstrated that using the $l_{1/2}$ norm yields more accurate and sparser abundance results than using the l_1 norm. We apply the $l_{1/2}$ norm to the abundance vector $\hat{a}_{ik}$, which is formulated as

$$Loss_{Sp} = \sum_{i=1}^{P}\sum_{k=1}^{R} \sqrt{|\hat{a}_{ik}|} \quad (14)$$

where $\hat{a}_{ik}$ represents the reference abundance fractional proportion of the kth endmember at the ith pixel in the HSI.

The minimum volume regularizer has already been proven to be beneficial for extracting endmembers [36]. Moreover, to make the estimated endmembers close to the observed spectrum, we design a more reasonable minimum volume constraint, denoted by

$$Loss_{Mv} = \frac{1}{BR}\sum_{k=1}^{R}\left\|\hat{e}_k - \bar{e}\right\|_2^2 \tag{15}$$

where $\bar{e} = (1/R)\sum_{k=1}^{R} e_k$ denotes the centroid vector. A geometrical explanation of this concept is shown in Figure 5. During each iteration, by minimizing $Loss_{Mv}$, the endmembers are pulled from the initial values (i.e., the vertices of the initial data simplex) to the vertices of the real data simplex.

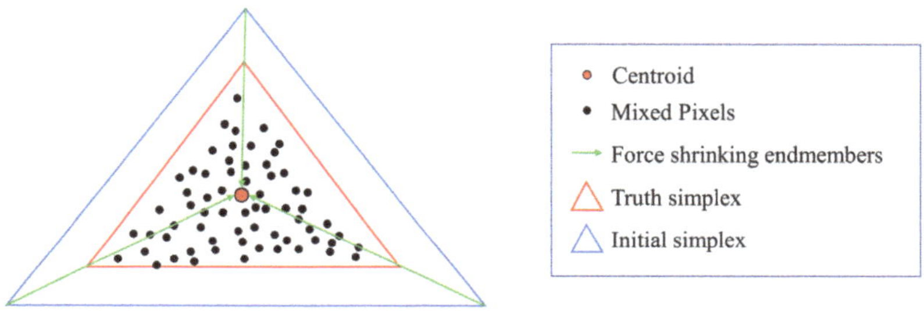

Figure 5. Geometric interpretation of minimum volume regularization. Each vertex of the simplex is considered as an endmember, and the initial endmembers are oriented toward the centroid of the simplex determined by the real endmembers.

To summarize, the overall loss function of SSANet is expressed as

$$Loss = Loss_{AE} + \lambda_1 Loss_{Sp} + \lambda_2 Loss_{Mv} \tag{16}$$

where λ_1 and λ_2 represent the regularization parameters.

4. Experiments

To validate the accuracy and validity of SSANet for HU, we conducted experiments using one synthetic data [26] and four widely used real hyperspectral scenes (Samson [37], Jasper Ridge [38], Houston [39], and Urban [40]), as shown in Figure 6. We chose seven representative unmixing methods (including classical methods and the most advanced methods) for comparison: VCA-FCLS [32,41], SGCNMF [42], DAEU [43], MTAEU [44], CNNAEU [45], CyCU-Net [23], and MiSiCNet [24]. VCA-FCLS is a baseline method, SGCNMF is based on non-negative matrix factorization, and the others are AE-based methods. DAEU uses only spectral information, whereas MTAEU, CNNAEU, CyCU-Net, and MiSiCNet use spectral–spatial information.

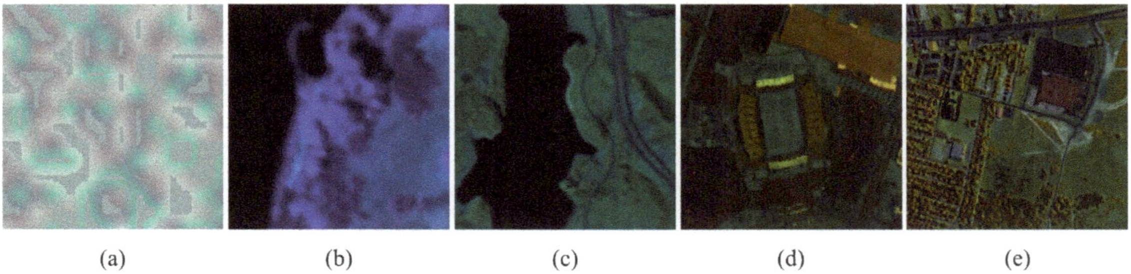

Figure 6. RGB images of the synthetic and real HSIs adopted in the experiments. (**a**) Synthetic data. (**b**) Samson. (**c**) Jasper Ridge. (**d**) Houston. (**e**) Urban.

4.1. Dataset Description

4.1.1. Synthetic Data

We created simulated data according to the approach adopted by Fang et al. [39]. Its size was 104 × 104 pixels, distributed over 200 spectral bands, with four endmembers. Each pixel in this image was a mixture that consisted of four endmembers. We generated these mixed pixels by multiplying four endmembers and four abundance maps according to the LMM. First, we created abundance maps that we decomposed into 8 × 8 homogeneous blocks, which we randomly chose as one of the endmember categories. Then, we degraded blocks by adopting a spatial low-pass filter of 9 × 9. Next, we added zero-mean Gaussian noise with various signal-to-noise ratios (SNRs) to the obtained synthetic dataset. Because of the different noise variances in different bands, we assigned different SNR values to different bands and obtained band-related SNR values from the baseline Indian Pines image. We assumed that the obtained SNR vector s was centralized and normalized; then, we could acquire the synthetic SNR n based on the rule $n = \beta s + r$, where β is the fluctuation amplitude of band-related SNR values and r is the center value that defines the total SNR of all bands. To investigate the robustness of our approach to various noise levels, we simulated three datasets with various noise values (SNR = 20, 30, 40 dB) by fixing $\beta = 5$ and varying r.

4.1.2. Samson Data

Samson data have three constituent materials: soil, trees, and water. This dataset was captured by the Samson sensor. The image contains 156 spectral channels ranging from 0.4–0.9 µm. Because the original image is large, we selected a subimage of the original data with a size of 95 × 95 pixels.

4.1.3. Jasper Ridge Data

Jasper Ridge data have four main materials: trees, water, soil, and roads. This dataset was obtained by the AVIRIS sensor. The original HSI covers 512 × 614 pixels in size and is spread over 224 spectral channels, covering wavelengths from 0.38 to 2.5 µm. It has a spatial resolution of 20 m/pixel. We selected an area of interest of 100 × 100 pixels and removed bands (1–3, 108–112, 154–166, and 220–224) to alleviate the influences of the atmosphere and water vapor. Finally, the Jasper Ridge dataset had 198 remaining bands.

4.1.4. Houston Data

Houston data have four dominant materials: parking lot 1, running track, healthy grass, and parking lot 2. The data were originally used in the 2013 IEEE GRSS data fusion competition. The original HSI contains 349 × 1905 pixels, distributed over 144 channels ranging from 0.35 to 1.05 µm. Its spatial resolution is 2.5 m/pixel. We selected a subimage containing 170 × 170 pixels. The subimage is centered on Robertson Stadium on the Houston campus.

4.1.5. Urban Data

Urban data have four constituent materials: asphalt, grass, tree, and roof. This dataset, collected by the HYDICE sensor, is characterized by a complex distribution. Its pixel resolution is 307 × 307, and there are 210 spectral bands ranging from 0.4 to 2.5 µm. It has a spatial resolution of 2 m/pixel. After we removed the contaminated bands, 162 bands remained.

4.2. Experimental Settings

4.2.1. Evaluation Metrics

We selected two commonly used evaluation metrics, the root mean square error (RMSE) and SAD, to assess the proposed method. These two indices are defined as

$$RMSE(\hat{\boldsymbol{a}}_i, \boldsymbol{a}_i) = \sqrt{\frac{1}{P}\sum_{i=1}^{P}\|\hat{\boldsymbol{a}}_i - \boldsymbol{a}_i\|_2^2} \tag{17}$$

$$SAD(\hat{\boldsymbol{e}}_k, \boldsymbol{e}_k) = arccos(\frac{\hat{\boldsymbol{e}}_k^T \boldsymbol{e}_k}{\|\hat{\boldsymbol{e}}_k\|_2 \|\boldsymbol{e}_k\|_2}) \tag{18}$$

where $\boldsymbol{e}_k$ and $\hat{\boldsymbol{e}}_k$ are the real endmember and extracted endmember, respectively, and $\boldsymbol{a}_i$ and $\hat{\boldsymbol{a}}_i$ are the real abundance and predicted abundance, respectively.

For both evaluation metrics, the lower the value, the better the corresponding unmixing results.

4.2.2. Hyperparameter Settings

In our experiments, we assumed that the number of endmembers R was known in advance, as determined by HySime [46]. In the training phase, we initialized the decoder with the endmembers extracted by VCA. We implemented our proposed SSANet in the environment of PyTorch 1.6 with an i7-8550U CPU. We applied the Adam optimizer to optimize the parameters. The selection of specific parameters for the proposed SSANet is displayed in Table 1. Figure 7 shows the convergence curves of the proposed SSANet during the learning process.

Table 1. Hyperparameter settings for the proposed SSANet.

Parameter	λ_1	λ_2	Epoch	Batch Size	Encoder Learning Rate	Decoder Learning Rate
Synthetic data	1×10^{-2}	1×10^{-2}	50	32	1×10^{-5}	1×10^{-5}
Samson	5×10^{-2}	0.5	50	128	1×10^{-3}	1×10^{-3}
Jasper Ridge	5×10^{-2}	0.5	50	128	1×10^{-3}	1×10^{-3}
Houston	5×10^{-2}	0.5	50	256	1×10^{-4}	1×10^{-5}
Urban	5×10^{-2}	0.5	50	64	1×10^{-3}	1×10^{-3}

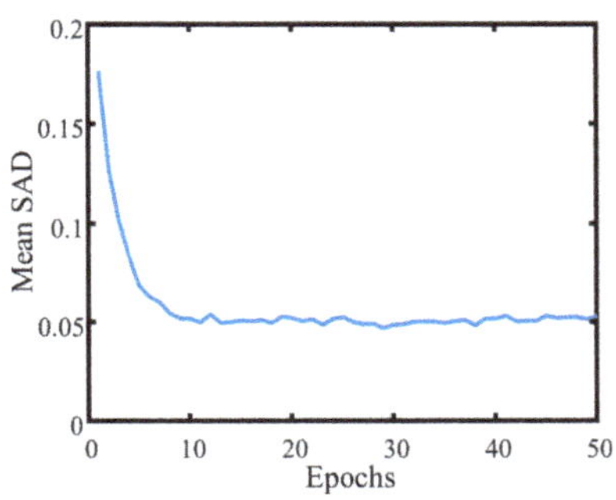

Figure 7. Convergence curves during 50 epochs.

4.3. *Comparison of SSANet with Other Methods*

4.3.1. Experiments with Synthetic Data

To study the robustness of SSANet to noise, we added zero-mean Gaussian noise with SNRs of 20, 30, and 40 dB to the synthetic dataset. Figure 8 shows the quantitative analysis results with varying SNR levels. Generally, SSANet achieved better (i.e., lower) SAD and RMSE results than the other methods, at both a low and high SNR. SGSNMF performed well when the noise intensity was relatively low. At high noise levels, the performance of SGSNMF deteriorated severely. CNNAEU and CyCU-Net could not obtain the desired performance at various noise levels. The reason is that, despite the introduction of spatial information, CNNAEU and CyCU-Net led to a noise-sensitivity problem because of insufficient spectral feature representation capability. For MiSiCNet, the image prior aimed to solve the degradation problem. As a result, MiSiCNet achieved relatively good results under low noise conditions. Other methods, such as DAEU and MTAEU, often obtained satisfactory results because of the introduction of abundance sparsity and spectral-spatial priors, respectively. The performance of SSANet did not degrade severely as noise levels increased. The overall performance at various noise levels verified the robustness of SSANet to noise, which mainly resulted from the advantage of the combination of the attention mechanism and associated physical properties. The visualization results of the abundances and endmembers for the synthetic data (SNR 40 dB) are shown in Figures 9 and 10, respectively. The experimental results indicated that our method successfully obtained relatively good results.

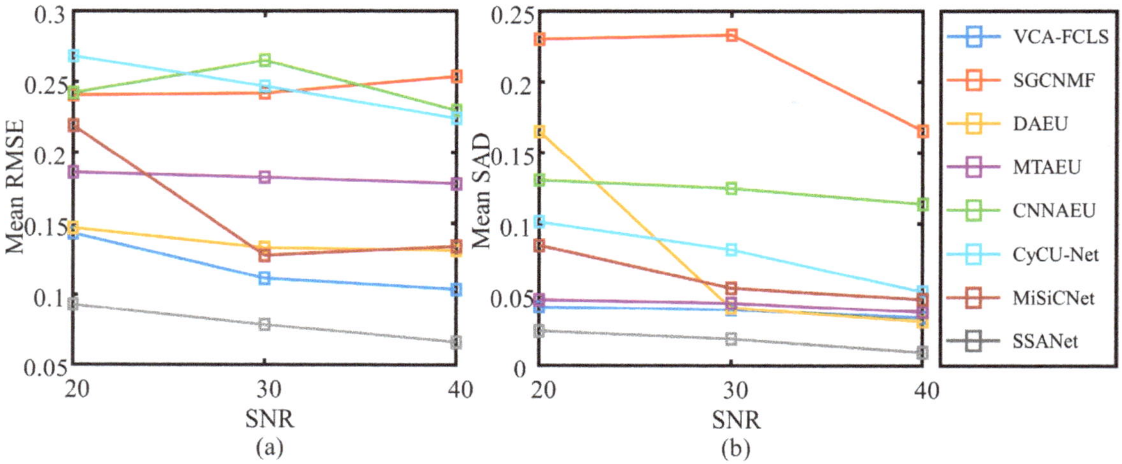

Figure 8. Experimental results of SSANet with various noise values (20, 30, and 40 dB) for the synthetic dataset. (**a**) Mean RMSE. (**b**) Mean SAD.

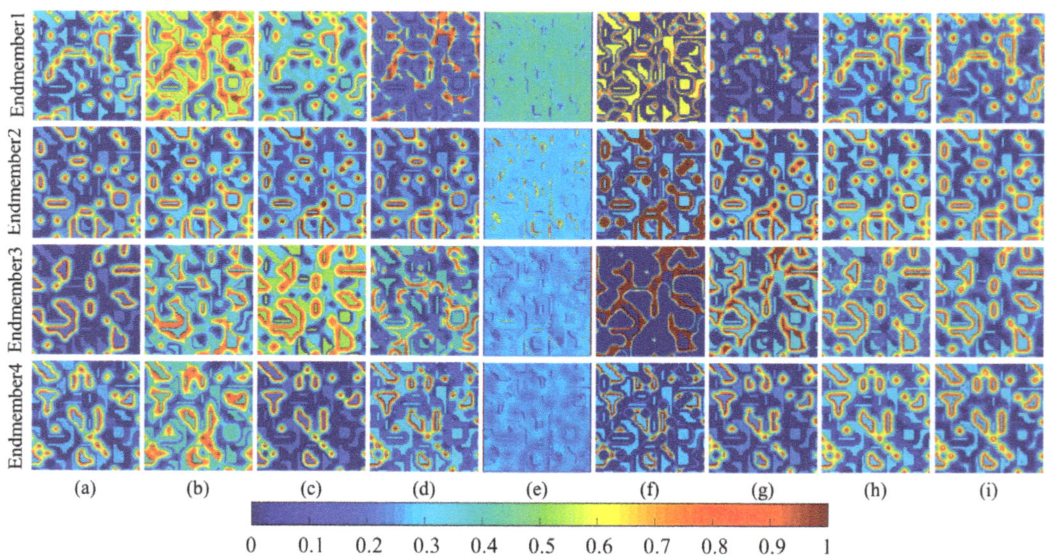

Figure 9. Visualization results of the abundances of the synthetic data (SNR 40 dB). (**a**) VCA-FCLS. (**b**) SGCNMF. (**c**) DAEU. (**d**) MTAEU. (**e**) CNNAEU. (**f**) CyCU-Net. (**g**) MiSiCNet. (**h**) SSANet. (**i**) Ground truth (GT).

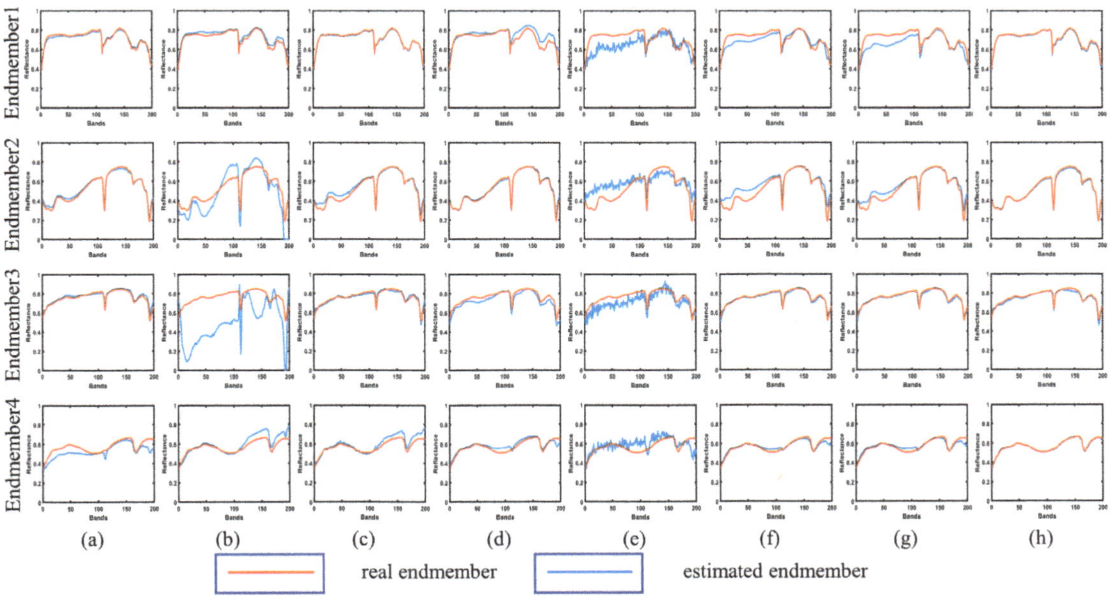

Figure 10. Visualization results of the endmembers of the synthetic data (SNR 40 dB). (**a**) VCA-FCLS. (**b**) SGCNMF. (**c**) DAEU. (**d**) MTAEU. (**e**) CNNAEU. (**f**) CyCU-Net. (**g**) MiSiCNet. (**h**) SSANet.

4.3.2. Experiments with Samson Data

The quantitative results for Samson are shown in Tables 2 and 3. Notably, our proposed SSANet outperformed the other methods in terms of the mean SAD and mean RMSE. Additionally, compared with the suboptimal results, these two metrics lowered by 16% and 69%, respectively. Figures 11 and 12 show the abundances and endmembers estimated by all the methods. Figure 11 shows that VCA-FCLS and SGCNMF performed relatively poorly, confusing soil and trees. By contrast, the DL-based methods confused nothing and distinguished each material more accurately, which demonstrates the advantage of the DL methods. However, the abundance results of these methods at the junction of two different materials were not ideal, whereas our method retained rich edge information and appeared much clearer visually. This may be the result of a moderate application of sparsity regularization, in addition to spatial attention. As shown by Figure 12, all methods achieved good performance. However, because SSANet took into account the geometric information of endmembers, in addition to the utilization of spectral attention to enhance the effective spectral bands, it made the extracted water endmember greatly superior to that of the competing methods. The superior performance further validated the effectiveness and reliability of SSANet.

Table 2. RMSE ($\times$100) and mean RMSE ($\times$100) of abundances acquired by various unmixing approaches on Samson data. Annotation: bold red text indicates the best results and bold blue text indicates the suboptimal results.

Methods		VCA-FCLS	SGCNMF	DAEU	MTAEU	CNNAEU	CyCU-Net	MiSiCNet	SSANet
RMSE	Soil	26.50	17.86	**11.02**	13.36	19.9	18.27	18.18	**4.06**
	Tree	25.11	24.49	9.89	**9.49**	25.01	19.19	17.91	**3.41**
	Water	42.35	35.77	10.71	**7.08**	27.91	15.78	31.31	**1.90**
Mean RMSE		31.32	26.04	10.54	**9.98**	24.27	17.75	22.47	**3.12**

Table 3. SAD (×100) and mean SAD (×100) of endmembers acquired by various unmixing approaches on Samson data. Annotation: bold red text indicates the best results and bold blue text indicates the suboptimal results.

Methods		VCA-FCLS	SGCNMF	DAEU	MTAEU	CNNAEU	CyCU-Net	MiSiCNet	SSANet
SAD	Soil	2.36	0.98	1.53	3.20	6.13	1.06	1.03	0.92
	Tree	4.33	4.60	4.52	6.21	4.01	2.50	3.54	3.55
	Water	15.04	22.97	3.39	4.98	16.09	5.37	40.08	2.96
Mean SAD		7.24	9.51	3.15	4.80	8.74	2.97	14.88	2.48

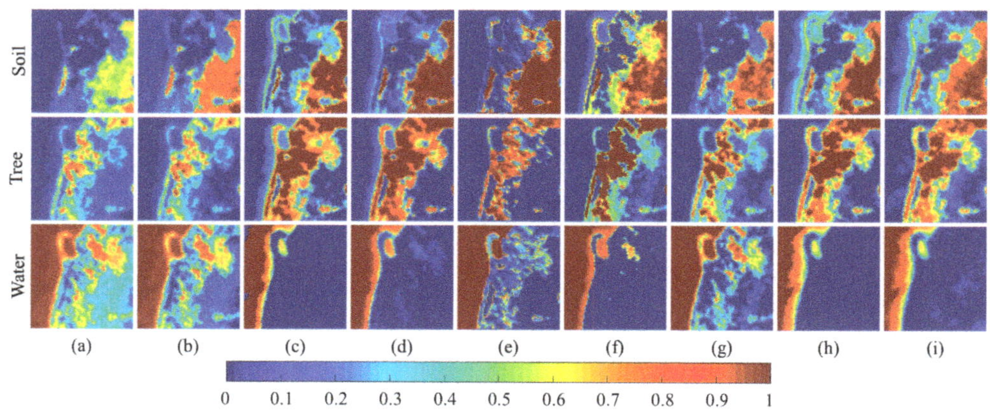

Figure 11. Visualization results of the abundances of Samson data. (**a**) VCA-FCLS. (**b**) SGCNMF. (**c**) DAEU. (**d**) MTAEU. (**e**) CNNAEU. (**f**) CyCU-Net. (**g**) MiSiCNet. (**h**) SSANet. (**i**) GT.

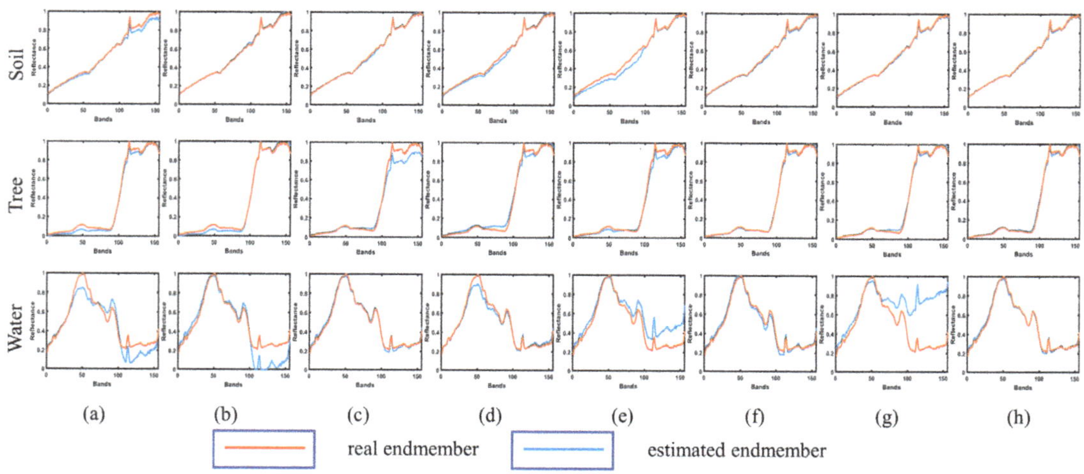

Figure 12. Visualization results of the endmembers of Samson data. (**a**) VCA-FCLS. (**b**) SGCNMF. (**c**) DAEU. (**d**) MTAEU. (**e**) MTAEU. (**f**) CyCU-Net. (**g**) MiSiCNet. (**h**) SSANet.

4.3.3. Experiments with Jasper Ridge Data

Tables 4 and 5 show the quantitative results for Jasper Ridge. The visualization results of abundances and endmembers are presented in Figures 13 and 14, respectively. As shown in Table 4, for RMSE of each material, our SSANet lowered by 56%, 51%, 45%, and 57%, respectively, compared with the suboptimal results. Table 5 shows that although SSANet did not achieve the best results for each material, it ranked first with respect to the mean

SAD. Figure 14 also shows that the endmembers obtained by SSANet were close to the GT. In Figure 13, the abundance maps generated by SSANet look much sharper. In the Jasper dataset, roads occupy a small portion of the scene. For material roads, estimating the abundances and endmembers is more challenging than for other materials because of the complex distribution. Numerous methods estimate unsatisfactory abundances and fail to completely separate roads, whereas SSANet separated roads more accurately because of the application of the abundance sparsity and the geometric feature of endmembers. Additionally, in both a heavily mixed area (soil) and homogeneous area (water), SSANet obtained superior separation results because of its powerful learning capability that fully integrated useful spectral and spatial information.

Table 4. RMSE (×100) and mean RMSE (×100) of abundances acquired by various unmixing approaches on Jasper Ridge data. Annotation: bold red text indicates the best results and bold blue text indicates the suboptimal results.

Methods		VCA-FCLS	SGCNMF	DAEU	MTAEU	CNNAEU	CyCU-Net	MiSiCNet	SSANet
RMSE	Road	14.48	11.99	19.03	20.83	44.82	11.75	24.94	5.11
	Soil	12.69	14.82	15.90	26.99	37.48	14.09	22.13	6.18
	Tree	15.63	15.80	16.32	21.75	23.64	9.66	9.60	5.27
	Water	18.73	26.27	8.05	5.19	30.65	10.04	11.42	2.25
Mean RMSE		15.39	17.22	14.95	18.69	34.15	11.38	17.02	4.70

Table 5. SAD (×100) and mean SAD (×100) of endmembers acquired by various unmixing approaches on Jasper Ridge data. Annotation: bold red text indicates the best results and bold blue text indicates the suboptimal results.

Methods		VCA-FCLS	SGCNMF	DAEU	MTAEU	CNNAEU	CyCU-Net	MiSiCNet	SSANet
SAD	Road	9.01	14.39	29.57	11.64	15.07	3.85	32.97	2.10
	Soil	22.34	22.45	6.03	15.80	9.52	3.70	6.63	7.52
	Tree	14.81	20.76	3.20	4.61	9.17	3.23	4.32	6.56
	Water	54.59	27.79	3.40	7.06	3.51	15.40	29.04	4.06
Mean SAD		25.19	21.35	10.55	9.78	9.32	6.44	18.24	5.06

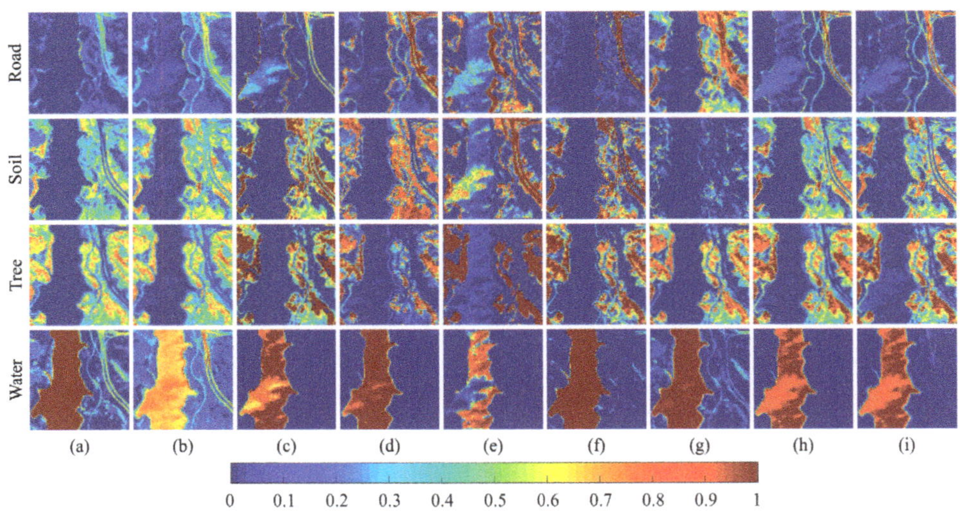

Figure 13. Visualization results of the abundances of Jasper Ridge data. (**a**) VCA-FCLS. (**b**) SGCNMF. (**c**) DAEU. (**d**) MTAEU. (**e**) CNNAEU. (**f**) CyCU-Net. (**g**) MiSiCNet. (**h**) SSANet. (**i**) GT.

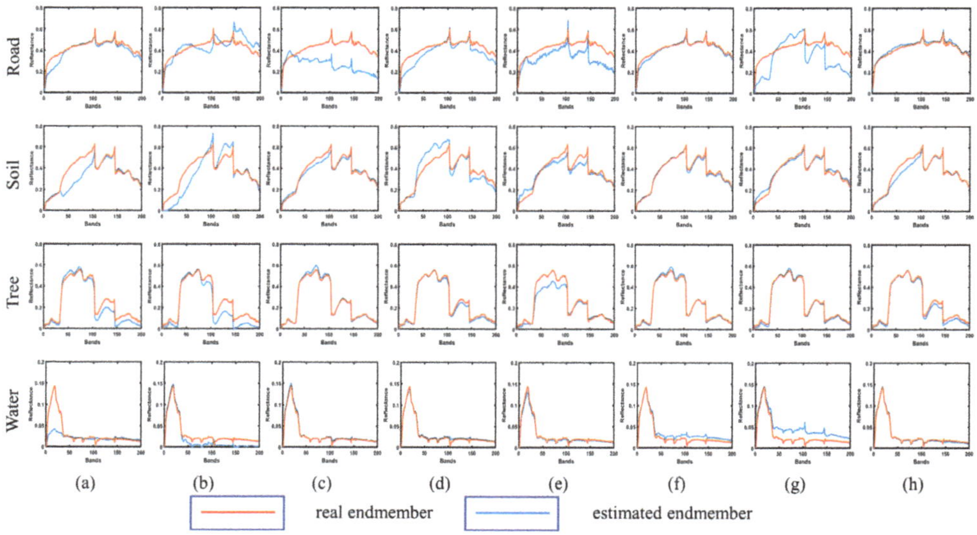

Figure 14. Visualization results of the endmembers of Jasper Ridge data. (**a**) VCA-FCLS. (**b**) SGCNMF. (**c**) DAEU. (**d**) MTAEU. (**e**) CNNAEU. (**f**) CyCU-Net. (**g**) MiSiCNet. (**h**) SSANet.

4.3.4. Experiments with Houston Data

The qualitative analysis results for the Houston dataset are shown in Tables 6 and 7. Figures 15 and 16 show the qualitative analysis results of the abundance maps and endmembers acquired, respectively. Clearly, with respect to both the RMSE and SAD, the results obtained by methods based on spectral–spatial information (MTAEU, MiSiCNet, and SSANet) were better than those obtained by methods that used only spectral information (DAEU and CyCU-Net). These results provide further confirmation that the full utilization of spectral-spatial features is advantageous for enhancing the precision of HU. Although SSANet did not acquire the best SAD results for each endmember, its mean SAD was the optimal result. Moreover, SSANet achieved the best results for all abundances with respect to the RMSE. Importantly, Figure 15 shows that all other methods performed poorly in terms of distinguishing similar materials (i.e., parking lot1 and parking lot2); however, it was relatively easier for our method to distinguish spectrally similar materials, which was facilitated by the attention mechanism selecting useful spectral–spatial features and suppressing useless features. In conclusion, we demonstrated the good performance of SSANet in real scenes with similar substances based on the combined RMSE and SAD evaluation.

Table 6. RMSE ($\times$100) and mean RMSE ($\times$100) of abundances acquired by various unmixing approaches on Houston data. Annotation: bold red text indicates the best results and bold blue text indicates the suboptimal results.

Methods		VCA-FCLS	SGCNMF	DAEU	MTAEU	CNNAEU	CyCU-Net	MiSiCNet	SSANet
RMSE	Running Track	**7.74**	9.47	15.19	21.45	14.33	40.12	10.36	**5.96**
	Grass Healthy	12.66	**7.02**	15.52	6.84	16.23	13.18	**5.99**	9.24
	Parking Lot1	24.76	23.86	30.66	22.05	43.68	25.07	**14.21**	**12.49**
	Parking Lot2	25.68	25.60	**15.81**	22.04	43.46	47.64	16.77	**14.62**
Mean RMSE		17.71	16.49	19.29	18.09	29.42	31.50	**11.83**	**10.58**

Table 7. SAD (×100) and mean SAD (×100) of endmembers acquired by various unmixing approaches on Houston data. Annotation: bold red text indicates the best results and bold blue text indicates the suboptimal results.

Methods		VCA-FCLS	SGCNMF	DAEU	MTAEU	CNNAEU	CyCU-Net	MiSiCNet	SSANet
SAD	Running Track	16.34	36.79	20.56	23.39	42.56	33.73	**7.24**	**15.23**
	Grass Healthy	11.85	12.54	7.00	**4.19**	**0.97**	7.30	9.05	8.48
	Parking Lot1	2.59	4.06	2.73	4.17	7.46	**2.55**	**1.10**	3.00
	Parking Lot2	26.64	12.49	**5.90**	6.82	**3.14**	10.50	19.40	9.39
Mean SAD		14.35	16.47	**9.05**	9.64	13.53	13.52	9.20	**9.02**

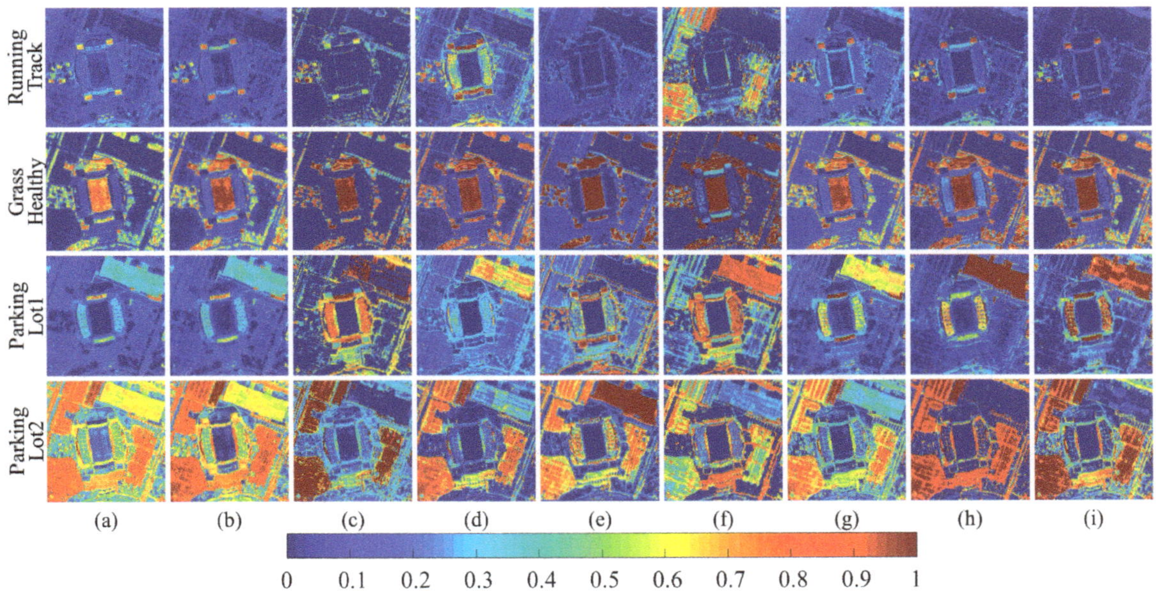

Figure 15. Visualization results of the abundances of Houston data. (**a**) VCA-FCLS. (**b**) SGCNMF. (**c**) DAEU. (**d**) MTAEU. (**e**) CNNAEU. (**f**) CyCU-Net. (**g**) MiSiCNet. (**h**) SSANet. (**i**) GT.

4.3.5. Experiments with Urban Data

Tables 8 and 9 show the quantitative metric comparisons for the Urban dataset. Figures 17 and 18 visualize the results of the abundances and endmembers, respectively. A feature of this dataset is its complex distribution, and mixed pixels are broadly distributed in this scene. It is worth noting that SSANet had the finest mean and individual RMSE, and the mean RMSE was 11% lower than that of the suboptimal method. Additionally, the individual SAD obtained by SSANet was also competitive. Figure 17 shows that the endmember mixed phenomenon appeared for VCA-FCLS and SGCNMF, which resulted in poor results. CyCU-Net and MiSiCNet achieved poor qualitative and quantitative performance. Although DAEU, MTAEU, and CNNAEU were able to distinguish each material, there were some errors in the details, which were related to the absence of useful adjacency information and a sparsity prior. Therefore, SSANet adopted a spatial attention that assigned weights to neighboring pixels, in addition to the sparsity regularizer to make the abundance maps look smooth and realistic. Figure 18 shows that the proposed SSANet acquired similar visual endmember maps to GT. However, because the roof endmember accounted for a small percentage of this large-scale scene, there were some gaps in the roof endmember obtained by SSANet; however, the overall results remained competitive. The superior unmixing results confirmed the reliability of SSANet in highly mixed scenes.

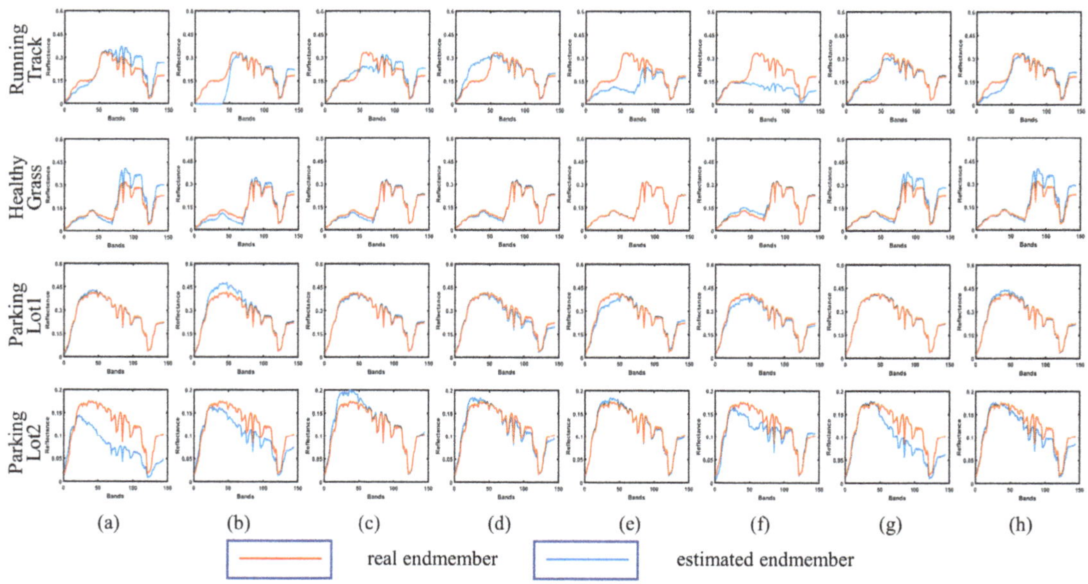

Figure 16. Visualization results of the endmembers of Houston data. (**a**) VCA-FCLS. (**b**) SGCNMF. (**c**) DAEU. (**d**) MTAEU. (**e**) CNNAEU. (**f**) CyCU-Net. (**g**) MiSiCNet. (**h**) SSANet.

Table 8. RMSE (×100) and mean RMSE (×100) of abundances acquired by various unmixing approaches on Urban data. Annotation: bold red text indicates the best results and bold blue text indicates the suboptimal results.

Methods		VCA-FCLS	SGCNMF	DAEU	MTAEU	CNNAEU	CyCU-Net	MiSiCNet	SSANet
RMSE	Asphalt	27.54	39.26	16.59	**15.35**	23.56	33.41	37.70	**13.73**
	Grass	40.10	33.83	15.21	**15.06**	29.81	44.90	31.64	**13.51**
	Tree	45.85	25.48	11.19	**9.39**	20.08	39.59	24.53	**7.58**
	Roof	17.08	18.93	8.68	**8.55**	13.70	15.15	15.64	**8.23**
Mean RMSE		32.64	29.37	12.92	**12.09**	21.79	33.27	27.38	**10.76**

Table 9. SAD (×100) and mean SAD (×100) of endmembers acquired by different unmixing approaches on Urban data. Annotation: bold red text indicates the best results and bold blue text indicates the suboptimal results.

Methods		VCA-FCLS	SGCNMF	DAEU	MTAEU	CNNAEU	CyCU-Net	MiSiCNet	SSANet
SAD	Asphalt	20.95	102.34	11.48	8.13	**6.02**	20.66	76.25	**7.51**
	Grass	26.03	44.42	6.85	**5.06**	10.05	34.99	39.12	**3.69**
	Tree	34.59	9.28	**3.39**	6.95	13.99	20.88	9.88	**3.80**
	Roof	82.28	16.45	30.91	14.83	**6.29**	9.86	**4.52**	26.31
Mean SAD		40.96	43.12	13.16	**8.74**	**9.09**	21.60	32.45	10.33

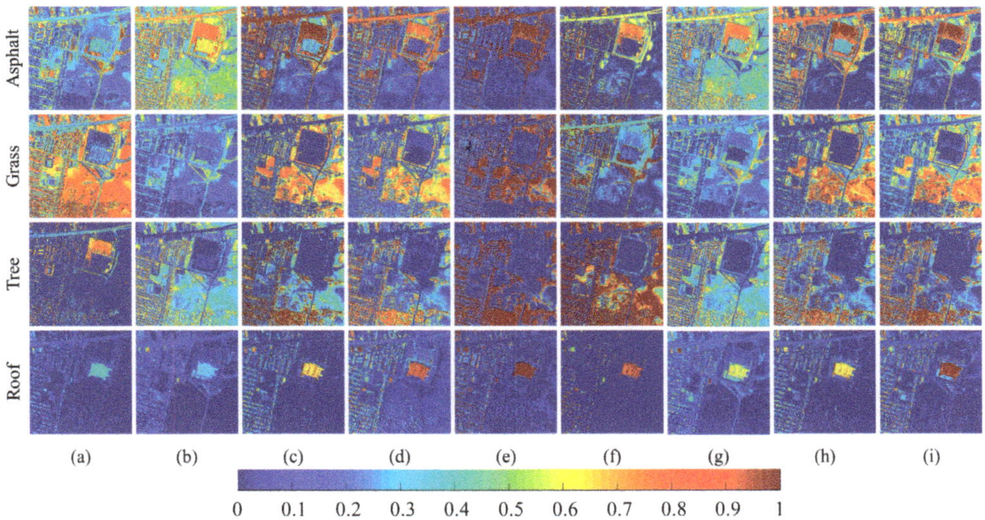

Figure 17. Visualization results of the abundances of Urban data. (**a**) VCA-FCLS. (**b**) SGCNMF. (**c**) DAEU. (**d**) MTAEU. (**e**) CNNAEU. (**f**) CyCU-Net. (**g**) MiSiCNet. (**h**) SSANet. (**i**) GT.

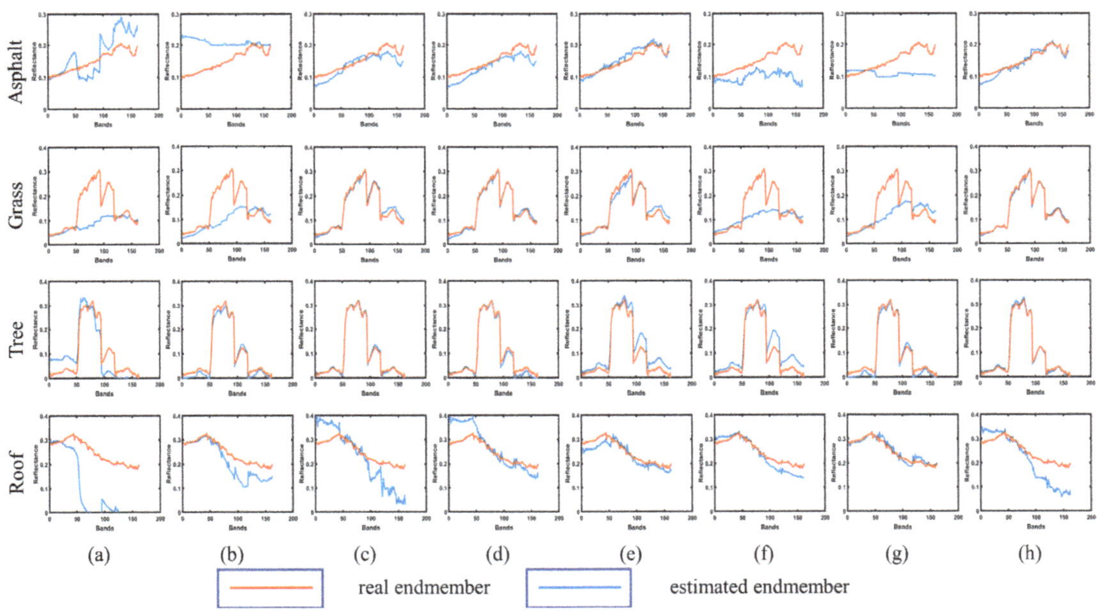

Figure 18. Visualization results of the endmembers of Urban data. (**a**) VCA-FCLS. (**b**) SGCNMF. (**c**) DAEU. (**d**) MTAEU. (**e**) CNNAEU. (**f**) CyCU-Net. (**g**) MiSiCNet. (**h**) SSANet.

4.4. Discussion

Through the qualitative and quantitative analysis of four real hyperspectral scenes, our SSANet vastly improved the unmixing performance. Because the distribution of real scenes may not have fulfilled the prior distribution assumption, VCA-FCLS and SGCNMF performed relatively poorly on real datasets compared with the DL-based methods, which also indicates the advantage of using the DL methods for the unmixing task. DAEU is an AE framework that does not contain spatial information; therefore, the overall performance of DAEU was not favorable; however, DAEU obtained satisfactory results in the abundance estimation because its special design took advantage of abundance sparsity in the form of adaptive thresholds. Additionally, the lack of ASC led to the poor performance of CyCU-Net in the reconstruction process. MTAEU and CNNAEU used spatial correlation but their objective functions simply used the SAD reconstruction term and did not impose regularizers on endmembers and abundances, which led to greater variances in endmember extraction and abundance estimation. MiSiCNet considered spatial information and used the geometric information of endmembers. The utilization of geometric properties allowed MiSiCNet to achieve competitive performance in endmember estimation, but it did not leverage the relevant properties of abundance, thus limiting unmixing performance. Although MTAEU, CNNAEU, and MiSiCNet combined spectral–spatial priors to make the unmixing performance relatively good, their limited performance can be attributed to their inability to combine useful spectral–spatial priors and the failure to consider both the geometric property of the endmember and the abundance sparsity. For the aforementioned problem, in our approach, we used SSAM to enhance useful information and weaken useless information, in addition to imposing a minimum volume regularizer and sparse regularizer on the endmembers and abundances, respectively. Therefore, our unmixing method obtained good unmixing accuracy. In conclusion, the overall experimental performance on four real-world HSIs illustrated the effectiveness and superior performance of our method.

4.5. Ablation Experiments

4.5.1. Ablation Study on Objective Functions

We selected the Jasper Ridge scene as an example to evaluate the contribution of the various parts of the objective function. Table 10 shows the results of the quantitative analysis of the ablation study. We observed that using the SAD reconstruction loss solely ensured the fulfillment of the HU task, but with limited accuracy. Incorporating appropriate regularization greatly improved the unmixing performance. Using the sparsity term exploited an inherent property of real scenes and guaranteed the sparsity of the abundance results. Moreover, we introduced the minimum simplex volume constraint to exploit the geometric information of the HSI. This term was beneficial for endmember extraction. To summarize, all these regularizations appear to be associated with achieving the best results and the optimal performance was obtained by combining all of them.

Table 10. Mean RMSE (×100) and mean SAD (×100) results of ablation experiments with various losses. Annotation: bold black text indicates the best results.

	$Loss_{AE}$	$Loss_{AE} + Loss_{Sp}$	$Loss_{AE} + Loss_{Mv}$	$Loss_{AE} + Loss_{Sp} + Loss_{Mv}$
Mean RMSE	14.53	6.27	9.54	4.70
Mean SAD	23.58	6.75	14.96	5.06

4.5.2. Ablation Study on Network Modules

In order to test whether both SSAM and SEAM improve the results, ablation experiments in the Jasper Ridge scene are shown in this section. We compared SSANet with SSANet without SSAM (SSANet-None), SSANet only with SEAM (SSANet-SEAM), and SSANet only with SAAM (SSANet-SAAM). The results are shown in Table 11. It can be seen from Table 11 that the SSANet after removing SEAM and SAAM yielded the worst unmixing performance. By introducing either SEAM or SAAM into the proposed AE model, the integrated SSANet had a certain improvement in the estimation of endmembers and abundances. Consequently, it was necessary to combine SEAM and SAAM to achieve superior performance.

Table 11. Mean RMSE (×100) and mean SAD (×100) results of ablation experiments with various network modules. Annotation: bold black text indicates the best results.

	None	SEAM	SAAM	SEAM + SAAM
Mean RMSE	6.87	6.48	5.46	4.70
Mean SAD	5.91	5.37	5.24	5.06

4.6. Processing Time

Table 12 shows the consumption time of all the unmixing approaches applied to the Jasper Ridge dataset in seconds. We ran all the experiments on a computer with a 3.6 GHz Intel Core i7-7820X CPU and NVIDIA GeForce RTX 1080 16GB GPU. We implemented VCA-FCLS and SGCNMF in MATLAB R2016a; implemented DAEU, MTAEU, and CNNAEU on the TensorFlow platform; and implemented CyCU-Net, MiSiCNet, and SSANet on the PyTorch platform. The proposed SSANet is not the quickest, but its time consumption was relatively satisfactory.

Table 12. Consumption time (in seconds) for all the unmixing approaches.

Methods	VCA-FCLS	SGCNMF	DAEU	MTAEU	CNNAEU	CyCU-Net	MiSiCNet	SSANet
Time(s)	1.75	26.82	15.35	23.26	1152.97	23.74	92.39	71.53

5. Conclusions

In this article, we present a convolutional AE unmixing network called SSANet, which effectively uses spectral–spatial information in HSIs. First, we propose a learnable SSAM, which refines spectral–spatial features by sequentially overlaying the SEAM and SAAM. This module strengthens high-information features and weakens low-information features by weighting the learning of features. Second, we use the sparsity of abundances and the geometric properties of endmembers by adding a sparsity constraint term and a minimum volume constraint term to the loss function to achieve sparse abundance results and accurate endmembers. We verify the effectiveness and robustness of SSANet in experiments by comparing it with several classical and advanced HU approaches in synthetic and real scenes.

Author Contributions: Conceptualization, J.W., J.X. and Q.C.; methodology, J.W., J.X. and Q.C.; software, J.W.; validation, Z.L. and W.Y.; formal analysis, H.X.; investigation, Q.X.; resources, Q.C.; data curation, Z.L.; writing—original draft preparation, J.W. and Q.C.; writing—review and editing, J.X. and M.N.; visualization, H.X.; supervision, Q.X. and M.N.; project administration, W.Y.; funding acquisition, J.X. All authors have read and agreed to the published version of the manuscript.

Funding: This research was funded by the National Natural Science Foundation of China under Grant 62072391 and Grant 62066013, and the Fundamental Research Funds for the Central Universities, CHD, 300102343518.

Data Availability Statement: The hyperspectral image datasets used in this study are freely available at http://rslab.ut.ac.ir/data, accessed on 3 September 2022.

Acknowledgments: All authors would sincerely thank the reviewers and editors for their suggestions and opinions for improving this article.

Conflicts of Interest: The authors declare no conflict of interest.

References

1. Bioucas-Dias, J.; Plaza, A.; Camps-Valls, G.; Scheunders, P.; Nasrabadi, N.M.; Chanussot, J. Hyperspectral remote sensing data analysis and future challenges. *IEEE Geosci. Remote Sens. Mag.* **2013**, *1*, 6–36. [CrossRef]
2. Mei, X.; Ma, Y.; Li, C.; Fan, F.; Huang, J.; Ma, J. Robust GBM hyperspectral image unmixing with superpixel segmentation based low rank and sparse representation. *Neurocomputing* **2018**, *275*, 2783–2797. [CrossRef]
3. Zou, J.; Lan, J.; Shao, Y. A hierarchical sparsity unmixing method to address endmember variability in hyperspectral image. *Remote Sens.* **2018**, *10*, 738. [CrossRef]

4. Zhong, Y.; Wang, X.; Zhao, L.; Feng, R.; Zhang, L.; Xu, Y. Blind spectral unmixing based on sparse component analysis for hyperspectral remote sensing imagery. *ISPRS J. Photogramm. Remote Sens.* **2016**, *119*, 49–63. [CrossRef]
5. Keshava, N.; Mustard, J. Spectral unmixing. *IEEE Signal Process. Mag.* **2002**, *19*, 44–57. [CrossRef]
6. Wang, P.; Shen, X.; Ni, K.; Shi, L.X. Hyperspectral sparse unmixing based on multiple dictionary pruning. *Int. J. Remote Sens.* **2022**, *43*, 2712–2734. [CrossRef]
7. Karoui, M.S.; Deville, Y.; Hosseini, S.; Ouamri, A. Blind spatial unmixing of multispectral images: New methods combining sparse component analysis, clustering and non-negativity constraints. *Pattern Recogn.* **2012**, *45*, 4263–4278. [CrossRef]
8. Xu, X.; Li, J.; Wu, C.S.; Plaza, A. Regional clustering-based spatial preprocessing for hyperspectral unmixing. *Remote Sens. Environ.* **2018**, *204*, 333–346. [CrossRef]
9. Rasti, B.; Hong, D.F.; Hang, R.L.; Ghamisi, P.; Kang, X.D.; Chanussot, J.; Benediktsson, J.A. Feature extraction for hyperspectral imagery: The evolution from shallow to deep: Overview and toolbox. *IEEE Geosci. Remote Sens. Mag.* **2020**, *8*, 60–88. [CrossRef]
10. Pattathal, V.A.; Sahoo, M.M.; Porwal, A.; Karnieli, A. Deep-learning-based latent space encoding for spectral unmixing of geological materials. *ISPRS J. Photogramm. Remote Sens.* **2022**, *183*, 307–320. [CrossRef]
11. Palsson, B.; Sveinsson, J.R.; Ulfarsson, M.O. Blind hyperspectral unmixing using autoencoders: A critical comparison. *IEEE J. Sel. Top. Appl. Earth Obs. Remote Sens.* **2022**, *15*, 1340–1372. [CrossRef]
12. Min, A.Y.; Guo, Z.Y.; Li, H.; Peng, J.T. JMnet: Joint metric neural network for hyperspectral unmixing. *IEEE Trans. Geosci. Remote Sens.* **2022**, *60*, 1–12. [CrossRef]
13. Qu, Y.; Qi, H.R. uDAS: An untied denoising autoencoder with sparsity for spectral unmixing. *IEEE Trans. Geosci. Remote Sens.* **2019**, *57*, 1698–1712. [CrossRef]
14. Zhao, Z.G.; Hu, D.; Wang, H.; Yu, X.C. Minimum distance constrained sparse autoencoder network for hyperspectral unmixing. *J. Appl. Remote Sens.* **2020**, *14*, 048501.
15. Jin, Q.; Ma, Y.; Mei, X.; Ma, J. TANet: An Unsupervised two-stream autoencoder network for hyperspectral unmixing. *IEEE Trans. Geosci. Remote Sens.* **2022**, *60*, 1–15. [CrossRef]
16. Li, H.-C.; Feng, X.-R.; Zhai, D.-H.; Du, Q.; Plaza, A. Self-supervised robust deep matrix factorization for hyperspectral unmixing. *IEEE Trans. Geosci. Remote Sens.* **2021**, *60*, 1–14. [CrossRef]
17. Ozkan, S.; Kaya, B.; Akar, G.B. Endnet: Sparse autoencoder network for endmember extraction and hyperspectral unmixing. *IEEE Trans. Geosci. Remote Sens.* **2018**, *57*, 482–496. [CrossRef]
18. Xiong, F.; Zhou, J.; Tao, S.; Lu, J.; Qian, Y. SNMF-Net: Learning a deep alternating neural network for hyperspectral unmixing. *IEEE Trans. Geosci. Remote Sens.* **2021**, *60*, 1–16. [CrossRef]
19. Qian, Y.; Xiong, F.; Qian, Q.; Zhou, J. Spectral mixture model inspired network architectures for hyperspectral unmixing. *IEEE Trans. Geosci. Remote Sens.* **2020**, *58*, 7418–7434. [CrossRef]
20. Tulczyjew, L.; Kawulok, M.; Longépé, N.; Saux, B.L.; Nalepa, J. A multibranch convolutional neural network for hyperspectral unmixing. *IEEE Geosci. Remote Sens. Lett.* **2022**, *19*, 1–5. [CrossRef]
21. Shi, C.; Wang, L. Incorporating spatial information in spectral unmixing: A review. *Remote Sens. Environ.* **2014**, *149*, 70–87. [CrossRef]
22. Hong, D.F.; Gao, L.R.; Yao, J.; Yokoya, N.; Chanussot, J.; Heiden, U.; Zhang, B. Endmember-guided unmixing network (EGU-Net): A general deep learning framework for self-supervised hyperspectral unmixing. *IEEE Trans. Neural Netw. Learn. Syst.* **2021**, *33*, 6518–6531. [CrossRef] [PubMed]
23. Gao, L.R.; Han, Z.; Hong, D.F.; Zhang, B.; Chanussot, J. CyCU-Net: Cycle-consistency unmixing network bylearning cascaded autoencoders. *IEEE Trans. Geosci. Remote Sens.* **2022**, *60*, 1–14. [CrossRef]
24. Rasti, B.; Koirala, B.; Scheunders, P.; Chanussot, J. Misicnet: Minimum simplex convolutional network for deep hyperspectral unmixing. *IEEE Trans. Geosci. Remote Sens.* **2022**, *60*, 1–15. [CrossRef]
25. Ayed, M.; Hanachi, R.; Sellami, A.; Farah, I.R.; Mura, M.D. A deep learning approach based on morphological profiles for Hyperspectral Image unmixing. In Proceedings of the 2022 6th International Conference on Advanced Technologies for Signal and Image Processing (ATSIP), Sfax, Tunisia, 24–27 May 2022; pp. 1–6.
26. Fang, Y.; Wang, Y.; Xu, L.; Zhuo, R.; Wong, A.; Clausi, D.A. Bcun: Bayesian fully convolutional neural network for hyperspectral spectral unmixing. *IEEE Trans. Geosci. Remote Sens.* **2022**, *60*, 1–14. [CrossRef]
27. Zhao, M.; Wang, M.; Chen, J.; Rahardja, S. Perceptual loss-constrained adversarial autoencoder networks for hyperspectral unmixing. *IEEE Geosci. Remote Sens. Lett.* **2022**, *19*, 1–5. [CrossRef]
28. Hadi, F.; Yang, J.; Ullah, M.; Ahmad, I.; Farooque, G.; Xiao, L. DHCAE: Deep hybrid convolutional autoencoder approach for robust supervised hyperspectral unmixing. *Remote Sens.* **2022**, *14*, 4433. [CrossRef]
29. Hua, Z.; Li, X.; Feng, Y.; Zhao, L. Dual branch autoencoder network for spectral-spatial hyperspectral unmixing. *IEEE Geosci. Remote Sens. Lett.* **2021**, *19*, 1–5. [CrossRef]
30. Bioucas-Dias, J.M.; Plaza, A.; Dobigeon, N.; Parente, M.; Du, Q.; Gader, P.; Chanussot, J. Hyperspectral unmixing overview: Geometrical, statistical, and sparse regression-based approaches. *IEEE J. Sel. Top. Appl. Earth Obs. Remote Sens.* **2012**, *5*, 354–379. [CrossRef]
31. Shi, C.; Liao, D.; Zhang, T.; Wang, L. Hyperspectral image classification based on expansion convolution network. *IEEE Trans. Geosci. Remote Sens.* **2022**, *60*, 1–16. [CrossRef]

32. Nascimento, J.M.; Dias, J.M. Vertex component analysis: A fast algorithm to unmix hyperspectral data. *IEEE Trans. Geosci. Remote Sens.* **2005**, *43*, 898–910. [CrossRef]
33. Qi, L.; Li, J.; Wang, Y.; Lei, M.; Gao, X. Deep spectral convolution network for hyperspectral image unmixing with spectral library. *Signal Process.* **2020**, *176*, 107672. [CrossRef]
34. Hua, Z.; Li, X.; Jiang, J.; Zhao, L. Gated autoencoder network for spectral–spatial hyperspectral unmixing. *Remote Sens.* **2021**, *13*, 3147. [CrossRef]
35. Qian, Y.T.; Jia, S.; Zhou, J.; Robles-Kelly, A. Hyperspectral unmixing via l-1/2 sparsity-constrained nonnegative matrix factorization. *IEEE Trans. Geosci. Remote Sens.* **2011**, *49*, 4282–4297. [CrossRef]
36. Miao, L.; Qi, H. Endmember extraction from highly mixed data using minimum volume constrained nonnegative matrix factorization. *IEEE Trans. Geosci. Remote Sens.* **2007**, *45*, 765–777. [CrossRef]
37. Azar, S.G.; Meshgini, S.; Beheshti, S.; Rezaii, T.Y. Linear mixing model with scaled bundle dictionary for hyperspectral unmixing with spectral variability. *Signal Process.* **2021**, *188*, 13.
38. Zhu, F.Y.; Wang, Y.; Xiang, S.M.; Fan, B.; Pan, C.H. Structured sparse method for hyperspectral unmixing. *ISPRS J. Photogramm. Remote Sens.* **2014**, *88*, 101–118. [CrossRef]
39. Debes, C.; Merentitis, A.; Heremans, R.; Hahn, J.; Frangiadakis, N.; van Kasteren, T.; Liao, W.; Bellens, R.; Pižurica, A.; Gautama, S. Hyperspectral and LiDAR data fusion: Outcome of the 2013 GRSS data fusion contest. *IEEE J. Sel. Top. Appl. Earth Obs. Remote Sens.* **2014**, *7*, 2405–2418. [CrossRef]
40. Zhu, F.Y.; Wang, Y.; Fan, B.; Xiang, S.M.; Meng, G.F.; Pan, C.H. Spectral unmixing via data-guided sparsity. *IEEE Trans. Image Process.* **2014**, *23*, 5412–5427. [CrossRef]
41. Heinz, D.C. Fully constrained least squares linear spectral mixture analysis method for material quantification in hyperspectral imagery. *IEEE Trans. Geosci. Remote Sens.* **2001**, *39*, 529–545. [CrossRef]
42. Wang, X.; Zhong, Y.; Zhang, L.; Xu, Y. Spatial group sparsity regularized nonnegative matrix factorization for hyperspectral unmixing. *IEEE Trans. Geosci. Remote Sens.* **2017**, *55*, 6287–6304. [CrossRef]
43. Palsson, B.; Sigurdsson, J.; Sveinsson, J.R.; Ulfarsson, M.O. Hyperspectral unmixing using a neural network autoencoder. *IEEE Access* **2018**, *6*, 25646–25656. [CrossRef]
44. Palsson, B.; Sveinsson, J.R.; Ulfarsson, M.O. Spectral-spatial hyperspectral unmixing using multitask learning. *IEEE Access* **2019**, *7*, 148861–148872. [CrossRef]
45. Palsson, B.; Ulfarsson, M.O.; Sveinsson, J.R. Convolutional autoencoder for spectral–spatial hyperspectral unmixing. *IEEE Trans. Geosci. Remote Sens.* **2021**, *59*, 535–549. [CrossRef]
46. Bioucas-Dias, J.M.; Nascimento, J.M.P. Hyperspectral subspace identification. *IEEE Trans. Geosci. Remote Sens.* **2008**, *46*, 2435–2445. [CrossRef]

remote sensing

MDPI

Article

3D-UNet-LSTM: A Deep Learning-Based Radar Echo Extrapolation Model for Convective Nowcasting

Shiqing Guo, Nengli Sun, Yanle Pei and Qian Li *

The College of Meteorology and Oceanography, National University of Defense Technology, Changsha 410005, China
* Correspondence: liqian@nudt.edu.cn

Abstract: Radar echo extrapolation is a commonly used approach for convective nowcasting. The evolution of convective systems over a very short term can be foreseen according to the extrapolated reflectivity images. Recently, deep neural networks have been widely applied to radar echo extrapolation and have achieved better forecasting performance than traditional approaches. However, it is difficult for existing methods to combine predictive flexibility with the ability to capture temporal dependencies at the same time. To leverage the advantages of the previous networks while avoiding the mentioned limitations, a 3D-UNet-LSTM model, which has an extractor-forecaster architecture, is proposed in this paper. The extractor adopts 3D-UNet to extract comprehensive spatiotemporal features from the input radar images. In the forecaster, a newly designed Seq2Seq network exploits the extracted features and uses different convolutional long short-term memory (ConvLSTM) layers to iteratively generate hidden states for different future timestamps. Finally, the hidden states are transformed into predicted radar images through a convolutional layer. We conduct 0–1 h convective nowcasting experiments on the public MeteoNet dataset. Quantitative evaluations demonstrate the effectiveness of the 3D-UNet extractor, the newly designed forecaster, and their combination. In addition, case studies qualitatively demonstrate that the proposed model has a better spatiotemporal modeling ability for the complex nonlinear processes of convective echoes.

Keywords: radar echo extrapolation; sequence-to-sequence (Seq2Seq) network; 3D-Unet; convective nowcasting

Citation: Guo, S.; Sun, N.; Pei, Y.; Li, Q. 3D-UNet-LSTM: A Deep Learning-Based Radar Echo Extrapolation Model for Convective Nowcasting. *Remote Sens.* **2023**, *15*, 1529. https://doi.org/10.3390/rs15061529

Academic Editor: Gwanggil Jeon

Received: 26 January 2023
Revised: 5 March 2023
Accepted: 6 March 2023
Published: 10 March 2023

1. Introduction

Convective nowcasting usually refers to forecasting the evolution trends of convective systems for lead times of up to a few hours, which is significant for protecting lives and property and supporting outdoor activities [1–3]. However, it is still challenging due to the obvious suddenness, rapid changes, and inherent uncertainty of convection systems.

In most cases, extrapolation-based forecasts have higher skills for lead times of up to 1–2 h. Spatiotemporal extrapolation techniques use statistical models or data-driven models to extrapolate radar or satellite images into the imminent future. After obtaining extrapolation results, convective nowcasting can be conducted with radar echo reflectivity values $\geq$ 35 dBZ [4] or cloud-top brightness temperatures below a certain threshold [5], and convective precipitation fields can also be estimated with the Z-R relation [6] and nonlinear mapping algorithms [7,8].

Traditional extrapolation techniques are usually based on statistical models, and most of them follow the framework of Lagrangian persistence, which utilizes the motion field calculated from recent images to extrapolate the latest available image under the assumption that the intensity and motion are constant [9]. These methods can be roughly divided into object-based extrapolation [10–12] and region-based extrapolation approaches [9,13–15]. Object-based extrapolation first identifies a convective storm cell and then extrapolates its trajectory based on the calculated motion vectors; this technique is mainly suitable for nowcasting convective storms with high intensity and stability. Region-based extrapolation focuses on the image and extrapolates all grid values without specific classifications. However, the performance of traditional extrapolation techniques is poor when they are used to forecast rapidly changing weather systems, especially for severe convection storms with abrupt intensity, location and size changes [2,16].

Recently, the continuous development of deep learning has contributed significantly to the modeling capabilities of extrapolation techniques. Deep neural networks (DNNs) are capable of modeling nonlinear processes in observation images, thus depicting complicated and rapidly developing weather phenomena such as the initiation, dissipation, and rotation of clouds. On the other hand, a data-driven solution enables DNNs to learn local weather patterns from massive historical observations, making them more suitable for regional convective forecasts. Furthermore, many studies have demonstrated that deep learning-based extrapolation methods perform better than traditional statistical extrapolation techniques [17–20]. Among those methods, the commonly used DNNs are convolutional recurrent neural networks (ConvRNNs) and convolutional neural networks (CNNs) [17,18].

ConvRNNs can explicitly model the temporal dependencies of consecutive observation images by recursively applying stacked ConvRNN units along the time direction, transmitting and updating the inside states. In prior works, most deep learning practitioners used ConvRNNs to address extrapolation-based nowcasting for convective storms and precipitation. For example, Shi et al. [17] proposed convolutional long short-term memory (ConvLSTM) to extrapolate radar images; this approach uses convolution operations instead of full connections in its state transitions. Shi et al. [21] then designed a more reasonable encoding-forecasting structure and proposed the trajectory-gated recurrent unit (TrajGRU) model to address the location invariance problem existing in ConvLSTM. To memorize spatial and temporal information simultaneously, Wang et al. [22] presented a general framework called the predictive RNN (PredRNN), which makes the states flow in two directions. Tuyen et al. [23] designed RainPredRNN, which could reduce the number of calculated operations based on PredRNN. In addition, Jing et al. [24] exploited radar images at three altitudes to extrapolate those at one and addressed the blurry prediction problem with adversarial training. A generative adversarial network (GAN) architecture was also applied by Ravuri et al. [19] to generate more sharp future radar images via a ConvRNN. Moreover, since observation images can be considered video sequences continuously recorded with a fixed "camera", other advanced ConvRNN models for video prediction [25,26] can also be applied to convective nowcasting.

Although it has already been concluded that a simple convolutional architecture can outperform recurrent architectures on diverse sequence modeling tasks [27,28], ConvRNNs are more generally used for spatiotemporal sequence forecasting than those using CNNs. In the past two years, the application of CNNs to extrapolation-based convective nowcasting has attracted increasing attention. Unlike ConvRNN-based extrapolation methods that explicitly model time, CNN-based approaches consider the forecasting task as an image-to-image translation problem, which aims to directly transform multiple concatenated past images into a future image/image sequence through layer-by-layer mapping [18,29–31]. Among the numerous available CNN models, UNet [32] can combine high-level and low-level features through skip connections to exploit more comprehensive information for future image generation, leading to increasing applications in radar-based nowcasting. For example, Agrawal et al. [18] used UNet to provide three pixel-level binary classifications that indicated whether the future rainfall intensity in the given pixel exceeded corresponding thresholds. Instead of predicting classes, UNet was applied in [20,33–35] to extrapolate radar images directly. Han et al. [20] demonstrated that UNet achieved comparable extrapolation performance to a ConvRNN-based model. The recent successes of UNet in the above applications indicate that the role of CNNs in extrapolation-based convective nowcasting needs to be reconsidered.

However, these two types of DNNs still have some limitations. First, it is not easy for standard ConvRNN models to tailor their predictions at different timestamps. One reason is that their sequence-to-sequence (Seq2Seq) structures use the same weights to generate the hidden states of all timestamps. Second, CNN models mainly emphasize spatial features while weakening the temporal variations between the input images, leading to difficulty learning relatively long-range temporal dependencies. Even though a few studies have noticed that 3D convolutions can extract spatiotemporal representations [36,37], they still follow the image-to-image translation paradigm and rarely explicitly model the temporal correlations among the extracted features in the prediction stage.

To leverage the advantages of the UNet and ConvRNN models while avoiding the above limitations, we develop a radar echo extrapolation model called 3D-UNet-LSTM for convective nowcasting, which combines 3D-UNet and a newly designed Seq2Seq network in an extractor-forecaster architecture. We first adopt 3D-UNet as the extractor to extract the spatiotemporal features of the input radar reflectivity images while retaining more detailed information, such as textures. In the forecaster, the Seq2Seq network uses different unstacked ConvLSTM layers to iteratively generate hidden states for different future timestamps. Finally, these hidden states are mapped to predicted images via a convolution layer.

The remainder of this paper is organized as follows. Section 2 describes the data used in this paper, and Section 3 illustrates the proposed model, the loss function, and the evaluation metrics in detail. The experimental results are presented in Section 4. Finally, a summary and discussions are given in Section 5. Appendix A briefly introduces some prior knowledge related to our work.

2. Data

The radar reflectivity data used in this paper are provided by an open meteorological database named MeteoNet [38], which covers two geographical areas, the northwest zone (NW) and southeast zone (SE) of France in Figure 1, and spans 3 years, 2016 to 2018, with 5-min intervals.

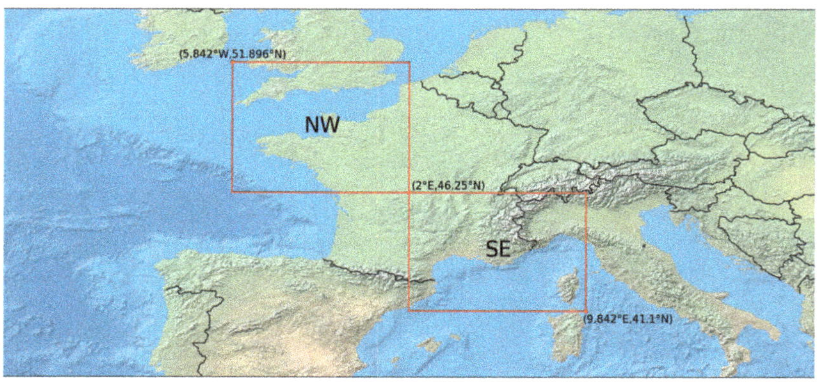

Figure 1. The geographical regions used for the radar reflectivity data (red rectangle).

The data were collected using the Doppler radar network of METEO FRANCE, and 3D reflectivity maps were obtained by each radar scanning the sky. The radar's spatial resolution is 0.01 degrees, and the projection system used is EPSG:4326.

To build our dataset, we first generate 1.5-h radar image sequences (each sequence has 19 radar images) every 25 min. Next, sequence samples are selected if the total number of pixels with reflectivity values $\geq$ 35 dBZ in one of their last 12 images exceeds 2000, and a total of 12,503 sequence samples are collected. To reduce the computational and memory cost and maintain adequate spatial resolution, the images in each sequencing sample are resized from 565 $\times$ 784 to 104 $\times$ 160 through bilinear interpolation, with a spatial resolution of approximately 0.05 degrees. Finally, to test the generalization ability of the proposed model, we ensure that the training, validation, and test subsets do not overlap in time, the details of which are shown in Table 1.

Table 1. The divided subsets for training, validation, and testing.

	Period	Sample Number		Total
		NW	SE	
Training	2016.1–2018.5	5504	4865	10,369
Validation	2018.6–2018.7	480	517	997
Test	2018.8–2018.10	308	829	1137

In addition, the reflectivity (in dBZ) can be approximated to a rainfall intensity R (mm/h) by using the Marshall-Palmer relation:

$$dBZ = 10\log a + 10b\log R \tag{1}$$

where a = 200 and b = 1.6.

3. Methodology

Consecutive radar images can directly show the evolution of convective systems. In this section, we propose a DNN model called 3D-UNet-LSTM to extrapolate future radar reflectivity images. The locations and intensities of convective systems over a very short term can be foreseen according to the extrapolated results. M consecutive radar images are given to predict the subsequent N radar images. In the implementation, we use the radar images in the past 0.5 h to forecast those in the next 1 h (i.e., M = 7, N = 12). We describe the architecture of 3D-UNet-LSTM in Section 3.1 and introduce the loss function and evaluation metrics in Section 3.2 and Section 3.3, respectively.

3.1. 3D-UNet-LSTM

The proposed 3D-UNet-LSTM is an end-to-end trainable model with an extractor-forecaster architecture, as illustrated in Figure 2. In the extractor part, we use 3D-UNet [39] to extract the comprehensive spatiotemporal features of consecutive radar images. It is composed of multiple 3D convolutional layers with kernel sizes of $2 \times 3 \times 3$, each of which is followed by a rectified linear unit (ReLU) activation function. Like UNet, the extractor contains a downsampling path, a symmetrical upsampling path and skip connections. Since skip connections require the temporal and spatial sizes of the features before each downsampling operation to be consistent with those observed after the symmetrical upsampling operation, we add a zero image before the 7 consecutive radar images and stack them along the temporal dimension as the model input. In the downsampling path, the temporal and spatial sizes of the input sequence are progressively halved by using three 3D convolutional layers with strides of 2, each followed by two 3D convolutional layers, and spatiotemporal features with different representation levels are extracted. In the upsampling path, the high-level features gradually return to the original size via three transposed 3D convolutional layers, each followed by two 3D convolutional layers. Furthermore, low-level features are received from the downsampling path through skip connections, bringing detailed information to the more comprehensive representations. Batch normalization (BN) [40] is used after the last convolutional layer to mitigate the vanishing gradient effect during backward propagation. After that, the comprehensive spatiotemporal features of the radar image sequence are output.

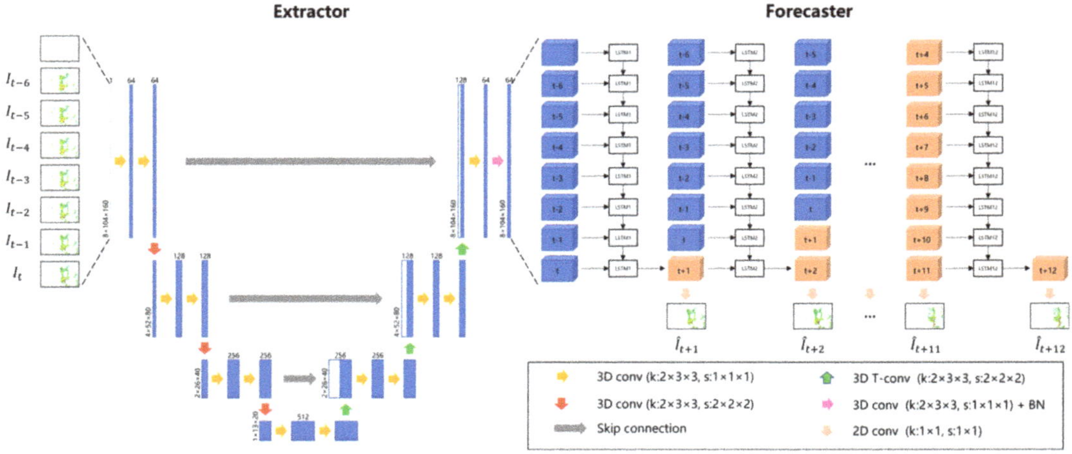

Figure 2. The 3D-UNet-LSTM architecture. 'k' and 's' represent the kernel size and the stride for a convolution, respectively.

The forecaster part is designed to further exploit the spatiotemporal features extracted by the extractor and output the predicted radar images. This part, a Seq2Seq network is presented to explicitly model time and extrapolate the hidden states step-by-step. ConvLSTM is selected as the basic unit due to its simplicity and effectiveness. For the Seq2Seq structure, considering the two common structures in Figure A1 that use shared parameters to generate hidden states for the predictions over all future timestamps, their ability to make corresponding adjustments according to the specific situations encountered at different timestamps in the future may be limited. To alleviate this problem, we utilize N ConvLSTM layers that have different parameters to individually generate the hidden states for future timestamps in an iterative way, as shown in Figure 2, each ConvLSTM layer has a step length of 8 with a convolutional kernel size of 3 × 3 and 64 hidden state channels thereby exploiting the long-term spatiotemporal information of the inputs and obtaining a hidden state correlated with a specific future timestamp. The hidden state output by the previous ConvLSTM layer is concatenated behind the inputs of the last 7 timestamps of this layer. Then, these are fed into the next layer to output the hidden state of the next future timestamp. In addition to utilizing different layers to tailor the predictions for different timestamps, the iterative design can ensure that the previous features, whether extracted by the extractor or generated by specific ConvLSTM layers, can be reused multiple times, thus, it is also helpful in improving the quality of long-term forecasts. Finally, the hidden state at each future timestamp is converted to a corresponding radar reflectivity image through a 2D convolutional layer with a kernel size of 1 × 1.

3.2. Loss Function

In many spatiotemporal sequence forecasting tasks, such as video prediction and traffic flow prediction, where the pixel values of images are relatively evenly distributed, the mean absolute error (*MAE*) and mean squared error (*MSE*) are used as the loss functions to train DNN models. However, for radar reflectivity images, the proportion of low-intensity pixels is much larger than that of high-intensity pixels [21]. Training the extrapolation model with the original *MAE* and *MSE* losses will make it focus on predicting low-intensity pixels (indicating no weather echoes and weak echoes), limiting the forecasting effect in areas with relatively strong echoes associated with hazardous convection. To achieve better forecasting performance for strong echoes, we introduce a balanced reconstruction loss function $L_{\text{B-rec}}$ that assigns greater weights to the errors of higher reflectivity values in the calculation process:

$$L_{\text{B-rec}} = \frac{1}{NHW} \sum_{n=1}^{N} \sum_{i=1}^{H} \sum_{j=1}^{W} \left\{ weight_{t+n,i,j} \times \left[\left| I_{t+n,i,j} - \hat{I}_{t+n,i,j} \right| + \left(I_{t+n,i,j} - \hat{I}_{t+n,i,j} \right)^2 \right] \right\} \quad (2)$$

$$weight_{t+n,i,j} = \begin{cases} 1, & I_{t+n,i,j} < 15dBZ \\ 2, & 15dBZ \leq I_{t+n,i,j} < 35dBZ \\ 5, & 35dBZ \leq I_{t+n,i,j} \end{cases} \quad (3)$$

where $I_{t+n,i,j}$ denotes the observed reflectivity value of the (i,j)th pixel of the future image at timestamp $t+n$, and $\hat{I}_{t+n,i,j}$ denotes the corresponding predicted value. $weight_{t+n,i,j}$ is the weight assigned to each pixel according to the range of its observed reflectivity. H and W are the height and width of the radar images, respectively. As in previous work [20,21,41], the values of $weight_{t+n,i,j}$ are determined based on experience. The prediction errors of high reflectivity values are given larger weights compared to those of low reflectivity values, but the difference between weights is only 2–3 times. Finally, the weights are determined by experiment. We verify the effectiveness of the balanced reconstruction loss function in Section 4.

3.3. Evaluation Metrics

To quantitatively evaluate the nowcasting performance of extrapolation models, we apply the probability of detection (*POD*), false-alarm ratio (*FAR*), bias score (*BIAS*), critical success index (*CSI*), root mean square error (*RMSE*) and correlation coefficient (*CC*) and design a temporally weighted average *CSI* (*twaCSI*) measure. These metrics can be computed based on a given threshold τ, representing a corresponding echo intensity level. *CSI* can provide a ratio of correct predictions. For its calculation, the observed image and predicted image are first binarized by a threshold τ. A pixel value greater than τ is set to 1; otherwise, it is set to 0. Then, *TP*, *FN*, and *FP*, which denote the numbers of true positives (prediction = 1, observation = 1), false negatives (prediction = 0, observation = 1) and false positives (prediction = 1, observation = 0), respectively, are obtained. The *CSI* is computed as

$$CSI^{\tau} = \frac{TP}{TP + FN + FP} \quad (4)$$

Furthermore, considering it becomes more challenging to forecast radar images with increasing lead time, we design $twaCSI^{\tau}$ to evaluate the temporal sequence of predicted radar images. It emphasizes the *CSI* scores of the images predicted at later timestamps by assigning them heavier weights; this step is defined as

$$twaCSI^{\tau} = \frac{\sum_{n=1}^{N} n \cdot CSI_{t+n}^{\tau}}{\sum_{n=1}^{N} n} \quad (5)$$

where CSI_{t+n}^{τ} is the *CSI* score of the predicted image at timestamp $t+n$.

POD and *FAR* would emphasize the amount of missed events and false alarms. Also, including *BIAS* will give an idea about the deviation of predictions.

$$POD^{\tau} = \frac{TP}{TP + FN} \quad (6)$$

$$FAR^{\tau} = \frac{FP}{TP + FP} \quad (7)$$

$$BIAS^{\tau} = \frac{TP + FP}{TP + FN} \quad (8)$$

when $BIAS > 1$, the forecast result is stronger than the real; when $BIAS < 1$, the forecast result is weaker; when $BIAS = 1$, the forecast deviation is 0, which is the highest prediction skill. In addition, for each predicted image, we utilize $RMSE^{\tau}$ and CC^{τ} to present the prediction error and consistency in the area where the observed reflectivities are greater than τ. Denoting the sets of observed values larger than τ and the corresponding predicted values as s and $\hat{s}$, respectively, $RMSE^{\tau}$ and CC^{τ} are calculated as follows:

$$RMSE^{\tau} = \sqrt{\frac{1}{|s|}\sum_{i=1}(s_i - \hat{s_i})^2} \tag{9}$$

$$CC^{\tau} = \frac{\mathrm{Cov}(s,\hat{s})}{\sqrt{\mathrm{Var}(s)\cdot\mathrm{Var}(\hat{s})}} \tag{10}$$

where $|s|$ represents the number of values in set s.

Specifically, we select 18 dBZ (0.5 mm/h, indicating rain or not [21]) and 35 dBZ (used to identify strong convections [10]) as the thresholds.

4. Experiments and Results

To evaluate the effectiveness and superiority of the proposed 3D-UNet-LSTM model, extrapolation-based 0–1 h nowcasting experiments are conducted. For comparison, six baseline models and a state-of-the-art model are reimplemented, including the Eulerian persistence model (hereafter called Persistence), which assumes that future radar images do not differ from the most recent observed image, a conventional model based on optical flow (Rainymotion [14]), five deep learning models including three four-layer ConvRNN models (ConvLSTM [17], PredRNN [22], SA-ConvLSTM [26]), a U-Net [32] model, and a state-of-the-art model (RainPredRNN [23]). In those models, ConvLSTM adopts the "same-side" structure, and PredRNN and SA-ConvLSTM apply the "opposite-side" structure.

We first separately train the 3D-UNet-LSTM model and the other deep learning models on the training set and validation set following the settings in Section 4.1 and then compare the performance of Persistence, Rainymotion and the well-trained models on the whole test set in Section 4.2. Then, to verify the effectiveness of the model design, Section 4.3 compares the 3D-UNet-LSTM model with two variations, including 3D-UNet. Next, in Section 4.4, we further investigate the impact of the balanced loss and adversarial loss functions on the performance of DNNs in accurately predicting convective echoes. Finally, two representative cases are studied in Section 4.5.

4.1. Implementation Details for Training

The radar reflectivity images are first normalized to [0, 1] and then fed into the DNN models. For a fair comparison, all models are trained with the balanced reconstruction loss function on the training set via the adaptive moment estimation (ADAM) optimizer [42] with an initial learning rate of 10^{-4}. The batch size of each training iteration is set to 4. To prevent overfitting, the training process is stopped if the $twaCSI^{35}$ obtained on the validation set is not improved for 20 epochs. All experiments are implemented in TensorFlow [43] and executed on a TITAN RTX GPU (24 GB).

4.2. Quantitative Evaluation of Eight Models on the Test Set

We quantitatively evaluate the overall 0–1 h nowcasting performance of the proposed 3D-UNet-LSTM model, RainPredRNN and six baseline models with the *CSI*, *twaCSI*, *CC* and *RMSE* scores (averaged over all 1137 samples) obtained on the test set. The *twaCSI* results and the mean *CSI*, *CC* and *RMSE* values obtained for all lead times at thresholds of 18 and 35 dBZ are tabulated in Table 2. Persistence has the poorest scores for all metrics. The optical flow based Rainymotion approach obviously performs better than Persistence with the help of the calculated motion field. The six well-trained DNN models significantly outperform the above two traditional models, which demonstrates the powerful modeling capability of deep learning. Among the ConvRNN models, although PredRNN achieves the same performance as ConvLSTM in terms of the *CSI* and *twaCSI*, it obtains higher *CC* and lower *RMSE* scores at both thresholds that the nowcasting values of PredRNN are more precise and closely aligned with the ground truth than those of ConvLSTM. RainPredRNN performs better than PredRNN with the help of the ST-LSTM unit and setting appropriate hyperparameters. Another SA-ConvLSTM obtains similar CSI^{18} and $twaCSI^{18}$ scores compared to those of ConvLSTM, PredRNN and RainPredRNN. Yet, it is superior to both when the threshold is set to 35 dBZ, particularly for $twaCSI^{35}$, implying that SA-ConvLSTM has a better nowcasting performance at longer lead times for echoes with high-intensity levels. The UNet model, which does not have a special design for time series modeling, obtains even better scores for all metrics than the above three advanced ConvRNN models at the thresholds of 18 dBZ and 35 dBZ, which is noteworthy, as it shows the high potential of the UNet architecture for extrapolation-based convective nowcasting. The proposed 3D-UNet-LSTM model yields the best nowcasting scores among the eight models, which verifies its superiority. Greater improvements in the *CSI* and *twaCSI* are achieved at the 35 dBZ threshold than at the 18 dBZ threshold because we focus more on improving the prediction accuracy for convective echoes, especially at longer lead times. In addition, the best *CC* and *RMSE* scores obtained at both thresholds indicate that the predicted radar reflectivities of 3D-UNet-LSTM are more precise and, thus better for estimating future rainfall intensities.

Table 2. Overall performance of the eight models on the test set.

Method	CSI↑		twaCSI↑		CC↑		RMSE↓	
	18 dBZ	35 dBZ	18 dBZ	35 dBZ	18 dBZ	35 dBZ	18 dBZ	35 dBZ
Persistence	0.4181	0.2068	0.3591	0.1554	0.2644	0.0355	16.92	21.34
Rainymotion	0.5149	0.2675	0.4581	0.2107	0.3616	0.0694	14.01	17.69
ConvLSTM	0.5814	0.3244	0.5421	0.2786	0.4350	0.1007	10.70	12.89
PredRNN	0.5898	0.3278	0.5468	0.2755	0.4500	0.1256	10.58	12.78
RainPredRNN	0.5906	0.3314	0.5483	0.2868	0.4624	0.1363	10.45	12.63
SA-ConvLSTM	0.5811	0.3349	0.5444	0.2933	0.4422	0.1110	10.47	12.50
UNet	<u>0.5938</u>	<u>0.3550</u>	<u>0.5497</u>	<u>0.2998</u>	<u>0.4707</u>	<u>0.1570</u>	<u>10.41</u>	<u>12.03</u>
3D-UNet-LSTM	**0.5990**	**0.3742**	**0.5512**	**0.3201**	**0.4853**	**0.1760**	**9.72**	**11.34**

The best and second-best scores are marked in bold and underlined, respectively, ↑ which means that higher is better, while ↓ lower is better.

The *POD*, *FAR* and *BIAS* values obtained for all lead times at thresholds of 18 and 35 dBZ are tabulated in Table 3. For the forecasting of medium and strong echoes, the *BIAS* score of our proposed model is greater than 1, and the overall forecast results are strong. The reason is that the model is designed to focus more on strong echoes. The model has the best *POD* and *FAR* scores at the thresholds of 35 dBZ (strong echo).

Table 3. Evaluation scores of our proposed model with others.

Method	POD↑		FAR↓		BIAS	
	18 dBZ	35 dBZ	18 dBZ	35 dBZ	18 dBZ	35 dBZ
Persistence	0.5664	0.3202	0.4205	0.6727	0.9845	1.0220
Rainymotion	0.6525	0.3585	0.3170	0.5315	0.9546	0.7718
ConvLSTM	0.7887	0.4776	0.3230	0.5085	1.1795	0.9820
PredRNN	0.7888	0.4651	**0.3129**	<u>0.4923</u>	1.1622	0.9072
RainPredRNN	0.7953	0.4836	0.3206	0.5049	1.1584	1.0659
SA-ConvLSTM	<u>0.8012</u>	0.5021	0.3319	0.5133	1.2178	1.0384
UNet	0.8005	<u>0.5480</u>	<u>0.3145</u>	0.5136	1.1863	1.1500
3D-UNet-LSTM	**0.8238**	**0.5610**	0.3235	**0.4844**	1.2462	1.1489

The best and the second-best scores are marked in bold and underlined, respectively, ↑ which means that higher is better, while ↓ lower is better.

Beyond that, to directly show the convective nowcasting performance over time, the *CSI*, *CC* and *RMSE* curves produced by the eight models at the 35 dBZ threshold against different nowcasting lead times up to 60 min are plotted in Figure 3. The results show that the performance of all extrapolation models deteriorates with increasing lead times, which can be expected and mainly results from unavoidable error accumulation and increasing uncertainty in the forecasting process. RainPredRNN and PredRNN obtain similar performance on all metrics over time. In addition, we notice that although UNet achieves a better overall performance in terms of mean CSI^{35} and $RMSE^{35}$ in Table 2 than the three ConvRNN models and RainPredRNN, this is largely due to the contribution of its better scores for lead times between 5 and 30 min. Later, the performance of UNet gradually becomes comparable to that of SA-ConvLSTM and is finally exceeded by that approach for lead times beyond approximately 45 min. One reason for this phenomenon presumably is that UNet focuses on maintaining or changing spatial appearances for radar images but fails to capture the internal temporal dependencies; this appears to affect its long-term prediction effectiveness.

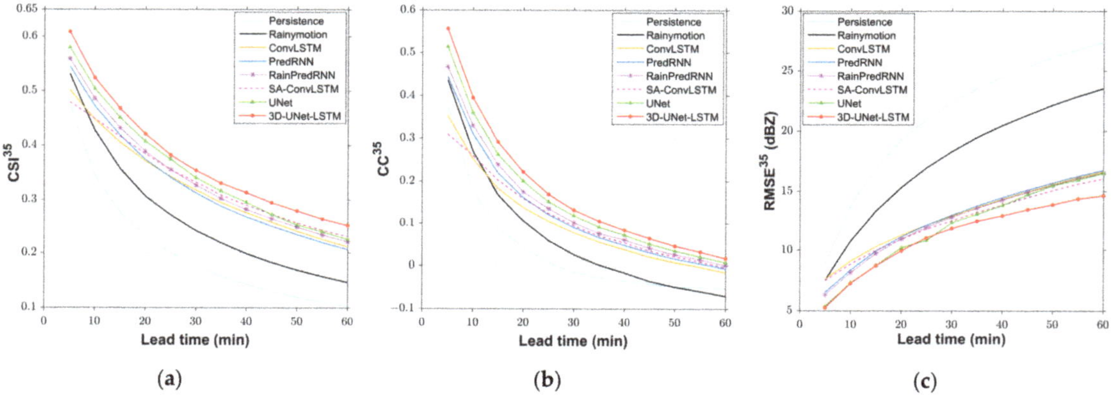

Figure 3. The (**a**) *CSI*, (**b**) *CC* and (**c**) *RMSE* curves produced by the eight models at the 35 dBZ threshold against different lead times. All values are the scores averaged over all cases in the test set at the corresponding lead time.

In contrast, the proposed 3D-UNet-LSTM produces the best CSI^{35} value for any lead time in one hour and achieves a score of more than 0.25 for 60-min nowcasts, while those of other deep learning models are in the range of 0.21 to 0.23. The same is true for $RMSE^{35}$; the proposed model remains competitive over the whole period, and its superiority becomes increasingly obvious at lead times after 30 min. For 60-min nowcasts, it reduces the average error by almost 2 dBZ compared with UNet. In terms of CC^{35}, the prediction results of the proposed model exhibit consistency with the observation values, especially at shorter lead times. Although its performance drops sharply as the lead time increases, our model still achieves the highest CC^{35} scores compared to other models. In general, 3D-UNet-LSTM has better early performance than UNet and consistently outperforms SA-ConvLSTM at long lead times, demonstrating its effective spatiotemporal modeling ability and better overall performance for convective nowcasting.

4.3. Evaluation of the Model Design

To evaluate the effectiveness of the 3D-UNet-LSTM model design, we first design two variations of the model, one that removes the forecaster and retains the 3D-UNet extractor only and another that replaces the forecaster with a two-layer ConvLSTM network (this variation model is referred to as '3D-UNet + ConvLSTM'). Then, the overall performance of the original ConvLSTM, UNet, 3D-UNet-LSTM and these two variations are compared, as shown in Table 4. When only the 3D-UNet extractor is retained, it still outperforms ConvLSTM and UNet in terms of the metrics at the 35 dBZ threshold, indicating that the 3D-UNet extractor has good potential for convective nowcasting. However, as we attempt to use a common ConvLSTM network to further leverage the features extracted by 3D-UNet and generate future hidden states according to the shared parameters, the nowcasting performance decreases considerably, becoming even worse than that of the original ConvLSTM. In contrast, when utilizing our designed forecaster to produce future hidden states with different parameters, the model obtains better scores than those of 3D-UNet, demonstrating the effectiveness of the forecaster.

Table 4. Quantitative evaluation of the model design.

Method	CSI↑		twaCSI↑		CC↑		RMSE↓	
	18 dBZ	35 dBZ	18 dBZ	35 dBZ	18 dBZ	35 dBZ	18 dBZ	35 dBZ
ConvLSTM	0.5814	0.3244	0.5421	0.2786	0.4350	0.1007	10.70	12.89
UNet	<u>0.5938</u>	0.3550	<u>0.5497</u>	0.2998	0.4707	0.1570	10.41	12.03
3D-UNet	0.5897	<u>0.3642</u>	0.5439	<u>0.3099</u>	<u>0.4735</u>	<u>0.1687</u>	<u>10.27</u>	<u>11.76</u>
3D-UNet + ConvLSTM	0.5567	0.3097	0.5197	0.2648	0.4208	0.1087	10.96	13.03
3D-UNet-LSTM	**0.5990**	**0.3742**	**0.5512**	**0.3201**	**0.4853**	**0.1760**	**9.72**	**11.34**

The best and second-best scores are marked in bold and underlined, respectively, ↑ which means that higher is better, while ↓ lower is better.

We also draw the CSI^{35}, CC^{35} and $RMSE^{35}$ curves of these methods for different lead times in Figure 4. It can be seen that by combining 3D-UNet and the forecaster, our model has better performance than the other approaches for nearly all lead times. The superiority of its design is more obvious for longer lead times.

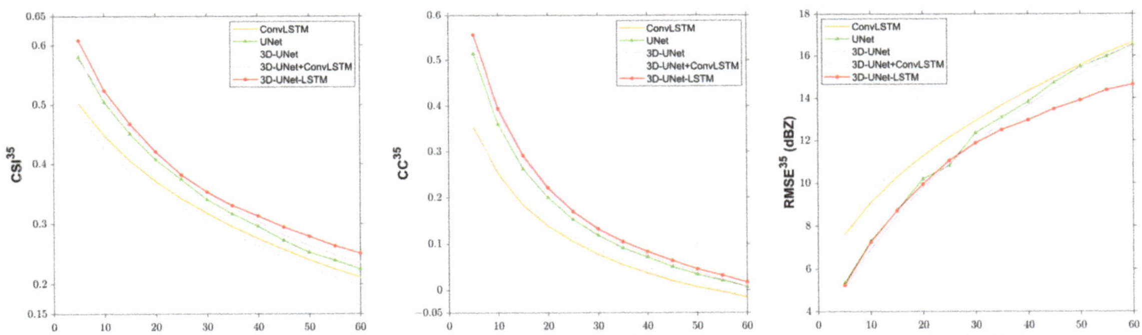

(a) (b) (c)

Figure 4. The (**a**) *CSI*, (**b**) *CC* and (**c**) *RMSE* curves at the 35 dBZ threshold against different lead times for the evaluation of the model design.

4.4. Evaluation of Different Loss Functions

In the following, we train the 3D-UNet-LSTM model with different loss functions and test their effects on the prediction accuracy for convective echo regions. These loss functions are the reconstruction loss (the sum of the *MAE* and *MSE*) widely used in video prediction tasks [22,26], the sum of the reconstruction loss and adversarial loss, which has been applied to address the blurring problem for echo prediction [24], the balanced reconstruction loss [21] applied in this paper, and the sum of the balanced reconstruction loss and adversarial loss [37,44]. The scaling factor of the adversarial loss is set to 0.03 to ensure that it can exert a certain degree of influence on the model training process. When the scaling factor is set to 0.003, its influence is quite slight. The results are shown in Table 5. We can see that without using any weights for reflectivities, the reconstruction loss slightly improves the CSI^{18} and $twaCSI^{18}$ scores but yields much poorer performance than that of the balanced loss functions in terms of other metrics, especially CSI^{35} and $twaCSI^{35}$. As we add an adversarial term to the reconstruction loss, these gaps are slightly narrowed. Regarding the balanced loss functions, the balanced reconstruction loss applied in this paper obtains the best scores for all evaluation metrics at the 35 dBZ threshold.

Table 5. Quantitative evaluation of different loss functions.

Loss Function	CSI↑		twaCSI↑		CC↑		RMSE↓	
	18 dBZ	35 dBZ	18 dBZ	35 dBZ	18 dBZ	35 dBZ	18 dBZ	35 dBZ
L_{rec}	**0.6045**	0.3302	**0.5575**	0.2636	0.4460	0.1114	11.26	13.86
$L_{\text{rec}} + 0.03L^{g}_{\text{adv}}$	0.5950	0.3392	0.5463	0.2794	0.4535	0.1433	11.08	13.37
$L_{\text{B-rec}}$	<u>0.5990</u>	**0.3742**	0.5512	**0.3201**	**0.4853**	**0.1760**	**9.72**	**11.34**
$L_{\text{B-rec}} + 0.003L^{g}_{\text{adv}}$	0.5978	<u>0.3716</u>	<u>0.5520</u>	<u>0.3161</u>	<u>0.4760</u>	<u>0.1622</u>	<u>10.12</u>	<u>11.57</u>
$L_{\text{B-rec}} + 0.03L^{g}_{\text{adv}}$	0.5884	0.3639	0.5385	0.3058	0.4635	0.1529	10.76	12.37

The best and second-best scores are marked in bold and underlined, respectively; ↑ which means that higher is better, which ↓ means that lower is better.

Regarding its combination with an adversarial loss, the convective nowcasting performance deteriorates with increasing scaling factors for the adversarial term. It can be concluded that compared with the original reconstruction loss, the balanced loss can significantly improve the convective nowcasting performance of a deep learning model. It seems that adding an adversarial loss to the reconstruction loss can slightly improve the prediction accuracy for convective echoes. However, for the balanced reconstruction loss, adding an adversarial loss term is of no help for further increasing the prediction precision.

4.5. Representative Case Study

To qualitatively evaluate the performance of the proposed model, we select two representative cases from the test set and visually examine the nowcasts produced by different models. The images of two cases, including radar observations and nowcasts, are presented in Figure 5 and Figure 6, respectively, and are displayed every 15 min to show the evolutions of convective systems.

Figure 5 shows a representative case of local strong convective growth over northwest France at a forecasting time of T = 7 August 2018, 11:55 UTC. In the input radar images, it can be seen that an isolated convective cell is located in the west at time T - 30 min, moving northeast together with other dispersed echoes, and the formation of a new strong small-scale convective cell occurs in Region B at forecasting time T. For the ground-truth observations in the next hour, the echoes continue to move in the northeast direction, and during this period, the new convective cell gradually grows and appears to merge with the older cell. Comparing the nowcasting results of each model with the ground truth, one can observe that all models can capture the movements of most echoes. However, the optical flow-based Rainymotion method simply advects the radar echoes. It fails to forecast the subsequent growth and evolution of the newly formed convective cell because it cannot completely model nonlinear processes. In contrast, all deep learning models successfully forecast that the newly formed convective cell will grow at time T + 30 min but underestimate its intensity. This under-forecasting problem, also called blurry prediction, is common when utilizing deterministic deep learning models for radar echo extrapolation, especially with longer lead times; this is mainly because a DNN model tends to average all probable outcomes to a blurry prediction in a case in which it has difficulty dealing with future uncertainty [45]. Nonetheless, the 30-min nowcast obtained by the 3D-UNet-LSTM model is closer to the ground truth in terms of the horizontal extent of the convection than those derived from other models. For the 60-min nowcasts, the forecasted intensities of the old convective cell in the results of other deep learning models deviate considerably from the ground truth, while the 3D-UNet-LSTM model and 3D-UNet model can maintain their intensity values at relatively high levels ($\geq$ 40 dBZ). It is noted that only the 3D-UNet-LSTM model forecasts a further growth trend in the size of the newly formed convective cell from time T + 30 min to T + 60 min, and its 60-min nowcasting result also successfully depicts the merging phenomenon of the two isolated convective echoes that occur in regions A and B one hour later.

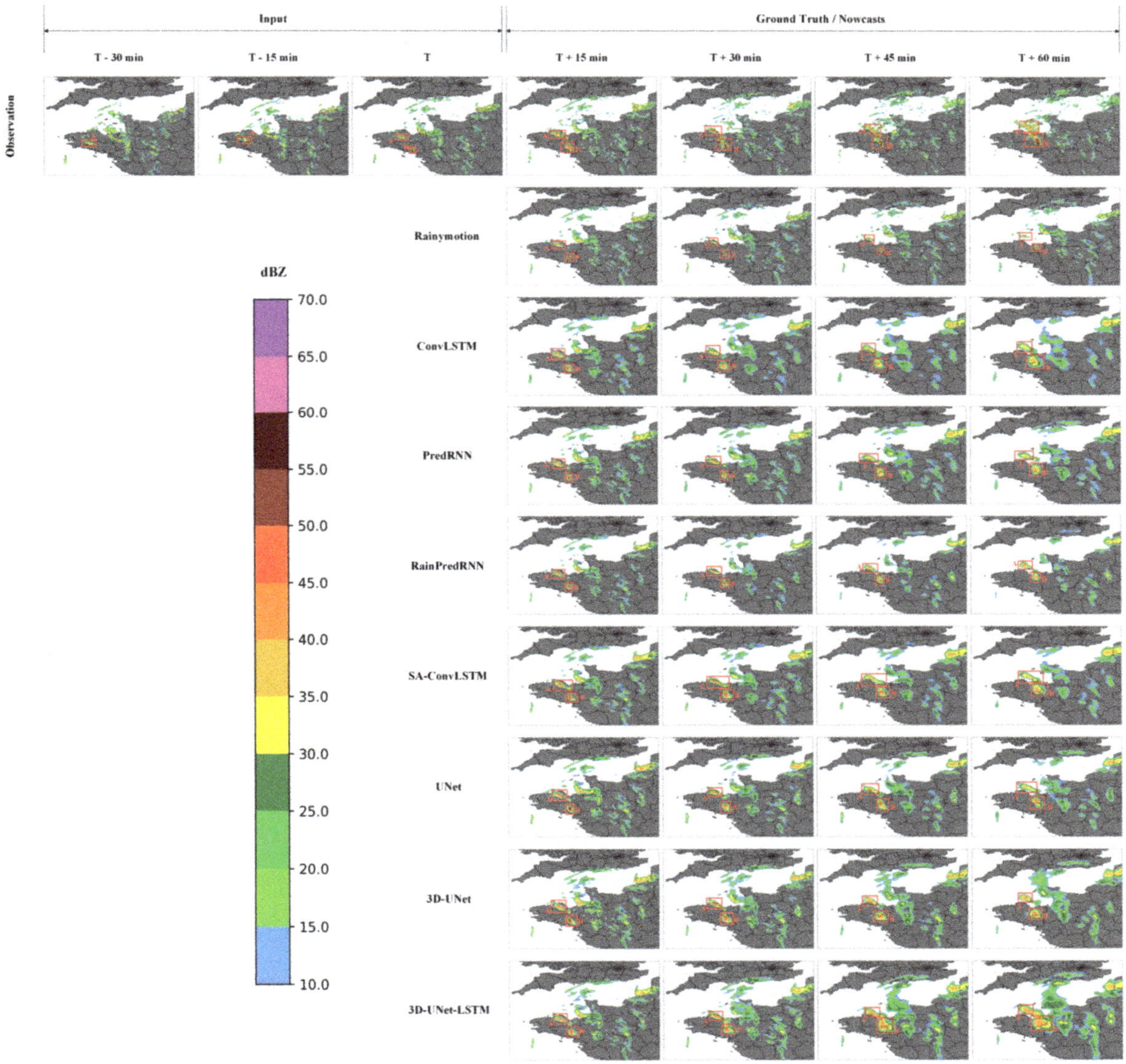

Figure 5. A representative case of local strong convective growth in the northwestern quarter of France at a forecasting time of T = 7 August 2018, 11:55 UTC. Letters A–B represents different regions where the proposed 3D-UNet-LSTM performs well.

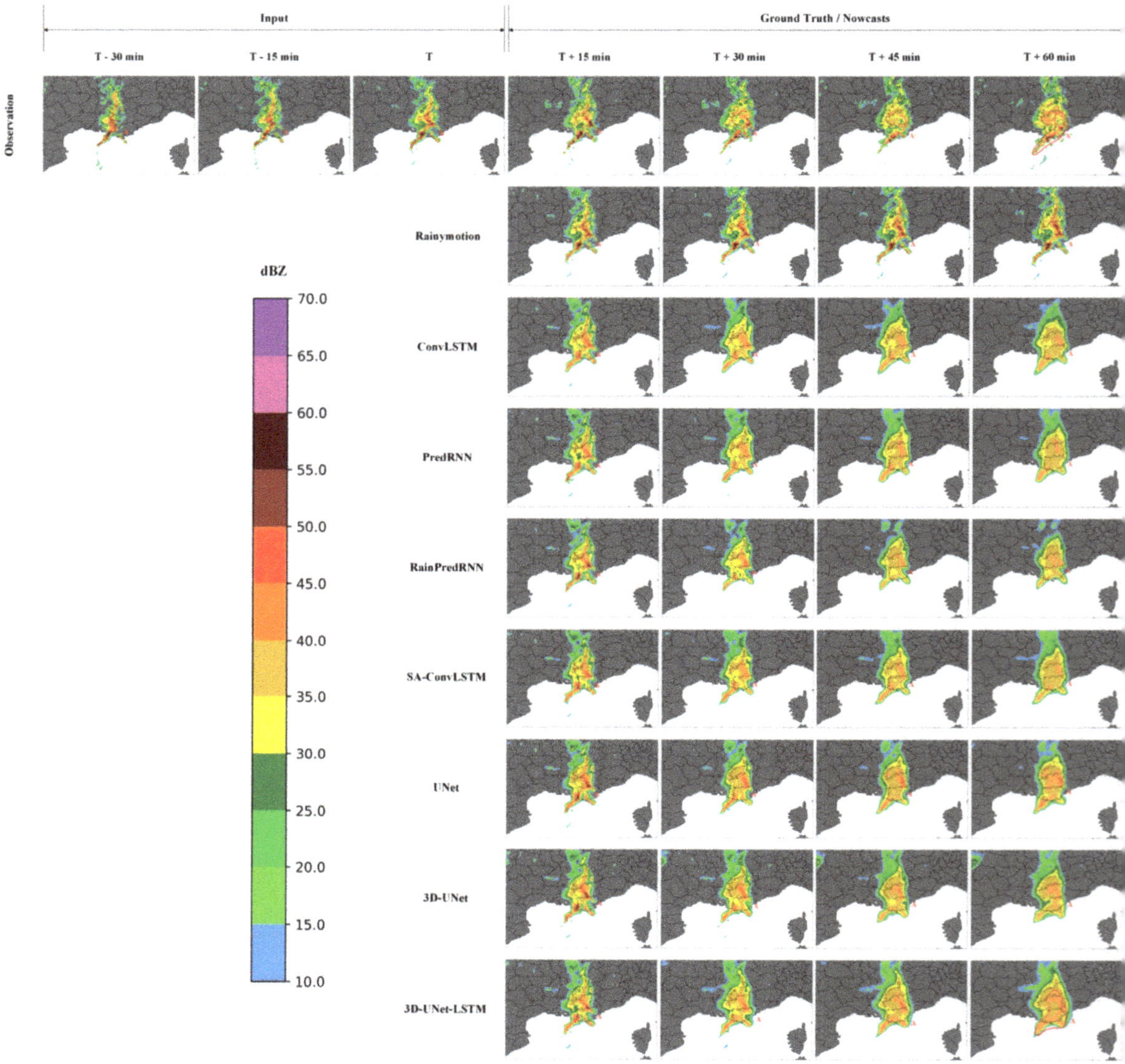

Figure 6. A representative case of squall line evolution in the southeastern quarter of France at a forecasting time of T = 13 August 2018, 05:00 UTC. Letter A represents the region where the proposed 3D-UNet-LSTM performs well.

Another representative case is shown in Figure 6, which describes the evolution of a severe squall line that occurs in southeast France at a forecasting time of T = 13 August 2018, 05:00 UTC. It is clear from the radar observations that a squall line is moving eastward while the convective area behind it gradually becomes larger, and it finally develops into a bow echo at time T + 60 min. As in the first case, all models provide relatively accurate moving directions for the quasi-linear convective system. The 30-min nowcasts obtained from all models, especially UNet, achieve good agreement with the radar observations, presumably because the system evolves relatively slowly during the first half hour after forecasting time T. However, for the 60-min nowcasts, it is difficult for the optical flow-based Rainymotion method to predict the subsequent convective evolution. Although the deep learning models successfully forecast that the convective area will expand in the future, significant differences remain between their 60-min nowcast performances. For example, one can observe that the three ConvRNN models give misleading information that high-impact meteorological hazards (reflectivity $\geq$ 40 dBZ) tend to decrease. Although UNet and 3D-UNet effectively preserve their intensities, neither they nor the ConvRNN models can forecast the bow echo structure at time T + 60 min. It is noted that the proposed 3D-UNet-LSTM yields a more trustable 60-min nowcast in Region A with a realistic bow echo structure (the region with reflectivity $\geq$ 40 dBZ in Figure 6) and a reasonable intensity distribution than those of other models. Bow echo is bowed toward the direction of movement. There are general weaknesses in reflectivity behind the bow. Only the nowcasting results of the proposed approach depict the squall line-to-bow echo transition clearly, indicating that 3D-UNet-LSTM has a better spatiotemporal modeling ability for the complex nonlinear processes of convective echoes.

5. Conclusions

In this paper, we propose a novel deep learning model called 3D-UNet-LSTM to precisely extrapolate radar reflectivity images for convective nowcasting. This model combines a well-known CNN named 3D-UNet and a newly designed Seq2Seq network in an extractor-forecaster architecture. We first apply 3D-UNet as the extractor to extract the comprehensive spatiotemporal representations of input radar images. Then, in the forecaster, the extracted features are further leveraged by the Seq2Seq network to individually generate hidden states for different future timestamps with different ConvLSTM layers. These hidden states are finally transformed into predicted radar images by a convolutional layer.

We conduct comparative experimental studies on a test set. The quantitative evaluation results show that 3D-UNet-LSTM outperforms conventional methods and state-of-the-art deep learning models regarding the prediction of convective echoes, particularly with long lead times. In addition, the evaluation of the model design demonstrates the effectiveness of the 3D-UNet extractor and the newly designed forecaster, as well as their combination. It is noteworthy that UNet-based models, especially 3D-UNet, achieve comparable or even superior performance to that of some ConvRNN-based models. We also verify the effectiveness of the utilized balanced loss function on the model performance for precisely forecasting strong echoes. Finally, representative case studies qualitatively illustrate that the 3D-UNet-LSTM model can better model the nonlinear processes of the evolutions of convective echoes and produce more reasonable and location-accurate nowcasts.

Although the quantitative and qualitative comparison and analysis verify the superiority and effectiveness of 3D-UNet-LSTM for extrapolation-based convective nowcasting, some limitations remain. We think these should be noted and discussed. First, like other deep learning models, the proposed model has difficulty forecasting convective initiation, which is still challenging for the meteorological community. One main reason is that the input reflectivity images cannot provide a DNN with sufficient early signals and characteristics of convective initiation. From there, adding relevant radar variables to supplement input reflectivities may be a promising direction. Second, the loss function has much room for improvement and introducing an additional classification network and an effective classification loss seems to be a good solution. Thirdly, we are currently working on only one benchmark dataset and will try to conduct studies using different benchmark data. In future work, we will carry out research on these three aspects.

Author Contributions: Conceptualization, Q.L., N.S. and S.G.; methodology, N.S. and S.G.; validation, S.G. and N.S.; investigation, N.S.; writing—original draft preparation, S.G.; writing—review and editing, N.S. and Y.P.; supervision, Q.L. All authors have read and agreed to the published version of the manuscript.

Funding: This research was funded by the National Natural Science Foundation of China (Grant No. U2242201, 42075139, 41305138), the China Postdoctoral Science Foundation (Grant No. 2017M621700), Hunan Province Natural Science Foundation (Grant No. 2021JC0009, 2021JJ30773) and Fengyun Application Pioneering Project (FY-APP-2022.0605).

Data Availability Statement: Meteonet data [38] is available at https://meteonet.umr-cnrm.fr/ (accessed on 6 April 2022).

Acknowledgments: The authors would like to thank the anonymous reviewers for providing professional and insightful comments about this manuscript. Finally, we thank the contributors of the Meteonet dataset for collecting, processing, and sharing their data.

Conflicts of Interest: The authors declare no conflict of interest.

Appendix A

Existing studies that have applied ConvRNNs or CNNs to conduct extrapolation-based convective nowcasting have included some important research directions, such as developing effective networks and designing loss functions. Two key issues need to be considered when designing a ConvRNN-based model: the basic ConvRNN unit and the Seq2Seq structure. In this appendix, we briefly introduce the typical ConvLSTM unit and the common Seq2Seq structures related to our method, as well as a typical adversarial loss function that is evaluated in experiments.

Appendix A.1. ConvLSTM Unit

The ConvLSTM unit is the basic component of a ConvLSTM model [17]. It receives the current input X_t, previous hidden state H_{t-1}, and temporal cell state C_{t-1} to generate a new hidden state H_t through a gate-controlled mechanism. This process can be formulated as

$$i_t = \sigma(W_{xi} * X_t + W_{hi} * H_{t-1} + b_i) \tag{A1}$$

$$f_t = \sigma\left(W_{xf} * X_t + W_{hf} * H_{t-1} + b_f\right) \tag{A2}$$

$$C_t = f_t \circ C_{t-1} + i_t \circ tanh(W_{xc} * X_t + W_{hc} * H_{t-1} + b_c) \tag{A3}$$

$$o_t = \sigma(W_{xo} * X_t + W_{ho} * H_{t-1} + b_o) \tag{A4}$$

$$H_t = o_t \circ tanh(C_t) \tag{A5}$$

where W and b represent the trainable 2D convolution kernel and bias, respectively. σ is the sigmoid activation function. $*$ and $\circ$ are the 2D convolution operation and the Hadamard product, respectively. The information flow is controlled by an input gate i_t, a forget gate f_t and an output gate o_t.

Appendix A.2. Structure

Two Seq2Seq structures were commonly used in prior works on RNN-based radar echo extrapolation, including the "same-side" structure (Figure A1a) [19,21], in which the inputs and predictions are on the same side, and the "opposite-side" structure (Figure A1b) [22,26], in which the predictions are on the opposite side of the inputs. As we can see from Figure A1, both structures can conduct direct multistep prediction by leveraging the shared parameters to generate hidden states over all future timestamps. The "same-side" structure is more suitable for input–output transformation since the spatial and channel sizes of the inputs and predictions are allowed to be different, while the "opposite-side" structure requires them to be consistent and can reduce the difficulty of training.

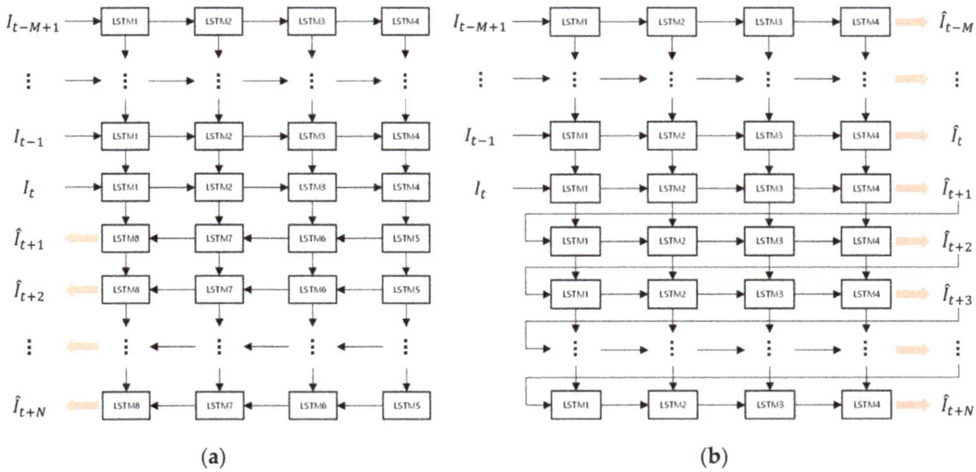

Figure A1. Two commonly used Seq2Seq structures for RNN-based radar echo extrapolation (choosing ConvLSTM as the basic unit). (**a**) The "same-side" structure; (**b**) The "opposite-side" structure.

Appendix A.3. Adversial Loss Function

A GAN [46] is a kind of architecture that is mostly used for image synthesis. A regular GAN-based architecture consists of a generator and a discriminator. The generator outputs images, and the discriminator is trained to distinguish whether its input is produced by the generator or derived from the training dataset (binary classification). At the same time, when training the generator with an adversarial loss function to fool the discriminator, the quality of its output images is improved.

In recent years, some studies have treated the extrapolation model as the generator and trained it in a GAN-based architecture with suitably designed adversarial loss functions to improve the textures of predicted images [19,24,44,47,48]. In that context, a simple yet effective adversarial loss function [48] can be defined as:

$$L_{\text{adv}}^{\text{g}} = \text{E}_x[1 - D(\{x, G(x)\})] \tag{A6}$$

$$L_{\text{adv}}^{\text{d}} = \text{E}_{x,y}[1 - D(\{x, y\})] + \text{E}_x[D(\{x, G(x)\})] \tag{A7}$$

where $L_{\text{adv}}^{\text{g}}$ and $L_{\text{adv}}^{\text{d}}$ denote the loss functions of the generator G and discriminator D, respectively. The generator G takes radar images x as input and generates predicted images $G(x)$, intended to have the same echo distribution as y, the training (ground-truth) data. $D(\cdot)$ is the output of the discriminator D. $\{\}$ represents the concatenation operation.

References

1. Sun, J.; Xue, M.; Wilson, J.W.; Zawadzki, I.; Ballard, S.P.; Onvlee-Hooimeyer, J.; Joe, P.; Barker, D.M.; Li, P.-W.; Golding, B.; et al. Use of NWP for Nowcasting Convective Precipitation: Recent Progress and Challenges. *Bull. Am. Meteorol. Soc.* **2014**, *95*, 409–426; [CrossRef]
2. Wilson, J.W.; Feng, Y.; Chen, M.; Roberts, R.D. Nowcasting Challenges during the Beijing Olympics: Successes, Failures, and Implications for Future Nowcasting Systems. *Weather Forecast.* **2010**, *25*, 1691–1714. [CrossRef]
3. Li, P.-W.; Wong, W.-K.; Cheung, P.; Yeung, H.-Y. An overview of nowcasting development, applications, and services in the Hong Kong Observatory. *J. Meteorol. Res.* **2014**, *28*, 859–876. [CrossRef]
4. Mecikalski, J.R.; Bedka, K.M. Forecasting Convective Initiation by Monitoring the Evolution of Moving Cumulus in Daytime GOES Imagery. *Mon. Weather Rev.* **2006**, *134*, 49–78. [CrossRef]
5. Cancelada, M.; Salio, P.; Vila, D.; Nesbitt, S.W.; Vidal, L. Backward Adaptive Brightness Temperature Threshold Technique (BAB3T): A Methodology to Determine Extreme Convective Initiation Regions Using Satellite Infrared Imagery. *Remote Sens.* **2020**, *12*, 337. [CrossRef]
6. Marshall, J.S.; Langille, R.C.; Palmer, W.M.K. Measurement of rainfall by radar. *J. Atmos. Sci.* **1947**, *4*, 186–192. [CrossRef]

7. Peng, X.; Li, Q.; Jing, J. CNGAT: A Graph Neural Network Model for Radar Quantitative Precipitation Estimation. *IEEE Trans. Geosci. Remote Sens.* **2022**, *60*, 1–14. [CrossRef]
8. Akbari Asanjan, A.; Yang, T.; Hsu, K.; Sorooshian, S.; Lin, J.; Peng, Q. Short-Term Precipitation Forecast Based on the PERSIANN System and LSTM Recurrent Neural Networks. *J. Geophys. Res. Atmos.* **2018**, *123*, 12543–12563. [CrossRef]
9. Germann, U.; Zawadzki, I. Scale-Dependence of the Predictability of Precipitation from Continental Radar Images. Part I: Description of the Methodology. *Mon. Weather Rev.* **2002**, *130*, 2859–2873. [CrossRef]
10. Dixon, M.; Wiener, G. TITAN: Thunderstorm Identification, Tracking, Analysis, and Nowcasting—A Radar-based Methodology. *J. Atmos. Ocean. Technol.* **1993**, *10*, 785–797. [CrossRef]
11. Johnson, J.T.; MacKeen, P.L.; Witt, A.; Mitchell, E.D.W.; Stumpf, G.J.; Eilts, M.D.; Thomas, K.W. The Storm Cell Identification and Tracking Algorithm: An Enhanced WSR-88D Algorithm. *Weather Forecast.* **1998**, *13*, 263–276. [CrossRef]
12. Walker, J.R.; MacKenzie, W.M.; Mecikalski, J.R.; Jewett, C.P. An Enhanced Geostationary Satellite–Based Convective Initiation Algorithm for 0–2-h Nowcasting with Object Tracking. *J. Appl. Meteorol. Climatol.* **2012**, *51*, 1931–1949. [CrossRef]
13. Rinehart, R.E.; Garvey, E.T. Three-dimensional storm motion detection by conventional weather radar. *Nature* **1978**, *273*, 287–289. [CrossRef]
14. Ayzel, G.; Heistermann, M.; Winterrath, T. Optical flow models as an open benchmark for radar-based precipitation nowcasting (rainymotion v0.1). *Geosci. Model Dev.* **2019**, *12*, 1387–1402. [CrossRef]
15. Pulkkinen, S.; Nerini, D.; Pérez Hortal, A.A.; Velasco-Forero, C.; Seed, A.; Germann, U.; Foresti, L. Pysteps: An open-source Python library for probabilistic precipitation nowcasting (v1.0). *Geosci. Model Dev.* **2019**, *12*, 4185–4219. [CrossRef]
16. Hwang, Y.; Clark, A.J.; Lakshmanan, V.; Koch, S.E. Improved Nowcasts by Blending Extrapolation and Model Forecasts. *Weather Forecast.* **2015**, *30*, 1201–1217. [CrossRef]
17. Shi, X.; Chen, Z.; Wang, H.; Yeung, D.-Y.; Wong, W.-K.; Woo, W.-c. Convolutional LSTM network: A machine learning approach for precipitation nowcasting. In Proceedings of the Advances in Neural Information Processing Systems, Montreal, QC, Canada, 7–12 December 2015; pp. 802–810.
18. Agrawal, S.; Barrington, L.; Bromberg, C.; Burge, J.; Gazen, C.; Hickey, J. Machine learning for precipitation nowcasting from radar images. *arXiv* **2019**, arXiv:1912.12132. [CrossRef]
19. Ravuri, S.; Lenc, K.; Willson, M.; Kangin, D.; Lam, R.; Mirowski, P.; Fitzsimons, M.; Athanassiadou, M.; Kashem, S.; Madge, S.; et al. Skilful precipitation nowcasting using deep generative models of radar. *Nature* **2021**, *597*, 672–677. [CrossRef] [PubMed]
20. Han, L.; Liang, H.; Chen, H.; Zhang, W.; Ge, Y. Convective Precipitation Nowcasting Using U-Net Model. *IEEE Trans. Geosci. Remote Sens.* **2022**, *60*, 1–8. [CrossRef]
21. Shi, X.; Gao, Z.; Lausen, L.; Wang, H.; Yeung, D.-Y.; Wong, W.-K.; Woo, W.-C. Deep learning for precipitation nowcasting: A benchmark and a new model. In Proceedings of the Advances in Neural Information Processing Systems, Long Beach, CA, USA, 4–9 December 2017; pp. 5618–5628.
22. Wang, Y.; Long, M.; Wang, J.; Gao, Z.; Yu, P.S. Predrnn: Recurrent neural networks for predictive learning using spatiotemporal lstms. In Proceedings of the Advances in Neural Information Processing Systems, Long Beach, CA, USA, 4–9 December 2017; pp. 879–888.
23. Tuyen, D.N.; Tuan, T.M.; Le, X.-H.; Tung, N.T.; Chau, T.K.; Van Hai, P.; Gerogiannis, V.C.; Son, L.H. RainPredRNN: A New Approach for Precipitation Nowcasting with Weather Radar Echo Images Based on Deep Learning. *Axioms* **2022**, *11*, 107. [CrossRef]
24. Jing, J.; Li, Q.; Peng, X. MLC-LSTM: Exploiting the Spatiotemporal Correlation between Multi-Level Weather Radar Echoes for Echo Sequence Extrapolation. *Sensors* **2019**, *19*, 3988. [CrossRef] [PubMed]
25. Villegas, R.; Yang, J.; Hong, S.; Lin, X.; Lee, H. Decomposing motion and content for natural video sequence prediction. *arXiv* **2017**, arXiv:1706.08033. [CrossRef]
26. Lin, Z.; Li, M.; Zheng, Z.; Cheng, Y.; Yuan, C. Self-attention convlstm for spatiotemporal prediction. In Proceedings of the AAAI Conference on Artificial Intelligence, New York, NY, USA, 7–12 February 2020; pp. 11531–11538.
27. Bai, S.; Kolter, J.Z.; Koltun, V. An empirical evaluation of generic convolutional and recurrent networks for sequence modeling. *arXiv* **2018**, arXiv:1803.01271. [CrossRef]
28. Chafik Bakkay, M.; Serrurier, M.; Kivachuk Burda, V.; Dupuy, F.; Citlali Cabrera-Gutierrez, N.; Zamo, M.; Mader, M.-A.; Mestre, O.; Oller, G.; Jouhaud, J.-C.; et al. Precipitaion Nowcasting using Deep Neural Network. *arXiv* **2022**, arXiv:2203.13263. [CrossRef]
29. Prudden, R.; Adams, S.; Kangin, D.; Robinson, N.; Ravuri, S.; Mohamed, S.; Arribas, A. A review of radar-based nowcasting of precipitation and applicable machine learning techniques. *arXiv* **2020**, arXiv:2005.04988. [CrossRef]
30. Klein, B.; Wolf, L.; Afek, Y. A Dynamic Convolutional Layer for short rangeweather prediction. In Proceedings of the 2015 IEEE Conference on Computer Vision and Pattern Recognition (CVPR), Boston, MA, USA, 7–12 June 2015; pp. 4840–4848.
31. Ayzel, G.; Heistermann, M.; Sorokin, A.; Nikitin, O.; Lukyanova, O. All convolutional neural networks for radar-based precipitation nowcasting. *Procedia Comput. Sci.* **2019**, *150*, 186–192. [CrossRef]
32. Ronneberger, O.; Fischer, P.; Brox, T. U-net: Convolutional networks for biomedical image segmentation. In Proceedings of the International Conference on Medical Image Computing and Computer-Assisted Intervention, Munich, Germany, 5–9 October 2015; pp. 234–241.
33. Ayzel, G.; Scheffer, T.; Heistermann, M. RainNet v1.0: A convolutional neural network for radar-based precipitation nowcasting. *Geosci. Model Develop.* **2020**, *13*, 2631–2644. [CrossRef]

34. Trebing, K.; Stańczyk, T.; Mehrkanoon, S. SmaAt-UNet: Precipitation nowcasting using a small attention-UNet architecture. *Pattern Recognit. Lett.* **2021**, *145*, 178–186. [CrossRef]
35. Pan, X.; Lu, Y.; Zhao, K.; Huang, H.; Wang, M.; Chen, H. Improving Nowcasting of Convective Development by Incorporating Polarimetric Radar Variables Into a Deep-Learning Model. *Geophys. Res. Lett.* **2021**, *48*, e2021GL095302. [CrossRef]
36. Che, H.; Niu, D.; Zang, Z.; Cao, Y.; Chen, X. ED-DRAP: Encoder–Decoder Deep Residual Attention Prediction Network for Radar Echoes. *IEEE Geosci. Remote Sens. Lett.* **2022**, *19*, 1–5. [CrossRef]
37. Larvor, G.; Berthomier, L.; Chabot, V.; Le Pape, B.; Pradel, B.; Perez, L. MeteoNet, An Open Reference Weather Dataset by Meteo-France. 2020. Available online: https://meteonet.umr-cnrm.fr/ (accessed on 6 April 2022).
38. Wang, C.; Wang, P.; Wang, P.; Xue, B.; Wang, D. Using Conditional Generative Adversarial 3-D Convolutional Neural Network for Precise Radar Extrapolation. *IEEE J. Sel. Top. Appl. Earth Obs. Remote Sens.* **2021**, *14*, 5735–5749. [CrossRef]
39. Çiçek, Ö.; Abdulkadir, A.; Lienkamp, S.S.; Brox, T.; Ronneberger, O. 3D U-Net: Learning dense volumetric segmentation from sparse annotation. In Proceedings of the International Conference on Medical Image Computing and Computer-Assisted Intervention, Athens, Greece, 17–21 October 2016; pp. 424–432.
40. Ioffe, S.; Szegedy, C. Batch normalization: Accelerating deep network training by reducing internal covariate shift. In Proceedings of the International Conference on Machine Learning, Lille, France, 6–11 July 2015; pp. 448–456.
41. Niu, D.; Huang, J.; Zang, Z.; Xu, L.; Che, H.; Tang, Y. Two-Stage Spatiotemporal Context Refinement Network for Precipitation Nowcasting. *Remote Sens.* **2021**, *13*, 4285. [CrossRef]
42. Kingma, D.P.; Ba, J. Adam: A method for stochastic optimization. *arXiv* **2014**, arXiv:1412.6980. [CrossRef]
43. Abadi, M.; Agarwal, A.; Barham, P.; Brevdo, E.; Chen, Z.; Citro, C.; Corrado, G.S.; Davis, A.; Dean, J.; Devin, M. Tensorflow: Large-scale machine learning on heterogeneous distributed systems. *arXiv* **2016**, arXiv:1603.04467. [CrossRef]
44. Liu, H.B.; Lee, I. MPL-GAN: Toward Realistic Meteorological Predictive Learning Using Conditional GAN. *IEEE Access* **2020**, *8*, 93179–93186. [CrossRef]
45. Oprea, S.; Martinez-Gonzalez, P.; Garcia-Garcia, A.; Castro-Vargas, J.A.; Orts-Escolano, S.; Garcia-Rodriguez, J.; Argyros, A. A Review on Deep Learning Techniques for Video Prediction. *IEEE Trans. Pattern Anal. Mach. Intell.* **2020**, *44*, 2806–2826. [CrossRef] [PubMed]
46. Goodfellow, I.J.; Pouget-Abadie, J.; Mirza, M.; Xu, B.; Warde-Farley, D.; Ozair, S.; Courville, A.; Bengio, Y. Generative adversarial nets. In Proceedings of the 27th International Conference on Neural Information Processing Systems, Montreal, QC, Canada, 8–13 December 2014; pp. 2672–2680.
47. Tian, L.; Li, X.; Ye, Y.; Xie, P.; Li, Y. A Generative Adversarial Gated Recurrent Unit Model for Precipitation Nowcasting. *IEEE Geosci. Remote Sens. Lett.* **2020**, *17*, 601–605. [CrossRef]
48. Veillette, M.; Samsi, S.; Mattioli, C. Sevir: A storm event imagery dataset for deep learning applications in radar and satellite meteorology. In Proceedings of the Advances in Neural Information Processing Systems, Online, 6–12 December 2020; pp. 22009–22019.

remote sensing

Article

Center-Ness and Repulsion: Constraints to Improve Remote Sensing Object Detection via RepPoints

Lei Gao [1], Hui Gao [1,2,*], Yuhan Wang [3], Dong Liu [4] and Biffon Manyura Momanyi [1]

1 School of Computer Science and Engineering, University of Electronic Science and Technology of China, Chengdu 611700, China
2 Kash Institute of Electronics and Information Industry, Kash 844000, China
3 School of Information and Software Engineering, University of Electronic Science and Technology of China, Chengdu 611700, China
4 Sichuan Huakun Zhenyu Intelligent Technology Co., Ltd., Chengdu 610095, China
* Correspondence: huigao@uestc.edu.cn

Abstract: Remote sensing object detection is a basic yet challenging task in remote sensing image understanding. In contrast to horizontal objects, remote sensing objects are commonly densely packed with arbitrary orientations and highly complex backgrounds. Existing object detection methods lack an effective mechanism to exploit these characteristics and distinguish various targets. Unlike mainstream approaches ignoring spatial interaction among targets, this paper proposes a shape-adaptive repulsion constraint on point representation to capture geometric information of densely distributed remote sensing objects with arbitrary orientations. Specifically, (1) we first introduce a shape-adaptive center-ness quality assessment strategy to penalize the bounding boxes having a large margin shift from the center point. Then, (2) we design a novel oriented repulsion regression loss to distinguish densely packed targets: closer to the target and farther from surrounding objects. Experimental results on four challenging datasets, including DOTA, HRSC2016, UCAS-AOD, and WHU-RSONE-OBB, demonstrate the effectiveness of our proposed approach.

Keywords: remote sensing object detection; point representation; sample quality assessment; aerial target recognition; center-ness quality

Citation: Gao, L.; Gao, H.; Wang, Y.; Liu, D.; Momanyi, B.M. Center-Ness and Repulsion: Constraints to Improve Remote Sensing Object Detection via RepPoints. *Remote Sens.* **2023**, *15*, 1479. https://doi.org/10.3390/rs15061479

Academic Editor: Gwanggil Jeon

Received: 6 February 2023
Revised: 4 March 2023
Accepted: 5 March 2023
Published: 7 March 2023

1. Introduction

With the improvement of imaging quality, remote sensing images have been applied in many fields. As the basis of many remote sensing image applications, the quality of remote sensing object detection directly affects the effect of downstream applications. Generally speaking, object detection aims at identifying the categories of objects of interest and locating their position and can be divided into horizontal object detection and oriented object detection according to the expression of the bounding box. Since the seminal creative work: R-CNN [1] and its successive improvements [2,3], horizontal object detection has achieved significant progress. As a fundamental yet essential sub-task in object detection, the development of oriented object detection has fallen behind horizontal object detection since it requires a more sophisticated mechanism to locate objects precisely. Recently, remote sensing object detection has drawn increasing attention. However, a significant and recurrent problem is that remote sensing objects are often in multiple scales with arbitrary orientations [4–6] and in densely packed distributions with complex background contexts [7–9]. Based on the horizontal bounding box, oriented object detection utilizes an angle parameter to position large aspect ratio objects and small remote sensing objects in a crowded environment. Besides, oriented bounding boxes can minimize the error effect caused by the non-maximum suppression compared with horizontal bounding boxes.

The mainstreamed-oriented object detection approaches typically take the perspective that horizontal object detection is a special case for oriented object detection. Accordingly, most oriented object detectors are often inherited from the classical horizontal detectors with an extra orientation parameter θ. As shown in Figure 1, oriented object detectors utilize an extra parameter θ to describe the orientation information of the target object, in other words, five parameters (x, y, w, h, θ). The oriented bounding box provides a more precise localization of the objects. Especially for the large aspect ratio and small targets, the angle parameter θ and center point (x, y) play a more significant role in the positioning paradigm. Taking ship detection as an example, detecting a ship in Figure 1a using a horizontal bounding box has an inferior performance compared with using an oriented bounding box in Figure 1b as more than half the area of the horizontal bounding box does not belong to the ship.

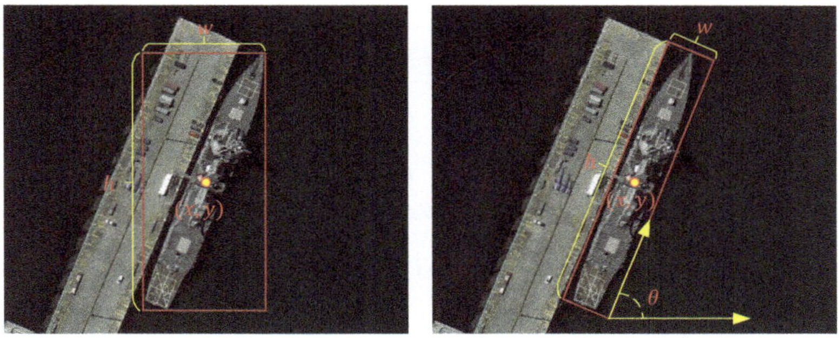

(x, y, w, h) (x, y, w, h, θ)

(**a**) Horizontal bounding box (**b**) Oriented bounding box

Figure 1. Horizontal bounding box (**a**) versus oriented bounding box (**b**), taking ship detection as an example. Point (x, y) denotes the coordinates of the center point of the target, while (w, h) denotes the width and height of the bounding boxes, respectively. The oriented bounding box, in particular, utilizes an extra parameter θ to represent the angle information making it better for locating aerial targets.

Most approaches treat oriented object detection as a problem of oriented object localization and the orientation regression-based methods [4,10,11] play the most important role in the research area. Benefiting from [12–14], these methods have achieved gratifying performance in research and application. However, the mechanism of angle-based regression methods has congenital drawbacks, including loss discontinuity and regression inconsistency [15–17]. These shortcomings are attributed to the periodicity of angular orientation and the specification of the oriented bounding box. For example, a bounding box rotated one degree clockwise or counterclockwise around the ground truth is equivalent under the Intersection over Union (IoU) evaluation metric. The transformation of five parameters (x, y, w, h, θ) and eight parameters $(x_1, y_1, x_2, y_2, x_3, y_3, x_4, y_4)$ also contains discontinuity of the loss problem caused by the order of the four points. The set $\{(x_i, y_i), i = 1, 2, 3, 4\}$ denotes four corner points of an oriented bounding box, respectively. Besides, some two-stage methods such as [4,9,18] design various complex modules to extract rotated features from the Region of Interest (RoI) and increase the computational complexity of the detectors.

Besides the discontinuity and complexity problems, orientated object detection has the challenge of precisely locating small and cluttered objects. This is especially true for aerial images, which are vital in remote sensing applications. To address this issue, SCRDet [9] proposed a pixel attention network and a channel attention network to suppress the noise and highlight object features. DRN [19] proposed a feature selection module and a dynamic refinement head to improve the receptive fields in accordance with the shapes and orientations of small and cluttered objects. However, these mainstream methods ignore spatial interaction among targets. While a vast majority of aerial images are taken from the bird's-view perspective, most targets are insufficiently covered by their surrounding targets. This fundamental feature of aerial targets is underutilized, and hence, spatial relative information should be considered in detector regression procedures.

Another challenge for oriented object detection is the design of sample assessment. As reported in [20–23], the selection, verification, and evaluation of samples can significantly improve the detectors' performance. ATSS [20] proved that the selection of positive and negative samples can improve the performance of detectors and proposed an adaptive sample assignment strategy. Chen et al. [21] discovered that joint inference with sample verification has a promising improvement over its foundation [24]. Hou et al. [22] considered shape information and measured the quality of proposals. Li et al. [23] proposed adaptive points assessment and assignment to improve the classification confidence and localization score. As pointed out in [25], the center-ness information plays a significant role in object localization. However, existing works do not have an effective measure of it.

As discussed above, the challenges associated with oriented object detection can be summarized as follows:

- The discontinuity of loss and the regression inconsistency caused by the expression of the oriented bounding box.
- The difficulty of locating small and cluttered objects precisely and the lack of spatial interaction among targets.
- Effective selection, verification, and assessment of samples and proposals, especially center-ness quality.

In this paper, we proposed repulsion and center-ness constraints based on RepPoints to improve remote sensing object detection. Firstly, we explore the representation of oriented objects in order to avoid the challenges caused by the oriented bounding box. As determined in RepPoints [21,24], point sets have demonstrated great potential while capturing vital semantic features produced by the multiple convolutional layers. In contrast to the conventional convolutional neural networks, RepPoints can have a weighted and wider reception field benefiting from [26]. To generate bounding boxes, a conversion function is applied to transform points into rectangles. For example, the conversion function MinAreaRect uses the oriented rectangle with minimum area to cover all the points in the learned point set over a target object. Secondly, as RepPoints only regresses the key points in the semantic feature maps but ignores measuring the quality of point sets, it attains an inferior performance for images with densely packed distributions and complex scenes. Therefore, we introduce the addition of a measuring strategy of center-ness to filter noisy samples located away from the center points of bounding boxes based on [23]. Thirdly, we design a novel loss function named oriented repulsion regression loss to illustrate the spatial interaction among targets. Specifically, we make the predicted bounding boxes closer to their corresponding ground truth boxes and farther from other ground truth boxes and predicted boxes, inspired by [27]. The main contributions of this paper are summarized as follows:

1. We utilize adaptive point sets to represent oriented bounding boxes to eliminate discontinuity and inconsistency and to capture key points with substantial semantic and geometric information.
2. We propose a center-ness constraint to measure the deviation of the point set to the center point in the feature map aiming to filter low-quality proposals and improve the localization accuracy.
3. We design a novel repulsion regression loss to effectively illustrate spatial information among remote sensing objects: closer to the target and farther from surrounding objects, especially helpful for small and cluttered objects.

In addition, to evaluate the effectiveness of our proposed method, we conducted a series of experiments on four challenging datasets, DOTA [28], HRSC2016 [29], UCAS-AOD [30], and WHU-RSONE-OBB [31], and obtained consistent and promising state-of-the-art results.

2. Related Work and Method

In this section, we first review the related studies of oriented object detection before providing sufficient information to illustrate our proposed methods.

2.1. Related Work

2.1.1. Oriented Object Detection

For several years, the representation of bounding boxes in object detection has been dominated by horizontal bounding boxes. With the increasing demand for object detection with arbitrary orientations, such as text localization and remote sensing object positioning, oriented object detection has drawn more attention. Recent advances in oriented object detection [4,9,16,32] are mainly derived from classical object detectors adapting horizontal object detectors with oriented bounding boxes to satisfy multi-oriented object detection. Generally, anchor-based oriented object detection can be divided into four categories: (1) generating rotated proposal regions directly and classifying the class of selected regions [4,10]; (2) regressing the angle parameter θ in a five parameter representation (x, y, w, h, θ) directly or based on horizontal proposal regions [5,9,33–35]; (3) using shape mask predicted by the mask branch to locate the object region [36]; and (4) transforming regression of the angle parameter into classification problem to address the periodicity of the angle and boundary discontinuity [16,17]. Although the anchor-based methods have achieved promising results, there are still some limitations for anchor-based detectors, such as various hyperparameters, complex post-processing, and overlapping calculation.

To further improve the efficiency of oriented object detection, some modifications have been made to anchor-free detectors for horizontal object detection, including key point-based methods [37,38], pixel-based methods [25], and point set-based methods [21,24]. Many superior methods have emerged verifying the effectiveness of the representation mentioned above. For example, O^2-Det [39] uses a pair of corresponding middle lines to locate rotated objects. In terms of overlapping calculation and boundary discontinuity, Yang et al. [40,41] transform the regression of the rotated bounding box to the Wasserstein Distance or Kullback–Leibler Divergence of 2-D Gaussian distributions, which achieves desirable results in oriented object detection.

2.1.2. Sample Assignment for Object Detection

Conventional object detection methods select positive and negative samples based on the fixed IoU threshold, i.e., MaxIoU strategy, which adopts IoU values as the only matching metric. Nevertheless, IoU-based assignment methods ignore the quality of training samples caused by the noise in the surroundings [42]. Various excellent adaptive sample assignment strategies have been proposed recently, which convert sample assignment into an optimization problem to select high-quality training samples. ATSS [20] uses a dynamic IoU threshold based on the statistical characteristic from the ground truth for the sample selection. FreeAnchor [43] enables the network to autonomously learn which anchor to match with the ground truth under the maximum likelihood principle. PPA [44] models the anchor assignment as a probabilistic procedure and calculates the scores of all anchors based on a probability distribution to determine the positive samples. DAL [45] defines a matching degree and sensitive loss to measure the localization potential of anchors, which enhances the correlation between classification and regression. SASM [22] utilizes the mean and standard deviation of the objects to capture shape information and add loss weights to each positive sample based on the quality.

In this paper, we divide the assignment into two phases: the initial stage and the refinement stage. In the initial stage, we utilize an IoU-based sample assignment, while we add a series of quality assessment strategies in the refinement stage, including center-ness constraint to filter noisy samples that can significantly enhance the effectiveness of adaptive points learning.

2.2. Overview of the Proposed Method

To alleviate boundary discontinuity, we adopt the adaptive point set proposed by [24] as a sophisticated representation of oriented bounding boxes instead of directly regressing the five parameters (x, y, w, h, θ). As a fine-grained representation, a point set enables the detectors to capture key points with substantial semantic information and geometric structure, which helps locate small and densely packed objects with arbitrary orientations. To converge from the ground truth boxes, a differentiable conversion function is applied to get oriented bounding boxes from the representative points. In the backward process, the coordinates are updated through the loss designed to adaptively cover an oriented object. To improve the effectiveness of adaptive point sets, we suggest a center-ness quality assessment strategy based on [23] for an additional constraint on the selected positive samples, which can make adaptive points concentrate more on the object rather than the background. To further address the issue of the localization of small and cluttered objects, we design a repulsion constraint in the form of a loss function, which makes the proposal bounding boxes closer to their ground truth boxes while farther from the other surrounding ground truth or proposal boxes. The assignment of samples is divided into two phases. In the initial stage, the detector selects positive samples according to the IoU values. To improve the qualities of the selected samples, we design an assessment module to score each sample, where the center-ness constraint score is calculated to filter low-quality samples alongside the orientation, classification, and localization quality measurement strategies. In the refinement stage, only high-quality samples selected by the assessment module are used to calculate loss values. Figure 2 illustrates an overview of our proposed anchor-free oriented object detector based on Reppoint.

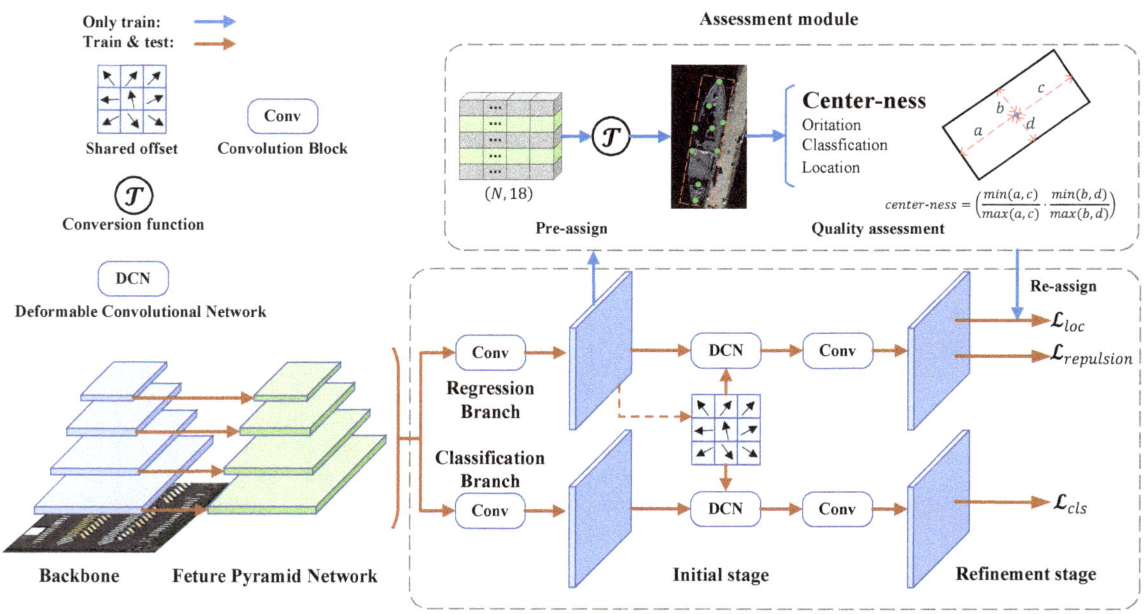

Figure 2. The pipeline of our proposed object detector. The proposed method is an anchor-free detector based on Reppoint [24] with adaptive point sets as the representation of an oriented bounding box, where a classical backbone with FPN [12] network is employed to encode multi-scale features. Deformable Convolutional Network (DCN) is utilized to capture shape-aware features. To cope with the harmony of the classification branch and the regression branch, the offset parameter is shared in the DCN block.

2.3. Deformable Convolutional Network

Traditional object detectors mainly use Convolutional Neural Networks (CNN) for feature encoding. However, the fixed receptive field of CNN leads to the defect that CNN can not capture information in the neighboring area. In the remote sensing images, objects are often sharply variable shapes, e.g., square tennis court and slender ship. While the defect appears to be more apparent, we alleviate it by adopting the Deformable Convolutional Network (DCN) [26] both in the classification and regression branches to capture shape-aware features of the objects. The process of DCN can be formulated as shown in Equation (1).

$$y(\mathbf{p_0}) = \sum_{\mathbf{p_n} \in \mathcal{R}} w(\mathbf{p_n}) \cdot x(\mathbf{p_o} + \mathbf{p_n} + \Delta\mathbf{p_n}), \tag{1}$$

where $w(\cdot)$ denotes the filter weights, $\mathcal{R} = \{(-1,-1),(-1,0),\cdots,(1,0),(1,1)\}$ is receptive field size and dilation taking a 3×3 kernel with dilation 1 as an example. $\{\Delta\mathbf{p_n}|n = 1,\cdots,N\}, N = |\mathcal{R}|$ is the offset set of each point in the receptive field, and is calculated as shown in Equation (2).

$$\Delta\mathbf{p_n} = Conv(F_i) - \mathcal{R}, \quad i \in \{1,\cdots,5\}, \tag{2}$$

where F_i denotes the i-th scale feature map, and $\mathcal{R}$ is the standard CNN receptive field. The function $Conv(\cdot)$ denotes a series of CNN layers and the dimension of its output is $w \times h \times 18$, where w and h are the width and height of F_i, respectively.

As shown in Figure 3, benefitting from the offset parameters, DCN gains the ability to aggregate information from the wider neighboring areas. As the offsets and the convolutional kernels are learned simultaneously during training, DCN can obtain dynamic and adaptive features of objects and is more sensitive to the variable shapes. More importantly, the inherent characteristic of DCN, i.e., learnable offset, perfectly fits the adaptive point set, which provides a more accurate localization of the oriented objects.

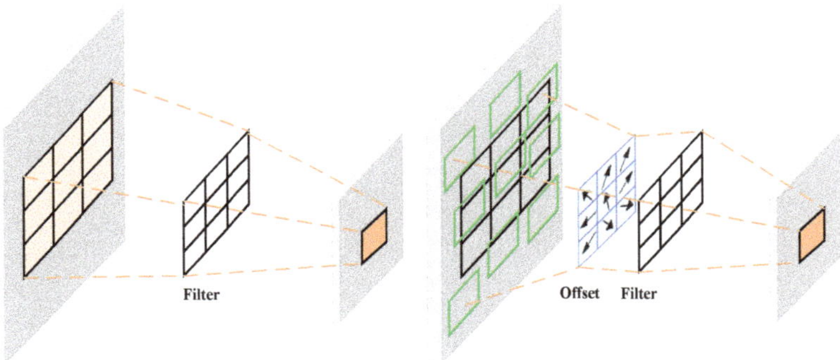

(**a**) Standard CNN (**b**) DCN

Figure 3. Illustration of standard CNN in (**a**) and DCN in (**b**). DCN utilizes an additional offset learned from the feature map to obtain a wider receptive field compared with CNN.

2.4. Center-Ness Constraint for Oriented Object Detection

Sample selection plays a critical role in the performance of detectors. Conventional IoU-based sample selection strategies overlook the shape information of the selected samples, which introduces many noisy samples and deteriorates the unbalance of positive and negative samples. In our proposed method, we divide the sample assignment into two phases: the initial stage and the refinement stage. In the refinement stage, all selected samples are assessed through our designed center-ness constraint alongside other strategies proposed by [23]. The center-ness constraint is first suggested in FCOS [25], aiming to remove redundant and meaningless proposal bounding boxes for horizontal object detection. Simply applying it in oriented object detection will introduce additional inconsistency between the distribution of the center-ness score and the oriented box. Concretely, the horizontal center-ness quality can not fit the oriented bounding box, as shown in Figure 4a. To modify this defect, we re-formulate the center-ness calculation process and make it fit the oriented bounding box appropriately. A horizontal bounding box can be simply expressed by (x, y, w, h), where (x, y), w, h denote the center point, width, and height of the horizontal bounding box, respectively. The center-ness score can be directly calculated by the offsets of the center point to the four edges.

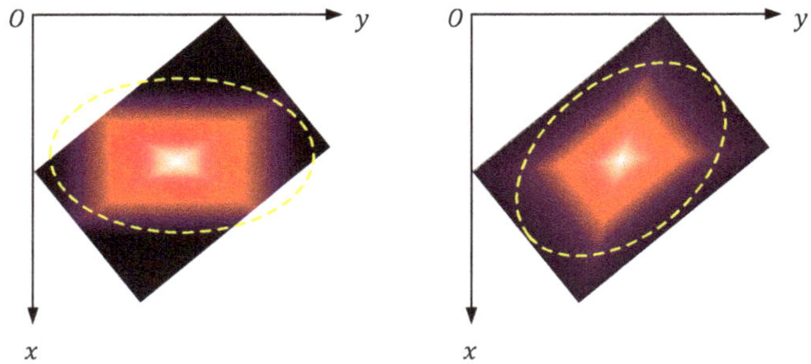

(**a**) Horizontal center-ness (**b**) Oriented center-ness

Figure 4. Heatmap of the horizontal and oriented center-ness. The distributions of horizontal and oriented center-ness scores are indicated by the ellipses with a yellow dotted outline.

In our proposed method, we utilize a point set $\mathcal{P}$ with nine points to represent an oriented bounding box, which is defined in Equation (3).

$$\mathcal{P} = \{(x_i, y_i) | i \in \{1, \cdots, 9\}\}, \tag{3}$$

where each (x_i, y_i) in $\mathcal{P}$ is calculated by the corresponding offset $\Delta \mathbf{p_n}$ and point (x, y) in the feature map projected to the original size of the input image. The process can be expressed as shown in Equation (4).

$$(x_i, y_i) = (x, y) + \Delta \mathbf{p_i}. \tag{4}$$

To simplify the computation procedure, the point set $\mathcal{P}$ is converted into a rotated rectangle through the $MinAeraRect(\cdot)$ function to measure the center-ness quality. $MinAreaRect$ uses the oriented rectangle with minimum area to cover all the points in $\mathcal{P}$. Equation (5) demonstrates how this conversion is formulated.

$$(x_1, y_1, x_2, y_2, x_3, y_3, x_4, y_4) = \text{MinAeraRect}(\mathcal{P}), \tag{5}$$

where $(x_1, y_1, x_2, y_2, x_3, y_3, x_4, y_4)$ denotes the four corner points of an oriented bounding box.

Since the vanilla center-ness proposed by FCOS [25] is measured w.r.t the axis-aligned edges, which can not be directly applied in oriented object detection, as shown in Figure 4, we suggest a distance function in the form of the cross product between the feature map point (x, y) and two adjacent corner points (c_x^1, c_y^1) and (c_x^2, c_y^2), as shown in Equation (6).

$$\begin{aligned} crossdist(c_x^1, c_y^1, c_x^2, c_y^2 | x, y) &= \frac{\mathbf{v_1} \times \mathbf{v_2}}{\|\mathbf{v_1}\|} \\ &= \frac{|(c_x^2 - c_x^1)(c_y^1 - y) - (c_x^1 - x)(c_y^2 - c_y^1)|}{\sqrt{(c_x^2 - c_x^1)^2 + (c_y^2 - c_y^1)^2}} \end{aligned} \tag{6}$$

With this formula, we can obtain four distance values between the four corner points $(x_1, y_1, x_2, y_2, x_3, y_3, x_4, y_4)$, as defined in Equation (7), given an oriented bounding box and feature map point (x, y).

$$\begin{aligned} a &= crossdist(x_1, y_1, x_2, y_2 | x, y) \\ b &= crossdist(x_2, y_2, x_3, y_3 | x, y) \\ c &= crossdist(x_3, y_3, x_4, y_4 | x, y) \\ d &= crossdist(x_4, y_4, x_1, y_1 | x, y) \end{aligned} \tag{7}$$

The oriented center-ness quality is then calculated as shown in Equation (8).

$$Q_{centerness} = \left(\frac{min(a, c)}{max(a, c)} \cdot \frac{min(b, d)}{max(b, d)} \right)^{\frac{1}{\gamma}}, \tag{8}$$

where γ is a hyper-parameter to control the sensitivity of the center-ness quality. As shown in Figure 5, the oriented center-ness constraint measured by the function above sufficiently evaluates the quality of an oriented bounding box. $Q_{centerness}$ ranges from 0 to 1, depending on whether the feature point is on the edges or at the center point, respectively. The closer the feature point is to the center point of an oriented bounding box, the higher the quality score. The goal of oriented center-ness is to remove redundant and low-quality bounding boxes generated in the initial stage, which will reduce the computational cost in the post-processing steps, e.g., NMS.

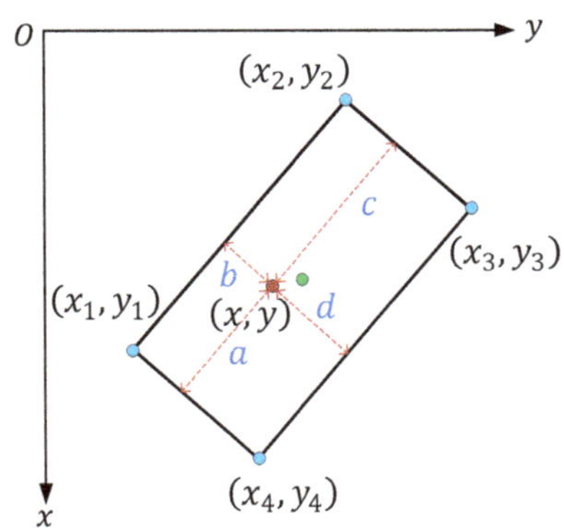

Figure 5. Illustration of the oriented center-ness. The four blue dots, red dots, and green dots denote the corner points of an oriented bounding box, the feature map point (x, y), and the center point of the oriented bounding box, respectively.

Based on the quality measurement strategy Q, we re-assign the samples selected in the initial stage according to the quality scores. Only the top k samples are selected for each ground truth. To retrieve high-quality samples, a ratio σ is utilized to control the number of samples. The value of k is calculated as shown in Equation (9).

$$k = \begin{cases} \sigma * N_t, & N_t \geq 2 \\ N_t, & N_t < 2 \end{cases} \tag{9}$$

where N_t denotes the number of proposals for each oriented object.

2.5. Repulsion Constraint for Oriented Object Detection

To address the issue of locating small and cluttered objects, we propose a repulsion constraint to discriminate the densely distributed objects. As mentioned before, the vast majority of aerial images are taken from the bird's-view and small objects are mostly in crowded scenes such as a parking lot. To locate them precisely, we should consider the spatial relative information, which means narrowing the gap between a proposal bounding box and its corresponding ground truth box and being away from other surrounding proposal and ground truth boxes. As illustrated in Figure 6, we utilize an IoU-based loss function to realize the repulsion constraint. A perfect proposal bounding box should have a maximum IoU to its ground truth while keeping IoUs within the surrounding ground truth and proposal bounding boxes.

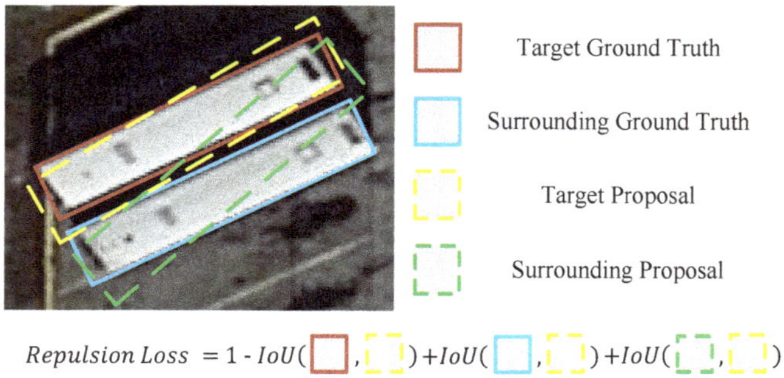

Figure 6. Visualization of repulsion constraint in the form of the loss function.

Inspired by [27], we divide the oriented repulsion loss into three components, defined as shown in Equation (10).

$$L_{repulsion} = L_{attr} + \alpha * L_{rgt} + \beta * L_{rp}, \tag{10}$$

where L_{attr} aims to narrow the gap between predicted boxes and ground truth boxes, while L_{rgt} and L_{rp} are designed to minimize the intersection among the surrounding ground truth and predicted boxes, respectively. Hyper-parameters α and β are used to balance the loss weight.

In practice, there is an accommodation relationship among objects of different categories, e.g., aircraft and airports. For simplicity, we only consider the repulsion constraint for the objects from the same category. Let $\mathbb{P}_+$ and $\mathbb{G}$ denote the sets of all positive samples and all ground truth boxes, respectively.

Given a ground truth box $G \in \mathbb{G}$, we assign the proposal containing the maximum rotated IoU to it, denoted by $P^G_{attr} = \operatorname{argmax}_{P \in \mathbb{P}_+} \mathrm{rIoU}(G, P)$. Then, L_{attr} can be calculated as shown in Equation (11).

$$L_{attr} = \frac{\sum_{G \in \mathbb{G}} \mathrm{rIoU}(G, P^G_{attr})}{|\mathbb{G}|}, \tag{11}$$

where $\mathrm{rIoU}(\cdot)$ is used to calculate the IoU between the two oriented boxes.

L_{rgt} is designed to repel a predicted box from its neighboring ground truth box. Here, we use intersection over ground truth: $\mathrm{IoG}(P, G) = \frac{\mathrm{area}(P \cap G)}{\mathrm{area}(G)} \in (0, 1)$ to describe the spatial relationship between a predicted box and its neighboring ground truth box. For each $G \in \mathbb{G}$, we define L_{rgt} as shown in Equation (12).

$$L_{rgt} = \frac{\sum_{P \in \mathbb{P}_+ \setminus P^G_{attr}} \mathrm{Smooth}_{\ln}(\mathrm{IoG}(P, G))}{|\mathbb{P}_+|}, \tag{12}$$

where $\mathrm{Smooth}_{\ln}$ function is applied to adjust the sensitivity of L_{rgt}. Equation (13) provides a definition of $\mathrm{Smooth}_{\ln}$.

$$\mathrm{Smooth}_{\ln} = \begin{cases} -\ln(1-x), & x \le \sigma \\ \frac{x-\sigma}{1-\sigma} - \ln(1-\sigma), & x > \sigma \end{cases} \tag{13}$$

NMS is an essential post-processing step in most detectors to select or merge the primary predicted bounding boxes. Especially for small and cluttered objects, NMS has a significant effect on the detection results. To alleviate the detectors' sensitivity to NMS, we use an additional constraint L_{rp} to minimize the overlap of two predicted boxes P_i and P_j, which are designated to different ground truth boxes. Equation (14) defines the definition of L_{rp}.

$$L_{rp} = \frac{\sum_{i \neq j} \text{Smooth}_{\ln}(\text{rIoU}(P_i, P_j))}{\sum_{i \neq j} \mathbf{1}[\text{rIoU}(P_i, P_j) \geq 0] + \epsilon}, \tag{14}$$

where $\mathbf{1}(\cdot)$ denotes the identity function and ϵ is introduced in case divided by 0.

Benefiting from the repulsion constraint, the loss $L_{repulsion}$ preserves the independence among predicted boxes, while preventing them from shifting toward nearby ground truth boxes, which makes the detector more robust to small and cluttered objects.

Eventually, the loss function of our proposed detector is formulated as shown in Equation (15).

$$L = L_{cls} + \lambda_1 L_{loc} + \lambda_2 L_{repulsion}, \tag{15}$$

where L_{cls} denotes the object classification loss, L_{loc} denotes regression loss for object localization, and $L_{repulsion}$ is repulsion constraint loss. In the experiment, we use focal loss [13] for classification and GIoU loss [46] for oriented polygon regression.

3. Results

In this section, we first introduce four challenging datasets that we use to verify the effectiveness of our proposed method, then describe the details of our experiment settings, and finally illustrate our results on the datasets.

3.1. Datasets

DOTA [28] is one of the largest datasets for oriented object detection in aerial images; it contains 15 categories: plane (PL), baseball diamond (BD), ground track field (GTF), small vehicle (SV), large vehicle (LV), bridge (BR), tennis court (TC), storage tank (ST), ship (SH), soccer ball field (SBF), harbor (HA), roundabout (RA), helicopter (HC), swimming pool (SP), and basketball court (BC). Labeled objects are in a wide range of scales, shapes, and orientations. DOTA contains 2806 images and 188,282 instances collected from different sensors and platforms. Each images size ranges from 800 × 800 to 20,000 × 20,000 pixels. The proportions of the training set, validation set, and testing set in DOTA are 1/2 (1411 images), 1/6 (458 images), and 1/3 (937 images), respectively. In our experiments, both the training and validation sets are utilized to train the proposed detector and the testing set without annotations for evaluation. All the images used for training were split into patches of 1024 × 1024 pixels with a stride of 200 pixels. Data augmentation operations, including random resizing and flipping, were employed in the training stage to avoid overfitting.

HRSC2016 [29] is a dataset for ship recognition that contains a large number of deformed strip and oriented ship objects collected from several famous harbors. The entire dataset contains 1061 images with sizes ranging from 300 × 300 to 1500 × 900. For a fair comparison, the training and validation sets (436 images and 181 images, 617 images in total) are used for training, while the testing set (444 images) is used for evaluation. All images are resized to 800 × 512 pixels for training and testing.

UCAS-AOD [30] is an aerial image dataset that labels airplanes and cars with oriented bounding boxes. The dataset contains 1510 images with approximately 1280 × 659 pixels (510 images for car detection and 1000 images for airplane detection). There are 14,596 instances in total. The entire dataset is randomly divided into the training set, validation set, and testing set with a ratio of 5:2:3, i.e., 755 images, 302 images, and 453 images, respectively.

WHU-RSONE-OBB [31] is a large-scale object detection dataset with oriented bounding boxes that contains 5977 images ranging from 600 × 600 pixels to 1372 × 1024 pixels. WHU-RSONE-OBB is a high spatial resolution remote sensing image dataset with spatial resolution ranging from 0.5 m to 0.8 m. Objects are of three: airplanes, storage tanks, and ships. Likewise, the training set (4781 images) and the validation set (598 images) were employed for training while the testing set (598 images) was used for evaluation. All images were resized to 1024 × 1024 pixels for both training and testing.

3.2. Implementation Details

We implement our proposed method based on MMRotate [47], an open-source toolbox for rotated object detection based on PyTorch, and utilize ResNet-50 and ResNet-101 [48] as the backbone with FPN [12]. The FPN block consists of P_3 to P_7 pyramid levels in the experiments. The SGD optimizer was selected during training with an initial learning rate of 0.008. The number of warming-up iterations was 500. At each decay step, the learning rate was decreased by a factor of 0.1. The momentum and weight decay of SGD were set to 0.9 and 10^{-4}, respectively. We trained the detector with 40 epochs, 120 epochs, 120 epochs, and 40 epochs for DOTA, HRSC2016, UCAS-AOD, and WHU-RSONE-OBB, respectively. In Equation (8), we set the sensitivity of center-ness to $\gamma = 4$. We set the balance weight to $\alpha = 0.5$ and $\beta = 0.5$ empirically in Equation (10). Meanwhile, the weights for L_{loc} and $L_{repulsion}$ were set to $\lambda_1 = 1.0$ and $\lambda_2 = 0.25$ in Equation (15), respectively.

We conducted all the experiments on a server with 2 NVIDIA RTX 3090 GPUs with a total batch size of four (two images per GPU) for training and a single NVIDIA RTX 3090 GPU for inference.

3.3. Comparisons with State-of-the-Art Methods

To verify the effectiveness of our proposed method, we conducted a series of experiments on DOTA, HRSC, UCAS-AOD, and WHU-RSONE-OBB. We adopted mean average precision (mAP) as the evaluation criteria for oriented object detection results, which can be calculated as shown in Equation (16).

$$mAP = \frac{1}{n}\sum_{i}^{n} AP_i \quad (16)$$

where AP_i denotes the value of the area under the precision–recall curve for the i-th class and n is the number of categories in one dataset.

Results on DOTA. As shown in Table 1, we report all the experimental results on the single-scale DOTA dataset to make fair comparisons with previous methods. The proposed method based on RepPoints obtains 76.93% mAP and 76.79% mAP with the backbone ResNet-50 and Resnet-101, respectively. It outperformed other methods with the same backbones. Using the tiny version of Swin-Transformer [49] with FPN, we achieved the best performance with 77.79% mAP. Notably, our results for the small vehicle (SV), which is a typical class of small and cluttered objects, consistently achieved the best performances under three different backbones, which demonstrates the effectiveness of our proposed method for small and cluttered objects.

Results on HRSC2016. Ship detection is a vital application direction of remote sensing images, where ships have large aspect ratios. Experiments on HRSC2016 have also verified the superiority of our proposed method. As shown in Table 2, our proposed method obtained 90.29% mAP, outperforming other methods listed in the table.

Table 1. Comparisons with state-of-the-art methods on the DOTA dataset. All the reported results were performed on the single-scale DOTA. The results with red color denote the best results in each column.

Type	Methods	Backbone	PL	BD	BR	GTF	SV	LV	SH	TC	BC	ST	SBF	RA	HA	SP	HC	mAP
Single-stage	RetinaNet-O [13]	R-50	88.67	77.62	41.81	58.17	74.58	71.64	79.11	90.29	82.18	74.32	54.75	60.60	62.57	69.67	60.64	68.43
	DAL [45]	R-101	88.61	79.69	46.27	70.37	65.89	76.10	78.53	90.84	79.98	78.41	58.71	62.02	69.23	71.32	60.65	71.78
	RSDet [15]	R-152	90.10	82.00	53.80	68.50	70.20	78.70	73.60	91.20	87.10	84.70	64.30	68.20	66.10	69.30	63.70	74.10
	R^3Det [34]	R-152	89.49	81.17	50.53	66.10	70.92	78.66	78.21	90.81	85.26	84.23	61.81	63.77	68.16	69.83	67.17	73.74
	S^2A-Net [5]	R-50	89.11	82.84	48.37	71.11	78.11	78.39	87.25	90.83	84.90	85.64	60.36	62.60	65.26	69.13	57.94	74.12
	R^3Det-DCL [17]	R-152	89.78	83.95	52.63	69.70	76.84	81.26	87.30	90.81	84.67	85.27	63.50	64.16	68.96	68.79	65.45	75.54
Two-stage	Faster RCNN [3]	R-50	88.44	73.06	44.86	59.09	73.25	71.49	77.11	90.84	78.94	83.90	48.59	62.95	62.18	64.91	56.18	69.05
	CAD-Net [33]	R-101	87.80	82.40	49.40	73.50	71.10	63.50	76.60	90.90	79.20	73.30	48.40	60.90	62.00	67.00	62.20	69.90
	CenterMap [50]	R-50	88.88	81.24	53.15	60.65	78.62	66.55	78.10	88.83	77.80	83.61	49.36	66.19	72.10	72.36	58.70	71.74
	SCRDet [9]	R-101	89.98	80.65	52.09	68.36	68.36	60.32	72.41	90.85	87.94	86.86	65.02	66.68	66.25	68.24	65.21	72.61
	FAOD [18]	R-101	90.21	79.58	45.49	76.41	73.18	68.27	79.56	90.83	83.40	84.68	53.40	65.42	74.17	69.69	64.86	73.28
	RoI-Trans. [4]	R-101	88.65	82.60	52.53	70.87	77.93	76.67	86.87	90.71	83.83	82.51	53.95	67.61	74.67	68.75	61.03	74.61
	MaskOBB [51]	R-50	89.61	85.09	51.85	72.90	75.28	73.23	85.57	90.37	82.08	85.05	55.73	68.39	71.61	69.87	66.33	74.86
	Gliding Vertex [52]	R-101	89.64	85.00	52.26	77.34	73.01	73.14	86.82	90.74	79.02	86.81	59.55	70.91	72.94	70.86	57.32	75.02
	ReDet [32]	ReR-50	88.79	82.64	53.97	74.00	78.13	84.06	88.04	90.89	87.78	85.75	61.76	60.39	75.96	68.07	63.59	76.25
	Oriented R-CNN [53]	R-101	88.86	83.48	55.27	76.92	74.27	82.10	87.52	90.90	85.56	85.33	**65.51**	66.82	74.36	70.15	57.28	76.28
Anchor-free	CenterNet-O [14]	DLA-34 [14]	81.00	64.00	22.60	56.60	38.60	64.00	64.90	90.80	78.00	72.50	44.00	41.10	55.50	55.00	57.40	59.10
	PIoU [54]	DLA-34	80.90	69.70	24.10	60.20	38.30	64.40	64.80	90.90	77.20	70.40	46.50	37.10	57.10	61.90	64.00	60.50
	O^2-DNet [39]	H-104	89.31	82.14	47.33	61.21	71.32	74.03	78.62	90.76	82.23	81.36	60.93	60.17	58.21	66.98	61.03	71.04
	DRN [19]	H-104	89.71	82.34	47.22	64.10	76.22	74.43	85.84	90.57	86.18	84.89	57.65	61.93	69.30	69.63	58.48	73.23
	CFA [7]	R-101	89.26	81.72	51.81	67.17	79.99	78.25	84.46	90.77	83.40	85.54	54.86	67.75	73.04	70.24	64.96	75.05
	Oriented RepPoints [23]	R-101	89.53	84.07	59.86	71.76	79.95	80.03	87.33	90.84	87.54	85.23	59.15	66.37	75.23	73.75	57.23	76.52
	Ours	R-50	88.39	84.00	54.68	73.58	80.89	80.38	87.60	90.90	85.33	86.93	64.48	69.85	74.72	72.32	59.98	76.93
	Ours	R-101	88.50	83.84	54.35	71.11	80.93	80.25	87.64	90.90	85.11	87.00	64.07	70.12	75.12	72.85	60.15	76.79
	Ours	Swin-T	88.90	84.13	55.24	75.68	81.84	82.98	87.75	90.90	86.12	86.45	64.17	69.10	76.90	73.47	63.25	77.79

Table 2. Results on HRSC2016. The best result is bolded.

Methods	Backbone	mAP
RRD [55]	VGG16	84.30
RoI-Trans. [4]	R-101-FPN	86.20
R^3Det-KLD [41]	R-101-FPN	87.45
CenterNet-O [14]	DLA-34	87.89
Gliding Vertex [52]	R-101-FPN	88.20
RetinaNet-O [13]	R-101-FPN	89.18
PIOU [54]	DLA-34	89.20
R^3Det [34]	R-101-FPN	89.26
R^3Det-DCL [17]	R-101-FPN	89.46
FPN-CSL [16]	R-101-FPN	89.62
DAL [45]	R-101-FPN	89.77
Ours	R-50-FPN	**90.29**

Results on UCAS-AOD. The UCAS-AOD dataset contains a mass of small and cluttered objects, which is competent to evaluate the effectiveness of our proposed method. All the experimental results are shown in Table 3 with our proposed method obtaining the best performance with 90.11% mAP. Although YOLOv7 [56] performs better in airplane detection, it lacks the ability to capture small and densely packed targets, such as cars, in remote sensing images.

Table 3. Results on UCAS-AOD. The best result is bolded in each column.

Methods	Car	Airplane	mAP
YOLOv3-O [57]	74.63	89.52	82.08
RetinaNet-O [13]	84.64	90.51	87.57
Faster RCNN [3]	86.87	89.86	88.36
RoI Trans. [4]	87.99	89.90	88.95
DAL [45]	89.25	90.49	89.87
YOLOv7-O [56]	83.35	**96.53**	89.94
Oriented RepPoints [23]	89.51	90.70	90.11
Ours	**89.73**	90.78	**90.26**

Results on WHU-RSONE-OBB. To further verify the effectiveness of the proposed method, we conducted a series of experiments on the WHU-RSONE-OBB dataset. As shown in Table 4, our proposed method achieved the best AP values for plane and ship with 92.83% mAP.

Table 4. Result on WHU-RSONE-OBB. The best result is bolded in each column.

Methods	Airplane	Storage-Tank	Ship	mAP
Faster-RCNN [3]	94.86	56.34	76.38	75.86
CNN-SOSF [58]	95.21	74.61	75.20	81.67
YOLOv3-O [57]	97.76	87.09	78.65	87.84
CNN-AOOF [31]	98.57	88.31	79.20	88.69
YOLOv7-O [56]	98.65	**95.69**	79.02	91.12
Ours	**99.57**	90.54	**88.38**	**92.83**

Model size and efficiency. The parameter size and the inference speed are shown in Table 5. Our proposed model requires additional memory to compute the repulsion loss during the training stage. However, as center-ness and repulsion constraints are only calculated in the training stage, the inference speed is not affected by these two constraints during the inference stage.

Table 5. Model size and efficiency. For a fair comparison, all the models utilized ResNet-50 as the backbone with a single NVIDIA RTX 2080S GPU.

Method	Backbone	Param	Inf Time (fps)
RetinaNet-O [13]	R-50	36.42 M	17.2
S^2A-Net [5]	R-50	38.6 M	15.5
Gliding Vertex [52]	R-50	41.14 M	16.4
RoI-Trans. [4]	R-50	55.13 M	16.5
R^3Det [34]	R-50	41.9 M	12.4
Ours	R-50	36.61M	16.8

3.4. Visualization of Results

To have an intuitive view of our proposed method, we selected some images from the testing set of the DOTA dataset to show the promising performance, as shown in Figure 7.

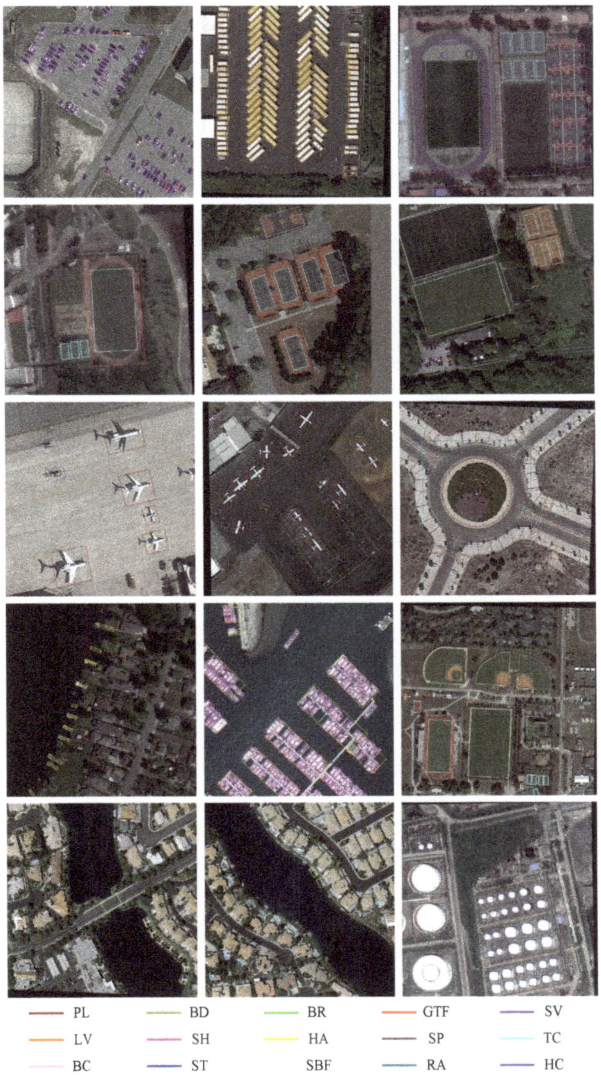

Figure 7. Visualization of the example detection results on DOTA testing set.

4. Discussion

In this section, we first demonstrate the superiority of the adaptive point set to represent the oriented bounding box. Secondly, we verify the effectiveness of our proposed center-ness quality assessment and repulsion constraint through a series of ablation studies. Thirdly, we explore the relationship among different categories via the confusion matrix on the DOTA validation set. Then, we further discuss how center-ness and repulsion constraints improve the distribution of localization scores. Finally, we discuss the limitation of the methods and possible future improvements.

4.1. Superiority of Adaptive Point Set

To examine the superiority of the adaptive point set to represent oriented boxes, we compare RepPoints with the anchor-based methods RoI-Trans [4] and R^3Det [34] on the HRSC2016 dataset. RoI-Trans proposes a transformation module to effectively mitigate the misalignment between RoIs and targets, while R^3Det utilizes a feature refinement module to reconstruct features. As shown in Table 6, the adaptive point set obtained nearly one percent enhancement with no bells and whistles, which displays its inherent superiority for the representation of oriented boxes.

Table 6. Comparisons between anchor-based orientation regression methods and our adaptive-point set-based method. The best result is bolded.

Methods	Backbone	mAP
RoI-Trans. [4]	R-101	86.20
R3Det [34]	R-101	89.26
RepPoints(adaptive point set)	R-50	**90.02**

4.2. Effectiveness of Center-Ness and Repulsion Constraints

To investigate the effectiveness of center-ness quality assessment and repulsion constraint, we compared them against the baseline method [23] without using them. Table 7 shows the experimental results.

Table 7. Performance evaluation on center-ness quality assessment and repulsion constraint. PL, SV, and SH denote the categories of plane, small vehicle, and ship, respectively. All the experiments adopt ResNet-50 with FPN as the backbone. '✓' and '✗' in the *Center-ness* and *Repulsion* columns denote the results with or without the corresponding constraint, respectively. We adopted ConvexGIoULoss for regression loss if the repulsion constraint is not applied. The best result is bolded in each column.

Center-ness	Repulsion	PL	SV	SH	mAP	Δ
✗	✗	87.02	80.18	87.28	75.97	-
✓	✗	88.30	80.78	87.51	76.05	0.08
✗	✓	**88.66**	80.73	87.54	76.31	0.34
✓	✓	88.39	**80.88**	**87.60**	**76.93**	**0.96**

Obviously, both center-ness and repulsion constraints improve the accuracy of the detector, especially the repulsion constraint, which considers the spatial correlation information and obtained a 0.34 mAP improvement compared with the baseline. Meanwhile, APs of three classic small and cluttered objects, plane, small vehicle, and ship, obtained consistent improvements. Although the center-ness constraint only has a slight improvement, with the collaboration of the repulsion constraint, the detector obtained a promising improvement with 0.96 mAP. This is because the center-ness constraint enforces the adaptive points to concentrate more on the center of objects, which is helpful to the localization tasks.

4.3. Correlation between Localization and Classification

To further explore how our proposed center-ness and repulsion constraints improve the quality of the proposals, we statistically analyze the correlation between the localization scores (IoU) and classification confidence of the predicted boxes. The closer the center of the distribution to the upper left corner is, the higher the quality of the predicted boxes the detector generates. In application scenarios, all the predicted boxes are filtered during the post-processing stage where NMS and IoU-thresholds are usually adopted. For a fair comparison, we only selected predicted boxes with no less than the IoU value of 0.5. All the experiments were conducted on the validation set of the DOTA dataset.

The experimental results are visualized in Figure 8. Obviously, the quality of the predicted boxes generated by the detector is more stable under the application of our two proposed constraints, compared to the baseline with no sample assessment strategy to filter low-quality samples. Furthermore, the center of quality distribution tends to move towards a higher degree under two constraints compared with simply applying one constraint.

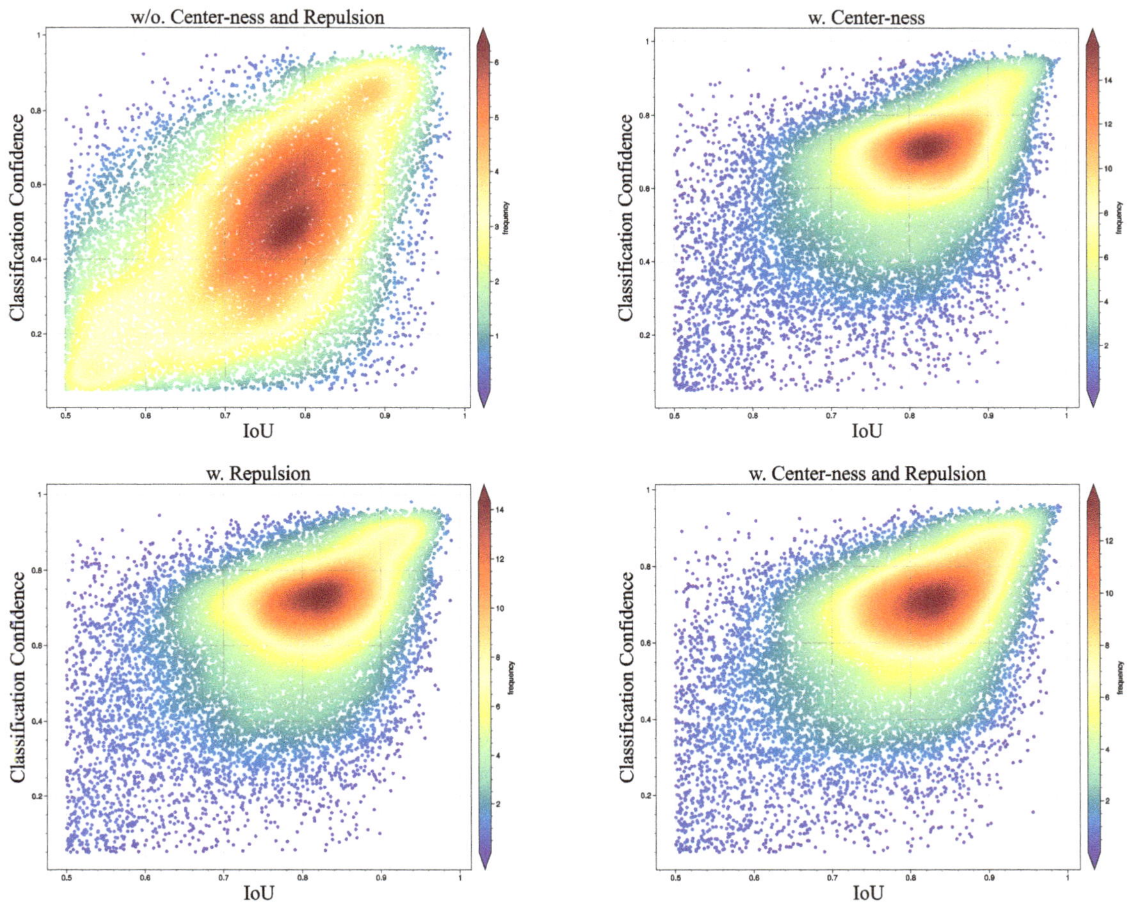

Figure 8. The correlation between the localization scores and classification confidence of predicted oriented boxes under four conditions: no center-ness and no repulsion constraints, center-ness constraint only, repulsion constraints only, and both center-ness and repulsion constraints. All the experiments adopt ResNet-50 with FPN as the backbone. The baseline is Rotated RepPoints [24].

4.4. Relationship among Categories

The confusion matrix is a standard format for expressing accuracy evaluation, which can visualize the detection results and discover the relevant information among categories. We provide the confusion matrix on the DOTA validation set to explore the detailed classification accuracy.

As shown in Figure 9, the detector is inferior at distinguishing between ground track and soccer ball fields, as they usually have similar shapes. Furthermore, in most scenarios, a soccer ball field is located within a ground track field. Moreover, we noticed that the detector mostly misidentifies the background targets as small and clustered targets such as small vehicles, which is mainly influenced by the complex scene environment. Meanwhile, the detector mostly misses objects such as ground track fields because they usually have the same color as the environment, and the iconic features are occluded by surrounding vegetation.

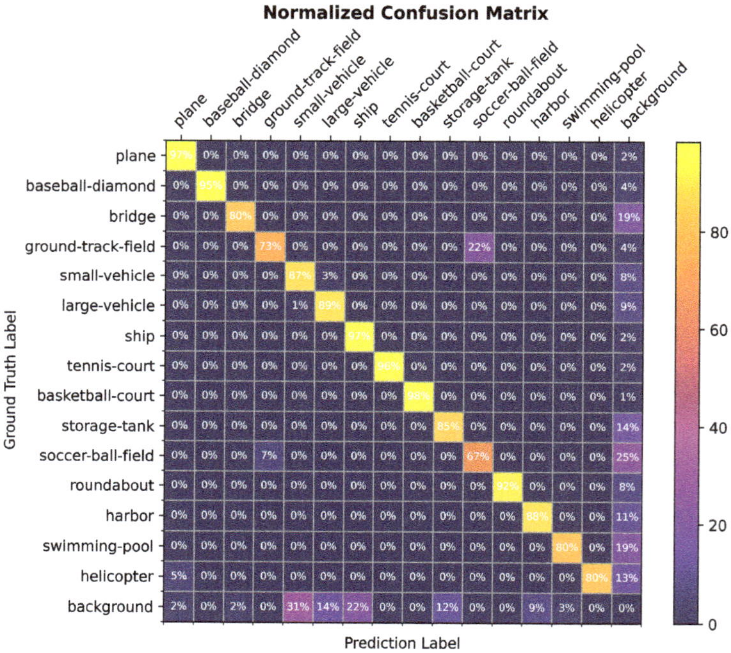

Figure 9. Normalized confusion matrix of detection results on the DOTA validation set.

4.5. Failure Analysis

During the validation stage, we noticed some failure cases, as shown in Figure 10. The yellow ellipses are the targets missed by the detector. Obviously, the detector missed the ship full of containers and classified the containers on board as ships in the left image. While in the right image, the detector missed the black car and truck covered by a gray patch. In the first case, the container on the ship completely covers the texture features of the ship, which is such an abstract situation. Although humans can make correct judgments through prior knowledge, it is difficult to obtain the hidden global semantic information for the detector. In the second case, similar colors with background and image noises (the irregular patch) lead to the omission. As we adopt the adaptive point set for the representation of oriented boxes, backgrounds with similar color and image noises may lead to the absence of some key points of objects. In the future, we may explore the attention mechanism similar to [59,60] for feature fusion to address this issue.

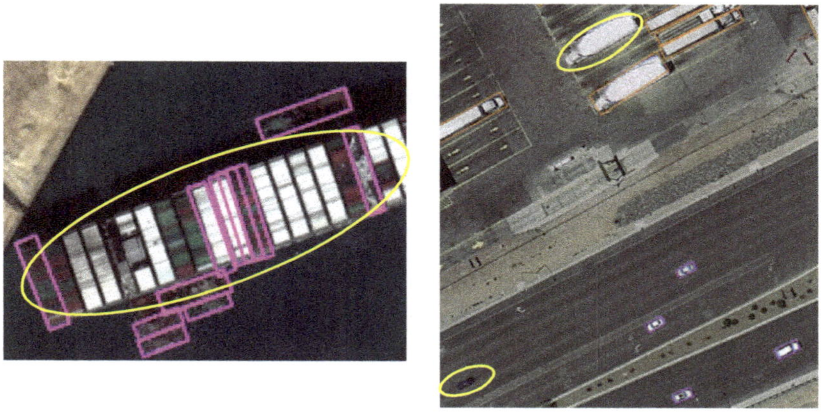

Figure 10. Failure cases for ship detection and small vehicle. Failures are marked by yellow ellipses.

4.6. Limitations and Future Directions

As mentioned before, we have verified the effectiveness of our proposed method through a series of carefully designed experiments on four challenging datasets. However, there are still some unsolved issues in our proposed method.

Since the measurement and assessment of samples are only carried out during the training process, it will not affect the speed of inference. Nevertheless, DCN requires more parameters than conventional CNN to obtain an adaptive receptive field, which leads to slower convergence of DCN during training.

In addition, there are usually hundreds or thousands of objects in one image under crowded scenarios, which leads to a sharp rise in computation cost in repulsion loss, especially computing rotated IoU values. In the experiments, we use a small trick to reduce the computation cost, where we use horizontal IoU values to exclude ulterior targets. In the future, we will try to exploit the Gaussian approximation methods proposed by [40,41] to simplify the calculation of the rotation IoU.

Finally, we notice that objects of some categories have a dependency relationship with each other, e.g., airplanes parking in airports and soccer ball fields inside ground track fields. We can utilize the prior knowledge of relationships between classes to improve the design of the repulsion loss.

5. Conclusions

In this work, we have presented an effective method for remote sensing object detection utilizing the adaptive point set to represent rotated boxes, which is able to capture key points with substantial semantic and geometric information. To improve the quality of sample selection and assignment, we introduce the center-ness constraint to assess the proposals and acquire high-quality samples. Furthermore, the repulsion constraint in the form of a loss function is designed to enhance the robustness of detecting small and clustered objects. Therefore, the extensive experiments on the four challenging datasets demonstrate the effectiveness of our proposed method.

Author Contributions: Conceptualization, L.G.; methodology, L.G. and H.G.; software, L.G. and Y.W.; validation, L.G. and Y.W.; formal analysis, L.G. and Y.W.; investigation, L.G., Y.W. and D.L.; resources, L.G. and H.G.; data curation, D.L.; writing—original draft preparation, L.G., Y.W. and B.M.M.; writing—review and editing, L.G. and B.M.M.; visualization, D.L.; supervision, H.G.; project administration, L.G. All authors have read and agreed to the published version of the manuscript.

Funding: This research work was supported by Sichuan Science and Technology Program (No. 2023YFG0021, No. 2022YFG0038, and No. 2021YFG0018), and by Xinjiang Science and Technology Program (No. 2022D01B185).

Data Availability Statement: DOTA, HRSC2016, UCAS-AOD, and WHU-RSONE-OBB are available at https://captain-whu.github.io/DOTA/index.html (accessed on 17 January 2023), https://www.kaggle.com/datasets/guofeng/hrsc2016 (accessed on 17 January 2023), https://github.com/fireae/UCAS-AOD-benchmark (accessed on 17 January 2023), and https://pan.baidu.com/share/init?surl=_Gdeedwo9dcEJqIh4eHHMA (password: 1234) (accessed on 17 January 2023), respectively. The source code is available at https://github.com/luilui97/Centerness-Repulsion-Object-Detection-OBB, (accessed on 6 March 2023).

Acknowledgments: We sincerely appreciate the constructive comments and suggestions of the anonymous reviewers, which have greatly helped to improve this paper.

Conflicts of Interest: The authors declare no conflict of interest.

Abbreviations

The following abbreviations are used in this manuscript:

CNN	Convolution Neural Network
DCN	Deformable Convolution Network
IoU	Intersection over Union
FPN	Linear Feature Pyramid Networks
SGD	Stochastic Gradient Descent
mAP	Mean Average Precision
NMS	Non-Maximum Suppression
RoI	Region of Interests

References

1. Girshick, R.; Donahue, J.; Darrell, T.; Malik, J. Rich Feature Hierarchies for Accurate Object Detection and Semantic Segmentation. In Proceedings of the IEEE Conference on Computer Vision and Pattern Recognition (CVPR), Columbus, OH, USA, 23–28 June 2014.
2. Girshick, R.B. Fast R-CNN. In Proceedings of the IEEE International Conference on Computer Vision (ICCV), Santiago, Chile, 7–13 December 2015; pp. 1440–1448.
3. Ren, S.; He, K.; Girshick, R.B.; Sun, J. Faster R-CNN: Towards Real-Time Object Detection with Region Proposal Networks. *IEEE Trans. Pattern Anal. Mach. Intell.* **2017**, *39*, 1137–1149. [CrossRef] [PubMed]
4. Ding, J.; Xue, N.; Long, Y.; Xia, G.; Lu, Q. Learning RoI Transformer for Oriented Object Detection in Aerial Images. In Proceedings of the IEEE/CVF Conference on Computer Vision and Pattern Recognition (CVPR), Long Beach, CA, USA, 15–20 June 2019; pp. 2849–2858.
5. Han, J.; Ding, J.; Li, J.; Xia, G. Align Deep Features for Oriented Object Detection. *IEEE Trans. Geosci. Remote Sens.* **2022**, *60*, 1–11. [CrossRef]
6. Ding, J.; Xue, N.; Xia, G.; Bai, X.; Yang, W.; Yang, M.Y.; Belongie, S.J.; Luo, J.; Datcu, M.; Pelillo, M.; et al. Object Detection in Aerial Images: A Large-Scale Benchmark and Challenges. *IEEE Trans. Pattern Anal. Mach. Intell.* **2022**, *44*, 7778–7796. [CrossRef]
7. Guo, Z.; Liu, C.; Zhang, X.; Jiao, J.; Ji, X.; Ye, Q. Beyond bounding-box: Convex-hull feature adaptation for oriented and densely packed object detection. In Proceedings of the IEEE/CVF Conference on Computer Vision and Pattern Recognition, Nashville, TN, USA, 20–25 June 2021; pp. 8792–8801.
8. Yang, F.; Fan, H.; Chu, P.; Blasch, E.; Ling, H. Clustered Object Detection in Aerial Images. In Proceedings of the IEEE International Conference on Computer Vision (ICCV), Seoul, Republic of Korea, 27 October–2 November 2019; pp. 8310–8319.
9. Yang, X.; Yang, J.; Yan, J.; Zhang, Y.; Zhang, T.; Guo, Z.; Sun, X.; Fu, K. SCRDet: Towards More Robust Detection for Small, Cluttered and Rotated Objects. In Proceedings of the IEEE International Conference on Computer Vision (ICCV), Seoul, Republic of Korea, 27 October–2 November 2019; pp. 8231–8240.
10. Azimi, S.M.; Vig, E.; Bahmanyar, R.; Körner, M.; Reinartz, P. Towards Multi-class Object Detection in Unconstrained Remote Sensing Imagery. In *Lecture Notes in Computer Science, Proceedings of the ACCV, Perth, Australia, 2–6 December 2018*; Springer: Berlin/Heidelberg, Germany, 2018; Volume 11363, pp. 150–165.
11. Jiao, J.; Zhang, Y.; Sun, H.; Yang, X.; Gao, X.; Hong, W.; Fu, K.; Sun, X. A Densely Connected End-to-End Neural Network for Multiscale and Multiscene SAR Ship Detection. *IEEE Access* **2018**, *6*, 20881–20892. [CrossRef]
12. Lin, T.; Dollár, P.; Girshick, R.B.; He, K.; Hariharan, B.; Belongie, S.J. Feature Pyramid Networks for Object Detection. In Proceedings of the IEEE Conference on Computer Vision and Pattern Recognition (CVPR), Honolulu, HI, USA, 21–26 July 2017; pp. 936–944.
13. Lin, T.; Goyal, P.; Girshick, R.B.; He, K.; Dollár, P. Focal Loss for Dense Object Detection. In Proceedings of the IEEE International Conference on Computer Vision (ICCV), Venice, Italy, 22–29 October 2017; pp. 2999–3007.
14. Zhou, X.; Wang, D.; Krähenbühl, P. Objects as Points. *arXiv* **2019**, arXiv:1904.07850.

15. Qian, W.; Yang, X.; Peng, S.; Yan, J.; Guo, Y. Learning Modulated Loss for Rotated Object Detection. In Proceedings of the AAAI Conference on Artificial Intelligence, Vancouver, BC, Canada, 2–9 February 2021; pp. 2458–2466.
16. Yang, X.; Yan, J. Arbitrary-Oriented Object Detection with Circular Smooth Label. In *Lecture Notes in Computer Science, Proceedings of the European Conference on Computer Vision (ECCV), Glasgow, UK, 23–28 August 2020*; Springer: Berlin/Heidelberg, Germany, 2020; Volume 12353, pp. 677–694.
17. Yang, X.; Hou, L.; Zhou, Y.; Wang, W.; Yan, J. Dense Label Encoding for Boundary Discontinuity Free Rotation Detection. In Proceedings of the IEEE Conference on Computer Vision and Pattern Recognition (CVPR), Nashville, TN, USA, 20–25 June 2021; pp. 15819–15829.
18. Li, C.; Xu, C.; Cui, Z.; Wang, D.; Zhang, T.; Yang, J. Feature-Attentioned Object Detection in Remote Sensing Imagery. In Proceedings of the 2019 IEEE International Conference on Image Processing (ICIP), Taipei, Taiwan, 22–25 September 2019; pp. 3886–3890.
19. Pan, X.; Ren, Y.; Sheng, K.; Dong, W.; Yuan, H.; Guo, X.; Ma, C.; Xu, C. Dynamic Refinement Network for Oriented and Densely Packed Object Detection. In Proceedings of the IEEE Conference on Computer Vision and Pattern Recognition (CVPR), Seattle, WA, USA, 13–19 June 2020; pp. 11204–11213.
20. Zhang, S.; Chi, C.; Yao, Y.; Lei, Z.; Li, S.Z. Bridging the Gap Between Anchor-Based and Anchor-Free Detection via Adaptive Training Sample Selection. In Proceedings of the 2020 IEEE/CVF Conference on Computer Vision and Pattern Recognition (CVPR), Seattle, WA, USA, 13–19 June 2020; pp. 9756–9765. [CrossRef]
21. Chen, Y.; Zhang, Z.; Cao, Y.; Wang, L.; Lin, S.; Hu, H. RepPoints v2: Verification Meets Regression for Object Detection. In Proceedings of the NeurIPS, Virtual, 6–12 December 2020.
22. Hou, L.; Lu, K.; Xue, J.; Li, Y. Shape-Adaptive Selection and Measurement for Oriented Object Detection. In Proceedings of the Thirty-Sixth AAAI Conference on Artificial Intelligence (AAAI 2022), Thirty-Fourth Conference on Innovative Applications of Artificial Intelligence (IAAI 2022), The Twelveth Symposium on Educational Advances in Artificial Intelligence (EAAI 2022), Virtual Event, 22 February–1 March 2022; pp. 923–932.
23. Li, W.; Chen, Y.; Hu, K.; Zhu, J. Oriented RepPoints for Aerial Object Detection. In Proceedings of the IEEE Conference on Computer Vision and Pattern Recognition (CVPR), New Orleans, LA, USA, 18–24 June 2022; pp. 1819–1828.
24. Yang, Z.; Liu, S.; Hu, H.; Wang, L.; Lin, S. RepPoints: Point Set Representation for Object Detection. In Proceedings of the IEEE International Conference on Computer Vision (ICCV), Seoul, Republic of Korea, 27 October–2 November 2019; pp. 9656–9665.
25. Tian, Z.; Shen, C.; Chen, H.; He, T. FCOS: Fully Convolutional One-Stage Object Detection. In Proceedings of the IEEE International Conference on Computer Vision (ICCV), Seoul, Republic of Korea, 27 October–2 November 2019; pp. 9626–9635.
26. Dai, J.; Qi, H.; Xiong, Y.; Li, Y.; Zhang, G.; Hu, H.; Wei, Y. Deformable Convolutional Networks. In Proceedings of the IEEE International Conference on Computer Vision (ICCV), Venice, Italy, 22–29 October 2017; pp. 764–773.
27. Wang, X.; Xiao, T.; Jiang, Y.; Shao, S.; Sun, J.; Shen, C. Repulsion Loss: Detecting Pedestrians in a Crowd. In Proceedings of the IEEE Conference on Computer Vision and Pattern Recognition (CVPR), Salt Lake City, UT, USA, 18–22 June 2018; pp. 7774–7783.
28. Xia, G.; Bai, X.; Ding, J.; Zhu, Z.; Belongie, S.J.; Luo, J.; Datcu, M.; Pelillo, M.; Zhang, L. DOTA: A Large-Scale Dataset for Object Detection in Aerial Images. In Proceedings of the IEEE Conference on Computer Vision and Pattern Recognition (CVPR), Salt Lake City, UT, USA, 18–22 June 2018; pp. 3974–3983.
29. Liu, Z.; Yuan, L.; Weng, L.; Yang, Y. A High Resolution Optical Satellite Image Dataset for Ship Recognition and Some New Baselines. In Proceedings of the 6th International Conference on Pattern Recognition Applications and Methods (ICPRAM), Porto, Portugal, 24–26 February 2017; pp. 324–331.
30. Zhu, H.; Chen, X.; Dai, W.; Fu, K.; Ye, Q.; Jiao, J. Orientation robust object detection in aerial images using deep convolutional neural network. In Proceedings of the 2015 IEEE International Conference on Image Processing (ICIP), Quebec City, QC, Canada, 27–30 September 2015; pp. 3735–3739.
31. Dong, Z.; Wang, M.; Wang, Y.; Liu, Y.; Feng, Y.; Xu, W. Multi-Oriented Object Detection in High-Resolution Remote Sensing Imagery Based on Convolutional Neural Networks with Adaptive Object Orientation Features. *Remote Sens.* **2022**, *14*, 950. [CrossRef]
32. Han, J.; Ding, J.; Xue, N.; Xia, G. ReDet: A Rotation-Equivariant Detector for Aerial Object Detection. In Proceedings of the IEEE Conference on Computer Vision and Pattern Recognition (CVPR), Nashville, TN, USA, 20–25 June 2021; pp. 2786–2795.
33. Zhang, G.; Lu, S.; Zhang, W. CAD-Net: A Context-Aware Detection Network for Objects in Remote Sensing Imagery. *IEEE Trans. Geosci. Remote Sens.* **2019**, *57*, 10015–10024. [CrossRef]
34. Yang, X.; Yan, J.; Feng, Z.; He, T. R3Det: Refined Single-Stage Detector with Feature Refinement for Rotating Object. In Proceedings of the AAAI Conference on Artificial Intelligence, Vancouver, BC, Canada, 2–9 February 2021; pp. 3163–3171.
35. Yang, X.; Yan, J.; Liao, W.; Yang, X.; Tang, J.; He, T. Scrdet++: Detecting small, cluttered and rotated objects via instance-level feature denoising and rotation loss smoothing. *IEEE Trans. Pattern Anal. Mach. Intell.* **2022**, *45*, 2384–2399. [CrossRef] [PubMed]
36. Li, Y.; Huang, Q.; Pei, X.; Jiao, L.; Shang, R. RADet: Refine Feature Pyramid Network and Multi-Layer Attention Network for Arbitrary-Oriented Object Detection of Remote Sensing Images. *Remote Sens.* **2020**, *12*, 389. [CrossRef]
37. Zhou, X.; Zhuo, J.; Krähenbühl, P. Bottom-Up Object Detection by Grouping Extreme and Center Points. In Proceedings of the IEEE Conference on Computer Vision and Pattern Recognition (CVPR), Long Beach, CA, USA, 16–20 June 2019; pp. 850–859. [CrossRef]

38. Duan, K.; Bai, S.; Xie, L.; Qi, H.; Huang, Q.; Tian, Q. CenterNet: Keypoint Triplets for Object Detection. In Proceedings of the 2019 IEEE/CVF International Conference on Computer Vision (ICCV), Seoul, Republic of Korea, 27 October–2 November 2019; pp. 6568–6577. [CrossRef]
39. Wei, H.; Zhang, Y.; Chang, Z.; Li, H.; Wang, H.; Sun, X. Oriented objects as pairs of middle lines. *ISPRS J. Photogramm. Remote Sens.* **2020**, *169*, 268–279. [CrossRef]
40. Yang, X.; Yan, J.; Ming, Q.; Wang, W.; Zhang, X.; Tian, Q. Rethinking Rotated Object Detection with Gaussian Wasserstein Distance Loss. In Proceedings of the International Conference on Machine Learning (ICML), Virtual Event, 18–24 July 2021; pp. 11830–11841.
41. Yang, X.; Yang, X.; Yang, J.; Ming, Q.; Wang, W.; Tian, Q.; Yan, J. Learning High-Precision Bounding Box for Rotated Object Detection via Kullback-Leibler Divergence. In Proceedings of the NeurIPS, Virtual, 6–14 December 2021; pp. 18381–18394.
42. Li, H.; Wu, Z.; Zhu, C.; Xiong, C.; Socher, R.; Davis, L.S. Learning From Noisy Anchors for One-Stage Object Detection. In Proceedings of the 2020 IEEE/CVF Conference on Computer Vision and Pattern Recognition (CVPR), Seattle, WA, USA, 13–19 June 2020; pp. 10585–10594. [CrossRef]
43. Zhang, X.; Wan, F.; Liu, C.; Ji, R.; Ye, Q. FreeAnchor: Learning to Match Anchors for Visual Object Detection. In Proceedings of the Advances in Neural Information Processing Systems 32: Annual Conference on Neural Information Processing Systems 2019 (NeurIPS 2019), Vancouver, BC, Canada, 8–14 December 2019; pp. 147–155.
44. Kim, K.; Lee, H.S. Probabilistic Anchor Assignment with IoU Prediction for Object Detection. In *Lecture Notes in Computer Science, Proceedings of the Computer Vision–ECCV 2020—16th European Conference, Glasgow, UK, 23–28 August 2020*; Vedaldi, A., Bischof, H., Brox, T., Frahm, J., Eds.; Springer: Berlin/Heidelberg, Germany, 2020; Volume 12370, pp. 355–371. [CrossRef]
45. Ming, Q.; Zhou, Z.; Miao, L.; Zhang, H.; Li, L. Dynamic Anchor Learning for Arbitrary-Oriented Object Detection. In Proceedings of the AAAI Conference on Artificial Intelligence, Vancouver, BC, Canada, 2–9 February 2021; pp. 2355–2363.
46. Rezatofighi, H.; Tsoi, N.; Gwak, J.; Sadeghian, A.; Reid, I.D.; Savarese, S. Generalized Intersection Over Union: A Metric and a Loss for Bounding Box Regression. In Proceedings of the IEEE Conference on Computer Vision and Pattern Recognition (CVPR), Long Beach, CA, USA, 16–20 June 2019; pp. 658–666. [CrossRef]
47. Zhou, Y.; Yang, X.; Zhang, G.; Wang, J.; Liu, Y.; Hou, L.; Jiang, X.; Liu, X.; Yan, J.; Lyu, C.; et al. MMRotate: A Rotated Object Detection Benchmark using PyTorch. In Proceedings of the 30th ACM International Conference on Multimedia, Lisboa, Portugal, 10–14 October 2022.
48. He, K.; Zhang, X.; Ren, S.; Sun, J. Deep Residual Learning for Image Recognition. In Proceedings of the 2016 IEEE Conference on Computer Vision and Pattern Recognition (CVPR), Las Vegas, NV, USA, 27–30 June 2016; pp. 770–778. [CrossRef]
49. Liu, Z.; Lin, Y.; Cao, Y.; Hu, H.; Wei, Y.; Zhang, Z.; Lin, S.; Guo, B. Swin Transformer: Hierarchical Vision Transformer using Shifted Windows. In Proceedings of the 2021 IEEE/CVF International Conference on Computer Vision (ICCV), Montreal, QC, Canada, 10–17 October 2021; pp. 9992–10002. [CrossRef]
50. Wang, J.; Yang, W.; Li, H.; Zhang, H.; Xia, G. Learning Center Probability Map for Detecting Objects in Aerial Images. *IEEE Trans. Geosci. Remote Sens.* **2021**, *59*, 4307–4323. [CrossRef]
51. Wang, J.; Ding, J.; Guo, H.; Cheng, W.; Pan, T.; Yang, W. Mask OBB: A Semantic Attention-Based Mask Oriented Bounding Box Representation for Multi-Category Object Detection in Aerial Images. *Remote Sens.* **2019**, *11*, 2930. [CrossRef]
52. Xu, Y.; Fu, M.; Wang, Q.; Wang, Y.; Chen, K.; Xia, G.; Bai, X. Gliding Vertex on the Horizontal Bounding Box for Multi-Oriented Object Detection. *IEEE Trans. Pattern Anal. Mach. Intell.* **2021**, *43*, 1452–1459. [CrossRef] [PubMed]
53. Xie, X.; Cheng, G.; Wang, J.; Yao, X.; Han, J. Oriented R-CNN for Object Detection. In Proceedings of the IEEE International Conference on Computer Vision (ICCV), Montreal, QC, Canada, 10–17 October 2021; pp. 3500–3509.
54. Chen, Z.; Chen, K.; Lin, W.; See, J.; Yu, H.; Ke, Y.; Yang, C. PIoU Loss: Towards Accurate Oriented Object Detection in Complex Environments. In *Lecture Notes in Computer Science, Proceedings of the European Conference on Computer Vision (ECCV), Glasgow, UK, 23–28 August 2020*; Springer: Berlin/Heidelberg, Germany, 2020; Volume 12350, pp. 195–211.
55. Liao, M.; Zhu, Z.; Shi, B.; Xia, G.; Bai, X. Rotation-Sensitive Regression for Oriented Scene Text Detection. In Proceedings of the 2018 IEEE Conference on Computer Vision and Pattern Recognition (CVPR), Salt Lake City, UT, USA, 18–22 June 2018; pp. 5909–5918. [CrossRef]
56. Wang, C.Y.; Bochkovskiy, A.; Liao, H.Y.M. YOLOv7: Trainable bag-of-freebies sets new state-of-the-art for real-time object detectors. *arXiv* **2022**, arXiv:2207.02696.
57. Redmon, J.; Farhadi, A. Yolov3: An incremental improvement. *arXiv* **2018**, arXiv:1804.02767.
58. Dong, Z.; Wang, M.; Wang, Y.; Zhu, Y.; Zhang, Z. Object Detection in High Resolution Remote Sensing Imagery Based on Convolutional Neural Networks With Suitable Object Scale Features. *IEEE Trans. Geosci. Remote Sens.* **2020**, *58*, 2104–2114. [CrossRef]
59. Sun, L.; Cheng, S.; Zheng, Y.; Wu, Z.; Zhang, J. SPANet: Successive Pooling Attention Network for Semantic Segmentation of Remote Sensing Images. *IEEE J. Sel. Top. Appl. Earth Obs. Remote Sens.* **2022**, *15*, 4045–4057. [CrossRef]
60. Yin, P.; Zhang, D.; Han, W.; Li, J.; Cheng, J. High-Resolution Remote Sensing Image Semantic Segmentation via Multiscale Context and Linear Self-Attention. *IEEE J. Sel. Top. Appl. Earth Obs. Remote Sens.* **2022**, *15*, 9174–9185. [CrossRef]

remote sensing

MDPI

Article

SquconvNet: Deep Sequencer Convolutional Network for Hyperspectral Image Classification

Bing Li, Qi-Wen Wang, Jia-Hong Liang *, En-Ze Zhu and Rong-Qian Zhou

College of Engineering, Shantou University, Shantou 515063, China
* Correspondence: 19jhliang1@stu.edu.cn

Abstract: The application of Transformer in computer vision has had the most significant influence of all the deep learning developments over the past five years. In addition to the exceptional performance of convolutional neural networks (CNN) in hyperspectral image (HSI) classification, Transformer has begun to be applied to HSI classification. However, for the time being, Transformer has not produced satisfactory results in HSI classification. Recently, in the field of image classification, the creators of Sequencer have proposed a Sequencer structure that substitutes the Transformer self-attention layer with a BiLSTM2D layer and achieves satisfactory results. As a result, this paper proposes a unique network called SquconvNet, that combines CNN with Sequencer block to improve hyperspectral classification. In this paper, we conducted rigorous HSI classification experiments on three relevant baseline datasets to evaluate the performance of the proposed method. The experimental results show that our proposed method has clear advantages in terms of classification accuracy and stability.

Keywords: hyperspectral image (HSI) classification; transformer; convolutional neural network (CNN); Sequencer; long short-term memory network (LSTM)

Citation: Li, B.; Wang, Q.-W.; Liang, J.-H.; Zhu, E.-Z.; Zhou, R.-Q. SquconvNet: Deep Sequencer Convolutional Network for Hyperspectral Image Classification. *Remote Sens.* **2023**, *15*, 983. https://doi.org/10.3390/rs15040983

Academic Editor: Gwanggil Jeon

Received: 6 January 2023
Revised: 7 February 2023
Accepted: 8 February 2023
Published: 10 February 2023

1. Introduction

Recent improvements in hyperspectral imaging sensors have resulted in hyperspectral images (HSI) that are rich in hundreds of contiguous and narrow spectral bands/depth. Due to its extensive spatial-spectral data, HSI has been used for a variety of purposes, including target detection [1], forestry [2,3], satellite calibration [4], identifying post-fire severity [5], and mineral identification [6]. Similarly, classification of hyperspectral land-cover information is one of the most significant application directions and has garnered a great deal of attention.

Two of the key distinguishing features of HSI are its high spatial correlation and an abundance of spectral information. A high spatial correlation from homogeneous areas can give secondary supplemental information for accurate mapping [7]. The ground material comprises a significant number of representative features that enable precise identification [8], taking advantage of the rich spectral information found in the continuous spectral bands. Contrarily, the curse of dimensionality is also brought on by an abundance of spectral information, which may have an impact on the performance of the classification [9–11]. Utilizing dimensionality reduction techniques is a crucial step for HSI classification, in order to improve the classification performance. The most used dimensionality reduction approach is Principal Component Analysis (PCA) [12]. In addition, other key dimensionality reduction techniques in hyperspectral classification include Factor Analysis (FA) [13], Linear Discriminant Analysis (LDA) [14,15], and Independent Discriminant Analysis (IDA) [16,17]. Early attempts at HSI classification included Support Vector Machines (SVMs) [18], Random Forest (RF) [19], K-mean clustering (KNN) [20], and Markov Random Field (MRF) [21]. However, because these techniques don't concentrate on spatial correlation and local consistency, they struggle to fully utilize spatial feature information, which leads to subpar classification performance. Recent advances in deep learning-based techniques, such as deep neural networks, have had a major impact on computer vision and have also been introduced to HSI classification. A deep belief network (DBN) [22] and stacked autoencoders (SAE) [12] are two methods that call for flattening the input data patches into one-dimensional features. However, both techniques alter the original spatial data, which leads to subpar performances. Hu et al. suggested five convolutional layers to create a 1D-CNN [23] for HSI classification. It accepts spectral data as input and can successfully extract discriminative features from the spectral data. The data conveyed by the first few principal components after dimensionality reduction are used by a 2D-CNN method [24], to extract spatial features. Chen et al. [25], introduced 3D-CNN to HSI classification, in order to extract both spatial and spectral information. The spectral-spatial residual network (SSRN) [26], which was inspired by Resnet [27], creates a deeper structure and utilizes identity mapping to connect additional three-dimensional convolutional layers. The deep pyramidal residual network (DPRN) has also been proposed for HSI data [28]. According to [29], a hybrid-CNN model (HybridSN) is proposed, that may overcome the failure of 2D-CNN to extract discriminative features from the spectral dimension, and that scales back the complexity of a single 3D-CNN. In addition to the convolution neural network, various networks with exceptional performance have been introduced to HSI classification, such as the completely convolution network (FCN) [30,31], the generative adversarial network (GAN) [32,33], the graph convolutional network (GCN) [34], etc.

Additionally, Transformer, the most well-liked neural network currently, has been introduced into hyperspectral classification [35]. These include a Spatial-Spectral Transformer (SST) [36], an upgraded transformer (SAT) [37], a restructured transformer encoder with a cross-layer model (SpectralFormer) [38], and a Spectral-Spatial Feature Tokenization Transformer (SSFTT) [39]. However, the performance of these Transformer-based methods is inferior to that of CNN-based methods.

Transformer still has certain limitations regarding the extraction of local spectral and local information disparities, which causes performance bottlenecks. Convolutional neural networks perform well in HSI classification, although there are still a number of problems. The first is that the ground's irregular shape prevents the convolution kernel from being able to capture all of its features [40]. The second is caused by the fact that the small convolutional kernels prevent convolutional neural networks' receptive fields from matching the hyperspectral features across the whole bandwidth [37]. Recently, Yuki and Masato [41] proposed Sequencer, a unique and straightforward architecture that uses LSTM for image classification. Sequencer uses a BiLSTM2D layer to replace the multi-head attention layer in the transformer encoder to create Sequencer block. Experiments reveal that self-attention is not required for modeling remote dependencies, and that competitive performance can be attained using the LSTM instead. As a supplement to convolutional neural networks, we have developed Sequencer, a Sequencer made up of vertical and horizontal bidirectional LSTMs, based on the context of the aforementioned problems and inspired by Sequencer [41]. Similar to the convolutional layer, we take a pixel as the center, regard the vertical and horizontal directions as sequences, and simultaneously expand the pixel to form a spatially significant receptive field. Contrary to the convolutional layer, however, the timed information capacity of LSTM gives the Sequencer the ability to blend spatial information memory, which we feel can be employed as a supplement to the convolutional layer's shortcomings. As a result, we suggest SquconvNet, a network integrating CNN with Sequencer2D block for HSI Classification, as being inspired by Sequencer. The proposed network in this study consists of three modules: the Spectral-Spatial Feature Extraction (SSFE) module, the Sequencer module, and the Auxiliary Classification (AC) module. The dimensionally reduced 3D-Patches input will be passed through the Sequencer module first, to capture long-term feature information, then the SSFE module will extract spatial features, and finally the AC module will further improve the classification performance. The LSTM shows a strong performance in utilizing long-range information to compensate for CNN's shortcomings, and our proposed model has demonstrated good results on three standard datasets.

The following is a summary of this paper's significant contributions and work:

(1) We introduce the BiLSTM2D layer and Sequencer module for the first time, and combine them with CNN to compensate for CNN's shortcomings and improve the performance of HSI classification.
(2) A supplementary classification module comprised of two convolutional layers and a fully connected layer is proposed, with the dual purpose of decreasing the network parameters and assisting the network in classification.
(3) Using three typical baseline datasets, we performed qualitative and quantitative evaluation studies (IP, UP, SA). The experimental findings show that, in terms of classification accuracy and stability, our proposed model verifies its superiority.

Next, Section 2 introduces in detail the illustration of the proposed SquconvNet architecture and its three modules. Section 3 describes the baseline datasets and presents an analysis of the experimental results. Ablation experiments and time loss are discussed in Section 4. Ultimately, the conclusion is drawn in Section 5.

2. Materials and Methods

2.1. Overview of SquconvNet

The proposed SquconvNet's general structure is shown in Figure 1. It is composed of three modules: the Spectral-Spatial Feature Extraction Module, the Sequencer2D Module, and the Auxiliary Classification Module.

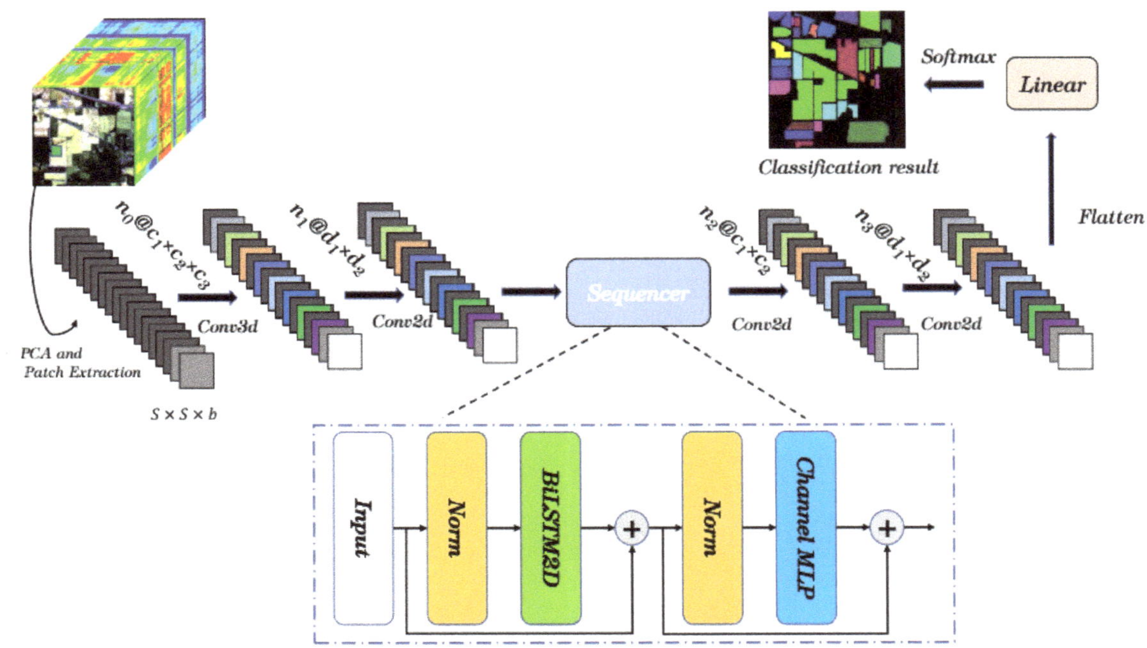

Figure 1. Illustration of the proposed SquconvNet.

2.2. Spectral-Spatial Feature Extraction Module

The Spectral-Spatial Feature Extraction (SSFE) module is where the proposed SquconvNet starts. HybridSN [29] and SSFTT [39] served as the inspiration for the SSFE module's design. Here, we adopt a comparable structure to them and improve its properties. The SSFE module primarily consists of a 3D-convolution layer and a 2D-convolution layer to reduce the amount of computation. Each convolution layer is followed by a batch normalization (BN) layer, a Relu non-linear activation, and another layer.

Firstly, an original HSI dataset is defined as $X \in R^{W \times H \times D}$, where D is the number of spectral bands, W is the width, and H is the height. Each HSI pixel forms a one-hot vector $Y = (y_1, y_2, \ldots, y_C) \in R^{1 \times 1 \times C}$, where C is the classes of land-cover. However, since D bands make up the HSI data, they only add a lot of unnecessary calculations and non-useful information. To eliminate redundant spectral information and preserve the same spatial dimensions, the principle component analysis (PCA) is performed on the HSI data, reducing the number of bands from D to b. Next, 3D-patches $P \in R^{s \times s \times b}$ are created from $X_{PCA} \in R^{M \times N \times b}$, where $s \times s$ is the window size of 3D-patch. Besides, when a single pixel is extracted, the edge pixels perform padding of $\frac{s-1}{2}$. Then, the true label is determined by the original label of the center pixel.

Secondly, each 3D-Patch, of size $s \times s \times b$, is used as the input to the SSFE module, to extract spectral-spatial features. In the operation of the 3D convolution layer, the value at the (x, y, z) position on the jth feature cube of the ith layer is calculated by:

$$v_{i,j}^{x,y,z} = f\left(b_{i,j} + \sum_k \sum_{h=-\alpha}^{\alpha} \sum_{w=-\beta}^{\beta} \sum_{c=-\gamma}^{\gamma} p_{i,j,k}^{h,w,c} v_{i-1,k}^{x+h,y+w,z+c}\right) \tag{1}$$

where $f(\cdot)$ defines the activation function; $b_{i,j}$ defines the bias; $2\alpha + 1$, $2\beta + 1$ and $2\gamma + 1$ respectively represent the height, width, and depth of the convolution kernel; $p_{i,k,k}^{h,w,c}$ is the weight parameter of the jth convolution kernel in the ith layer, and the kth feature of the previous layer at position (h, w, c); $v_{i-1,k}^{x+h,y+w,z+c}$ represents the value at the position $(x + h,\ y + w,\ z + c)$.

Similarly, for the 2D convolution layer, its formula can be expressed as:

$$v_{i,j}^{x,y} = f\left(b_{i,j} + \sum_k \sum_{h=-\alpha}^{\alpha} \sum_{w=-\beta}^{\beta} p_{i,j,k}^{h,w} v_{i-1,k}^{x+h,y+w}\right) \tag{2}$$

The 3D convolution layer and 2D convolution layer in the two convolution models discussed above have different features. The convolution kernel of the 3D convolution layer is $k_1 \times k_2 \times k_3$, forming a rectangular body that can cover the spectrum- spatial information. The convolution kernel of the two-dimensional convolutional layer is $k_1 \times k_2$, which forms a rectangular body to extract spatial information. In other words, while 2D convolution layers are unable to extract spectral correlations, 3D convolution layers may extract both spectral and spatial information simultaneously. On the other hand, a 3D convolution layer typically has parameters that are much higher than a 2D convolution layer. Therefore, the use of 3D-convolution layers alone may lead to performance reduction due to an excessive number of parameters, and the use of 2D convolution layers alone may lead to an insufficient ability to extract spatial features, so a hybrid 3D-2D convolution layer is considered here, to extract spectral-spatial features.

Lastly, in our SSFE module, the dimensions of the 3D convolution kernels are $8 \times 3 \times 7 \times 7 \times 1$, where 1 is the number of spectral bands of the input data, 8 is the number of channels produced by the convolution, and $(3 \times 7 \times 7)$ is the size of the convolving kernel. The sizes of the 2D convolution kernels are $64 \times 3 \times 3 \times (8 \times (b - 2))$, where $8 \times (b - 2)$ is the number of spectral bands of the input data, 64 is the number of channels produced by the convolution, and (3×3) is the size of the convolving kernel. Assuming that the input patch size is $s \times s \times b$, then the output patch size is $(s - 8) \times (s - 8) \times 64$.

2.3. Sequencer Module

The Sequencer (SDB) module is used to extract the spatial features after the spectral-spatial features have been extracted by the SSFE module. We integrate the Sequencer into the HSI classification process and perform an adaptive transformation on it in order to address the proposed solution for the traditional image classification problems [41]. The Sequencer module's BiLSTM2D core, which consists of vertical BiLSTM, horizontal BiLSTM, and channel fusion, is its most significant component. Contrarily, the BiLSTM is made up of two standard LSTM. Figure 2 depicts the precise LSTM structure and describes the BiLSTM2D layer.

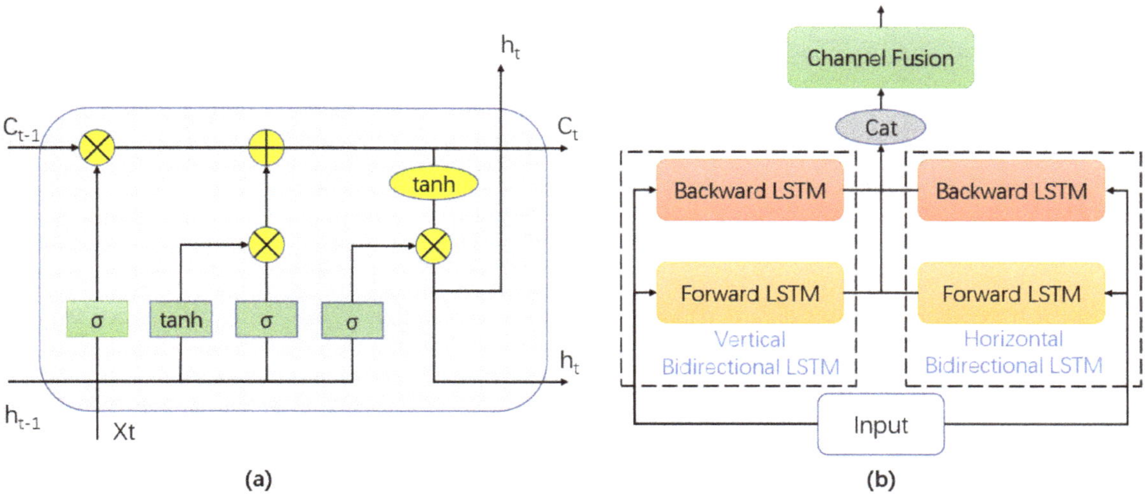

Figure 2. (**a**) The specific structure of LSTM layer (**b**) The figure outlines the BiLSTM2D layer.

The LSTM [42] has an input gate i_t, a forget gate f_t, and an output gate o_t. Where the input gate controls the storage of the input, the forget gate controls the forgetting of the previous cell state, and the output gate controls the cell output h_t of the current cell state c_t. As a review, the original LSTM is formulated as:

$$\begin{cases} i_t = \sigma(b_i + W_{xi}x_t + W_{hi}h_{t-1}) \\ f_t = \sigma\left(b_f + W_{xf}x_t + W_{hf}h_{t-1}\right) \\ o_t = \sigma(b_o + W_{xo}x_t + W_{ho}h_{t-1}) \\ c_t = c_{t-1} \odot f_t + i_t \odot \tanh(b_a + W_{xa}x_t + W_{ha}h_{t-1}) \\ h_t = o_t tanh \odot (c_t) \\ \tanh(x) = \frac{1-e^{-2x}}{1+e^{-2x}} \\ \sigma(x) = \frac{1}{1+e^{-x}} \end{cases} \tag{3}$$

where $\odot$ is the Hadamard product,b_k $(k = i,\ f, o)$ is the offset, W_{xj} *and* W_{hj} $(j = a,\ f, i, o)$ are the weight matrices.

A BiLSTM consists of two LSTMs, which is formulated as:

$$h = concatenate\left(LSTM_{for}(\vec{x}),\ LSTM_{back}\left(\overleftarrow{x}\right)\right) \tag{4}$$

where $\vec{x}$ is the input series, $\overleftarrow{x}$ is the rearrangement of $\vec{x}$ in reverse order, and h is a 2D dimensional vector output.

Consisting of a vertical BiLSTM and a horizontal BiLSTM, the BiLSTM2D layers are a technique for efficiently mixing 2D spatial information. Let $X \in R^{H\times W\times C}$ be the input of the Sequencer module, the BiLSTM2D can be formulated as:

$$H = concatenate(BiLSTM(X_{h,W,C}),\ BiLSTM(X_{H,w,C})), \hat{X} = FC(H) \tag{5}$$

where $\{X_{h,W,C} \in R^{W\times C}\}_{h=1}^{H}$ and $\{X_{H,w,C} \in R^{H\times C}\}_{w=1}^{W}$ can be viewed as a set of sequences, and $FC(\cdot)$ is the fully-connected layer with weight $W \in R^{C\times 4D}$.

In this process, X is the input 2D-patches, and its horizontal and vertical directions are treated separately, as sequences, as input to the LSTM.

Figure 3 demonstrates that the Transformer block contains multi-head attention, while the Sequencer module contains BiLSTM2D. In place of the multi-head self-attention in the Transformer block, the Sequencer module uses BisLSTM2D, as seen in Figure 3. Multi-head self-attention is thought to have had a significant role in the success of the Transformer. In contrast, multi-head self-attention is less memory and parameter efficient than LSTM, which is also equally capable of learning long-term dependencies. The Sequencer module is utilized in this case to extract additional discriminative spatial features. Specifically, the output, $O \in R^{s-8 \times s-8 \times 64}$, for the previous module does not change its size in this module.

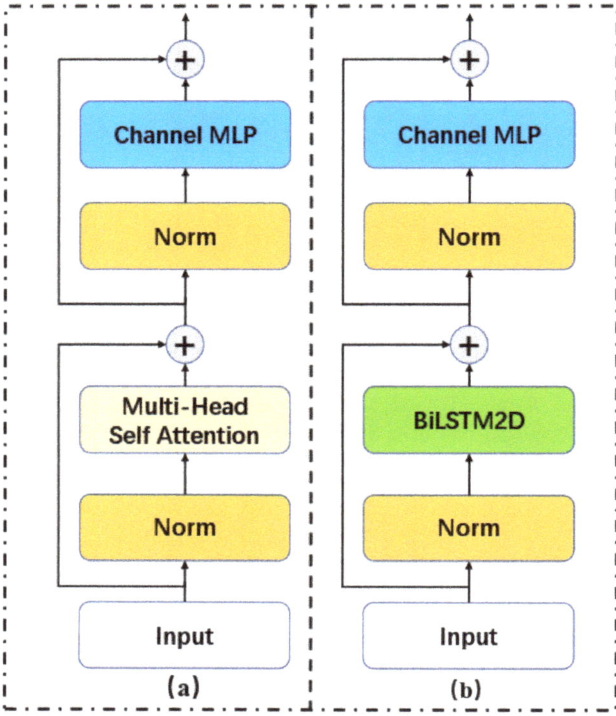

Figure 3. (**a**) The Transformer block consists of multi-head attention (**b**) The Sequencer module consists of BiLSTM2D.

2.4. Auxiliary Classification Module

We suggest the Auxiliary Classification (AC) module, which is based on further extracting feature information and minimizing the number of parameters in fully connected layers. The AC module, the final module in the proposed model, consists of two 2D-convolution layers, a flattened layer and a fully connected layer. A BN layer, followed by a relu non-linear activation function, comes after each convolution layer. Direct classification will not be as successful as it could be after the previous two modules, despite the fact that numerous discriminative features have been retrieved and the patch size is still quite high. As a result, two 2D convolutional layers are utilized to reduce the size of the patches and the number of parameters. These layers' convolutional kernel sizes are (7×7) and (3×3), with 128 and 256 kernels, respectively. The last input channel in the last fully connected layer is 256 if the initial input patch size is $(17 \times 17 \times b)$. Finally, the label will be expressed as the predicted category of the sample after passing the softmax function of the AC module the highest probability value.

3. Experiment and Analysis

3.1. Hyperspectral Image Datasets Description

To examine the effectiveness and stability of our suggested SquconvNet model, we take into account three publicly available standard hyperspectral image datasets: Indian

Pines (IP), University of Pavia (UP), and the Salians Scene (SA). The three datasets are summarized in Table 1.

Table 1. Summary of the Characteristics of the IP, the UP, and the SA Datasets.

Description	Datasets		
	IP	UP	SA
Spatial Size	145 × 145	610 × 340	512 × 217
Spectral Band	224	103	204
No of Classes	16	9	16
Total sample pixels	10,249	42,776	54,129
Sensor	AVIRIS	ROSIS	AVIRIS
Spatial Resolution (m)	20	1.3	3.7

3.1.1. Indian Pines Dataset (IP)

The IP dataset was acquired by the AVIRIS sensor in northwest Indiana, and consists of 145 × 145 pixels and 224 spectral reflectance bands in the wavelength range 400 nm to 2500 nm, 24 of which, covering the region of water absorption, have been eliminated. Figure 4 depicts the false-color image and the image of the real world. For training, we randomly chose 30% of the data, and for testing, we randomly chose the remaining 70%. The category names, training samples, test samples, and the number of samples per category are listed in Table 2.

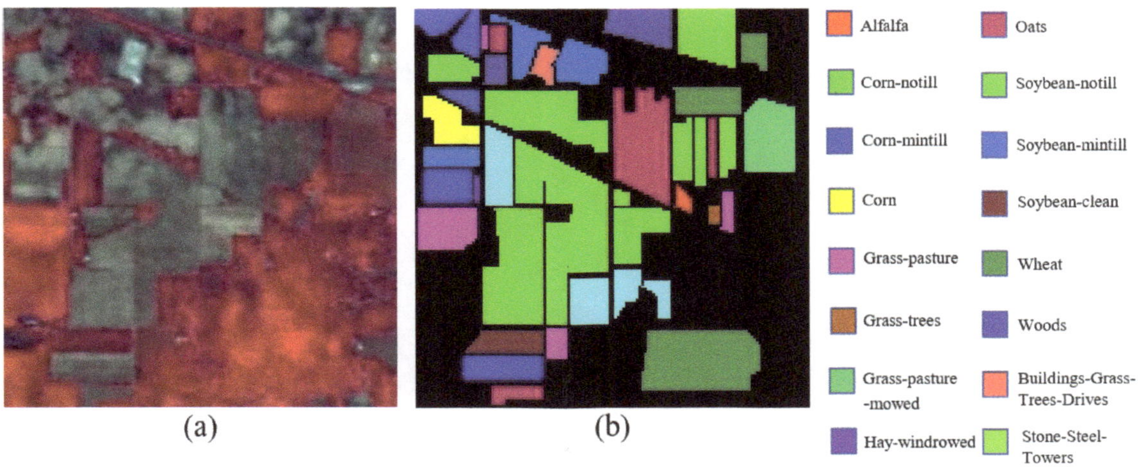

Figure 4. Indian Pines dataset. (**a**) False-color image; (**b**) Ground Truth.

Table 2. Training and Test Samples for the IP Dataset.

Category	Category Name	Training Samples	Test Samples	Number of Samples per Category
1	Alfalfa	14	32	46
2	Corn-notill	431	997	1428
3	Corn-mintill	250	580	830
4	Corn	71	166	237
5	Grass-pasture	145	338	483
6	Grass-trees	219	511	730
7	Grass-pasture-mowed	8	20	28

Table 2. *Cont.*

Category	Category Name	Training Samples	Test Samples	Number of Samples per Category
8	Hay-windrowed	143	335	478
9	Oats	6	14	20
10	Soybean-nottill	292	680	972
11	Soybean-mintill	736	1719	2455
12	Soybean-clean	178	415	593
13	Wheat	62	143	205
14	Woods	381	884	1265
15	Building-Grass-Trees-Drives	117	269	386
16	Stone-Steel-Towers	28	65	93
Total.		3081	7168	10,249

3.1.2. University of Pavia Dataset (UP)

The UP dataset includes imagery of 610 × 340 pixels, 103 spectral depths, and a wavelength range of 430∼ 860nm , and was collected by the ROSIS sensor during a flight campaign above Pavia University. There are 42,776 labeled pixels altogether, divided into nine classes of urban land-cover. Figure 5 displays the ground truth image and the false-color image. In the UP dataset, the entire set is divided into two separate datasets at random, with 10% of the samples utilized for training and the remaining 90% for classification evaluation. Table 3 provides further details on each category as well as general information.

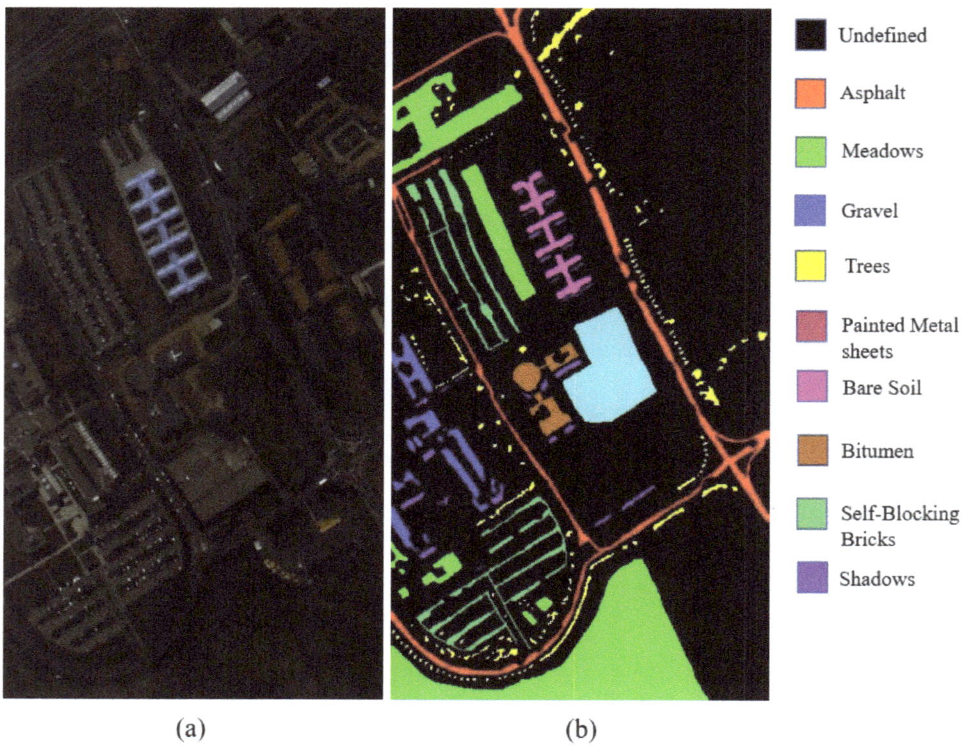

Figure 5. University of Pavia dataset. (**a**) False-color image; (**b**) Ground Truth.

Table 3. Training and Test Samples for UP.

Category	Category Name	Training Samples	Test Samples	Number of Samples per Category
1	Asphalt	663	5968	6631
2	Meadows	1865	16,784	18,649
3	Gravel	210	1889	2099
4	Trees	306	2758	3064
5	Painted metal sheets	134	1211	1345
6	Bare Soil	503	4526	5029
7	Bitumen	133	1197	1330
8	Self-Blocking Bricks	368	3314	3682
9	Shadows	95	852	947
Total.		4277	38,499	42,776

3.1.3. Salians Scene Dataset (SA)

The SA dataset, acquired by the AVRIS sensor, consists of 512 × 217 spatial sizes and 224 spectral depths in the wavelength range of 360 to 2500 nm; 20 of the spectral bands, spanning the region of water absorption, have been eliminated. With the category names, training samples, test samples, and the number of samples per category indicated in Table 4, the false-color image and ground truth map are shown in Figure 6 along with the false-color image itself. Of the samples, 10% are randomly chosen for training and the remaining 90% are used for the classification evaluation.

Table 4. Training and Testing Samples for SA.

Category	Category Name	Training Samples	Test Samples	Number of Samples per Category
1	Brocoli_green_weeds_1	201	1808	2009
2	Brocoli_green_weeds_2	372	3354	3726
3	Fallow	197	1779	1976
4	Fallow_rough_plow	139	1255	1394
5	Fallow_smooth	268	2410	2678
6	Stubble	396	3563	3959
7	Celery	358	3221	3579
8	Grapes_untrained	1127	10,144	11,271
9	Soil_vineyard_develop	620	5583	6203
10	Corn_senesced_green_weeds	328	2950	3278
11	Lettuce_romaine_4wk	107	961	1068
12	Lettuce_romaine_5wk	193	1734	1927
13	Lettuce_romaine_6wk	91	825	916
14	Lettuce_romaine_7wk	107	963	1070
15	Vineyard_untrained	727	6541	7268
16	Vineyard_vertical_trellis	181	1626	1807
Total.		5412	48,717	54,129

3.2. Experimental Settings

In order to make a fair comparison, both our proposed model and the compared methods were tested in the PyTorch environment on a GPU server equipped with an NVIDIA GeForce GTX 3060 12 GB. With a 256-miniature batch size, we decided to use the Adam optimizer, an optimizer with an adaptable learning rate, to improve the proposed model. According to classification performance, 1×10^{-3} is chosen as the initial learning rate. There are 100 training epochs applied to each dataset. The 3D patches of 17 × 17 × 30 for IP, and 17 × 17 × 15 for UP and SA, are used for a fair comparison. To test the performance of our experiment, four important and common measurements are used: each class accuracy, the Overall Accuracy (OA), the Average Accuracy (AA), and the Kappa Coefficient (Kappa/k). To reduce the error associated with the randomly selected training samples, each model is run ten times to compute the average accuracy and standard deviation.

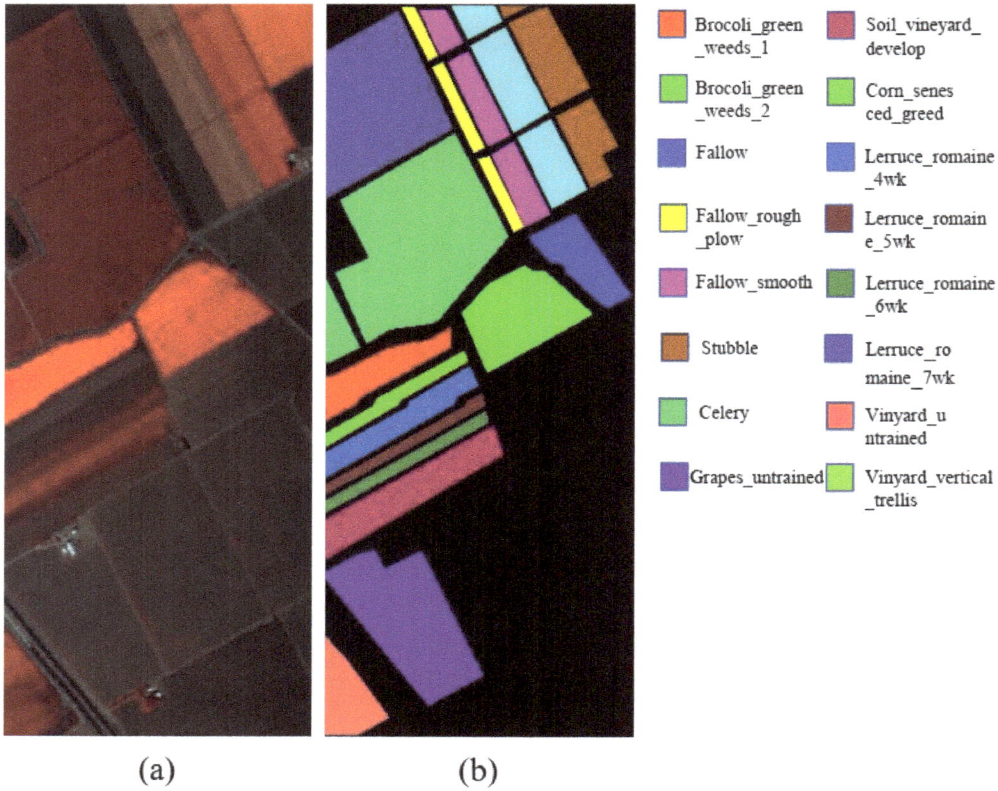

Figure 6. Salians Scene dataset. (**a**) False-color image; (**b**) Ground Truth.

3.3. Experimental and Evaluation on Three Datasets

For a better demonstration of the superiority and stability of the proposed SquconvNet, it is compared with some representative methods: Resnet [27], 3D-CNN [25], SSRN [26], HybridSN [29], SPRN [43], and SSFTT [39]. For the Resnet, we use an optimal method, that is consistent with our model. The 3D-CNN, SSRN, HybridSN, SPRN, and SSFTT are set up as described in their corresponding references.

3.3.1. Experiment on IP Dataset

The methods of each classification are shown in Table 5. The table highlights the optimal outcomes. Particularly, HybirdSN, 3D-CNN, Resnet, etc., fared worse than our suggested method in order of best average OA value, which was 99.87%. Additionally, the performance of our proposed method is the best. The differences between the mean and the suboptimal methods of the proposed method for the evaluation of the OA, AA, and Kappa are +0.1, +0.24, and +0.11, respectively, as shown in Table 5. Additionally, the standard deviation of our proposed method is also the smallest, demonstrating our method's higher level of stability. But it is important to remember that SSRN and SPRN's volatility is what led to their poor classification performance. In our ten studies, the best OA achieved by SSRN and SPRN were 99.87% and 99.72%, respectively. In comparison to Resnet and SSFTT, both methods have a higher upper bound, but due to the high sample imbalance in the IP dataset, their average effectiveness is very low. The deep learning-based methods discussed above have all achieved quite good results, particularly HybridSN, based on 3D-2D convolutional architecture. However, convolutional neural networks have difficulties in classifying when the ground is irregularly shaped. In terms of the convolutional architecture, our proposed SquconvNet complements the convolution layer; by transmitting "memory" information in the horizontal and vertical directions, it is possible to overcome, to a certain extent, the convolution kernel's inability to capture all the features in the convolution layer due to the uneven shape of the ground. For the SSFTT based on the convolution-Transformer framework, even if it can supplement the inadequacy of the global information extraction of the convolutional layer, it is also constrained by the issue that the Transformer finds it challenging to perform better on tiny data samples and has a constrained accuracy. The classification map of Ground Truth and all methods is shown in Figure 7a–h. The classification map illustrates how our proposed model produces a classification map that is not only smoother but also better in terms of texture and edge features. This demonstrates even more how effective Sequencer is at handling unusual ground forms. The proposed method SquconvNet, and the HybridSN-based classification map, outperform other methods in terms of visual performance. In conclusion, on the IP dataset, our proposed method of merging 3D-2DCNN and LSTM2D outperforms its rivals in terms of accuracy and stability.

Table 5. Results of the various methods for IP using 30% training data (Highest performance is in Boldface).

NO.	Resnet	3D-CNN	SSRN	HybridSN	SPRN	SSFTT	Proposed
1	99.69 ± 0.936	**100 ± 0**	99.69 ± 0.936	99.69 ± 0.936	97.5 ± 6.527	99.38 ± 1.248	**100 ± 0**
2	99.23 ± 0.237	99.16 ± 0.398	95.54 ± 12.69	99.64 ± 0.237	79.56 ± 33.04	99.32 ± 0.496	**99.87 ± 0.11**
3	99.8 ± 0.168	99.72 ± 0.520	98.24 ± 4.936	99.64 ± 0.598	87.77 ± 29.73	98.86 ± 0.712	**100 ± 0**
4	99.32 ± 0.351	**100 ± 0**	**100 ± 0**	**100 ± 0**	91.08 ± 22.41	99.31 ± 0.728	**100 ± 0**
5	99.76 ± 0.553	99.70 ± 0.400	99.67 ± 0.362	99.67 ± 0.598	98.25 ± 0.018	97.62 ± 3.56	**99.94 ± 0.12**
6	99.32 ± 0.351	99.86 ± 0.197	99.90 ± 0.158	99.98 ± 0.06	97.04 ± 7.517	99.65 ± 0.325	**100 ± 0**
7	98.5 ± 2.29	**100 ± 0**	99 ± 2	99.5 ± 1.5	94.5 ± 11.5	99.85 ± 0.447	**100 ± 0**
8	**100 ± 0**	**100 ± 0**	**100 ± 0**	99.91 ± 0.27	96.63 ± 6.803	74.76 ± 17.73	**100 ± 0**
9	94.29 ± 4.286	**99.29 ± 2.142**	97.86 ± 4.574	93.57 ± 7.458	80.71 ± 38.61	98.79 ± 0.746	**99.29 ± 2.142**
10	99.63 ± 0.165	99.57 ± 0.309	99.50 ± 0.466	99.84 ± 0.364	88.44 ± 29.58	98.79 ± 0.746	**99.62 ± 0.256**
11	99.76 ± 0.099	99.92 ± 0.118	91.83 ± 21.039	99.78 ± 0.18	86.95 ± 19.72	99.56 ± 0.37	**99.92 ± 0.063**
12	98.22 ± 0.763	98.91 ± 0.500	98.91 ± 0.571	99.69 ± 0.241	95.25 ± 11.66	97.73 ± 1.876	**99.11 ± 0.343**
13	99.51 ± 0.77	99.65 ± 0.472	99.86 ± 0.28	**100 ± 0**	97.48 ± 6.64	99.2 ± 0.94	**100 ± 0**
14	**100 ± 0**	**100 ± 0**	99.98 ± 0.069	99.94 ± 0.117	99.82 ± 0.543	99.83 ± 0.192	**100 ± 0**
15	99.96 ± 0.111	99.85 ± 0.342	99.18 ± 1.835	99.63 ± 0.409	92.89 ± 10.91	98.38 ± 2.313	**100 ± 0**
16	99.23 ± 1.033	98.31 ± 2.790	99.38 ± 0.754	99.08 ± 1.411	96.31 ± 6.123	94.58 ± 6.193	**100 ± 0**
OA (%)	99.6 ± 0.05	99.69 ± 0.083	97.09 ± 7.261	99.77 ± 0.095	90.57 ± 12.81	99.09 ± 0.434	**99.87 ± 0.034**
AA (%)	99.22 ± 0.287	99.62 ± 0.182	98.66 ± 2.481	99.35 ± 0.459	92.51 ± 10.77	96.94 ± 1.566	**99.86 ± 0.125**
k × 100	99.54 ± 0.064	99.64 ± 0.095	96.77 ± 8.027	99.74 ± 0.111	89.35 ± 14.46	98.96 ± 0.496	**99.85 ± 0.041**

3.3.2. Experimental on UP Dataset

The average OA, AA, and Kappa (k) on the PU dataset are shown in Table 6 along with their standard deviations. Overall, the proposed SquconvNet performs better in terms of OA, AA, and Kappa than all the other methods utilized for comparison. Our method performs best in eight of the nine categories, and the standard deviation for these eight categories is the lowest of any method. On the UP dataset, the proposed method achieved an excellent OA performance of 99.93%, an improvement of +0.24 over the less-than-ideal method SSRN. On the PU dataset, 3D-CNN, SSRN, HybridSN, and SSFTT all outperformed Resnet and SPRN. With the minimum standard deviation of all the methods, the proposed SquconvNet has also shown an addition in stability. The classification maps of the UP dataset using Ground Truth and several of the methods is shown in Figure 8. Our suggested model performs better on this dataset. This outcome, from the accuracy level, is difficult to explain. It is clear that the UP dataset's distribution is less homogeneous, and its ground shape is more erratic, than for the IP dataset. As a result, other methods have trouble in capturing discriminative features. However, the Sequencer created by LSTM performs better in the results, and is more resistant to ground shape anomalies. Additionally, the classification map's shape is smoother, includes less noise, and has clearer bounds. Resnet and SPRN, in contrast, find it difficult to extract the most discriminative feature information, and as a result have more salt-pepper noise and incorrectly categorized area blocks. Despite having nice visual effects, the other methods still contain a lot of point noise.

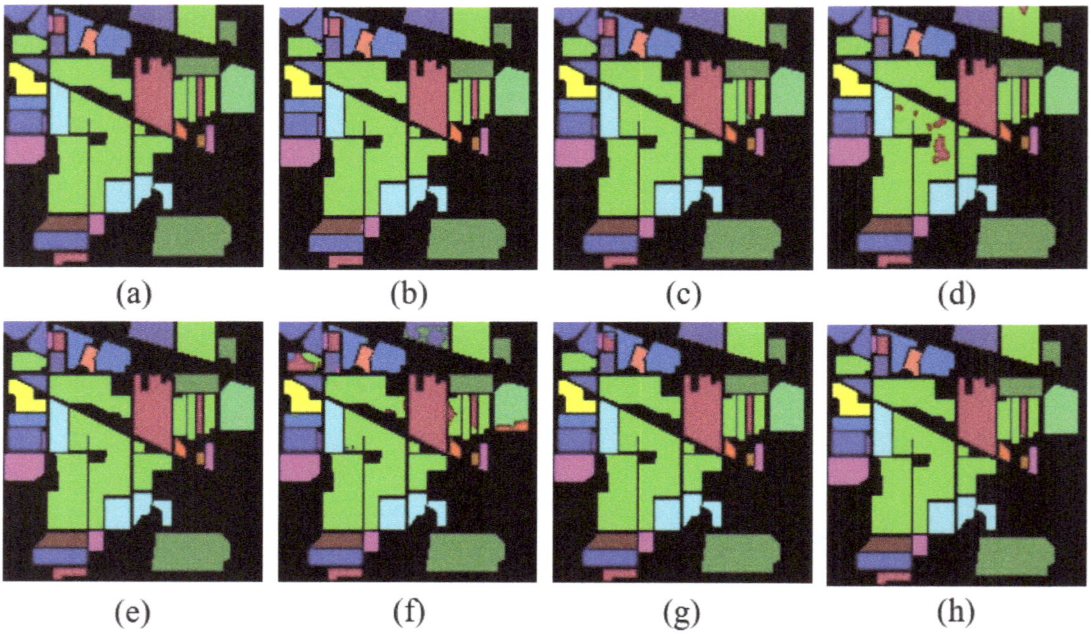

Figure 7. Classification map of various methods for IP (**a**) Ground Truth, (**b**) Resnet, (**c**) 3D-CNN, (**d**) SSRN, (**e**) HybridSN, (**f**) SPRN, (**g**) SSFTT, (**h**) SquconvNet.

Table 6. Results of various methods for UP using 10% training data (Highest performance is in Boldface).

NO.	Resnet	3D-CNN	SSRN	HybridSN	SPRN	SSFTT	Proposed
1	99.42 ± 0.558	99.81 ± 0.160	99.91 ± 0.190	99.98 ± 0.023	99.5 ± 0.948	99.61 ± 0.298	**99.99 ± 0.006**
2	99.75 ± 0.177	99.99 ± 0.003	99.98 ± 0.252	99.99 ± 0.005	96.14 ± 10.36	99.94 ± 0.121	**100 ± 0**
3	98.39 ± 1.395	99.32 ± 0.484	99.30 ± 0.395	99.15 ± 0.798	91.71 ± 22.28	98.98 ± 0.644	**99.48 ± 0.266**
4	99.41 ± 0.277	98.77 ± 0.260	**100 ± 0**	98.95 ± 1.052	99.99 ± 0.024	98.73 ± 0.542	99.80 ± 0.128
5	99.92 ± 0.105	99.92 ± 0.154	**100 ± 0**	99.65 ± 0.624	96.96 ± 7.054	99.37 ± 0.677	**100 ± 0**
6	99.77 ± 0.471	99.99 ± 0.012	99.9 ± 0.203	**100 ± 0**	96.96 ± 7.054	99.98 ± 0.024	**100 ± 0**
7	96.3 ± 3.819	99.89 ± 0.129	98.33 ± 3.506	99.48 ± 0.773	94.55 ± 10.92	99.63 ± 0.531	**99.99 ± 0.024**
8	96.94 ± 4.011	98.99 ± 0.420	99.97 ± 0.742	98.81 ± 0.659	92.05 ± 14.43	98.66 ± 0.995	**99.71 ± 0.158**
9	96.44 ± 1.714	95.56 ± 1.967	99.79 ± 0.321	94.52 ± 3.216	98.78 ± 0.44	97.46 ± 0.007	**99.85 ± 0.211**
OA (%)	99.19 ± 0.574	99.64 ± 0.069	99.69 ± 0.437	99.58 ± 0.165	96.55 ± 7.78	99.57 ± 0.129	**99.93 ± 0.026**
AA (%)	98.93 ± 0.763	99.09 ± 0.200	99.69 ± 0.402	98.88 ± 0.395	98.56 ± 1.431	99.15 ± 0.183	**99.86 ± 0.049**
k × 100	98.93 ± 0.763	99.53 ± 0.090	99.72 ± 0.438	99.44 ± 0.219	95.61 ± 9.772	99.43 ± 0.172	**99.90 ± 0.031**

3.3.3. Experiment on SA Dataset

Table 7 displays the classification results for several networks utilizing 10% training data on the Salinas Scene Dataset. Due to instabilities and limited abilities to extract features, Resnet and SPRN perform badly, as indicated in Table 7. In contrast, the 3D-CNN, SSRN, HybridSN, and SSFTT algorithms all extracted Spectral-Spatial features with the aid of 3D-convolution, and obtained better classification results. However, they can still be made more accurate and consistent. Our proposed method achieves a mean of 99.99 on OA, AA, and Kappa, while having a lower standard deviation, thanks to the combination of 3D-2D convolution and Sequencer2D block. The classification map of the SA dataset using Ground Truth and several methods is shown in Figure 9. With large noise levels and subsequent blocks of classification mistakes, the performance of the related classification maps obtained by Resnet and SPRN was subpar. Improved results were obtained, less point noise was present, and there was better continuity between different object classes with 3D-CNN, SSRN, and HybridSN. Overall, nevertheless, our suggested approach has less point noise and smoother bounds. Table 7 makes it clear that practically all of the compared methods attain good accuracies. In fact, because there are more data and the ground is flatter, it is a very simple dataset to classify. Using only 30% of the training dataset, the HybridSN authors achieved 100% accuracy. However, in order to reduce the expense of manual annotation, we anticipate using fewer training datasets. We used 10% of the training dataset in the experiment to get an accuracy rate that was extremely close to 100%. This is not due to the overfitting phenomenon, rather, the model we suggested is superior at extracting spectral-spatial features.

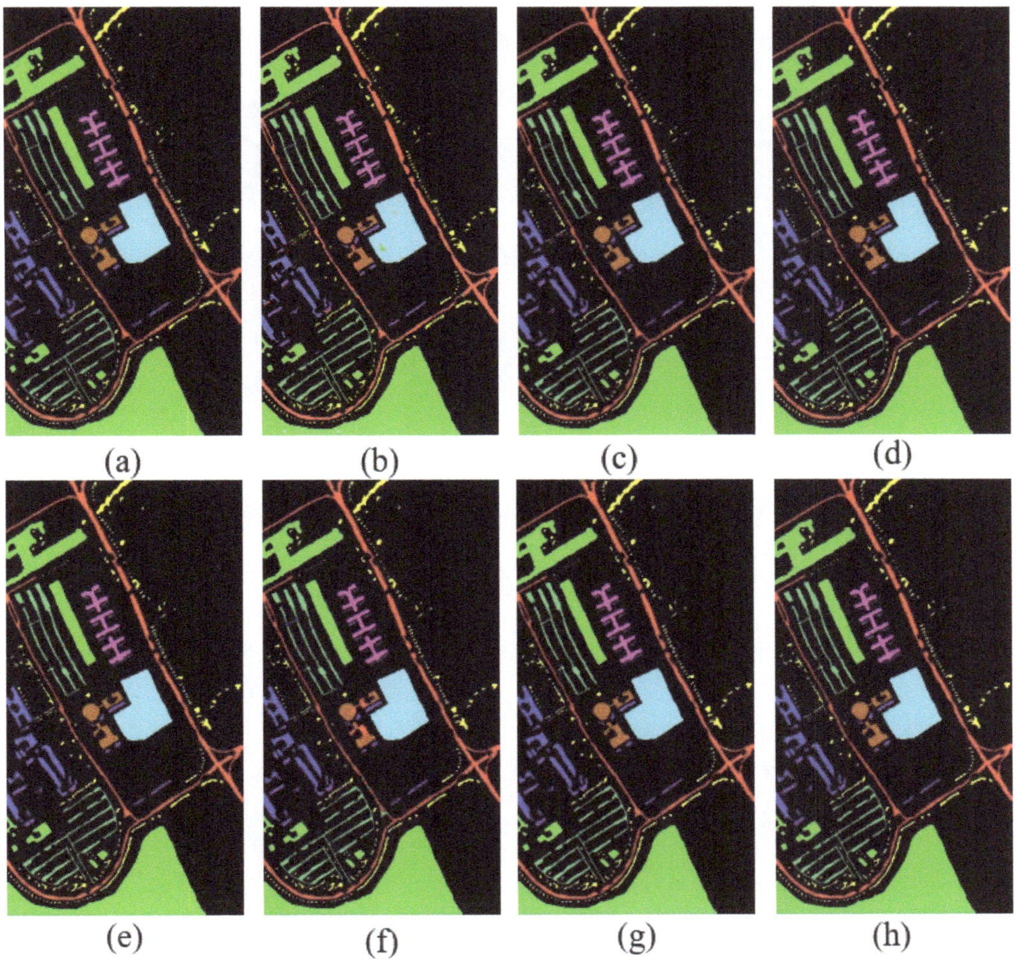

Figure 8. Classification map of various methods for UP (**a**) Ground Truth, (**b**) Resnet, (**c**) 3D-CNN, (**d**) SSRN, (**e**) HybridSN, (**f**) SPRN, (**g**) SSFTT, (**h**) SquconvNet.

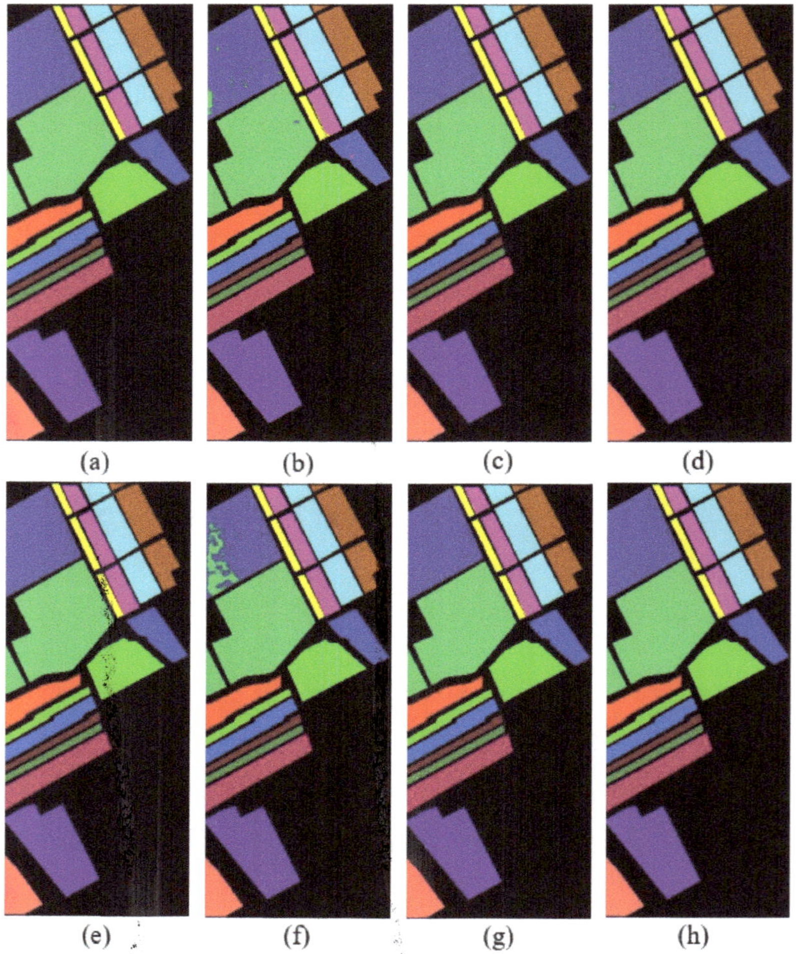

Figure 9. Classification map of various methods for SA (**a**) Ground Truth, (**b**) Resnet, (**c**) 3D-CNN, (**d**) SSRN, (**e**) HybridSN, (**f**) SPRN, (**g**) SSFTT, (**h**) SquconvNet.

Table 7. Results of various methods for SA using 10% training data (Highest performance is in Boldface).

NO.	Resnet	3D-CNN	SSRN	HybridSN	SPRN	SSFTT	Proposed
1	99.13 ± 2.41	**100 ± 0**	**100 ± 0**	**100 ± 0**	**100 ± 0**	**100 ± 0**	**100 ± 0**
2	99.76 ± 0.695	99.99 ± 0.009	**100 ± 0**	**100 ± 0**	**100 ± 0**	99.99 ± 0.009	**100 ± 0**
3	99.14 ± 1.068	**100 ± 0**	**100 ± 0**	**100 ± 0**	97.51 ± 7.252	99.93 ± 0.185	**100 ± 0**
4	99.39 ± 1.116	**100 ± 0**	99.95 ± 0.069	99.97 ± 0.096	99.41 ± 0.842	99.31 ± 1.131	**100 ± 0**
5	99.40 ± 0.905	99.32 ± 0.39	99.73 ± 0.216	99.78 ± 0.111	98.11 ± 0.525	99.42 ± 0.621	**99.93 ± 0.070**
6	**100 ± 0**	**100 ± 0**	**100 ± 0**	**100 ± 0**	**100 ± 0**	99.88 ± 0.149	**100 ± 0**
7	99.91 ± 0.107	99.95 ± 0.056	99.99 ± 0.020	99.99 ± 0.009	**100 ± 0**	99.91 ± 0.021	**100 ± 0**
8	84.2 ± 30.68	**100 ± 0**	99.98 ± 0.021	99.98 ± 0.032	92.8 ± 11.17	99.89 ± 0.148	99.99 ± 0.003
9	99.95 ± 0.112	**100 ± 0**	**100 ± 0**	**100 ± 0**	98.45 ± 0.046	99.99 ± 0.021	**100 ± 0**
10	99.78 ± 0.499	**100 ± 0**	99.93 ± 0.054	99.98 ± 0.028	99.92 ± 0.151	99.89 ± 0.148	99.96 ± 0.060
11	99.69 ± 0.299	99.88 ± 0.278	99.92 ± 0.205	99.32 ± 1.005	99.99 ± 0.03	99.73 ± 0.403	**100 ± 0**
12	99.84 ± 0.276	**100 ± 0**	**100 ± 0**	**100 ± 0**	99.93 ± 0.188	99.94 ± 0.124	**100 ± 0**
13	99.77 ± 0.536	99.92 ± 0.121	**100 ± 0**	**100 ± 0**	**100 ± 0**	99.21 ± 0.884	**100 ± 0**

Table 7. *Cont.*

NO.	Resnet	3D-CNN	SSRN	HybridSN	SPRN	SSFTT	Proposed
14	99.92 ± 0.146	99.95 ± 0.095	99.95 ± 0.139	99.97 ± 0.067	99.8 ± 0.384	99.5 ± 0.594	**99.99 ± 0.030**
15	99.26 ± 0.86	99.98 ± 0.020	99.74 ± 0.151	99.98 ± 0.060	95.1 ± 9.154	99.96 ± 0.025	**99.99 ± 0.006**
16	99.94 ± 0.186	**100 ± 0**	**100 ± 0**	**100 ± 0**	**100 ± 0**	99.72 ± 0.731	**100 ± 0**
OA (%)	96.44 ± 6.513	99.95 ± 0.022	99.94 ± 0.026	99.97 ± 0.028	97.46 ± 2.178	99.88 ± 0.038	**99.99 ± 0.007**
AA (%)	98.69 ± 2.198	99.94 ± 0.033	99.95 ± 0.026	99.94 ± 0.071	98.81 ± 0.825	99.77 ± 0.099	**99.99 ± 0.009**
k × 100	96.082 ± 7.15	99.95 ± 0.025	99.93 ± 0.710	99.96 ± 0.0297	97.17 ± 2.411	99.87 ± 0.044	**99.99 ± 0.007**

3.4. Learning Rate Experiment

An essential hyperparameter that influences how well the model fits, is the initial learning rate. In this experiment, unlike in other experiments, our dataset is divided into a 20% training set, a 10% validation set, and a 70% test set. And the results of this experiment are given by the validation set. Each of the following initial learning rates are set to 0.0001, 0.0005, 0.001, 0.005, 0.01, and 0.05 for the purposes of our experimental investigation. Figure 10 shows the classification outcomes for the IP datasets at various speeds. The best initial learning rate is 0.001 and the suboptimal initial learning rate is 0.0005, as seen in Figure 10. We set the initial learning rate for other experiments to 0.001 based on the classification results.

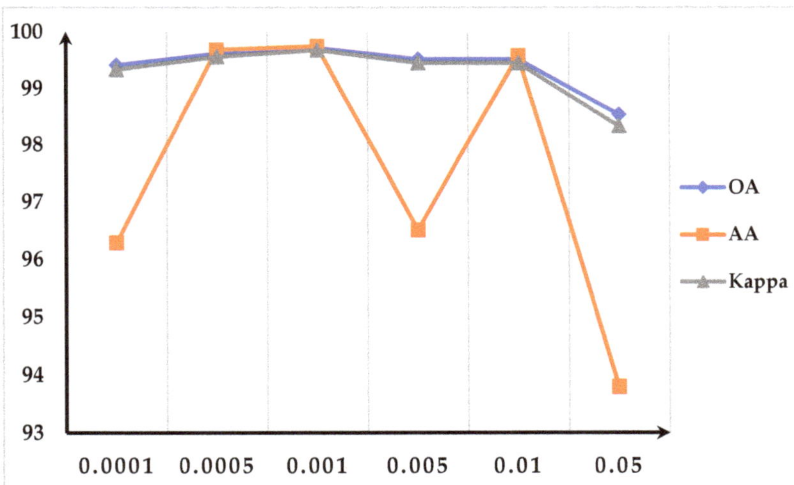

Figure 10. The OA, AA, and Kappa of IP at different learning rates.

4. Discussion

We initially talk about the effects of the three modules on the three datasets in this section (ablation experiment). Finally, we conduct comparison experiments for a number of relatively advanced methods, as well as the proposed method, to compare training time, testing time, and parameter number.

4.1. Discussion on the Ablation Experiment

To better explore the efficiency of each SquconvNet network component, a series of ablation experiments are conducted utilizing three datasets. SDB-CNN, 3D-2DCNN, SS-SDB, and our proposed SquconvNet are four combinations we set up based on three modules. An example of their combination, and their best classification results on the IP dataset, are shown in Table 8. The methods based on the SDB and SSFE modules produce the worst outcomes. The best accuracy is attained by the proposed method. Additionally, for each of the four methods, we try training with fewer data to explore the stability of the model methods. On the three standard datasets, Figures 11–13 show the overall accuracy of each of the four methods. On 5% of the training dataset for the IP dataset, SquconvNet outperforms SDB-CNN and SS-SDB. This is because there are not enough training samples for some classes to learn features with effective discriminative power due to the significant imbalance of samples (such as class.9) in the IP dataset. The experimental results also show that when the data are balanced and sufficient, the technique we suggest can produce the best results. It shows that, while our proposed method can withstand an imbalance caused by insufficient data, it loses effectiveness when the amount of data falls below a particular threshold. SDB-CNN has demonstrated a comparatively good performance while dealing with less data. When there is a small amount of data, we speculate that spatial information may be more significant than spectral information in our proposed model. Additionally, the model made up of 3D-CNN and 2D-CNN had the worst outcome. We speculate that in the case of small data, it might be brought about by the model's poor generalization ability, brought about by 3D-CNN's excessive emphasis on spectral information.

Table 8. Ablation analysis of the proposed methods on the IP dataset with 30% labeled samples.

Method	SSFE	SDB	AC	OA	AA	Kappa
SDB-CNN		√	√	99.79	99.81	99.76
3D-2DCNN	√		√	99.87	99.91	99.86
SS-SDB	√	√		99.69	99.75	99.65
SquconvNet	√	√	√	99.94	99.96	99.94

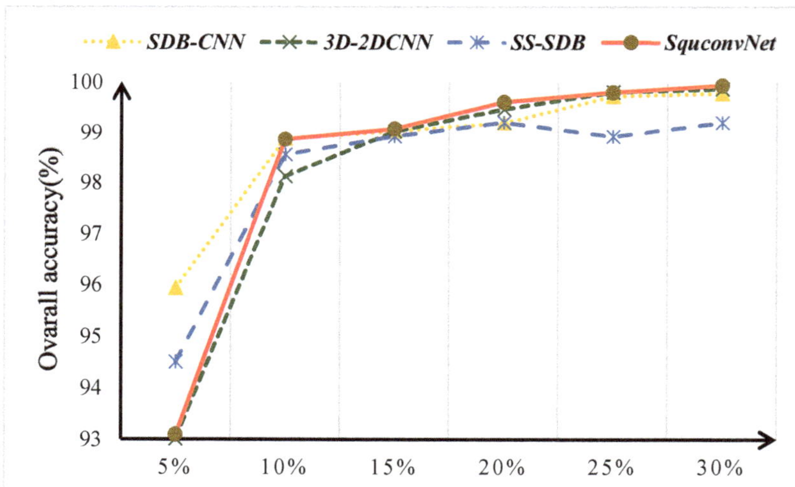

Figure 11. The Overall Accuracy of the different proposed models on the IP dataset at different training samples.

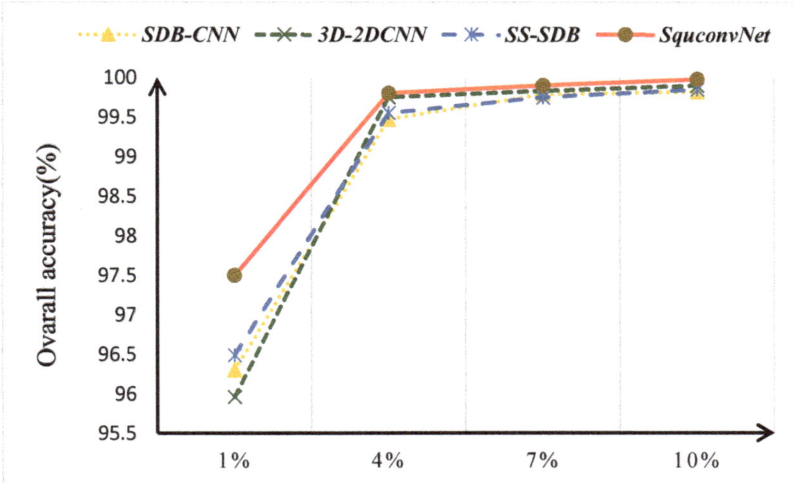

Figure 12. The Overall Accuracy of the different proposed models on the UP dataset at different training samples.

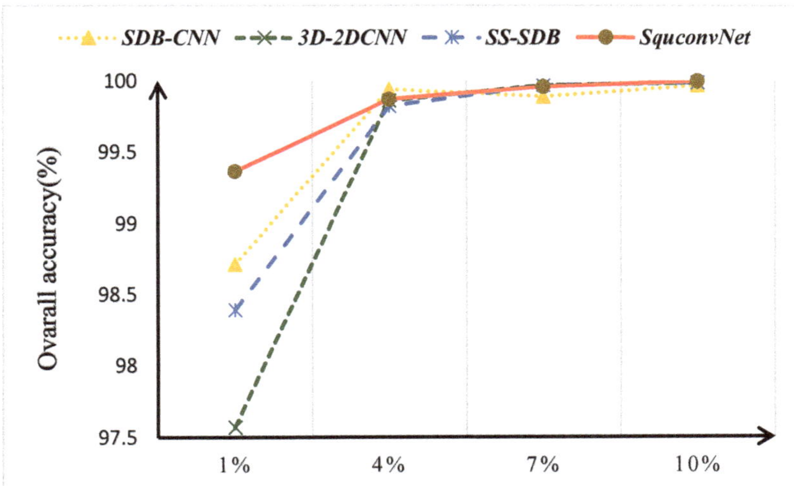

Figure 13. The Overall Accuracy of the different proposed models on the SA dataset at different training samples.

4.2. Discussion on the Time Cost

The training time, test time, and total number of parameters for the 3D-CNN, SSRN, HybridSN, SSFTT, and SquconvNet are listed in Table 9. The slowest training speed is shown by SSRN using deep residuals. Additionally, 3D-CNN, SSRN, and HybridSN all struggle, with lengthy training and testing periods. Furthermore, the last three fully connected layers of HybridSN have resulted in an excessive number of its overall parameters, overburdening the system with parameters. SSFTT provides speed benefits, albeit at the expense of accuracy. SquconvNet increases training speed by at least twelve times and testing speed by at least four times over SSRN. SquconvNet reduced the training time for UP and SA by a factor of three and a factor of nine, respectively, when compared to HybridSN. SquconvNet has around six times fewer parameters than HybridSN, and is two to three times larger than SSRN in terms of parameter number. Furthermore, there has been a significant improvement in the classification accuracy and stability. As a result, the proposed method is useful and has promising application possibilities.

Table 9. Training and Test time of different methods on three datasets.

Method		3D-CNN	SSRN	HybridSN	SSFTT	SquconvNet
IP	Train(s)	174.3	498.4	318.6	38.2	35.1
	Test(s)	1.80	1.67	3.4	0.32	0.37
	Params.	144 k	364 k	5122 k	427 k	878 k
UP	Train(s)	120.2	495.5	106.7	52.6	35.24
	Test(s)	4.48	6.0	3.97	1.75	1.33
	Params.	135 k	217 k	4845 k	427 k	807 k
SA	Train(s)	143.7	555.8	136.5	67.0	46.5
	Test(s)	5.41	7.76	4.98	2.24	1.71
	Params.	136 k	370 k	4846 k	427 k	809 k

5. Conclusions

This article suggests applying a hybrid SquconvNet to HSI classification that combines a 3D convolution layer, a 2D convolution layer, and a BiLSTM2D layer. The Spectral-Spatial Feature Extraction Module, the Sequencer Module, and the Auxiliary Classification Module make up the methodology. We suggest using the Sequencer based of LSTM as a supplement to the convolutional neural network in order to address its shortcomings. On three freely accessible and publicly accessible hyperspectral remote image datasets, we conduct numerous compared experiments and this method is shown to improve classification accuracy, classification speed, and stability effectively and efficiently in the experiments. When compared to conventional convolutional methods, our new method efficiently counters classification mistakes caused by erratic ground forms. The proposed method obtains average accuracies of 99.87%, 99.93%, and 99.99% on the three standard public datasets.

Additionally, we put forward the theory that, in the context of small data, spatial information may be more significant than spectral information in our proposed method. In the upcoming phase of our research, based on the SquconvNet, we intend to investigate the validity of this hypothesis under small-sample learning settings, as well as the viability of substituting 3D Sequencer for convolutional layers.

Author Contributions: Conceptualization, B.L. and Q.-W.W.; methodology, B.L. and J.-H.L.; software, Q.-W.W. and J.-H.L.; validation, B.L., Q.-W.W., J.-H.L., E.-Z.Z. and R.-Q.Z.; formal analysis, B.L., Q.-W.W., J.-H.L., E.-Z.Z. and R.-Q.Z.; writing—original draft preparation, J.-H.L., B.L. and Q.-W.W., writing—review and editing, B.L., J.-H.L., Q.-W.W., E.-Z.Z. and R.-Q.Z. All authors have read and agreed to the published version of the manuscript.

Funding: This research was funded by the Start-up research projects of Shantou University (NTF19016

Data Availability Statement: The datasets presented in this paper is available through https://www ehu.eus/ccwintco/index.php/Hyperspectral_Remote_Sensing_Scenes, accessed on 11 November 2022

Conflicts of Interest: The authors declare no conflict of interest.

References

1. Prasad, S.; Bruce, L.M. Limitations of Principal Components Analysis for Hyperspectral Target Recognition. *IEEE Geosci. Remote Sens. Lett.* **2008**, *5*, 625–629. [CrossRef]
2. Piiroinen, R.; Heiskanen, J.; Maeda, E.; Viinikka, A.; Pellikka, P. Classification of Tree Species in a Diverse African Agroforestry Landscape Using Imaging Spectroscopy and Laser Scanning. *Remote Sens.* **2017**, *9*, 875. [CrossRef]
3. Chen, S.Y.; Lin, C.S.; Tai, C.H.; Chuang, S.J. Adaptive Window-Based Constrained Energy Minimization for Detection of Newly Grown Tree Leaves. *Remote Sens.* **2018**, *10*, 96. [CrossRef]
4. Zhang, H.; Zhang, B.; Chen, Z.C.; Huang, Z.H. Vicarious Radiometric Calibration of the Hyperspectral Imaging Microsatellites SPARK-01 and-02 over Dunhuang, China. *Remote Sens.* **2018**, *10*, 120. [CrossRef]
5. Tane, Z.; Roberts, D.; Veraverbeke, S.; Casas, A.; Ramirez, C.; Ustin, S. Evaluating Endmember and Band Selection Techniques for Multiple Endmember Spectral Mixture Analysis using Post-Fire Imaging Spectroscopy. *Remote Sens.* **2018**, *10*, 389. [CrossRef]

6. Ni, L.; Wub, H. Mineral Identification and Classification by Combining Use of Hyperspectral VNIR/SWIR and Multispectral TIR Remotely Sensed Data. In Proceedings of the IGARSS 2019—2019 IEEE International Geoscience and Remote Sensing Symposium, Yokohama, Japan, 28 July–2 August 2019; pp. 3317–3320.
7. Li, J.; Bioucas-Dias, J.M.; Plaza, A. Spectral-Spatial Classification of Hyperspectral Data Using Loopy Belief Propagation and Active Learning. *IEEE Trans. Geosci. Remote Sens.* **2013**, *51*, 844–856. [CrossRef]
8. Zhang, L.; Zhong, Y.; Huang, B.; Gong, J.; Li, P. Dimensionality reduction based on clonal selection for hyperspectral imagery. *IEEE Trans. Geosci. Remote Sens.* **2007**, *45*, 4172–4186. [CrossRef]
9. Brown, A.J.; Sutter, B.; Dunagan, S. The MARTE VNIR Imaging Spectrometer Experiment: Design and Analysis. *Astrobiology* **2008**, *8*, 1001–1011. [CrossRef]
10. Brown, A.J.; Hook, S.J.; Baldridge, A.M.; Crowley, J.K.; Bridges, N.T.; Thomson, B.J.; Marion, G.M.; de Souza, C.R.; Bishop, J.L. Hydrothermal formation of Clay-Carbonate alteration assemblages in the Nil Fossae region of Mars. *Earth Planet. Sci. Lett.* **2010**, *297*, 174–182. [CrossRef]
11. Zhu, J.S.; Hu, J.; Jia, S.; Jia, X.P.; Li, Q.Q. Multiple 3-D Feature Fusion Framework for Hyperspectral Image Classification. *IEEE Trans. Geosci. Remote Sens.* **2018**, *56*, 1873–1886. [CrossRef]
12. Chen, Y.S.; Lin, Z.H.; Zhao, X.; Wang, G.; Gu, Y.F. Deep Learning-Based Classification of Hyperspectral Data. *IEEE J. Sel. Top. Appl. Earth Obs. Remote Sens.* **2014**, *7*, 2094–2107. [CrossRef]
13. Lavanya, A.; Sanjeevi, S. An Improved Band Selection Technique for Hyperspectral Data Using Factor Analysis. *J. Indian Soc. Remote Sens.* **2013**, *41*, 199–211. [CrossRef]
14. Bandos, T.V.; Bruzzone, L.; Camps-Valls, G. Classification of Hyperspectral Images With Regularized Linear Discriminant Analysis. *IEEE Trans. Geosci. Remote Sens.* **2009**, *47*, 862–873. [CrossRef]
15. Ye, Q.; Yang, J.; Liu, F.; Zhao, C.; Ye, N.; Yin, T. L1-Norm Distance Linear Discriminant Analysis Based on an Effective Iterative Algorithm. *IEEE Trans. Circuits Syst. Video Technol.* **2018**, *28*, 114–129. [CrossRef]
16. Villa, A.; Benediktsson, J.A.; Chanussot, J.; Jutten, C. Independent Component Discriminant Analysis for hyperspectral image classification. In Proceedings of the 2010 2nd Workshop on Hyperspectral Image and Signal Processing: Evolution in Remote Sensing, Reykjavik, Iceland, 14–16 June 2010; pp. 1–4.
17. Villa, A.; Benediktsson, J.A.; Chanussot, J.; Jutten, C. Hyperspectral Image Classification With Independent Component Discriminant Analysis. *IEEE Trans. Geosci. Remote Sens.* **2011**, *49*, 4865–4876. [CrossRef]
18. Melgani, F.; Bruzzone, L. Classification of hyperspectral remote sensing images with support vector machines. *IEEE Trans. Geosci. Remote Sens.* **2004**, *42*, 1778–1790. [CrossRef]
19. Gislason, P.O.; Benediktsson, J.A.; Sveinsson, J.R. Random Forests for land cover classification. *Pattern Recognit. Lett.* **2006**, *27*, 294–300. [CrossRef]
20. Haut, J.M.; Paoletti, M.; Plaza, J.; Plaza, A. Cloud implementation of the K-means algorithm for hyperspectral image analysis. *J. Supercomput.* **2017**, *73*, 514–529. [CrossRef]
21. Tarabalka, Y.; Fauvel, M.; Chanussot, J.; Benediktsson, J.A. SVM- and MRF-Based Method for Accurate Classification of Hyperspectral Images. *IEEE Geosci. Remote Sens. Lett.* **2010**, *7*, 736–740. [CrossRef]
22. Li, T.; Zhang, J.; Zhang, Y. Classification of hyperspectral image based on deep belief networks. In Proceedings of the 2014 IEEE International Conference on Image Processing (ICIP), Paris, France, 27–30 October 2014; pp. 5132–5136.
23. Hu, W.; Huang, Y.; Wei, L.; Zhang, F.; Li, H. Deep Convolutional Neural Networks for Hyperspectral Image Classification. *J. Sens.* **2015**, *2015*, 258619. [CrossRef]
24. Zhao, W.; Du, S. Spectral–Spatial Feature Extraction for Hyperspectral Image Classification: A Dimension Reduction and Deep Learning Approach. *IEEE Trans. Geosci. Remote Sens.* **2016**, *54*, 4544–4554. [CrossRef]
25. Chen, Y.; Jiang, H.; Li, C.; Jia, X.; Ghamisi, P. Deep Feature Extraction and Classification of Hyperspectral Images Based on Convolutional Neural Networks. *IEEE Trans. Geosci. Remote Sens.* **2016**, *54*, 6232–6251. [CrossRef]
26. Zhong, Z.; Li, J.; Luo, Z.; Chapman, M. Spectral–Spatial Residual Network for Hyperspectral Image Classification: A 3-D Deep Learning Framework. *IEEE Trans. Geosci. Remote Sens.* **2018**, *56*, 847–858. [CrossRef]
27. He, K.; Zhang, X.; Ren, S.; Sun, J. Deep Residual Learning for Image Recognition. In Proceedings of the 2016 IEEE Conference on Computer Vision and Pattern Recognition (CVPR), Las Vegas, NV, USA, 27–30 June 2016; pp. 770–778.
28. Paoletti, M.E.; Haut, J.M.; Fernandez-Beltran, R.; Plaza, J.; Plaza, A.J.; Pla, F. Deep Pyramidal Residual Networks for Spectral–Spatial Hyperspectral Image Classification. *IEEE Trans. Geosci. Remote Sens.* **2019**, *57*, 740–754. [CrossRef]
29. Roy, S.K.; Krishna, G.; Dubey, S.R.; Chaudhuri, B.B. HybridSN: Exploring 3-D–2-D CNN Feature Hierarchy for Hyperspectral Image Classification. *IEEE Geosci. Remote Sens. Lett.* **2020**, *17*, 277–281. [CrossRef]
30. Li, J.J.; Zhao, X.; Li, Y.S.; Du, Q.; Xi, B.B.; Hu, J. Classification of Hyperspectral Imagery Using a New Fully Convolutional Neural Network. *IEEE Geosci. Remote Sens. Lett.* **2018**, *15*, 292–296. [CrossRef]
31. Tun, N.L.; Gavrilov, A.; Tun, N.M.; Trieu, D.M.; Aung, H. Hyperspectral Remote Sensing Images Classification Using Fully Convolutional Neural Network. In Proceedings of the 2021 IEEE Conference of Russian Young Researchers in Electrical and Electronic Engineering (ElConRus), St. Petersburg, Moscow, 26–29 January 2021; pp. 2166–2170.
32. Bi, X.J.; Zhou, Z.Y. Hyperspectral Image Classification Algorithm Based on Two-Channel Generative Adversarial Network. *Acta Opt. Sin.* **2019**, *39*, 1028002. [CrossRef]

33. Xue, Z.X. Semi-supervised convolutional generative adversarial network for hyperspectral image classification. *IET Image Process.* **2020**, *14*, 709–719. [CrossRef]
34. Hong, D.; Gao, L.; Yao, J.; Zhang, B.; Plaza, A.; Chanussot, J. Graph Convolutional Networks for Hyperspectral Image Classification. *IEEE Trans. Geosci. Remote Sens.* **2021**, *59*, 5966–5978. [CrossRef]
35. Dosovitskiy, A.; Beyer, L.; Kolesnikov, A.; Weissenborn, D.; Zhai, X.; Unterthiner, T.; Dehghani, M.; Minderer, M.; Heigold, G.; Gelly, S. An image is worth 16x16 words: Transformers for image recognition at scale. *arXiv* **2020**, arXiv:2010.11929.
36. He, X.; Chen, Y.S.; Lin, Z.H. Spatial-Spectral Transformer for Hyperspectral Image Classification. *Remote Sens.* **2021**, *13*, 498. [CrossRef]
37. Qing, Y.H.; Liu, W.Y.; Feng, L.Y.; Gao, W.J. Improved Transformer Net for Hyperspectral Image Classification. *Remote Sens.* **2021**, *13*, 2216. [CrossRef]
38. Hong, D.F.; Han, Z.; Yao, J.; Gao, L.R.; Zhang, B.; Plaza, A.; Chanussot, J. SpectralFormer: Rethinking Hyperspectral Image Classification With Transformers. *Ieee Trans. Geosci. Remote Sens.* **2022**, *60*, 5518615. [CrossRef]
39. Sun, L.; Zhao, G.; Zheng, Y.; Wu, Z. Spectral–Spatial Feature Tokenization Transformer for Hyperspectral Image Classification. *IEEE Trans. Geosci. Remote Sens.* **2022**, *60*, 5522214. [CrossRef]
40. Zhu, J.; Fang, L.; Ghamisi, P. Deformable convolutional neural networks for hyperspectral image classification. *IEEE Geosci. Remote Sens. Lett.* **2018**, *15*, 1254–1258. [CrossRef]
41. Tatsunami, Y.; Taki, M. Sequencer: Deep LSTM for Image Classification. *arXiv* **2022**, arXiv:2205.01972.
42. Hochreiter, S.; Schmidhuber, J. Long short-term memory. *Neural Comput.* **1997**, *9*, 1735–1780. [CrossRef]
43. Zhang, X.; Shang, S.; Tang, X.; Feng, J.; Jiao, L. Spectral Partitioning Residual Network With Spatial Attention Mechanism for Hyperspectral Image Classification. *IEEE Trans. Geosci. Remote Sens.* **2022**, *60*, 5507714. [CrossRef]

MDPI
St. Alban-Anlage 66
4052 Basel
Switzerland
www.mdpi.com

Remote Sensing Editorial Office
E-mail: remotesensing@mdpi.com
www.mdpi.com/journal/remotesensing

www.ingramcontent.com/pod-product-compliance
Lightning Source LLC
Chambersburg PA
CBHW061612220926
51850CB00005B/490
9783725807710